To Bette and Gayle, for their help and understanding

INTRODUCTION TO HYDROLOGY

Fifth Edition

Warren Viessman, Jr.
University of Florida

Gary L. Lewis
Consulting Engineer

PEARSON

Prentice
Hall

Pearson Education International

Vice President and Editorial Director, ECS: *Marcia J. Horton*
Acquisitions Editor: *Laura Fischer*
Editorial Assistant: *Erin Katchmar*
Vice President and Director of Production and
 Manufacturing, ESM: *David W. Riccardi*
Executive Managing Editor: *Vince O'Brien*
Managing Editor: *David A. George*
Production Editor: *Donna King*
Director of Creative Services: *Paul Belfanti*
Creative Director: *Carole Anson*
Art Director: *Jayne Conte*
Art Editor: *Greg Dulles*
Cover Designer: *Bernadette Travis*

Manufacturing Manager: *Trudy Pisciotti*
Manufacturing Buyer: *Lynda Castillo*
Marketing Manager: *Holly Stark*

© 2012 by Pearson Education, Inc.
Pearson Education, Inc.
Upper Saddle River, NJ 07458

Printed in the United States of America

10 9 8 7 6 5 4 3 2 1

ISBN 13: 978-0-13-276360-8
ISBN 10: 0-13-276360-5

Pearson Education LTD. , London
Pearson Education Australia PTY, Limited
Pearson Education Singapore, Pte. Ltd
Pearson Education North Asia Ltd
Pearson Education Canada, Inc.
Pearson Educación de Mexico, S.A. de C.V.
Pearson Education -- Japan
Pearson Education Malaysia, Pte. Ltd
Pearson Education, Upper Saddle River, New Jersey

Contents

Preface

Given the burgeoning interest in environmental issues, it is believed that course offerings in hydrology will expand nationwide and that the need for contemporary elementary textbooks on the subject will increase. This fifth edition of *Introduction to Hydrology* has been redesigned to play an important role in meeting that need.

Water scientists and engineers of tomorrow must be equipped to deal with a diversity of issues such as the design and operation of data retrieval and storage systems; forecasting; developing alternative water use futures; estimating water requirements for natural systems; exploring the impacts of climate change; developing more efficient systems for applying water in all water-using sectors; and analyzing and designing water management systems incorporating technical, economic, environmental, social, legal, and political elements. A knowledge of hydrologic principles is a requisite for dealing with such issues.

In the early years of the twentieth century, water resources development and management were focused almost exclusively on water supply and flood control. Today, these issues are still important, but environmental protection, ensuring safe drinking water, and providing aesthetic and recreational experiences compete equally for attention and funds. Furthermore, an environmentally conscious public is pressing for greater reliance on improved management practices, with fewer structural components, to solve the nation's water problems. The notion of continually striving to provide more water has been replaced by one of husbanding this precious natural resource.

There is a growing constituency for allocating water for the benefit of fish and wildlife, for protection of marshes and estuary areas, and for other natural system uses. But estimating the quantities of water needed for environmental protection and for maintaining and/or restoring natural systems is difficult. Scientific data are sparse, and our understanding of the complex interactions inherent in ecosystems of all scales is rudimentary. And this is a critical issue, since the quantities of water involved in environmental protection can be substantial, and competition for these waters from traditional water users is keen. The nations of the world are facing major decisions regarding natural systems, decisions that are laden with significant economic and social impacts. Thus there is an urgency associated with developing a better understanding of ecologic systems and of their hydrologic components.

The fifth edition has been rewritten to acquaint future water scientists and managers with the basic elements of the hydrologic cycle. It reviews data sources, introduces statistical analyses in the context of hydrologic problem-solving, covers the components of the hydrologic budget, discusses hydrograph analysis and routing, and introduces groundwater hydrology, urban hydrology, hydrologic models, and hydrologic design. The book is designed to meet the needs of students who expect to become involved in programs that are concerned with the development, management, and protection of water resources. Many solved examples and problems serve to amplify the concepts presented in the text. Many appropriate Internet addresses are provided.

Numerous sources have been drawn upon to provide subject matter for the book, and the authors hope that suitable acknowledgment has been given to them. The authors also thank the following reviewers: Istvan Bogardi, Meteoroligia, Hungary; Praveen Kumar, University of Illinois; David B. Thompson, University of Texas; and Jose D. Salas, Colorado State University. Colleagues and students are also recognized for their helpful comments and reviews.

WARREN VIESSMAN, JR.
GARY L. LEWIS

CHAPTER 1

Introduction

OBJECTIVES

The purpose of this chapter is to:

- Define hydrology
- State the fundamental equation of hydrology
- Demonstrate how hydrologic principles can be applied to supplement decision-support systems for water and environmental management.

1.1 HYDROLOGY DEFINED

Hydrology is an earth science. It encompasses the occurrence, distribution, movement, and properties of the waters of the earth. A knowledge of hydrology is one of the key ingredients in decision-making processes where water is involved. The important role that water plays in ecosystem viability and performance must be better understood [1]–[6].*

1.2 THE HYDROLOGIC CYCLE

The hydrologic cycle is a global sun-driven process whereby water is transported from the oceans to the atmosphere to the land and back to the sea. The ocean is the earth's principal reservoir; it stores over 97 percent of the terrestrial water. Water is evaporated by the sun, incorporated into clouds as water vapor, falls to the land and sea as precipitation, and ultimately finds its way back to the atmosphere through a variety of hydrologic processes. The hydrologic cycle can be considered a closed system for the earth because the total amount of water in the cycle is fixed even though its distribution in time and space varies. There are many subcycles within the worldwide system, however, and they are generally open-ended. It is these subsystems that give rise to the many problems of water supply and allocation that confront hydrologists and water managers.

*Numbers in brackets indicate references at the end of the chapter.

The hydrologic cycle is usually described in terms of six major components: precipitation (P), infiltration (I), evaporation (E), transpiration (T), surface runoff (R), and groundwater flow (G). For computational purposes, evaporation and transpiration are sometimes lumped together as evapotranspiration (ET). Figure 1.1 defines these components and illustrates the paths they define in the hydrologic cycle.

Figures 1.1 and 1.2 illustrate that some precipitation evaporates before reaching the earth and remains in the atmosphere as water vapor. Water also evaporates after reaching the earth. Plants take up infiltrated water and groundwater and return a portion of it to the atmosphere through their leaves, a process known as *transpiration*. Some infiltrated water may emerge to surface-water bodies as interflow, while other portions may become groundwater flow. Groundwater may ultimately be discharged into streams or may emerge as springs. After an initial filling of interception and depression storages, and providing that the rate of precipitation exceeds that of infiltration, overland flow (surface runoff) begins. The magnitude and duration of a precipitation event determine the relative importance of each component of the hydrologic cycle during that event. During storm events, evaporation and transpiration may be minor considerations, for example, but during rain-free periods, ET becomes a dominant feature of the hydrologic cycle.

As water moves through the hydrologic cycle, its quality changes. Seawater is naturally desalted and purified by solar-driven evaporation. Rainwater falling from clouds may pick up contaminants as it descends through the atmosphere, as it is used by humans, or as it flows over land or moves through underground passages. Soils act as natural filters and can adsorb contaminants from water seeping through them toward

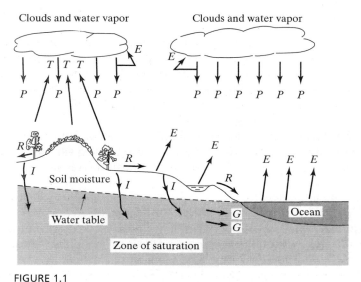

FIGURE 1.1

The hydrologic cycle: T, transpiration; E, evaporation; P, precipitation; R, surface runoff; G, groundwater flow; and I, infiltration.

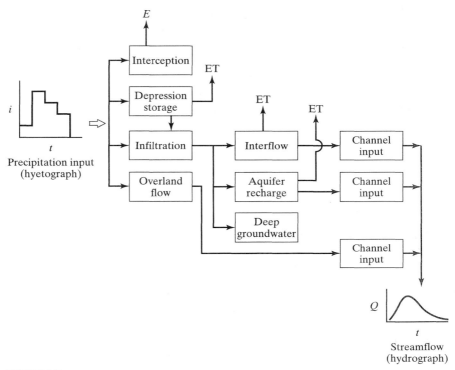

FIGURE 1.2

Distribution of precipitation input.

groundwater reserves or surface outlets. But soils and rocks are subjected to the dissolving processes of water, and these processes may also impart chemical constituents to water moving through them. A major problem for water managers is that of dealing with impaired water quality, especially where groundwater systems are involved. It is important to consider both the quantity and quality of water when decisions regarding its management are being contemplated.

The hydrologic cycle, while simple in concept, is in reality, very complex. Paths taken by precipitated droplets of water are many and varied before the sea is reached. The time scale may be on the order of seconds, minutes, hours, days, or even years.

1.3   THE HYDROLOGIC BUDGET

A water budget comprised of the components of the hydrologic cycle can be formulated. It is an accounting of the inflow, outflow, and storage of water in a designated hydrologic system.

The U.S. Geological Survey (USGS) has estimated the hydrologic budget for the coterminous United States (Fig. 1.3). Approximately 40,000 billion gallons per day (bgd) of water passes over the nation as vapor. Of this amount, about 4,200 bgd

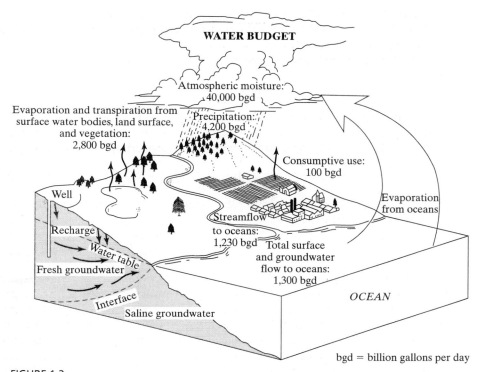

WATER BUDGET

Atmospheric moisture: 40,000 bgd

Evaporation and transpiration from surface water bodies, land surface, and vegetation: 2,800 bgd

Precipitation: 4,200 bgd

Consumptive use: 100 bgd

Well

Evaporation from oceans

Recharge

Water table

Fresh groundwater

Streamflow to oceans: 1,230 bgd

Total surface and groundwater flow to oceans: 1,300 bgd

OCEAN

Interface

Saline groundwater

bgd = billion gallons per day

FIGURE 1.3

Hydrologic budget of the coterminous United States (U.S. Geological Survey).

(roughly 30 in. or 76 cm per year) falls to the earth as precipitation. About two-thirds of this amount (2,750 bgd) is returned to the atmosphere by evaporation or by transpiration. The remaining 1,450 bgd is accounted for by storage; flows to Canada, Mexico, and the oceans; evaporation from surface-water storage; and consumptive use. Of the 1,450 bgd that could potentially be used, only about 675 bgd is considered to be available in 95 out of 100 years [5].

The development of an equation for the water budget of a hydrologic region is straightforward. For a designated time period, it provides for balancing the gains and losses of water in the region with the quantities of water stored in the region (a continuity equation).

With T, E, P, R, G, I, and ET defined as in Figs. 1.1 and 1.2, and letting ΔS stand for change in storage, a hydrologic budget can be derived. Inflows to the region are denoted as positive quantities and outflows as negative ones. Subscripts s and g indicate surface and underground components, respectively.

For surface flow, the hydrologic budget can be written as:

$$P + R_1 - R_2 + R_g - E_s - T_s - I = \Delta S_s \tag{1.1}$$

where precipitation, surface-water inflow, and groundwater appearing as surface water (R_g) are inflows; surface-water outflow, evaporation, and infiltration are outflows; and all variables are volumes per unit of time.

For underground flow, the hydrologic budget can be written as:

$$I + G_1 - G_2 - R_g - E_g - T_g = \Delta S_g \tag{1.2}$$

where infiltration and groundwater inflow are inflows and groundwater outflow, groundwater appearing as surface water, evaporation, and transpiration are outflows. The combined hydrologic budget for a region is derived by summing Eqs. 1.1 and 1.2:

$$P - (R_2 - R_1) - (E_s + E_g) - (T_s + T_g) - (G_2 - G_1) = \Delta(S_s + S_g) \tag{1.3}$$

If the subscripts are dropped and the quantities in parentheses are taken as net changes, the equation reduces to:

$$P - R - E - T - G = \Delta S \tag{1.4}$$

Equation 1.4 is the fundamental equation of hydrology. It is the basis for all hydrologic modeling. Various applications of this important equation are referred to in later chapters. One of its uses is in calculating the combined evaporation and transpiration, or evapotranspiration (ET), for a region when estimates of other variables in the equation can be reasonably made. For example, in large river basins (measured in thousands of square miles or kilometers), groundwater system boundaries often follow surface-water divides. In cases where this assumption can be considered valid, the groundwater flux into and out of the region can be assumed equal to zero ($G = 0$). In addition, over a long period of time (usually five or more years), seasonal excesses and deficits in storage often tend to balance out in large watersheds, and in such cases the average condition for ΔS may sometimes be assumed to be equal to zero. Under these two particular assumptions, the hydrologic equation becomes:

$$P - R - \text{ET} = 0 \tag{1.5}$$

By knowing P and R, a rough estimate of ET can be obtained.

To solve the hydrologic budget equation in terms of any one of its variables, reasonable estimates of the other variables must be made. This is not always possible or easily done. Sometimes data are lacking on the variables of concern and sometimes the data may not be in the right form for the use anticipated. Under certain conditions, simplifying assumptions can be made to minimize problems of data availability, but such circumstances can be counted on only in special cases. The challenge is to expand data-collection and -monitoring programs so that model applications to hydrologic problem-solving can be improved.

Example 1.1

The drainage area of the James River at Scottsville, Virginia, is 11,839 km². If the mean annual runoff is determined to be 144.4 m³/s and the average annual rainfall is

1.08 m, estimate the ET losses for the area. How does this compare with the lake evaporation of 1 m/yr measured at Richmond, Virginia?

Solution.

1. Assuming that $G = 0$ and $\Delta S = 0$, Equation 1.5 can be used to estimate ET.
2. Runoff is converted from m^3 to m/yr as follows:

$$R = [144.4 \times 86,400 \times 365]/[11,839 \times 10^6] = 0.38 \text{ m}$$

3. $\text{ET} = P - R = 1.08 - 0.38 = 0.7 \text{ m}$
4. The ET losses over the drainage basin are less than the measured lake ET losses at Richmond. (Note that ET losses over open bodies of water are due to E only.) This shows that the ET rate is less for the vegetated drainage basin than for the open, available body of water.

1.4   HYDROLOGIC MODELS

Hydrologic systems are generally analyzed by using mathematical models. These models may be empirical or statistical, or founded on known physical laws. They may be used for such simple purposes as determining the rate of flow that a roadway grate must be designed to handle, or they may be used to guide decisions about the best way to develop a river basin for a multiplicity of objectives. The choice of the model should be tailored to the purpose for which it is to be used. In general, the simplest model capable of producing information adequate to deal with the issue should be chosen.

Unfortunately, most water resources systems of practical concern have physical, social, political, environmental, and legal dimensions, and their interactions cannot be exactly described in mathematical terms. Furthermore, the historical data necessary for meaningful hydrologic analyses are often lacking or unreliable. And when one considers that hydrologic systems are generally probabilistic in nature, it is easy to understand that the modeler's task is not an easy one. In fact, it is often the case that the best that can be hoped for from a model is an enhanced understanding of the system being analyzed. But this in itself can be of great value, leading, for example, to the implementation of data-collection programs that can ultimately support reliable modeling efforts.

For the most part, mathematical models are designed to describe the way a system's elements respond to some type of stimulus (input). For example, a model of a groundwater system might be developed to demonstrate the effects on groundwater storage of various schemes for pumping. Equations 1.1 and 1.2 are mathematical models of the hydrologic budget. In later chapters, some hydrologic models that can be used as the basis for informed water management decisions are discussed.

1.5   APPLICATIONS

Land, air, and water are inextricably linked. Solving the challenging environmental problems faced by the inhabitants of planet Earth will require a thorough understanding of hydrologic processes and how they relate to the other ecosystem components with which they interact. Applications of hydrology abound, and the need for scientists, engineers, and others with competence in this field is clear.

Topics requiring expertise in hydrology are far-ranging and they include water supply development and management; treated wastewater disposal; floodplain management; wetlands protection; preservation, protection, and restoration of natural systems; solid waste landfill design; water resources management; habitat protection; and groundwater protection and development. Hydrologists have an important role to play in ensuring that future allocations of water are scientifically based and efficiently made so as to satisfy the needs of both human and natural systems.

SUMMARY

Hydrology is the science of water. It embraces the occurrence, distribution, movement, and properties of the waters of the earth. A mathematical accounting system may be constructed for the inputs, outputs, and water storages of a region so that a history of water movement over time can be estimated.

After reading this chapter, you should understand the hydrologic budget and be able to make a simple accounting of water transport in a region. You should also have gained an understanding of how hydrologic analyses can be used to facilitate design and management processes for water resources systems.

PROBLEMS

1.1 Two cm of runoff result from a storm on a drainage area of 100 km^2. Convert this amount to cubic meters and acre-feet (1 acre = 43,560 square feet).

1.2 Assume you are dealing with a vertical-walled reservoir having a surface area of 500,000 m^2 and that an inflow of 1.0 m^3/s occurs. How many hours will it take to raise the reservoir level by 30 cm?

1.3 The storage existing in a river reach at a given time is 13 acre-ft, and at the same time, the inflow to the reach is 450 cfs and the outflow is 500 cfs. One hour later, the inflow is 500 cfs and the outflow is 530 cfs. Calculate the change in storage during the hour in cubic meters and acre-feet.

1.4 The annual evaporation from a lake is found to be 125 cm. If the lake's surface area is 12 km^2, what is the daily evaporation rate in centimeters and inches?

1.5 If a vertical-walled reservoir having a surface area of 1 mi^2 receives an inflow of 12 cfs, how long will it take to raise the reservoir level by 6 in.?

1.6 If the mean annual runoff of a drainage basin of 10,000 km^2 is 140 m^3, and the average annual precipitation is 105 cm, estimate the ET losses for the area in 1 year. What are your assumptions? How reliable do you think this estimate is?

1.7 The evaporation rate from the surface of a 3,650-acre lake is 100 acre-ft/day. Estimate the depth change in the lake in feet during a 1-year period if the net inflow to the lake is 25 cfs. Is the depth increasing or decreasing? What is the change in centimeters?

1.8 The storage in a reach of river is 20,000 m^3 at a given time. Determine the storage 1 hr later if the average rates of inflow and outflow during the hour are 20 and 18 m^3/s, respectively.

REFERENCES

[1] P. B. Jones, G. D. Walker, R. W. Harden, and L. L. McDaniels, "The Development of the Science of Hydrology," Circ. No. 60-03, Texas Water Commission, Apr. 1963.

[2] W. D. Mead, *Notes on Hydrology*, Chicago: D. W. Mead, 1904.

[3] Ven Te Chow (ed.), *Handbook of Applied Hydrology*, New York: McGraw-Hill, 1964.

[4] O. E. Meinzer, *Hydrology*, Vol. 9 of *Physics of the Earth*, New York: McGraw-Hill, 1942. Reprinted by Dover, New York, 1949.

[5] Water Resources Council, *The Nation's Water Resources: 1975–2000*, U.S. Gov't. Print. Off., Washington, D.C., 1978.

[6] D. R. Maidment (ed.), *Handbook of Hydrology*, New York: McGraw-Hill, 1993.

CHAPTER 2

Hydrologic Measurements and Data Sources

OBJECTIVES

The purpose of this chapter is to:

- ■ Describe the principal sources of data for hydrologic investigations
- ■ Describe instruments used in measuring hydrologic variables
- ■ Indicate ways in which data are recorded and transmitted
- ■ Indicate the importance of real-time and continuous recording of hydrologic events.

2.1   UNITS OF MEASUREMENT

Stream and river flows are usually recorded as cubic meters per second (m^3/sec), cubic feet per second (cfs), or second-feet (sec-ft); groundwater flows and water supply flows are commonly measured in gallons per minute, hour, or day (gpm, gph, gpd), millions of gallons per day (mgd), or m^3 or liters per unit time; and flows used in agriculture or related to water storage are often expressed as acre-feet (acre-ft), acre-ft per unit time, inches (in.) or centimeters (cm) depth per unit time, or acre-inches per hour (acre-in./hr).

Volumes are often given as gallons, cubic feet, cubic meters, liters, acre-ft, second-foot-days, and inches or centimeters over an area. An acre-ft is equivalent to a volume of water 1 ft deep over an area of 1 acre of land (43,560 ft^3). A second-foot-day (cfs-day, sfd) is the accumulated volume produced by a flow of 1 cfs in a 24-hr period. A second-foot-hour (cfs-hr) is the accumulated volume produced by a flow of 1 cfs in 1 hr. Inches or centimeters of depth relate to a volume equivalent to that many inches or centimeters of water over the area of interest. In hydrologic mass balances, it is sometimes useful to note that 1 cfs-day = 2 acre-ft with sufficient accuracy for most calculations.

Precipitation depths are usually recorded in inches or centimeters, whereas precipitation rates are given in inches or centimeters per unit time. Evaporation, transpiration, and infiltration rates are also given as inches or centimeters of depth per unit time. Some useful constants, conversion factors, and physical properties of water are given in Appendix A.

2.2 HYDROLOGIC DATA

Data on hydrologic variables are fundamental to analyses, forecasting, and modeling. Such data may be found in numerous publications of state and federal agencies, research institutes, universities, and other organizations. Many of these sources are identified and referenced in this chapter [1]–[4].

General Climatological Data

The most readily available sources of data on temperature, solar radiation, wind, and humidity are climatological data bulletins published by the Environmental Data Service of the National Oceanic and Atmospheric Administration (NOAA) and monthly summary solar radiation data published by the National Climatic Data Center (www.ncdc.noaa.gov). The Environmental Data Service, in cooperation with the World Meteorological Organization (WMO), also publishes monthly climatic data for the world. A publication of the Environmental Sciences Service Administration, *Climatic Atlas of the United States*, summarizes wind, temperature, humidity, evaporation, precipitation, and solar radiation on a series of maps. State environmental, geologic, water resources, and agricultural agencies also collect and publish hydrologic data. They should also be consulted.

Rainfall and Snowfall Data

There are probably more records of precipitation (rainfall and snowfall) than of most other hydrologic variables. The principal federal source of data on precipitation is NOAA. "Climatological Data," published monthly and annually for each state or combination of states, the Pacific area, Puerto Rico, and the Virgin Islands by the Environmental Data Service, provides tables of monthly averages, departures from normal, and extremes of precipitation and temperature. The publication also includes tables of daily precipitation, temperature, snowfall, snow on ground, evaporation, wind, and soil temperature. "Hourly Precipitation Data" is issued monthly and annually for each state or combination of states and presents alphabetically by station, hourly and daily precipitation amounts for stations equipped with recording gauges. A station location map is also included. This publication is available from the Environmental Data Service. Another publication, "World Weather Records," is issued by geographic regions for 10-year periods. Data are listed by country or area name, station name, latitude and longitude, and elevation. Monthly and annual mean values of station pressure, sea-level pressure and temperature, and monthly and annual total precipitation are given in sequential order. Aside from NOAA, other federal and state agencies and universities publish precipitation data at varying intervals, often in a storm- or site-specific context. In addition, many municipalities and water and wastewater utilities also collect and maintain precipitation and other hydrologic data. Computerized precipitation data are available from the National Climatic Data Center in Asheville, North Carolina (www.ncdc.noaa.gov).

Streamflow Data   The principal sources of streamflow data for the United States are the U.S. Geological Survey (USGS), U.S. Natural Resources Conservation Service [(NRCS) formerly the Soil Conservation Service (SCS)], U.S. Forest Service, and U.S. Agricultural Research Service (ARS). See the following USGS Web sites: National Water Information System (http://water.usgs.gov/nwis), National Water Quality (http://water.usgs.gov/nawqa), and Water Modeling (http://water.usgs.gov/software/). See also the U.S. Environmental Protection Agency (EPA) Office of Water (www.epa.gov/OW) and the NRCS (www.nrcs.usda.gov). In addition, the U.S. Army Corps of Engineers (COE), the Tennessee Valley Authority (TVA), and the U.S. Bureau of Reclamation (USBR) make some streamflow measurements and tabulate streamflow data relative to their missions. State agencies, universities, and various research organizations also compile and publish streamflow data.

The USGS Water Supply Papers (WSPs) are the benchmark for referencing streamflow data. Furthermore, computerized data are also available from the USGS. *Publications of the Geological Survey*, published every five years and supplemented annually, is an excellent source of information on that agency's reports. The NRCS historically published data on streamflow from small watersheds in its "Hydrologic Bulletin" series, but much of these data have been republished by the ARS. Records from NRCS pilot watersheds are published in cooperation with the USGS. U.S. Forest Service streamflow data are published at irregular intervals in technical bulletins and professional papers.

Evaporation and Transpiration Data

Monthly and annual issues of "Climatological Data" published by NOAA include pan evaporation and related data. The ARS, agricultural colleges, and water utilities are other sources of information. Data on evapotranspiration are often published by university researchers working through their agricultural experiment stations.

Soils

The primary particles of soil that form its texture are sand, silt, and clay. Soil is generally categorized into A, B, and C horizons. The surface layer of soil is known as the A-horizon. This layer contains decaying organic matter. The B-horizon lies below the A-horizon and usually contains more clay than the A-horizon. The C-horizon contains the soil parent material from which the A- and B-horizons were formed. These horizons, in combination, constitute the soil profile, which may also contain subhorizons [5].

Soil classification systems are based on the intended use of the system. In the United States, the Pedalogical System, which extends down into the C-horizon, has been developed mainly for agricultural uses. This system is used by the U.S. Department of Agriculture and state soil scientists to provide soil surveys for numerous counties in the

United States. The system is especially tailored to suit agronomic purposes. The Engineering Unified Soil Classification System serves to classify soils for use in foundation and hydraulic structure designs. It is not applicable to agricultural soil classification. Soil data are provided mainly by the U.S. Department of Agriculture and its Soil Conservation Service [1],[6]–[9].

2.3   HYDROLOGIC MEASUREMENTS

Hydrologic measurements support water resources planning, management, design, and construction activities. They produce the database needed for developing and verifying hydrologic models. Contemporary techniques and instruments used for measuring hydrologic variables are described in this section [10]–[12].

Wind, Temperature, Humidity, and Solar Radiation

Measurements of wind, temperature, humidity, and solar radiation are needed to support hydrologic analyses. Wind is commonly measured using an anemometer, a device that has a wind-propelled element such as a cup or propeller whose speed is calibrated to reflect wind velocity. Wind direction is obtained using a vane that orients itself with the direction of the wind. Temperature measurements are made using standard thermometers of various types, while humidity is measured using a psychrometer. A psychrometer consists of two thermometers, one called a wet bulb and the other a dry bulb. Upon ventilation, the two thermometers give different measurements, and this difference is called the *wet-bulb depression*. Using appropriate tables, the dew point, vapor pressure, and relative humidity can be determined [13].

Solar radiation is an important component of the snowmelt process. Only a fraction of the radiant energy emitted by the sun reaches the earth, but the amount received is the ultimate source of the planet's energy. Daily amounts of incoming solar radiation (insolation) received at the outer limits of the earth's atmosphere can be calculated from the solar constant (generally considered to be 1.94 langleys per minute) for a specified latitude and time of the year. Insolation measurements are made using instruments called *pyrheliometers*. Data on insolation in the United States are provided by the National Weather Service and may be obtained through the National Climatic Data Center.

Figure 2.1 depicts a complete weather station incorporating measurements of precipitation, wind, temperature, barometric pressure, and humidity. Such stations can automatically report weather data from remote sites on either an event and/or timed basis to a central site. A station such as this can be used for marine weather forecasting, quantitative determinations of oncoming storms, determination of wind effect on tidal areas, establishment of a database for irrigation, and many other purposes.

Precipitation

Gauges for measuring rainfall and snowfall may be recording or nonrecording. The most common nonrecording gauge is the U.S. Weather Bureau standard 8-inch gauge. The gauge may be read at any desired interval but often this is daily. The gauge is calibrated so that a measuring stick, when inserted, shows the equivalent rainfall depth.

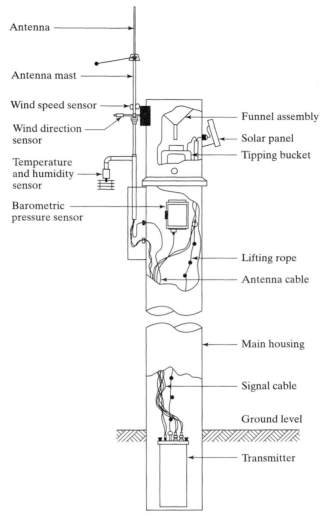

Antenna

Antenna mast

Wind speed sensor

Wind direction
sensor

Temperature
and humidity
sensor

Barometric
pressure sensor

Funnel assembly

Solar panel

Tipping bucket

Lifting rope

Antenna cable

Main housing

Signal cable

Ground level

Transmitter

FIGURE 2.1

Self-reporting weather station (courtesy of Sierra-Misco, Inc.,
Environmental Products, Berkeley, CA).

Such gauges are useful when only periodic volumes are required but they cannot be
used to indicate the time distribution of rainfall.

Recording gauges continuously sense the rate of rainfall and its time of occurrence. These gauges are usually either of the weighing-recording type or the tipping-
bucket type. Weighing-type gauges usually run for a period of 1 week, at which time
their charts must be changed. A mass curve of rainfall depth vs. time is the product.
Tipping-bucket gauges, on the other hand, sense each consecutive rainfall accumulation when it reaches a prescribed amount, usually 0.01 inch or 1 millimeter (mm) of

rain. A small calibrated bucket is located below the rainfall entry port of the gauge, and when it fills to the 0.01-inch increment, it tips over, bringing a second bucket into position. These two small buckets are placed on a swivel, and the buckets tip back and forth as they fill. Each time a bucket spills it produces an indication on a strip chart or other recording form. In this way a record of rainfall depth vs. time (intensity) is the outcome. For rain gauges to record snow accumulations, some modifications must be made. Usually these involve providing a melting agent so that the snow can be converted into measurable water.

Figure 2.2a is a diagram of a self-reporting rain-gauging station; the tipping bucket mechanism generates a digital input signal whenever 1 mm of rainfall drains through the funnel assembly. The signal from the gauge is automatically transmitted to a receiving station, where the station ID number and an accumulated amount of rainfall are recorded. The receiving station records the time at which the message was received, and rainfall rates for desired periods can thus be calculated. Figure 2.2b shows a similar gauge equipped to measure snow. In this case, a glycometh solution is used to melt the snow. The melt water overflows through a temperature-compensating mechanism and is measured by the tipping bucket, which operates the station's transmitter. Tipping-bucket gauges can easily be incorporated into real-time monitoring systems that can be used in a variety of forecasting and operating modes.

Rainfall measurements can also be made using satellite sensors and radar [3], [14],[15]. Radar is widely used to track the movement of rainfall events and it can also be used to make quantitative assessments. Information provided by radar covers reasonably large areas in real time and generally has a resolution of about 5 km on a 5–15-minute time scale [15]. The value of radar as a means to make precipitation measurements is related to the fact that it provides information covering large spaces while rain gauges provide only point observations.

At wavelengths over which weather radar typically operates (3 to 10 cm), large cloud droplets, raindrops, and snow particles are reflective. Back-scattered radiation (reflectivity) is highly correlated with the volume of precipitation illuminated by the radar screen [14],[15]. There are, however, a number of error sources associated with radar measurements. Accordingly, the best results will usually be obtained by combining radar measurements with those made by point gauges. In this way, the spatial and temporal accuracy of radar and the point-specific accuracy of precipitation gauges are combined to enhance the best features of both approaches.

For a large portion of the earth (the area covered by the oceans), satellite sensors provide the only means of measuring rainfall [3]. These sensors work under the principle that the atmosphere selectively transmits radiation at various wavelengths. Satellite systems are attractive because they provide complete global coverage and can yield spatially continuous data. Visible reflection and thermal infrared emission from cloud tops measured by satellites can be used to estimate rainfall rates. It has been found that the intensity of convective rainfall is correlated with the brightness and radiative temperature of cloud tops. Studies have shown that colder and brighter cloud tops are associated with more intense convection, which is also associated with rainfall intensity and volume. Validation and calibration of satellite methods can be achieved using radar estimates, rain gauge network data, and aviation weather reports [15]. So far, direct measurement of rainfall by satellites has been limited, but by

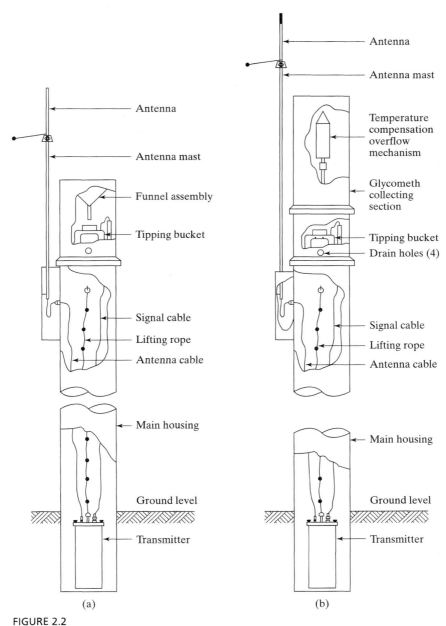

FIGURE 2.2

Self-reporting (a) rain and (b) snow stations (courtesy of Sierra-Misco, Inc., Environmental Products, Berkeley, CA).

combining satellite data with those from other sources, useful estimates for large regions can be obtained.

Snow

Snow measurements are made through the use of standard and recording rain gauges, seasonal storage precipitation gauges, snow boards, snow stakes, and remote sensing methods [16]. Rain gauges are usually equipped with shields to reduce the effect of wind. Snow boards, about 16 in. square, are laid on the snow so that new snowfall that occurs between observation periods will be accumulated above them. Care must be taken to ensure that adverse wind effects or other conditions do not produce an erroneous sample. Snow stakes are calibrated wooden posts driven into the ground for periodic observation of the snow depth or inserted into the snowpack to determine its depth.

Direct measurements of snow depth at a single station are generally not very useful in making estimates of the snow's distribution over large areas, since the measured depth may be unrepresentative due to drifting or blowing. To circumvent this problem, snow-surveying techniques have been developed. Such surveys provide information on snow depth and on the water equivalent, density, and quality of the snowpack at various points along the snow course.

Water equivalent is the depth of water that would weigh the same as that of the snow sample. In this way snow depth can be described in terms of inches or centimeters of water. Density is the percentage of snow volume that would be occupied by its water equivalent. The quality of snow relates to the ice content of the snowpack and is expressed as a decimal number. It is the ratio of the weight of the ice content to the total weight. Snow quality is usually about 0.95 except during periods of rapid melt, when it may drop to 0.70 to 0.80 or less.

A snow course includes a series of sampling locations, normally not less than 10 in number [3],[17]. The stations are spaced about 50–100 ft apart in a geometric pattern designed in advance. Points are permanently marked so that the same location will be surveyed each year. Snow survey data are obtained directly by foresters, and others, by aerial photographs and observations, and by automatic recording stations that telemeter information to a central processing location.

Evaporation and Transpiration

Evaporation pans have been widely used for estimating the amount of evaporation from free water surfaces. These devices, depicted in Fig. 2.3, are easy to use, but relating measurements taken from them to actual field conditions is difficult and the data they produce are often of questionable value for making areal estimates. A variety of pan types have been developed but the U.S. Weather Bureau Class A pan is the standard in the United States [18]. Pan evaporation observations have been used to estimate both free water (lake) evaporation and evapotranspiration from well-watered vegetation. Interestingly enough, field experiments have shown a high degree of correlation of pan data with evapotranspiration from surrounding vegetation when there is full cover and good water supply [16]. As in the case of precipitation gauges, pan data can be continuously recorded and transmitted to a central receiving station.

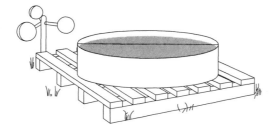

FIGURE 2.3

U.S. Weather Bureau Class A pan.

Evapotranspiration measurements are often made using lysimeters. These devices are containers placed in the field and filled with soil, on which some type of vegetative growth is maintained. The object is to study soil-water-plant relationships in a natural surrounding. The main feature of a weighing lysimeter is a block of undisturbed soil, usually weighing about 50 tons, encased in a steel shell that is $10 \times 10 \times 8$ ft. The lysimeter is buried so that only a plastic border marks the top of the contained soil. The entire block of soil and the steel casing are placed on an underground scale sensitive enough to record even the movement of a rabbit over its surface. The soil is weighed at intervals, often every 30 minutes around the clock, to measure changes in soil water level. The scales are set to counterbalance most of the dead weight of the soil and measure only the active change in weight of water in the soil.[19] The scales can accurately weigh about 400 grams (slightly under 1 pound), which is equivalent to 0.002 inches of water. The weight loss from the soil in the lysimeter represents water used by the vegetative cover, plus any soil evaporation. Added water is also weighed and thus an accounting of water content can be kept. Crops or cover are planted in the area surrounding the lysimeter to provide uniformity of conditions surrounding the instrument. Continuous records at the set weighing intervals provide almost continuous monitoring of conditions. The data obtained can be transmitted to any desired location for analysis and/or other use. Weighing lysimeters can produce accurate values of evapotranspiration over short periods of time, but they are expensive. Nonweighing types of lysimeters, which are less costly, have also been used, but unless the soil moisture content can be measured reliably by some independent method, the data obtained from them cannot be relied upon except for long-term measurements such as between precipitation events [19].

Soil Moisture

Soil moisture measurements are made in a variety of ways but the principal method employed is the gravimetric method. In that procedure, a soil sample is weighed and placed in a container of known weight. The weight of the wet soil sample is thus known. The soil sample is sealed in its container and taken to a laboratory where it is removed and oven-dried until all of the water has evaporated. The sample is then weighed again, and the loss in weight is equated to the weight, and thus the amount, of soil moisture that the field sample contained [17],[20]–[22].

A number of other methods for determining soil moisture are also used. They include radiological, electric resistance, time-domain reflectometry, nuclear magnetic resonance, and remote sensing. Details on these procedures are readily available in the literature [17],[20]–[22].

Infiltration

Commonly used methods for determining infiltration capacity are hydrograph analyses and infiltrometer studies. Infiltrometers are usually classified as rainfall simulators or flooding devices. In the former, artificial rainfall is simulated over a test plot and the infiltration is calculated from observations of rainfall and runoff, with consideration given to depression storage and surface detention [3],[17]. Flooding infiltrometers are usually rings or tubes inserted in the ground. Water is applied and maintained at a constant level and observations are made of the rate of replenishment required.

Estimates of infiltration based on hydrograph analyses have the advantage over infiltrometers of relating more directly to prevailing conditions of precipitation and field. They are, however, no better than the precision with which rainfall and runoff are measured. Of special importance in such studies is the areal variability of rainfall. Several methods have been developed and are in use [3],[15].

Streamflow

Measurements of open channel (natural and man-made) flow are made using standard measuring devices such as flumes and weirs; they are also made by calibrating special control sections along rivers and streams so that measurements of depth (stage) of flow can be related to discharge. Flow-measuring devices are designed so that sensing some parameter such as depth automatically translates the observation into units of flow (discharge). When a control section is used, observations of cross-sectional area for various depths must be obtained, and average flow velocities must be ascertained for various stages so that a section rating curve can be established. In the United States, the U.S. Geological Survey, the U.S. Bureau of Reclamation, the Natural Resources Conservation Service, and the Army Corps of Engineers have done extensive flow measuring and have been active in developing instruments and procedures for ascertaining rates of flow [11],[13].

Weirs. Weirs are common water-measuring devices. When they are properly installed and maintained they can be a very simple and accurate means for gauging discharge. The most often used weir types are the rectangular weir and the V-notch weir (Fig. 2.4). To be effective, weirs usually require a fall of about 0.5 ft or more in the channel in which they are placed. Basically, a weir is an overflow structure placed across an open channel. For a weir of specific size and shape with free-flow steady-state conditions and a proper weir-to-pool relationship, only one depth of water can exist in the upstream pool for a particular discharge. Flow rate is determined by measuring the vertical distance from the crest of the overflow part of the weir to the water surface in

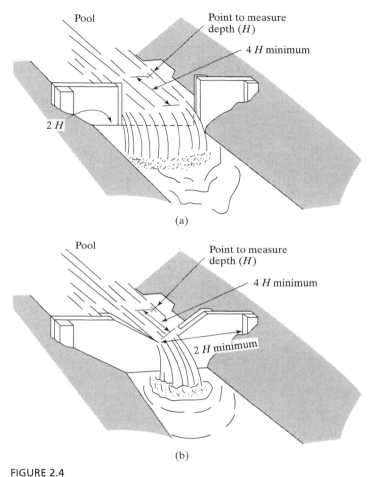

Pool

Point to measure
depth (H)

4 H minimum

2 H

(a)

Pool

Point to measure
depth (H)

4 H minimum

2 H minimum

(b)

FIGURE 2.4

Field installation of weirs: (a) rectangular and (b) V-notch. USDA
Cooperative Extension Service, Mountain States area.

the upstream pool. The weir's calibration curve then translates this recorded depth into
rate of flow at the device.

Parshall Flumes. A Parshall flume is a specially shaped open channel flow
section that can be installed in a channel section. Figure 2.5 depicts one of these
devices. The flume has several major advantages: It can operate with a relatively small
head loss; it is fairly insensitive to the approach velocity; it can be used even under
submerged conditions; and its flow velocity is usually sufficient to preclude sediment
deposits in the structure [11]. The Parshall flume, developed by the late Ralph L.
Parshall, is a particular form of venturi flume.

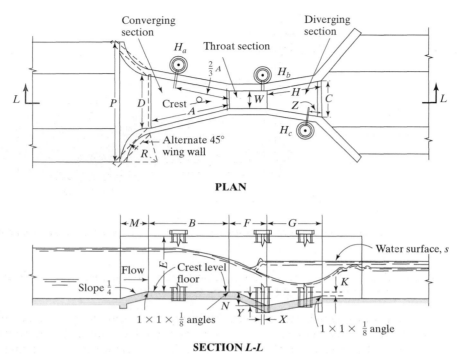

PLAN

SECTION L-L

FIGURE 2.5

Parshall flume (U.S. Soil Conservation Service).

The constricted throat of the flume produces a differential head that can be related to discharge. Thus as in the case of the weir, an observation of depth (head) is all that is required to determine the rate of flow at the control point. Weirs and flumes are generally best suited to gauging small streams and open channels, although large broad-crested weirs can be installed at dam sites as part of overflow structures. For major rivers, other measuring approaches, such as developing field ratings at a specified control section, must be relied upon.

Control Sections. Where the installation of a weir, flume, or some other flow-measuring device is impractical, it is sometimes possible to develop a rating curve at some location along a stream by taking measurements of depth, cross-sectional area, and velocity and calculating the rate of flow for a particular stage at the location. By doing this for a range of depths of flow, a station rating curve can be developed. Instruments required to develop such a curve are depth-sensing devices, surveying instruments, and velocity meters. The velocity meter is similar to an anemometer. It is placed at various positions in the channel and a velocity is recorded. By doing this at a number of locations, a velocity profile for a given depth can be developed. From this an average flow velocity can be computed, and using that determination and the cross-sectional area, discharge can be calculated as the product of the mean velocity and cross-sectional area. If observations can be made for a range of depths, a rating curve

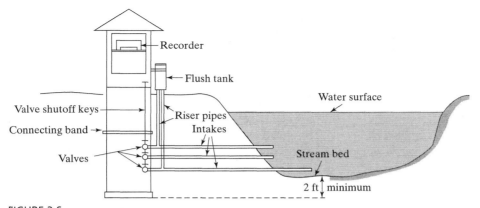

FIGURE 2.6

Recorder house and stilling well for a stream-gauging station (U.S. Bureau of Reclamation).

can be developed for the control section so that only measurements of depth will be needed to estimate rate of flow at some later time [11],[13].

Depth (Stage) Measurements. Most depth measurements are made using a float and cable arrangement in a stilling well or a bubbler gauge. In the first instance, a stilling well connected to the channel is used to house a float device that activates a recorder as it moves up and down (Fig. 2.6). Figure 2.7 illustrates a self-reporting stilling well liquid-level station. Data from the gauging station can be transmitted to any central location for analysis and/or other use. A bubbler-type installation makes use of dry air or nitrogen as a fluid for bubbling through an orifice into a channel bed. As the depth of flow changes, the change in head above the bubbler orifice causes a corresponding pressure change. This results in a fluid level change in the manometer connected to the gas supply, and this in turn is used to reflect stage variation over time.

The foregoing descriptions are of a few of the instruments used in hydrologic work. Both their reliability and the limitations associated with their use must be understood if they are to be reliably used and their outputs are to be considered credible.

2.4 DATA NETWORKS AND TELEMETRY

Hydrologic-Meteorologic Networks

Most modern-hydrologic-meteorologic networks are designed to provide real-time information for purposes such as scheduling hydropower, releasing flows for irrigation, developing and testing hydrologic system models, regulating reservoir discharges, allocating water from multiple sources, forecasting streamflow, tracking pollutant transport, and enforcing environmental regulations. As shown in Fig. 2.8, hydro-met networks may be designed to monitor physiographic, climatic, hydrologic, biologic, and chemical features, or combinations of these, in a region or river basin. They must have gauge densities and distributions that are sufficient to permit interpolation between gauge sites in a manner permitting valid conclusions to be drawn for the entire area

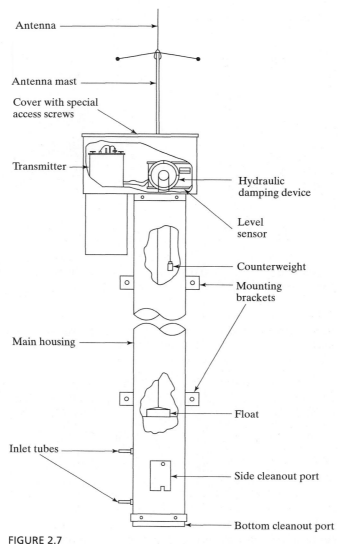

Antenna

Antenna mast

Cover with special
access screws

Transmitter

Hydraulic
damping device

Level
sensor

Counterweight

Mounting
brackets

Main housing

Float

Inlet tubes

Side cleanout port

Bottom cleanout port

FIGURE 2.7

Self-reporting stilling well liquid-level station (courtesy of Sierra-Misco,
Inc., Environmental Products, Berkeley, CA).

covered by the network. Typically, measurements are made of such variables as precipitation, solar radiation, temperature, relative humidity, barometric pressure, snow depth, soil moisture, wind, streamflow, and water quality. In any event, special basin or regional climatic factors must be given due consideration. Each hydro-met network is different in its purpose and setting, and thus its design must reflect both the spatial and temporally varying features at the locality to be monitored along with the objectives of the monitoring program [23],[24].

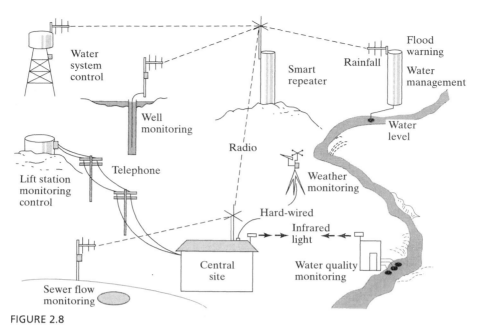

FIGURE 2.8

A telemetry monitoring system (courtesy of Sierra-Misco, Inc., Environmental Products, Berkeley, CA).

With the rapid technological development of computers, especially inexpensive personal computers (PCs), the opportunities for automated collection of all types of hydrologic and water quality data have increased substantially. PCs, used with analog-to-digital converters, pressure or liquid-level sensors, and the appropriate software can, for example, be used in hydrologic monitoring systems as flow-metering/data acquisition systems (Fig. 2.9) [22]. Such systems are highly versatile and relatively inexpensive. Computer systems can be custom-designed for almost any data acquisition application, and they are often less costly than other commercially available hardware systems designed for the same purpose. Computers can convert raw data into other more useful forms, store data for later use, and communicate with other computers if necessary. As such, they are a powerful and important component of modern-day hydrologic monitoring systems. Figure 2.10 illustrates the use of computers in a real-time telemetry system.

Modeling hydrologic systems requires an understanding of how these systems actually function; cleaning up a toxic waste discharge requires tracking the effects of remedial actions; enforcing environmental regulations requires knowledge of what has happened since the rules were implemented; and regulating reservoir releases to meet specified targets requires a continuous understanding of the state of the system being operated. The key to meeting such requirements lies in the products of carefully designed and managed monitoring networks. Developing such networks is no small task, however, as the number of variables that must be observed may be very large, the instruments to measure them costly to install and operate, and the data storage and management requirements extensive. Accordingly, a monitoring network's design must

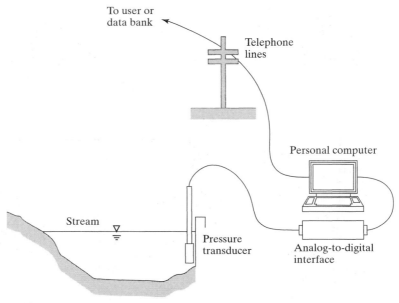

FIGURE 2.9

Personal computer used in stream gauging.

begin with a thorough understanding of its purpose so that the degree of resolution provided by its observations is adequate, but not excessive, for the task at hand. A good rule is to keep the network as simple as possible, within the constraints of what must be accomplished.

The purpose of monitoring is to gather information in a continuum such that the dynamics of the system can be ascertained. According to Dressing, objectives of monitoring for nonpoint source pollution control include developing baseline information, generating data for trend analysis, developing and/or verifying models, and investigating single incidents or events [25],[26]. These objectives are also valid for hydrologic monitoring in general, but they should be supplemented by the following objectives: planning, real-time system operation, enforcing regulatory programs, and environmental policy-making. The ultimate purpose of monitoring is to enhance decision-making, whether it be for development, management, regulatory, or research aims.

Telemetry Systems

Historically, many gauges were read periodically by an individual making the rounds of installations. This served well when the purpose of the data was to establish a base record of some variable such as rainfall. But in more modern times it has, under many circumstances, become necessary to continuously record rainfalls, streamflows, evaporation rates, etc. and to have these data available for the real-time operation of water management systems, and for forecasting hydrologic events. Some examples of activities requiring real-time hydrologic data are managing reservoirs, issuing flood

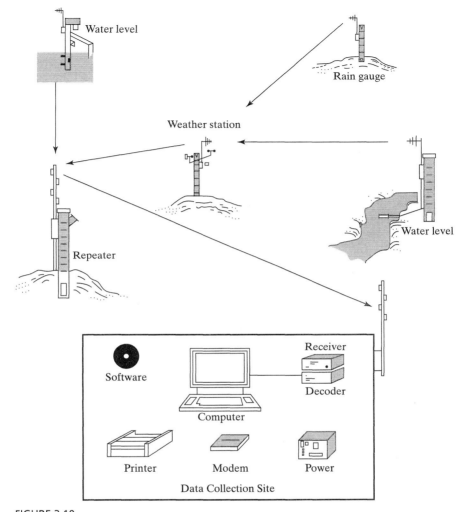

FIGURE 2.10

Computer use in a real-time telemetry system (courtesy of Sierra-Misco, Inc., Environmental Products, Berkeley, CA).

warnings, allocating water for various uses such as irrigation, monitoring streamflows to ensure that treaties and compacts are honored, and monitoring the quality and quantity of water for regulatory and environmental purposes. Accordingly, gauging stations capable of electronically transmitting their data to a central location for immediate use have now become common. The advantages of such stations include providing information to users in a time frame that meets management needs, reducing the costs of collecting data, and providing a continuous and synchronous record of hydrologic events. Figure 2.11 shows a stream gauge reporting station using radio transmission. Figure 2.12 illustrates a satellite data collection and transmitting operation [27]–[31].

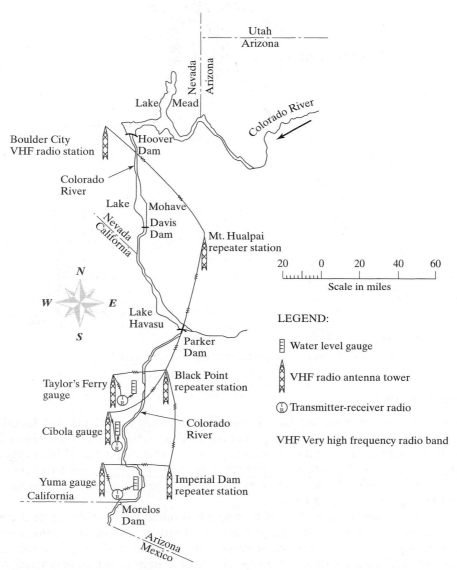

FIGURE 2.11

Stream gauge reporting system using radio transmission. Water stage information is requested from the gauging stations by VHF radio signal. In turn, this water stage information is obtained from the stream gauges and automatically encoded and transmitted to the Boulder City receiving station. All downstream releases from Hoover Dam are determined and integrated with this streamflow information in controlling the flow of the lower Colorado River (U.S. Bureau of Reclamation).

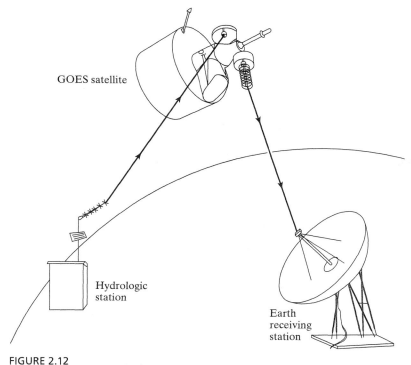

GOES satellite

Hydrologic
station

Earth
receiving
station

FIGURE 2.12

Hydrologic data collection by satellite (U.S. Geological Survey).

Remote Sensing

Since the 1960s, remote sensing has become a common hydrologic tool. Examples of aircraft and satellite data collection and transmission abound [32]–[36]. Figure 2.13 illustrates the use of aircraft and satellites in a snow survey system. Other types of surveys such as those for determining impervious areas, classifying land uses for assessing basin-wide runoff indices, determining lake evaporation, and prospecting groundwater can be depicted in similar fashion.

The principal value of remote sensing is its ability to provide regional coverage and at the same time provide point definition. Furthermore, satellite communications can be digitized and are thus compatible with the transfer of computerized information. Following the evolution of linkages between computer and communications technology, new software systems incorporating powerful data management systems have been developed. These systems facilitate the storage, compaction, and random access of large data banks of information. One data management option, geographic information systems (GISs), allows the overlaying of many sets of data (particularly satellite-derived data) for convenient analysis. Versatile color pictorial and graphic display systems are also becoming attractive as their costs have decreased [35].

With the advancement of satellite technology, the use of satellites as remote sensor platforms has spread. Currently available sensors can operate in a multitude of

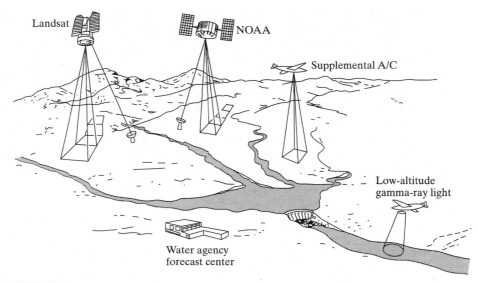

FIGURE 2.13

Satellite snow survey system (after Calabrese and Thome [15]).

electromagnetic radiation wavelengths, and the information content of their signals can include surface temperatures, radiation, atmospheric pollutants, and other types of meteorological data. As remote sensors are improved to permit attaining greater radiometric and geographic resolution, and as computer image-enhancing techniques become more sophisticated, this powerful water management tool will surely see even greater and more diversified use.

2.5   URBAN RUNOFF MONITORING

The fact that most people live in urban areas suggests that urban runoff, and its implications for population centers, requires special attention. The conception of an urban runoff monitoring system requires a thorough understanding of the prevailing hydrologic regime and of the objectives to be met by monitoring. Usually, both the quality and quantity of urban runoff are of concern. Because monitoring costs can be substantial, it is important that monitoring networks be designed and operated in the most cost-effective manner [25],[26].

In general, urban monitoring networks have spatial and temporal dimensions. Although monitoring at a specific point location may be all that is necessary under some circumstances, it is becoming more common to monitor what is happening in a regional setting. The temporal aspect is similar. While a snapshot at some point in time may sometimes suffice, the time variance of the conditions to be tracked is usually critical for effective analyses. Both the short-term and long-term variability of many targets of monitoring must be ascertained. For example, water quality in a stream can change rapidly with time while changes in lake levels, such as those experienced in the Great Lakes in the 1980s, are the result of long-term hydrologic variability.

Spatial variability is an important consideration. Infiltration rates may vary considerably within a region, rainfall intensities may be quite different within even short distances, water quality might be different in upstream and downstream locations, and so on. Topography, soils, vegetal cover, paving, structures, and other factors affecting the performance of a hydrologic system are unevenly distributed in space, and these differences must be recognized. The trick is to develop a monitoring system that can provide the needed data, recognize regional and temporal variabilities, and keep installation, operation, and maintenance costs to a minimum. To do this requires a comprehensive knowledge of the system to be monitored, an understanding of what the data obtained by the system will be used for, and a prescription for the amount, type, and precision of the data to be collected.

Selecting appropriate instruments, determining sampling frequency, and setting data formats must also be addressed. Questions such as "How much do we need to know?" and "When do we need to know it?" must be answered. The form and extensiveness of data must be tightly related to the monitoring objectives. Furthermore, it is sometimes necessary to monitor surrogates instead of the condition to be tracked [26]. For example, if lake eutrophication is the issue, phosphorous and chlorophyll "a" concentrations might be surrogate measures. Hydrologic, water quality, land use and treatment, topographic, soils, vegetative cover, meteorologic, and other types of data may be needed, and the quantity of data required to meet these needs can be extensive.

Most monitoring systems include quality control and quality assurance (QA/QC) elements. Quality control is a planned system of activities designed to produce a quality product (data in this case) that meets the needs of the user. Quality assurance is a planned system of activities designed to guarantee that the quality control program is being carried out properly. A quality management plan should be part of the overall monitoring plan. It should be prepared when the monitoring program is being developed [25].

2.6   GROUNDWATER MONITORING

Groundwater monitoring embraces the measurement of the amount and quality of flows in aquifers. Baseline information on the extent and quality of groundwater resources is needed to effectively manage them. Monitoring yields benchmarks, supports mapping trends over time, and facilitates ascertaining the effects of regulatory and other policies on local and regional groundwater systems.

Flow measurements are made using networks of monitoring wells that record changes in groundwater elevations and permit estimates of the slopes of piezometric surfaces (see Chapter 10). Water quality determinations are made at locations near potential sources of pollution, at points of water withdrawal, and over regions of concern [37]–[39]. In this way, threats from contaminated sites can be evaluated, the quality of drinking water sources can be assessed, and overall regional groundwater quality can be estimated. The data obtained at monitoring stations can be used with a variety of groundwater quantity and quality models to forecast trends over time.

These trends portray expected future conditions under either present or modified states of development.

Numerous state and federal programs are active in groundwater monitoring. The U.S. Geological Survey is one of the best sources of data on groundwater systems. Its Regional Aquifer System Analysis (RASA) characterizes the major aquifers in the United States [37]. The Safe Drinking Water Act requires quality assurance monitoring of groundwater sources used for drinking water supplies. And the Resource Conservation and Recovery Act (RCRA) mandates that operators of hazardous waste facilities monitor groundwater in the vicinity of their sites.

Groundwater monitoring programs are designed to range from general assessments of trends in groundwater quantity and quality to very sophisticated networks capable of providing oversight for industry compliance with groundwater regulations [40]–[42]. In most cases, management systems for handling the large volumes of data generated during the monitoring process are required to support the effort.

2.7   NATIONAL WATER DATA EXCHANGE

The National Water Data Exchange (NAWDEX) is designed to match water data user needs with available data. It is a national confederation of water-oriented organizations. The member organizations are linked so that their water data holdings may be easily exchanged for maximum use. The NAWDEX program is coordinated from a program office of the U.S. Geological Survey. The program office indexes the data held by NAWDEX members and participants to provide a central source of water data information. These data may be in computerized and noncomputerized form. A variety of services assist users in identifying, locating, and obtaining the data they need. NAWDEX provides access to a number of large data files including those of the USGS Water Data Storage and Retrieval (WATSTORE) System; the Storage and Retrieval (STORET) System of the EPA; the Environmental Data and Information Service (EDIS) of the National Oceanic and Atmospheric Administration (NOAA); the Water Resources Scientific Information Center (WRSIC) of the U.S. Department of the Interior; several state governmental organizations; and the Water Resources Document (WATDOC) Reference Center of the Inland Waters Directorate, Canadian Department of the Environment. Detailed information on NAWDEX services can be obtained by contacting National Water Data Exchange, U.S. Geological Survey, 421 National Center, Reston, VA 22092.

SUMMARY

Hydrologic data are the building blocks for modeling hydrologic processes. Many sources of data may be accessed to support model development and verification, statistical analyses, and other studies. The quality of data obtained relates to the attributes of measuring instruments and to the features of gauging sites. The capabilities and limitations of measuring devices must be understood.

Information obtained by monitoring hydrologic systems enhances the understanding of system interactions and facilitates the design and testing of hydrologic models. Monitoring is the key to ascertaining the effectiveness of measures taken to protect the environment and/or alter watershed performance.

PROBLEM

2.1 Develop a list of data sources in your state or locality, by visiting the library or through other channels.

REFERENCES

[1] USDA Soil Conservation Service, *SCS National Engineering Handbook*, "Hydrology," Section 4, U.S. Govt. Print. Off., Washington, D.C., 1985.

[2] J. F. Miller, "Annotated Bibliography of NOAA Publications of Hydrometeorological Interest," NOAA Technical Memorandum NWS HYDRO-22, National Oceanic and Atmospheric Administration, Washington, D.C., May 1975.

[3] D. R. Maidment (ed.), *Handbook of Hydrology*, New York: McGraw-Hill, 1993.

[4] D. K. Todd (ed.), *The Water Encyclopedia*, New York: Water Information Center, 1970.

[5] "Soil Survey Manual," U.S. Department of Agriculture Handbook 18, 1951.

[6] "Soil Classification: A Comprehensive System, 7th Approximation," U.S. Department of Agriculture, Soil Survey Staff, August 1960.

[7] "Earth Manual," U.S. Department of the Interior, Bureau of Reclamation, 1960. "Unified Soil Classification System," U.S. Army Corps of Engineers Waterways Experiment Station, Technical Manual 3-357, Vicksburg, MS, March 1953.

[8] "Engineering Soil Classification for Residential Developments," Federal Housing Administration, F.H.A. 373, August 1959.

[9] USDA Soil Conservation Service, "Urban Hydrology for Small Watersheds," *Technical Release No. 55* (TR-55), Washington, D.C., 1986.

[10] "Irrigation Water Measurement," Mountain States Regional Publication 1, revision of Extension Circular 132, Irrigation Water Measurement, University of Wyoming, Laramie, WY, June 1964.

[11] U.S. Bureau of Reclamation, *Water Measurement Manual*, 2nd ed., U.S. Gov't. Print. Off., Washington, D.C., 1967.

[12] *Stevens Water Resources Data Book*, 4th ed. Beaverton, OR: Leupold and Stevens, Jan. 1987.

[13] R. K. Linsley, Jr., M. A. Kohler, and J. L. H. Paulhus, *Applied Hydrology*, New York: McGraw-Hill, 1982.

[14] R. L. Bras, *Hydrology: An Introduction to Hydrologic Science*, New York: Addison-Wesley, 1990.

[15] American Society of Civil Engineers, *Hydrology Handbook*, 2nd ed., ASCE Manual of Engineering Practice No. 28, Washington, D.C., 1996.

[16] Corps of Engineers, "Snow Hydrology," NTIS PB 151 660, North Pacific Division, Corps of Engineers, Portland, OR, June 1956.

[17] Ven Te Chow (ed.), *Handbook of Applied Hydrology*, New York: McGraw-Hill, 1964.

[18] W. Brutsaert, *Evaporation into the Atmosphere*, London: D. Reidel, 1982.

[19] "Texas Lab Installs Weighing Lysimeters," *Irrigation Journal*, Vol. 37, No. 3, May/June 1987, Encino, CA.

[20] American Society for Testing and Materials, "Standard Test Method for Classification of Soils for Engineering Purposes, d 2487-83, *1985 Annual Book of ASTM*, Vol. 4.08, pp. 395–408, Philadelphia, 1985.

[21] W. H. Gardner, "Water Content," in *Methods of Soil Analysis, Part I—Physical and Mineralogical Methods,* A. Klute, ed., American Society of Agronomy Monograph 9, 2nd ed., pp. 493–544, 1986.

[22] T. J. Schmugge, T. J. Jackson, and H. L. McKim, "Survey of Methods for Soil Moisture Determination," *Water Resour. Res.*, Vol. 16, No. 6, pp. 961–979, 1980.

[23] P. J. Gabrielsen and A. J. Carmeli, "Operation of a Hydrologic-Meteorologic Monitoring Network in a Severe Winter Environment," in Reference 25, 1987, pp. 113–122.

[24] H. E. Post and T. J. Grizzard, "The Monitoring of Stream Hydrology and Quality Using Microcomputers," in Reference 25, 1987, pp. 199–208.

[25] S. J. Nix and P. E. Black, eds., "Proceedings of the Symposium on Monitoring, Modeling, and Mediating Water Quality," American Water Resources Association, Bethesda, MD, 1987.

[26] S. A. Dressing, "Nonpoint Source Monitoring and Evaluation Guide," in Reference 25, 1987, pp. 69–78.

[27] R. J. C. Burnash and T. M. Twedt, "Event-Reporting Instrumentation for Real-Time Flash Flood Warning," American Meteorological Society, Preprints, Conference on Flash Floods: Hydro-meteorological Aspects, May 1978.

[28] D. E. Colton and R. J. C. Burnash, "A Flash-Flood Warning System," American Meteorological Society, Preprints, Conference on Flash Floods: Hydro-meteorological Aspects, May 1978.

[29] R. J. C. Burnash, "Automated Precipitation Measurements," Aug. 1980.

[30] R. J. C. Burnash and R. L. Ferral, "A Systems Approach to Real Time Runoff Analysis with a Deterministic Rainfall-Runoff Model," International Symposium on Rainfall-Runoff Modeling, University of Mississippi, May 18–21, 1981.

[31] Hydrologic Services Division, National Weather Service, Western Region, "Automated Local Evaluation in Real Time: A Cooperative Flood Warning System for Your Community," Feb. 1981.

[32] R. J. C. Burnash and R. L. Ferral, "Examples of Benefits and the Technology Involved in Optimizing Hydrosystem Operation Through Real-Time Forecasting," Conference on Real-Time Operation of Hydrosystems, Waterloo, Ontario, June 24–26, 1981.

[33] M. Deutsch, D. R. Wiesnet, and A. Rango (eds.), *Satellite Hydrology*, Bethesda, MD: American Water Resources Association, 1981.

[34] J. F. Bartholic, "Agricultural Meteorology: Systems Approach to Weather and Climate Needs for Agriculture, Forestry, and Natural Resources," in *Proceedings, Thirty-Second Meeting, Agricultural Research Institute*, Bethesda, MD: Agricultural Research Institute, 1983, pp. 75–85.

[35] M. A. Calabrese and P. G. Thome, "NASA Water Resources/Hydrology Remote Sensing Program in the 1980's," in *Satellite Hydrology* (M. Deutsch, D. R. Wiesnet, and A. Rango, eds.), Bethesda, MD: American Water Resources Association, 1981, pp. 9–15.

[36] G. K. Moore, "An Introduction to Satellite Hydrology," in *Satellite Hydrology* (M. Deutsch, D. R. Wiesnet, and A. Rango, eds.), Bethesda, MD: American Water Resources Association, 1981, pp. 37–41.

[37] T. R. Henderson, J. Trauberman, and T. Gallagher, *Groundwater Strategies for State Action*, Environmental Law Institute, Washington, D.C., 1984.

[38] C. Everett, *Groundwater Monitoring*, General Electric Co., Schenectady, NY, 1980.

[39] S. Nacht, "Ground Water Monitoring System Considerations," *Ground Water Monitoring Review*, Vol. 3, No. 2, 1983.

[40] D. Miller, "Guidelines for Developing a Statewide Groundwater Monitoring Program," *Groundwater Monitoring Review*, Spring 1981.

[41] "Phase II Report to OMB, Implementation of the Requirements," U.S. EPA, Office of Solid Waste, Washington, D.C., March 10, 1983.

[42] W. G. Gray and J. L. Hoffman, "A Numerical Model Study of Groundwater Contamination from Price's Landfill, New Jersey: I and II," *Ground Water*, Vol. 21, No. 1, 1983.

Statistical Methods in Hydrology

OBJECTIVES

The purpose of this chapter is to:

- Present the commonly used methods for statistical analysis of hydrologic data that need to be understood before reading later chapters
- Introduce concepts of probability, frequency, recurrence interval, return period, and regression and correlation
- Introduce the basic tenets of probability theory as applied to random, hydrologic variables
- Describe common probability distributions and show how they are applied in hydrology
- Illustrate several methods for conducting frequency analyses, including the use of frequency factors that allow estimation of magnitudes of variables for given recurrence intervals
- Explain the widely used Bulletin No. 17B log–Pearson Type III method for analyzing magnitudes of extremes in hydrology
- Review the fundamentals of linear correlation and regression as applied in hydrology
- Provide the reader with references to other standard sources that expand discussions of statistics, probability, and risk beyond the scope of this book.

3.1   RANDOM VARIABLES AND STATISTICS

A *random variable* is one that demonstrates variability that isn't sufficiently explained by analytical measures of physical processes. Many hydrologic phenomena have this tendency, appearing at times to be fully subject to chance themselves, or driven by some other closely related factor. In practice, hydrologists often analyze problems as systems of connected random and deterministic processes. For example, precipitation is often evaluated statistically as a random variable because of the complexity of understanding and modeling the atmospheric processes that are known to drive the precipitation system. Runoff is also random, but when it is calculated as a function of precipitation, it is being viewed deterministically, using the rainfall–runoff analogs that are the nucleus of much of hydrology.

Methods of *statistical analysis* in hydrology provide ways to reduce and summarize observed data, to present information in precise and meaningful form, to determine the underlying characteristics of the observed phenomena, and to make predictions concerning future behavior [1]–[5]. These inferences include information about the central tendency, range, distribution within the range, variability around the central tendency, degree of uncertainty, and frequency of occurrence of values. The U.S. Geological Survey maintains updates of software for performing statistical computations of surface water data. A download of the program SWSTAT is available at http://water.usgs.gov/software/surface_water.html.

The random variables in the process under study are *continuous* if they may take on all values in the range of occurrence, including figures differing only by an infinitesimal amount; they are *discrete* if they are restricted to specific, incremental values. Distribution of the variables over the range of occurrence is defined in terms of the frequency or *probability* with which different values have occurred or might occur.

3.2   PROBABILITY DISTRIBUTIONS

Random variables, either discrete or continuous, are characterized by the distribution of probabilities attached to the specific values that the variable may assume [1],[2]. A random variable throughout its range of occurrence is generally designated by a capital letter, and a specific value or outcome of the random process is designated by a lowercase letter. For example, $P(X = x_1)$ is the probability that random variable X takes on the value x_1. A shorter version is $P(x_1)$. Figure 3.1 shows the probability distribution of

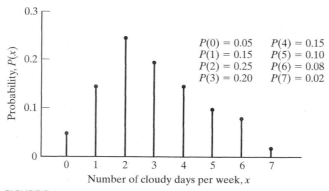

$P(0) = 0.05$	$P(4) = 0.15$
$P(1) = 0.15$	$P(5) = 0.10$
$P(2) = 0.25$	$P(6) = 0.08$
$P(3) = 0.20$	$P(7) = 0.02$

FIGURE 3.1

Probability distribution of cloudy days per week.

the number of cloudy days in a week. It is a discrete distribution because the number of days is exact; in the record from which the relative frequencies were taken, a day had to be described as cloudy or not. Observe that each of the seven events has a finite probability and the sum is 1; that is:

$$\sum_i P(x_i) = 1 \qquad (3.1)$$

The *cumulative distribution function*, CDF, is a graph of the probability that any outcome in X is less than or equal to a stated, limiting value x. The cumulative distribution function is denoted $F(x)$. Thus:

$$F(x) = P(X \le x) \qquad (3.2)$$

and the function increases monotonically from a lower limit of zero to an upper bound of unity. Figure 3.2 is the CDF of the number of cloudy days in a week derived from Fig. 3.1 by taking cumulative probabilities. The function shows that the probability is 90% that the number of cloudy days in the week will be 5 or less. Conversely, there is a 10 percent probability that it will be cloudy for 6 or 7 days. This complementary cumulative probability is sometimes called $G(x)$, where:

$$G(x) = 1 - F(x) = P(X \ge x) \qquad (3.3)$$

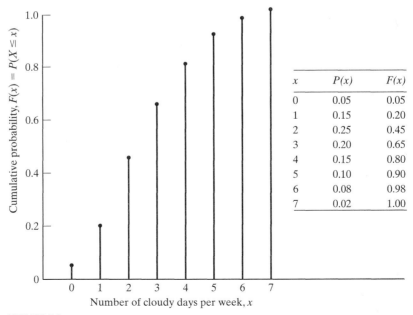

x	$P(x)$	$F(x)$
0	0.05	0.05
1	0.15	0.20
2	0.25	0.45
3	0.20	0.65
4	0.15	0.80
5	0.10	0.90
6	0.08	0.98
7	0.02	1.00

FIGURE 3.2

Cumulative distribution of cloudy days per week.

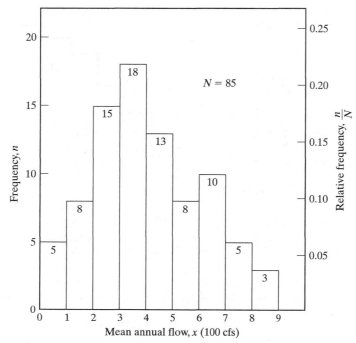

FIGURE 3.3

Frequency distribution of mean annual flows.

Continuous variables present a slightly different picture. Figure 3.3 is the *histogram* of an 85-year record of annual streamflows. The observations were grouped into nine intervals ranging from 0 to 900 cfs and the number falling in each interval was plotted as frequency on the left ordinate. A convenient alternative is to plot the relative frequency as shown by the right ordinate. The CDF for the streamflow record is shown in Fig. 3.4. As the number of observations increases, the continuous distribution will be developed by reducing the size of the intervals. In the limit, the broken curves of Figs. 3.3 and 3.4 will appear as those in Fig. 3.5.

The ordinates of Figs. 3.3 and 3.5a are different. Because relative frequency is synonymous with probability, it is convenient to reconstitute the histogram so that the area in each interval represents probability; the total area contained is thus unity. To do this, the ordinate in each interval, say n/N for relative frequency or probability, is divided by the interval width, Δx. The ratio $n/N\,\Delta x$ is literally the probability per unit length in the interval and therefore represents the average density of probability. The probability n/N in the interval is represented on the CDF (before the limiting process) as $\Delta F(x)$, or $F(x + \Delta x/2) - F(x - \Delta x/2)$. We then can define:

$$f(x) = \lim_{\Delta x \to 0} \frac{\Delta F(x)}{\Delta x} = \frac{dF(x)}{dx} \tag{3.4}$$

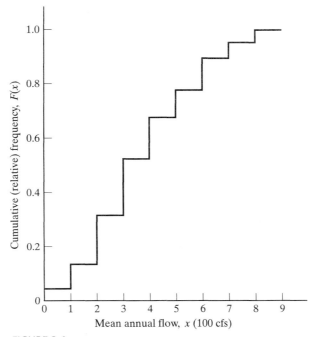

FIGURE 3.4

Cumulative frequency distribution of mean annual flows.

which is called the *probability density function*, PDF. This function is the density (or intensity) of probability at any point; $f(x)\,dx$ is described as the differential probability.

For continuous variables, $f(x) \geq 0$, since negative probabilities have no meaning. Also, the function has the property that:

$$\int_{-\infty}^{\infty} f(x)\,dx = 1 \tag{3.5}$$

which again is the requirement that the probabilities of all outcomes sum to 1. Furthermore, the probability that x will fall between the limits a and b is written:

$$P(a \leq X \leq b) = \int_{a}^{b} f(x)\,dx \tag{3.6}$$

Note that the probability that x takes on a particular value, say a, is zero; that is:

$$\int_{a}^{a} f(x)\,dx = 0 \tag{3.7}$$

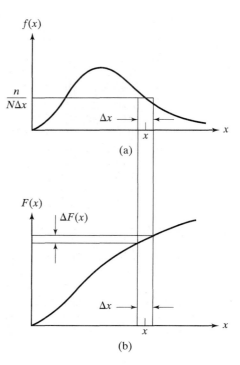

FIGURE 3.5

Continuous probability distributions: (a) probability
density function and (b) cumulative distribution
function.

which emphasizes that finite probabilities are defined only as areas under the PDF
curve between finite limits.

The CDF can now be defined in terms of the PDF as:

$$P(-\infty \leq X \leq x) = P(X \leq x) = F(x) = \int_{-\infty}^{x} f(u)\, du \qquad (3.8)$$

where u is used as a dummy variable to avoid confusion with the limit of integration.
The area under the CDF curve has no meaning, only the ordinates, or the difference in
ordinates. For example, $P(x_1 \leq X \leq x_2)$, which is equivalent to Eq. 3.6, can be evaluated as $F(x_2) - F(x_1)$.

Example 3.1

Table B.1 contains the area beneath a "standard normal" bell-shaped PDF curve.
Because the distribution is symmetrical, areas are provided only on one side of the
center. Use the distribution to determine the values of

1. $P(0 \leq z \leq 2)$.
2. $P(-2 \leq z \leq 2)$.
3. $P(z \geq 2)$.
4. $P(z \leq -1)$.

Solution

1. $P(0 \leq z \leq 2) = 0.4772$.
2. From symmetry, $P(-2 \leq z \leq 0) = P(0 \leq z \leq 2) = 0.4772$. Since $P(-2 \leq z \leq 2) = P(-2 \leq z \leq 0) + P(0 \leq z \leq 2)$, then $P(-2 \leq z \leq 2) = 0.4772 + 0.4772 = 0.9544$.
3. This is the area under the curve in the right tail beyond $z = 2.0$. Because the area right of center ($z = 0$) is 0.5000, $P(z \geq 2) = P(z \geq 0) - P(0 \leq z \leq 2)$, or $P(z \geq 2) = 0.5000 - 0.4772 = 0.0228$.
4. From the solution to (3), $P(z \leq -1) = P(z \leq 0) - P(-1 \leq z \leq 0)$. By symmetry, $P(-1 \leq z \leq 0) = P(0 \leq z \leq 1) = 0.3413$, and $P(z \leq -1) = 0.5000 - 0.3413 = 0.1587$.

3.3   DISTRIBUTION STATISTICS

Characteristics of random variable distributions are *central tendency*, the grouping of observations or probability about a central value; *variability*, the dispersion of the variate or observations; and *skewness*, the degree of asymmetry of the distribution [1],[3],[4],[6]. The theoretical functions shown in Fig. 3.6 exhibit approximately the same grouping about a central value, but f_2 has much greater variability than f_1, and f_2 possesses a pronounced right-skew while f_1 is symmetrical.

Sample Versus Population Statistics

In introducing the parameters of distributions, the usual sequence of statistical problems will be followed—that is, statistics are derived from the distribution of *sample* data and used as estimates of the parameters of the *population* distribution. Summation forms of integrals are used to compute moments for samples. For example, the mean of sample data is designated $\bar{x}$ and it is used as the best estimate of the population mean. By convention, Greek letters are used to denote population parameters.

Central Tendency

The familiar arithmetic average, the *mean*, is the most used measure of central tendency. It is estimated by the first moment about the origin for the sample data and

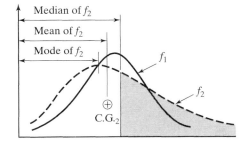

FIGURE 3.6

Symmetrical and skewed probability distribution for continuous variables.

is designated:

$$\bar{x} = \frac{1}{n}\sum_{i=1}^{n}x_i \tag{3.9}$$

The statistic $\bar{x}$ is only an estimate of the population mean μ.

Means of defining central tendency other than the arithmetic mean—for example, the *geometric mean* $\bar{x} = (x_1x_2x_3\cdots x_n)^{1/n}$ or *harmonic mean* $\bar{x} = n/\Sigma(1/x_i)$—are also used. Two additional measures of central tendency are the *median*, which is the middle value of the observed data and divides the distribution into equal areas, and the *mode*, which in discrete variables is the value occurring most frequently and in continuous variables is the peak value of probability density. All three are illustrated in Fig. 3.6.

Variability

Variability of a random variable can be represented by the total range of values or by the deviation about the mean; however, the parameter of statistical importance is the mean squared deviation as measured by the second moment about the mean. The parameter is termed the *variance* and is designated by:

$$\sigma^2 = \frac{1}{n}\sum_{i=1}^{n}(x_i - \mu)^2 \tag{3.10}$$

But the population mean μ is not known precisely and therefore it is necessary to compute instead:

$$s^2 = \frac{1}{n-1}\sum_{i=1}^{n}(x_1 - \bar{x})^2 \tag{3.11}$$

As an unbiased estimate of σ^2, the quantity s^2 is found using $n-1$ in place of n in Eq. 3.10. The reasoning for this substitution involves the loss of a degree of freedom by using $\bar{x}$ instead of μ, but a proof is beyond the scope of this text [1],[6]–[9].

The square root of the variance is a statistic known as the *standard deviation* (σ or s), in which form variability is measured in the same units as the variate and the mean, and hence is easier to interpret and manipulate. The *coefficient of variation* C_v, defined as σ/μ or $s/\bar{x}$, is an expression useful in comparing relative variability.

Skewness

A fully symmetrical distribution would exhibit the property that all odd moments equal zero. A skewed distribution, however, would have excessive weight to either side of the center and the odd moments would exist. The third moment α is:

$$\alpha = \frac{1}{n}\sum_{i=1}^{n}(x_i - \mu)^3 \tag{3.12}$$

An unbiased estimate of the third moment is computed by:

$$a = \frac{n}{(n-1)(n-2)} \sum_{i=1}^{n} (x_i - \bar{x})^3 \tag{3.13}$$

The *coefficient of skewness* is the ratio α/σ^3 and is estimated by:

$$C_s = \frac{a}{s^3} \tag{3.14}$$

For symmetrical distributions, the third moment is zero and $C_s = 0$; for right skewness (i.e., the long tail to the right side) $C_s > 0$, and for left skewness $C_s < 0$. The PDF for f_2 shown in Fig. 3.6 has a right or positive skew. The property of skewness is of questionable statistical value when it must be estimated from less than 50 sample data points.

Example 3.2

Determine the sample statistics and compare the three distributions of annual rainfall for the records shown in Table 3.1.

TABLE 3.1   Typical Annual Rainfall for Selected Cities

Year	Annual Rainfall (in.)		
	Anniston, AL	Los Angeles, CA	Richmond, VA
2000	48	9	43
1999	49	19	44
1998	55	19	38
1997	98	9	31
1996	43	8	47
1995	53	6	49
1994	56	15	52
1993	47	20	31
1992	69	11	51
1991	57	9	40
1990	61	18	41
1989	64	8	43
1988	99	23	37
1987	54	17	36
1986	40	23	34
1985	47	17	38
1984	58	10	36
1983	44	18	37
1982	44	5	43
1981	64	24	34
1980	44	19	53
1979	51	15	49
1978	71	21	47

Solution

Parameter	Anniston	Los Angeles	Richmond
Mean, $\bar{x}$	57.2 in.	14.9 in.	41.5 in.
Standard deviation, s	15.5 in.	5.9 in.	6.7 in.
Coefficient of variation, $C_v = s/\bar{x}$	0.27	0.40	0.16
Coefficient of skewness, $C_s = a/s^3$	1.69	−0.16	0.16

Comments

(1) Anniston's record shows a high annual average and a fairly large variability. In particular, Anniston's distribution has a pronounced right skew, caused principally by two very large observed values in this short period of record. (2) Los Angeles has a small annual average but a very large variability and a slightly negative skewness. (3) Richmond has the most uniform distribution: a relatively small variability and only a slight positive skewness.

3.4   PROBABILITY APPLICATIONS: FREQUENCY AND RETURN PERIOD

The laws of probability underlie any study of the statistical nature of repeated observations or trials. The probability of a single event, say E_1, is defined as the relative number of occurrences of the event after a long series of trials. Thus $P(E_1)$, the probability of event E_1, is n_1/N for n_1 occurrences of the same event in N trials if N is sufficiently large. The number of occurrences n_1 is the *frequency*, and n_1/N the *relative frequency*.

Often the probabilities and the rules governing their manipulation are known intuitively or from experience. In the familiar coin-tossing experiment, P (heads) $=$ P (tails) $= \frac{1}{2}$. Each outcome of a single toss (a trial) has a finite probability, and the sum of the probabilities of all possible outcomes is 1. Also, the outcomes are *mutually exclusive*; that is, if one occurs, say a head, then a tail cannot occur. In two successive tests, there are four possible outcomes—HH, TT, HT, TH—each with a probability of $\frac{1}{4}$. In this case, because each trial is independent of the other one, probabilities for each outcome are found by P (first trial) $\times P$ (second trial) $= \frac{1}{2} \times \frac{1}{2} = \frac{1}{4}$. Again, the sum of the probabilities of the possible outcomes is 1. Note that the probability of getting exactly one head and one tail during the experiment (without any regard to the order) is P (HT) $+ P$ (TH) $= \frac{1}{2}$.

Summarizing the rules of probability indicated by coin tossing, we find the following: [4],[5].

1. The probability of an event is nonnegative and never exceeds 1:

$$0 \le P(E_i) \le 1 \tag{3.15}$$

2. The sum of the probabilities of all possible outcomes in a single trial is 1:

$$\sum_i P(E_i) = 1 \qquad (3.16)$$

3. The probability of a number of *independent* and *mutually exclusive* events is the sum of the probabilities of the separate events:

$$P(E_1 \cup E_2) = P(E_1) + P(E_2) \qquad (3.17)$$

The probability statement, $P(E_1 \cup E_2)$, signifies the probability of the *union* of two events and is read "the probability of E_1 or E_2."

4. The probability of two *independent* events occurring simultaneously or in succession is the product of the individual probabilities:

$$P(E_1 \cap E_2) = P(E_1) \times P(E_2) \qquad (3.18)$$

$P(E_1 \cap E_2)$ is called probability of the *intersection* of two events or *joint probability* and is read "the probability of E_1 and E_2."

Consider the following example of events that are not independent or mutually exclusive: An urban drainage canal reaches flood stage each summer with relative frequency of 0.10; power failures in industries along the canal occur with probability of 0.20; experience shows that when there is a flood the chances of a power failure for whatever reason are raised to 0.40. The probability statements are

$P(\text{flood}) = P(F) = 0.10$ $P(\text{power failure}) = P(P) = 0.20$

$P(\text{no flood}) = P(\overline{F}) = 0.90$ $P(\text{no power failure}) = P(\overline{P}) = 0.80$

$P(\text{power failure given that a flood occurs}) = 0.40$

The last statement is called a *conditional probability*. It signifies the joint occurrence of events and is usually written $P(P \mid F)$. Rules 3 and 4 no longer are strictly applicable. If Rule 3 applied, $P(F \cup P) = P(F) + P(P) = 0.3$. If the events remained independent, the conditional probability $P(P \mid F)$ would equal the *marginal* probability $P(P)$. Thus the events are independent if the probability of either is not "conditioned by" or changed by knowledge that the other has occurred. For independent events, the joint probabilities would be

$$P(F \cap P) = 0.1 \times 0.2 = 0.02$$

$$P(F \cap \overline{P}) = 0.1 \times 0.8 = 0.08$$

$$P(\overline{F} \cap P) = 0.9 \times 0.2 = 0.18$$

$$P(\overline{F} \cap \overline{P}) = 0.9 \times 0.8 = 0.72$$

The probability of a flood or a power failure during the summer would be the sum of the first three joint probabilities above.

$$P(F \cup P) = P(F \cap P) + P(F \cap \overline{P}) + P(\overline{F} \cap P) = 0.28$$

The events are dependent, however, from the statement of conditional probability: When a flood occurs with $P(F) = 0.1$, a power failure will occur with probability 0.4, and true joint probability is $P(F) \times P(P \mid F) = 0.1 \times 0.4 = 0.04 = P(F \cap P)$. The probability of the union is then $P(F \cup P) = P(F) + P(P) - P(F \cap P) = 0.1 + 0.2 - 0.04 = 0.26$. Note the contrast:

$$P(F \cup P) = 0.30 \quad \text{for mutually exclusive events}$$

$$P(F \cup P) = 0.28 \quad \text{for joint but independent events}$$

$$P(F \cup P) = 0.26 \quad \text{otherwise}$$

The new, more general rule for the union of probabilities is:

5. $P(E_1 \cup E_2) = P(E_1) + P(E_2) - P(E_1 \cap E_2)$ \hfill (3.19)

and a sixth rule should be added for conditional probabilities:

6. $P(E_1 \mid E_2) = \dfrac{P(E_1 \cap E_2)}{P(E_2)}$ \hfill (3.20)

A very important concept of independence is expressed in a variation of Rule 6, namely, $P(E_1 \mid E_2) = P(E_1)$ if events E_1 and E_2 are independent. This further explains Rule 3 that $P(E_1) \times P(E_2) = P(E_1 \cap E_2)$ for independent events.

The example of flooding can be extended to show some interesting features about probabilities and risks associated with hydrologic phenomena. $P(F) = .10$ implies a 10 percent chance each year for the flood to "occur," meaning that the flood level will be exceeded. Because the probability of any single, exact value of a continuous variable is 0, "occur" can also mean the level will be reached or exceeded. In the long run, the level would be reached or exceeded on the average once in 10 years. Thus the average *return period** T in years is defined as:

$$T = \frac{1}{P(F)} = \frac{1}{1 - P(\overline{F})}$$ \hfill (3.21)

and the following general probability relations hold:

1. The probability that F will be equaled or exceeded in any year:

$$P(F) = \frac{1}{T}$$ \hfill (3.22)

---

*The terms *return period* and *recurrence interval* are used interchangeably to denote the reciprocal of the annual probability of exceedence.

2. The probability that F will not be exceeded in any year:

$$P(\overline{F}) = 1 - P(F) = 1 - \frac{1}{T} \tag{3.23}$$

3. The probability that F will not be equaled or exceeded in any of n successive years:

$$P_1(\overline{F}) \times P_2(\overline{F}) \times \cdots \times P_n(\overline{F}) = P(\overline{F})^n = \left(1 - \frac{1}{T}\right)^n \tag{3.24}$$

4. The probability R, called *risk*, that F will be equaled or exceeded at least once in n successive years:

$$R = 1 - \left(1 - \frac{1}{T}\right)^n = 1 - \{P(\overline{F})\}^n \tag{3.25}$$

Table 3.2 shows return periods associated with various levels of risk.

TABLE 3.2   Return Periods Associated with Various Degrees of Risk and Expected Design Life

Risk (%)	Expected design life (years)							
	2	5	10	15	20	25	50	100
75	2.00	4.02	6.69	11.0	14.9	18.0	35.6	72.7
50	3.43	7.74	14.9	22.1	29.4	36.6	72.6	144.8
40	4.44	10.3	20.1	29.9	39.7	49.5	98.4	196.3
30	6.12	14.5	28.5	42.6	56.5	70.6	140.7	281
25	7.46	17.9	35.3	52.6	70.0	87.4	174.3	348
20	9.47	22.9	45.3	67.7	90.1	112.5	224.6	449
15	12.8	31.3	62.0	90.8	123.6	154.3	308	616
10	19.5	48.1	95.4	142.9	190.3	238	475	950
5	39.5	98.0	195.5	292.9	390	488	976	1949
2	99.5	248	496	743	990	1238	2475	4950
1	198.4	498	996	1492	1992	2488	4975	9953

Example 3.3

If T is the recurrence interval for a flood with magnitude Q_a, find the probability (risk) that the peak flow rate will equal or exceed Q_a at least once in two consecutive years. Assume the events are independent.

Solution.   The solution is easily obtained by substitution into Eq. 3.25. To assist in understanding the equations, an alternative derivation follows.

The four possible outcomes for the two years are:

a: nonexceedance in both years
b: exceedance in the first year only
c: exceedance in the second year only
d: exceedance in both years

Because these four represent all possible outcomes, the probability of the union of a, b, c, and d is 1.0, or from Eq. 3.16, $P(a \cup b \cup c \cup d) = 1.0$. Exceedance in at least one year is satisfied by b, c, or d, but not a. Thus the risk of at least one exceedance is $P(b \cup c \cup d)$, which is the total less the probability of a. From Eqs. 3.16 and 3.17, we find that:

$$2\text{-year risk} = P(b \cup c \cup d) = 1 - P(a)$$

From Eq. 3.17, we find that:

$$P(a) = P(Q < Q_a \text{ in Year 1}) \times P(Q < Q_a \text{ in Year 2})$$

$$= \left(1 - \frac{1}{T}\right)\left(1 - \frac{1}{T}\right)$$

and
$$\text{Risk} = 1 - P(a) = 1 - \left(1 - \frac{1}{T}\right)^2$$

Example 3.4

What return period must a highway engineer use in designing a critical underpass drain to accept only a 10 percent risk that flooding will occur in the next 5 years?

Solution

$$R = 1 - \left(1 - \frac{1}{T}\right)^n$$

$$0.10 = 1 - \left(1 - \frac{1}{T}\right)^5$$

$$T = 48.1 \text{ years}$$

Note that this is the same result obtained by entering Table 3.2 with a 10 percent risk and a 5-yr design life.

3.5   TYPES OF PROBABILITY DISTRIBUTION FUNCTIONS

Many standard theoretical probability distributions have been used to describe hydrologic processes [1],[4],[5]. It should be emphasized that any theoretical distribution is not an exact representation of the natural process but only a description that approximates the underlying phenomenon and has proved useful in describing the observed data. Table 3.3 summarizes the common distributions, giving the PDF, mean, and variance of the functions. The distributions presented in the table have experienced wide application and are derived and discussed in many standard textbooks on statistics. In the material to follow, only aspects of the most used distributions are given.

The uses of binomial and Poisson discrete probability distributions in Table 3.3 are restricted generally to those random events in which the outcome can be described either as a success or failure. Furthermore, the successive trials are independent and the probability of success remains constant from trial to trial [8],[9]. In a sense, the common discrete distributions are counting or enumerating techniques.

The binomial distribution is frequently used to approximate other distributions, and vice versa. For example, with discrete values, when n is large and p small (such that $np < 5$ preferably), the binomial approaches the Poisson distribution. This is a single-parameter distribution ($\lambda = np$) and is very useful in describing arrivals in queueing theory. When p approaches $\frac{1}{2}$ and n grows large, the binomial becomes indistinguishable from the normal distribution described in the next section.

3.6   CONTINUOUS PROBABILITY DISTRIBUTION FUNCTIONS

Most hydrologic variables are assumed to be continuous random processes, and the common continuous distributions are used to fit historical sequences, as in frequency analysis (Section 3.7). Other applications are also important for continuous distributions. The elementary uniform distribution is the basis for computing random numbers so important in simulation studies. The whole body of material in the area of reliability and estimating depends on derived distributions like Student's t, chi-squared, and the F distribution. The explanations that follow concern the more common distributions applied in fitting hydrologic sequences. The reader is referred to standard texts for more detailed treatment [4],[6–9].

Normal Distribution

The normal distribution is a symmetrical, bell-shaped frequency function, also known as the Gaussian distribution or the natural law of errors. It describes many processes that are subject to random and independent variations. The whole basis for a large body of statistics involving testing and quality control is the normal distribution. Although it often does not perfectly fit sequences of hydrologic data, it has wide application, for example, in dealing with transformed data that do follow the normal distribution and in estimating sample reliability by virtue of the central limit theorem.

The normal distribution has two parameters, the mean μ and the standard deviation σ, for which $\bar{x}$ and s, derived from sample data, are substituted. By a simple transformation, the distribution can be written as a single-parameter function only. Defining $z = (x - \mu)/\sigma, dx = \sigma\,dz$, the PDF becomes:

$$f(z) = \frac{1}{\sqrt{2\pi}}\,e^{-z^2/2} \tag{3.26}$$

TABLE 3.3   Table of Common Distributions Used in Hydrology

Distribution of random variable X	Probability density function and CDF	Range	Mean $\bar{x}$ or μ	Variance s^2 or σ^2
Binomial	$P(x) = \dfrac{n!}{x!(n-x)!}p^x(1-p)^{n-x}$	$0 \le x < n$	np	$np(1-p)$
Poisson	$P(x) = \dfrac{\lambda^x e^{-\lambda}}{x!}$	$0 \le x \le \cdots$	λ	λ
Uniform	$f(x) = \dfrac{1}{b-a}$	$a \le x \le b$	$\dfrac{b+a}{2}$	$\dfrac{(b-a)^2}{12}$
Exponential	$f(x) = \dfrac{1}{a}e^{-x/a}$	$0 \le x \le \infty$	a	a^2
Normal	$f(x) = \dfrac{1}{\sigma\sqrt{2\pi}}e^{-(x-\mu)^2/2\sigma^2}$	$-\infty \le x \le \infty$	μ	σ^2
Lognormal $(y = \ln x)$	$f(y) = \dfrac{1}{x\sigma_y\sqrt{2\pi}}\exp\left[\dfrac{-(y-\mu_y)^2}{2\sigma_y^2}\right]$	$-\infty \le y \le \infty$ $(0 \le x \le \infty)$	μ_y	σ_y^2
Gamma	$f(x) = \dfrac{x^\alpha e^{-x/\beta}}{\beta^{\alpha+1}\Gamma(\alpha+1)}$	$0 \le x \le \infty$	$\beta(\alpha+1)$	$\beta^2(\alpha+1)$
Gumbel	$f(x) = \dfrac{1}{\alpha}\exp\left[-\dfrac{x-\xi}{\alpha} - \exp\left(-\dfrac{x-\xi}{\alpha}\right)\right]$ $F(x) = \exp\left[-\exp\left(-\dfrac{x-\xi}{\alpha}\right)\right]$ $x = \xi - \alpha\ln[-\ln F]$	$-\infty \le x \le \infty$	$\mu = \xi + 0.5772\alpha$	$\sigma^2 = \dfrac{\pi^2\alpha^2}{6} \approx 1.645\alpha^2$
Weibull	$f(x) = \left(\dfrac{k}{\alpha}\right)\left(\dfrac{x}{\alpha}\right)^{k-1}\exp\left[-\left(\dfrac{x}{\alpha}\right)^k\right]$ $F(x) = 1 - \exp[-(x/\alpha)^k]$	$x \ge 0; \alpha, k \ge 0$	$\mu = \alpha\Gamma\left(1+\dfrac{1}{k}\right)$	$\sigma^2 = \alpha^2\left\{\Gamma\left(1+\dfrac{2}{k}\right) - \left[\Gamma\left(1+\dfrac{1}{k}\right)\right]^2\right\}$
Extreme value	$f(x) = \alpha\exp\{-\alpha(x-u) - e^{-\alpha(x-u)}\}$	$-\infty \le x \le \infty$	$u + \dfrac{0.5772}{\alpha}$	$\dfrac{\pi^2}{6\alpha^2}$
Log–Pearson III $(y = \ln x)$	$f(x) = \dfrac{(y-\gamma)^\alpha}{\beta^2 x\Gamma(\alpha+1)}\exp\left[\dfrac{-(y-\gamma)}{\beta}\right]$	$-\infty \le y \le \infty$ $(0 \le x \le \infty)$	$\mu_y = \gamma + \beta(\alpha+1)$	$\sigma_y^2 = \beta^2(\alpha+1)$

and the CDF becomes:

$$F(z) = \frac{1}{\sqrt{2\pi}} \int_{-\infty}^{z} e^{-u^2/2} \, du \tag{3.27}$$

The variable z is called the *standard normal variate*; it is normally distributed with zero mean and unit standard deviation. Tables of areas under the standard normal curve, as given in Appendix B, Table B.1, serve all normal distributions after standardization of the variables. Given a cumulative probability, the deviate z is found in the table of areas and x is found from the inverse transform:

$$x = \mu + z\sigma \quad \text{or} \quad x = \bar{x} + zs \tag{3.28}$$

Example 3.5

Assume that the Richmond, Virginia, annual rainfall in Table 3.1 follows a normal distribution. Use the standard normal transformation to find the rain depth that would have a recurrence interval of 100 years.

Solution.

1. From example 3.2, the mean is 41.5 in. and the standard deviation is 6.7 in. This gives:

    $$x = 41.5 + z(6.7)$$

2. Equation 3.8 shows that the area under the PDF to the right of z is the exceedence probability of the event. For the 100-yr event, Eq. 3.21 gives the exceedence probability $P(z) = 1/T_r = 1/100 = 0.01$. From the figure accompanying Table B.1 in Appendix B, $F(z) = 0.5 - P(z) = 0.49$, and $z = 2.326$ by interpolating the table. The expected 100-yr rain depth is therefore:

    $$x = 41.5 + (2.326 \times 6.7) = 57.1 \text{ in.}$$

    The 100-yr event for a normal distribution is 2.326 standard deviations above the mean.

Lognormal Distribution

Many hydrologic variables exhibit a marked right skewness, partly due to the influence of natural phenomena having values greater than zero, or some other lower limit, and being unconstrained, theoretically, in the upper range. In such cases, frequencies will not follow the normal distribution, but their logarithms may follow a normal distribution [10]. The PDF shown in Table 3.3 for the lognormal comes from substituting $y = \ln x$ in the normal. With μ_y and σ_y as the mean and standard deviation, respectively, and with α as the expected value of the log-transformed variable, the following relations have been found to hold between the characteristics of the untransformed variate x and the transformed variate y [10],[11]:

$$\mu = \exp(\mu_y + \sigma_y^2/2) \tag{3.29}$$

$$\sigma^2 = \mu^2[\exp(\sigma_y^2) - 1] \tag{3.30}$$

$$\alpha = [\exp(3\sigma_y^2) - 3\exp(\sigma_y^2) + 2]C_v^3 \tag{3.31}$$

$$C_v = [\exp(\sigma_y^2) - 1]^{1/2} \tag{3.32}$$

$$C_s = 3C_v + C_v^3 \tag{3.33}$$

Also $\mu_y = \ln M$, where M is the median value *and* the geometric mean of the x's.

The lognormal is especially useful because the transformation opens the extensive body of theoretical and applied uses of the normal distribution. Since both the normal and lognormal are two-parameter distributions, it is necessary only to compute the mean and variance of the untransformed variate x and solve Eqs. 3.29 and 3.30 simultaneously. Information on three-parameter or truncated lognormal distributions can be found in the literature [10],[11].

Gamma (and Pearson Type III) Distribution

The gamma distribution has wide application in mathematical statistics and has been used increasingly in hydrologic studies. In greater use is the *Pearson Type III*. This distribution has been widely adopted as the standard method for flood frequency analysis in a form known as the *log–Pearson III* in which the transform $y = \log x$ is used to reduce skewness [12]–[14]. Although all three moments are required to fit the distribution, it is extremely flexible in that a zero skew will reduce the log–Pearson III distribution to a lognormal and the Pearson Type III to a normal. Tables of the cumulative function are available [14],[15] and will be explained in a later section. A very important property of gamma variates as well as normal variates (including transformed normals) is that the sum of two such variables retains the same distribution. This feature is important in generating synthetic hydrologic sequences [17],[18].

Gumbel's Extremal Distribution

The theory of extreme values considers the distribution of the largest (or smallest) observations occurring in each group of repeated samples. The distribution of the n_1 extreme values taken from n_1 samples, with each sample having n_2 observations, depends on the distribution of the $n_1 n_2$ total observations. Gumbel was the first to employ extreme value theory for analysis of flood frequencies [18]. Chow has demonstrated that the Gumbel distribution is a lognormal with constant skewness [19]. The CDF of the density function given in Table 3.3 is:

$$P(X \le x) = F(x) = \exp\{-\exp[-\alpha(x - u)]\} \tag{3.34}$$

which is a convenient form to evaluate the function. Parameters α and u are given as functions of the mean and standard deviation in Table 3.3. Tables of the double exponential are usually in terms of the reduced variate, $y = \alpha(x - u)$ [20]. Gumbel also has proposed another extreme value distribution that appears to fit instantaneous (minimum annual) drought flows [21],[22].

CDFs in Hydrology

Normal and Pearson distributions can often be used to describe hydrologic variables if the variable is the sum or mean of several other random variables [4]. The sum of a number of independent random variables is approximately normally distributed. For example, the annual rainfall is the sum of the daily rain totals, each of which is viewed as a random variable. Other examples include annual lake evaporation, annual pumpage from a well, annual flow in a stream, and mean monthly temperature.

The lognormal CDF has been successfully used in approximating the distribution of variables that are the product of powers of many other random variables [1],[19]. The logarithm of the variable is approximately normally distributed because the logarithm of products is a sum of transformed variables.

Examples of variables that have been known to follow a lognormal distribution include:

1. Annual series of peak flow rates.
2. Daily precipitation depths and streamflow volumes (also monthly, seasonal, and annual).
3. Daily peak discharge rates.
4. Annual precipitation and runoff (primarily in the western United States).
5. Earthquake magnitudes.
6. Intervals between earthquakes.
7. Yield stress in steel.
8. Sediment sizes in streams where fracturing and breakage of larger into smaller sizes are involved.

The Pearson Type III (a form of gamma) has been applied to a number of variables such as precipitation depths in the eastern United States and cumulative watershed runoff at any point in time during a given storm event. The transformed log–Pearson Type III is most used to approximate the CDF for annual flood peaks. If the skew coefficient C_s of the variable is zero, the CDF reverts to a lognormal. It has also been used with monthly precipitation depth and yield strengths of concrete members.

A useful CDF for values of annual extreme is the Gumbel or extreme value distribution. The mean of the distribution has a theoretical exceedance probability of 0.43 and a recurrence interval T of 2.33 years. Flood peaks in natural streams have exhibited strong conformance to this distribution, including means with 2.33-year recurrence intervals. Graph paper that produces a straight-line fit for Gumbel variables is available and useful for graphical tests of annual extremes. The CDF has been applied to peak annual discharge rates, peak wind velocities, drought magnitudes and intervals, maximum rainfall intensities of given durations, and other hydrologic extremes that are independent events.

3.7   FREQUENCY ANALYSIS

Unless stated otherwise, the *frequency* of a hydrologic event is the probability that some value of a discrete variable will occur or some value of a continuous variable will be equaled or exceeded in any given year. The latter is more appropriately called the exceedance probability or *exceedance frequency*, but is often termed the *frequency*. Note that frequency is a probability, not a regular incidence, and has no units of measure. The reciprocal of the exceedance frequency, as shown by Eq. 3.21, is the return period, having units of years.

Two methods of frequency analysis are described. One is a straightforward plotting technique to obtain the cumulative distribution and the other uses frequency factors. The cumulative distribution function provides a rapid means of determining the probability of an event equal to or less than some specified quantity. The inverse is used to obtain recurrence intervals. As a general rule, frequency analysis is cautioned when working with records shorter than 10 years and in estimating frequencies of expected hydrologic events greater than twice the record length.

Plotting Formulas

The frequency of an event can be obtained by use of "plotting position" formulas. When annual maximum values are being analyzed, the recurrence interval is approximated as the mean time in years, with N future trials, for the mth largest value to be exceeded once on the average. The mean number of exceedances for this condition can be shown to be:

$$\bar{x} = N\frac{m}{n+1} \tag{3.35}$$

where $\bar{x}$ = the mean number of exceedances
N = the number of future trials
n = the number of values
m = the rank of descending values, with largest equal to 1

If the mean number of exceedances $\bar{x} = 1$ and $N = T$, then:

$$T = \frac{n+1}{m} \tag{3.36}$$

indicating that the recurrence interval is equal to the number of years of record plus 1, divided by the rank of the event.

Several plotting position formulas are available [23],[24]. They give different results as noted in Table 3.4. The range in recurrence intervals obtained for 10 years of record is illustrated in the right-hand column. Most plotting position formulas do not account for the sample size or length of record. One formula that does account for sample size was given by Gringorten [24] and has the general form:

$$T = \frac{n+1-2a}{m-a} \tag{3.37}$$

TABLE 3.4   Plotting Position Formulas

Method	Solve for $P(X > x)$	For $m = 1$ and $n = 10$	
		P	T
California	$\dfrac{m}{n}$	.10	10
Hazen	$\dfrac{2m - 1}{2n}$	.05	20
Beard	$1 - (0.5)^{1/n}$	.067	14.9
Weibull	$\dfrac{m}{n + 1}$	.091	11
Chegadayev	$\dfrac{m - 0.3}{n + 0.4}$	.067	14.9
Blom	$\dfrac{m - \frac{3}{8}}{n + \frac{1}{4}}$	.061	16.4
Tukey	$\dfrac{3m - 1}{3n + 1}$	.065	15.5

where n = the number of years of record

m = the rank

a = a parameter depending on n as follows:

n	10	20	30	40	50
a	0.448	0.443	0.442	0.441	0.440

n	60	70	80	90	100
a	0.440	0.440	0.440	0.439	0.439

In general, $a = 0.4$ is recommended in the Gringorten equation. If the distribution is approximately normal, $a = \frac{3}{8}$ is used. A value of $a = 0.44$ is used if the data follow a Gumbel distribution.

The technique in all cases is to arrange the data in increasing or decreasing order of magnitude and to assign order number m to the ranked values. The most efficient formula for computing plotting positions for unspecified distributions [23], and the one now commonly used for most sample data, is the Weibull equation:

$$P = \frac{m}{n + 1} \tag{3.38}$$

When m is ranked from lowest to highest, P is an estimate of the probability of values being equal to or less than the ranked value, that is, $P(X \leq x)$; when the rank is from highest to lowest, P is $P(X \geq x)$. For probabilities expressed in percentages, the value is $100m/(n + 1)$. The probability that $X = x$ is zero for any continuous variable.

Example 3.6

Using the 23 years of annual precipitation depths for Los Angeles, California (see data in Table 3.1), estimate the exceedance frequencies and recurrence intervals of the highest ten values using the Weibull equation.

Solution

1. The ten highest flow rates are tabulated below. By ranking them from highest to lowest, the value P in Eq. 3.38 becomes the exceedance probability.
2. Each value can have only one recurrence interval; therefore repeat values have one calculated probability. The greatest value, 24 in., has a calculated recurrence interval of 24 years.

Year	Rain depth (in.)	Rank, m	$P, m/(n + 1)$	T_r (yr)
1981	24	1	0.042	24
1986	23	3	—	—
1988	23	3	0.125	8
1978	21	4	0.167	6
1993	20	5	0.208	4.8
1999	19	8	—	—
1998	19	8	—	—
1980	19	8	0.333	3
1990	18	10	—	—
1983	18	10	0.417	2.4

Annual and Partial-Duration Series

In earlier examples of frequency analysis, only the series of annual maximum or minimum occurrences in the hydrologic record have been described. These extremes constitute an *annual series* that is consistent with frequency analysis and the manipulation of annual probabilities of occurrence. All the observed data—say, all floods or all the daily streamflows—would constitute a *complete series*. Any subset of the complete series is a *partial series*. In selecting the maximum annual events from a record, it often happens that the second greatest event in one year exceeds the annual maximum in some other year. Analysis of the annual series neglects such events. The extreme values analyzed without regard for the period (i.e., year) of occurrence are termed the *partial-duration series*.

The theoretical differences in recurrence intervals based on annual and partial-duration series of the same length are shown in Table 3.5. The difference for intervals greater than 10 years is negligible. An example of the use of annual and partial-duration series is included in Chapter 4, Section 4.8.

TABLE 3.5   Relation Between the Partial
Duration Series and the Annual
Series

Recurrence interval (yr)	
Partial duration series	Annual series
0.5	1.2
1.0	1.6
1.5	2.0
2.0	2.5
5.0	5.5
10.0	10.5

Plotting Paper

Several theoretical cumulative distribution functions plot as straight lines on special graph paper developed for use with Eq. 3.38. This facilitates extrapolation and interpolation of the data. Arithmetic probability paper has an arithmetic ordinate and a probability abscissa scale (see example in Fig. 3.6). It can be used to plot the calculated apparent frequencies of a variable to evaluate whether a normal CDF is approximated by the data. A straight-line plot would identify a normal CDF.

The same paper, but with a logarithmic scale as the ordinate, tests the apparent fit to a lognormal distribution. A third type of paper contains an extreme-value probability scale versus either an arithmetic or a logarithmic scale (both types are available). This allows a test of whether the data approximate a Gumbel or log–Gumbel extreme-value CDF. Extrapolation using any of these graphical aids is not recommended beyond two times the period of record.

As an alternative to using commercial plotting paper, spreadsheets and the properties of various cumulative distribution functions tabulated in the appendixes can be used to construct any desired graph paper [24],[25]. The abscissa should be a probability scale, scaled to result in a straight line for any data fitting the CDF. The ordinate can be either an arithmetic scale or several log cycles for plotting the measured values of the variable at the apparent exceedance probability values to visualize how well the data follow the distribution.

Frequency Analysis by Frequency Factors

Chow [26] has proposed the use of:

$$x = \bar{x} + Ks \tag{3.39}$$

as the general equation for hydrologic frequency analysis, where K is the *frequency factor* and s is the standard deviation. K is a function of recurrence interval T and varies with the coefficient of skewness in skewed distributions and can be affected greatly by the number of years of record. For the normal distribution, and for the transformed variate in a lognormal distribution, the deviate z given in standard normal tables (Table B.1) is synonymous with the frequency factor.

For a normal distribution, the value of variable Q corresponding to a given recurrence interval T is:

$$Q = \overline{Q} + zs_Q \qquad (3.40)$$

Values of z for a given T can be obtained from Table B.1 by recognizing that the exceedance probability, $1/T$, is the area under the probability density function to the right of the corresponding value of z.

Example 3.7

The mean annual rainfall at Los Angeles is 14.9 in., and the standard deviation is 5.9 in. Find the 10-year rainfall depth, assuming that the annual rain is normally distributed.

Solution. The exceedance probability of the 10-yr annual rain is $1/10$, or 0.1. Because this is the area under the normal curve to the right of the normalized 10-yr rain depth, the value of z_{10} can be found from the standard normal curve shown at the top of Table B.1 in the appendix. From the figure, the area under the curve between the origin and z_{10} is $F(z) = 0.500 - 0.100 = 0.400$. By interpolating Table B.1, the corresponding value for z_{10} is 1.282. Thus $Q_{10} = 14.9 + 1.282(5.9) = 15.1$ in. This shows that the 10-yr rain depth (or 10-yr value for any other normally distributed variable) is about 1.3 standard deviations above the mean value.

Lognormal Many variables that plot as curves on normal probability paper plot as straight lines when the logarithms are plotted, or when the values themselves are plotted on logarithmic probability paper. If either occurs, frequency factors for the normal distribution (Table B.1) can be applied using:

$$\log Q = \overline{\log Q} + z(s_{\log s}) \qquad (3.41)$$

Simply stated, the logarithms of Q follow a normal distribution, and Eq. 3.40 is applied to the logarithms. The mean and standard deviation of the logarithms are both required. Note that the mean and standard deviation of logarithms are not the same as the logarithms of the arithmetic mean and standard deviation. This common error must be avoided.

Log–Pearson III (Bulletin 17B) All federal agencies in the United States currently use Bulletin 17B, *Guidelines for Determining Flood Flow Frequency*, for fitting log–Pearson Type III probability distributions to peak flow data for stream gauges. The guidelines were developed by the Federal Hydrology Subcommittee of the Interagency Advisory Committee on Water Data to provide a consistent and uniform technique for flood-frequency analyses among various federal agencies. Earlier versions of the bulletin (Bulletin 17, Bulletin 17A) contained discrepancies and incon-sistencies of some techniques. The initial version of 17B, published in 1981, had numer-ous typographical errors, which were corrected in the current version [14]. Among other changes, the current version contains:

■ Revised guidelines for estimating and using generalized skew.
■ A new procedure for weighting station skew and generalized skew.
■ A new test for detecting high outliers and a revised test for detecting low outliers.
■ Revised guidelines for the application of conditional probability adjustments.

The U.S. Geological Survey maintains updates of software for performing Bulletin 17B frequency analyses. A download of the program PEAKFQ is available at http://water.usgs.gov/software/surface_water.html.

Frequency factors for the Pearson Type III (logarithmic or arithmetic) are shown in Appendix B, Table B.2, for various recurrence intervals (or exceedance probabili-ties) and skew coefficients. As outlined by the Water Resources Council, the fitting technique involves transforming annual floods to logarithmic values ($y_i = \log x_i$) and finding the mean, standard deviation, and skew coefficients of the logarithms. Flood magnitudes (Q) are estimated from:

$$\log Q = \bar{y} + K s_y \tag{3.42}$$

which is the same form as Eq. 3.39. Note that $K = \phi(T, C_s)$, a function of both recur-rence interval and skewness. Because the skewness coefficient has a much greater vari-ability than the mean or standard deviation, Beard [27] has recommended that only average regional coefficients of skew be employed in flood analysis for a single station unless the record exceeds 100 years. In practice this may be impractical to attain, and it is best to compute all parameters and compare results with any other records, experi-ence, or regional studies. Regional skew coefficients are described further in Section 13.3 (see Fig. 13.9). The use of logarithms to reduce the skewness of an already skewed distribution helps. Hazen recommended that the computed skewness for Pearson III analysis be multiplied by a factor of $(1 + 8.5/n)$ to obtain an adjusted skewness when dealing wih small samples [28]. Chow has developed K versus T curves for the distribution [29].

If the skew coefficient falls between −1.0 and 1.0, approximate values of fre-quency factors for the Pearson Type III can be obtained from:

$$K = \frac{2}{C_s}\left\{\left[\left(z - \frac{C_s}{6}\right)\frac{C_s}{6} + 1\right]^3 - 1\right\} \tag{3.43}$$

where z is the standard normal deviate for the selected recurrence interval T, and C_s is the skew coefficient from Eq. 3.14.

From examining Tables B.1 and B.2, the reader can establish the fact that a Pearson III distribution with a skew C_s of zero is normal. For example, both tables yield a 100-year frequency factor of 2.326. Through logarithmic transformation, this also means that a log–Pearson III CDF with zero skew of logarithms is a lognormal distribution.

Example 3.8

For rainfall data developed in Example 3.2, fit distribution functions to the records of Richmond, Virginia, and Los Angeles, California.

Solution.   Both distributions exhibit small skewness and are approximately normal. For the purposes of illustration, the Richmond data are fitted with the normal and the Los Angeles data with a Pearson Type III.

1. The data are arrayed for plotting in Table 3.6. The points are plotted in Figs. 3.7 and 3.8 as exceedance probability (left-hand scale) versus inches of rainfall.
2. The theoretical normal of best fit is a straight line through $(\bar{x} - s)$ at 15.9 percent, $\bar{x}$ at 50 percent, and $(\bar{x} + s)$ at 84.1 percent, (see Table B.1). Thus for Richmond,

TABLE 3.6   Plotting Data for Example 3.8

m	Richmond	Los Angeles	$100m/(n + 1)$
1	53	24	4.2
2	52	23	8.3
3	51	23	12.5
4	49	21	16.7
5	49	20	20.8
6	47	19	25.0
7	47	19	29.2
8	44	19	33.3
9	43	18	37.5
10	43	18	41.7
11	43	17	45.8
12	41	17	50.0
13	40	15	54.7
14	38	15	58.3
15	38	11	62.5
16	37	10	66.7
17	37	9	70.8
18	36	9	75.0
19	36	9	79.7
20	34	8	83.3
21	34	8	87.5
22	31	6	91.7
23	31	5	95.8

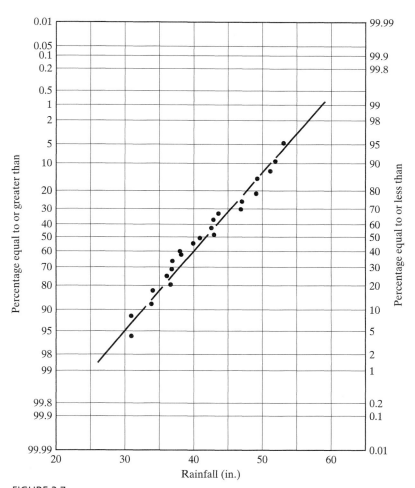

FIGURE 3.7

Annual rainfall for Richmond, Virginia, 1978–2000, plotted on normal probability paper.

x	Plotting position (right-hand scale)
$\overline{x} - s = 41.5 - 6.7 = 34.8$	15.9
$\overline{x} = 41.5$	50.0
$\overline{x} + s = 41.5 + 6.7 = 48.2$	84.1

3. The plotting positions (read as percent chance) come from Table B.2 for the computed skewness. Thus for Los Angeles,

Percent chance	99	95	90	80	50	20	10	4	2	1	0.5
$K(C_s = -0.16)$	−2.44	−1.69	−1.30	−0.83	0.03	0.85	1.26	1.69	1.97	2.21	2.43
$x = 14.9 + K(5.9)$	0.5	4.9	7.2	10.0	15.1	19.9	22.3	24.9	26.5	27.9	29.2

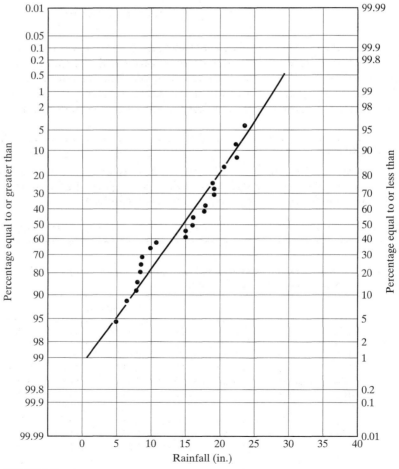

FIGURE 3.8

Annual rainfall for Los Angeles, California, 1978–2000.

4. To illustrate the use of this table, the 1 percent chance (100-yr) annual rainfall is 27.9 in.

Gumbel's Extreme Value   Equation 3.34 can be solved for the recurrence interval T and for the variate x, as follows:

$$\frac{1}{T} = 1 - F(x) = 1 - \exp\{-\exp[-\alpha(x - u)]\} \qquad (3.44)$$

$$x = u - \frac{1}{\alpha} \ln[\ln T - \ln(T - 1)] \qquad (3.45)$$

On substituting into Eq. 3.39, the general frequency equation, with u and α as defined in Table 3.3, the frequency factor for the extreme value distribution becomes:

$$K = -\frac{\sqrt{6}}{\pi}\left(0.5772 + \ln\ln\frac{T}{T - 1}\right) \qquad (3.46)$$

It should be noted that this expression for K is valid only in the limit, that is, as n approaches infinity. For a finite sample, K varies with the sample or length of record as shown in Table 3.7. K versus T curves have also been developed [23]. In Eq. 3.46, when $K = 0$, $T = 2.33$ years; thus in flood frequency analysis the recurrence interval of the mean annual flood is commonly designated as the 2.33-year event.

Example 3.9

The mean of the annual maximum discharges at a streamflow site with 25 years of record is 1000 cfs. The standard deviation is 400 cfs. Estimate the magnitude of the 50-year flood for a Gumbel extreme-value distribution.

Solution.   From Table 3.7, $K = 3.088$; $x = \bar{x} + Ks = 1000 + 3.088(400) = 2235$ cfs.

TABLE 3.7   Gumbel Extreme-Value Frequency Factors

	Recurrence interval								
Sample Size	2.33	5	10	20	25	50	75	100	1000
15	0.065	0.967	1.703	2.410	2.632	3.321	3.721	4.005	6.265
20	0.052	0.919	1.625	2.302	2.517	3.179	3.563	3.836	6.006
25	0.044	0.888	1.575	2.235	2.444	3.088	3.463	3.729	5.842
30	0.038	0.866	1.541	2.188	2.393	3.026	3.393	3.653	5.727
40	0.031	0.838	1.495	2.126	2.326	2.943	3.301	3.554	5.476
50	0.026	0.820	1.466	2.086	2.283	2.889	3.241	3.491	5.478
60	0.023	0.807	1.446	2.059	2.253	2.852	3.200	3.446	5.410
70	0.020	0.797	1.430	2.038	2.230	2.824	3.169	3.413	5.359
75	0.019	0.794	1.423	2.029	2.220	2.812	3.155	3.400	5.338
100	0.015	0.779	1.401	1.998	2.187	2.770	3.109	3.349	5.261
∞	−0.067	0.720	1.305	1.866	2.044	2.592	2.911	3.137	4.900

Example 3.10

The annual maximum discharge data in Table 3.8 have been obtained from the U.S. Geological Survey Water Resources Division for a small stream in Missouri. Rank the data and plot on extreme-value probability paper.

TABLE 3.8   Peak Annual Flows for a Missouri Stream

Water year	Annual maximum discharge (cfs)	Rank	$(n + 1)/m$
2000	2510	8	1.375
1999	4150	1	11.0
1998	2990	5	2.2
1997	2120	10	1.1
1996	3555	2	5.5
1995	2380	9	1.22
1994	2550	7	1.57
1993	2800	6	1.83
1992	3300	3	3.67
1991	3150	4	2.75

Solution.   The data are plotted in Fig. 3.9.

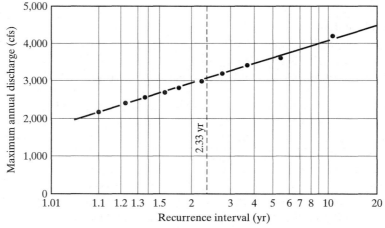

FIGURE 3.9

Annual floods on a small Missouri stream.

Steps in Frequency Analyses

The steps in a comprehensive frequency analysis of an annual or partial-duration series of values would include:

1. Evaluate the data homogeneity. Have there been changes in the basin that would cause the data to be split into two or more populations? Has the gauge been

moved during the period of record? Have there been changes in the gauge setting, such as datum shifts or buildings or trees, that may have affected the homogeneity of the data?

2. Rank the homogeneous sets of data in descending magnitude, select a plotting equation, and determine the plotting position P and recurrence interval ($1/P$); then plot the data on several types of probability graph paper to assess whether the data are linear when plotted on any background.

3. Using the probability distribution of the type that appears best fitted by the data, calculate the population statistics and plot the cumulative distribution function on the graph of plotted sample data.

4. Assess high outliers by assessing whether any might have a larger recurrence interval than the period of record. Remove the outliers, and recompute and replot the frequency curve.

5. Compare the higher-order population statistics (standard deviation, skew, and kurtosis) with regional values and adjust the frequency curve if appropriate.

6. Compare the estimates with values computed from other methods such as regional regression equations, maximum likelihood methods, L-moment methods, or rainfall-runoff models.

7. Be cautious about extrapolation of the frequency curve beyond about twice the period of record.

Only part of step 2 of this procedure was illustrated for an annual series of rain depths at Los Angeles in Example 3.6. Step 3 was illustrated in Example 3.8. The other steps are important and should not be neglected. The ten largest rain depths at Los Angeles appear to be uniformly distributed between 18 and 24 inches. In contrast, the Anniston, Alabama, data in Table 3.1 reveal that the gauge recorded a depth of 99 inches in 1988 and 98 inches in 1997. The next highest rain depth was 71 inches in 1978. Nothing in the table reveals anything about the homogeneity of the data, but the extraordinarily high values of these two measurements suggest that they may be *outliers*. Step 2 applied to these data would suggest that a 99-in. value would have a 24-yr recurrence interval. If 60 more years of data were collected and this was still the highest value, its apparent recurrence interval would be 84 years, which may be true even though only 23 years of data are being evaluated. Extreme values in an annual or partial-duration series often are considered outliers because an insufficient number of years have passed to accurately assess their recurrence intervals.

Effect of Record Length on Flood Prediction

The length of the period of record used in a frequency analysis significantly affects the results. Data from a 68-year record at one gauging station were analyzed in various groupings and subsets using log–Pearson III frequency procedures, with the following results [30]:

1. Use of less than all 68 years caused large increases in the estimated 10-, 50-, and 100-year floods. If only the most recent 10-year record rather than all 68 years

had been used, the estimated 100-year flood was increased by 211 percent. Even the 10-year flood was overestimated by 62 percent.

2. If only the most recent 20-, 30-, 40-, 50-, and 60-year records were available, the overestimate in Q_{100} ranged from 123 percent for 20 years of data to 2 percent for 60 years. The Q_{10} estimates were 51 percent and −5 percent different, respectively, for the 20- and 60-year subsets of data.

3. The choice of wet versus dry sequences during the 68 years also affected the results significantly. A 10-year record in the 1930s (dry period) resulted in Q_{100} that is 61 percent below the 68-year value. Adding 10 more years reduced this error only slightly. On the other hand, use of the 10-year wet cycle in the 1960s produced a Q_{100} that is 181 percent greater than the corresponding 68-year value.

4. Much of the difference is attributed to the sensitivity that the skew coefficient has to the number and type (wet versus dry) of years. For comparison, the 68-year skew of −0.546 changed to −1.827 when 1903–1912 floods were used, and 0.993 when 1943–1952 records were used.

Frequency Curve Confidence Limits

Approximate confidence limits can be placed on frequency curves. A method proposed by Beard [27] involves placing lines above and below the fitted curve to form a reliability band. Table 3.9 shows the factors by which the standard deviations of the variate must be multiplied to mark off a 90 percent reliability band above and below the frequency curve. The 5 percent level, for example, means that only 5 percent of future values should fall above the limit, and, similarly, only 5 percent should fall below the 95 percent limit. Nine of ten should fall within the band.

TABLE 3.9   Error Limits for Flood Frequency Curves

Years of record (n)	Exceedance frequency (%, at 5% level)						
	99.9	99	90	50	10	1	0.1
5	1.22	1.00	0.76	0.95	2.12	3.41	4.41
10	0.94	0.76	0.57	0.58	1.07	1.65	2.11
15	0.80	0.65	0.48	0.46	0.79	1.19	1.52
20	0.71	0.58	0.42	0.39	0.64	0.97	1.23
30	0.60	0.49	0.35	0.31	0.50	0.74	0.93
40	0.53	0.43	0.31	0.27	0.42	0.61	0.77
50	0.49	0.39	0.28	0.24	0.36	0.54	0.67
70	0.42	0.34	0.24	0.20	0.30	0.44	0.55
100	0.37	0.29	0.21	0.17	0.25	0.36	0.45
	0.1	1	10	50	90	99	99.9
	Exceedance frequency (%, at 95% level)						

Note: Tabular values are multiples of the standard deviation of the variate. Five percent error limits are added to the flood value from the fitted curve at the same exceedance frequency and the sum plotted. Ninety-five percent limits are subtracted from the flood value at the same exceedance frequency. Log values are added or subtracted before antilogging and plotting.

Example 3.11

The maximum annual instantaneous flows from the Maury River near Lexington, Virginia, for a 26-year period are listed in Table 3.10.

Plot the log–Pearson III curve of best fit and determine the magnitude of the flood to be equaled or exceeded once in 5, 10, 50, and 100 years. Using Table 3.9, also plot the upper and lower confidence limits.

TABLE 3.10   Maury River Peak Flow Rates

Water (year)	Discharge (cfs)	Water (year)	Discharge (cfs)	Water (year)	Discharge (cfs)
1975	6,730	1984	13,800	1993	6,680
1976	9,150	1985	40,000	1994	6,540
1977	6,310	1986	10,200	1995	5,560
1978	10,000	1987	13,400	1996	7,700
1979	15,000	1988	8,950	1997	8,630
1980	2,950	1989	11,900	1998	14,500
1981	8,650	1990	5,840	1999	23,700
1982	11,100	1991	20,700	2000	15,100
1983	6,360	1992	12,300		

Solution

1. The statistical calculations are summarized as follows:

	Arithmetic	Log
Mean, $\bar{x}$	11,606	4.001
Variance, s^2	53.87×10^6	0.0516
Skew coefficient, C_s	2.4	0.38

2. After forming an array and computing plotting positions, the data are plotted in Fig. 3.9.

3. Plotting data for log–Pearson III (Table 3.11) are developed from Table B.2. Confidence limits are plotted in Fig. 3.10 using Table 3.9.

TABLE 3.11   Plotting Data for Log–Pearson III Distribution

Chance (%)	T_r (yr)	$(C_s = 0.38)$ K	$(\bar{y} = 4.001)$ $(s_y = 0.227)$ $\bar{y} + Ks_y = \log Q$	Q
99	1.01	−2.044	3.537	3,443
95	1.05	−1.530	3.653	4,498
90	1.11	−1.234	3.721	5,260
80	1.25	−0.855	3.760	5,754
50	2	−0.062	3.987	9,705
20	5	0.818	4.187	15,380
10	10	1.315	4.300	19,950
4	25	1.874	4.426	26,690
2	50	2.251	4.512	32,510
1	100	2.601	4.591	39,030
0.5	200	2.930	4.666	46,360

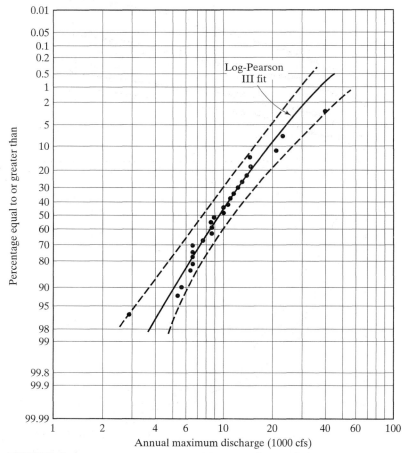

FIGURE 3.10

Maximum instantaneous annual flows, Maury River, Lexington, Virginia.

3.8 FLOW DURATION ANALYSIS

Figures 3.11–3.14 illustrate further applications of frequency analysis. Figures 3.11 and 3.12 represent standard point frequency analyses of the annual series of high and low flows for different durations. Figure 3.13 is a low-flow duration–frequency curve based on the same data as Fig. 3.12. Figure 3.14 is based on an analysis of the *complete* series of daily flows although all observed values are not plotted. Presented in this form, such a curve usually is called a *duration* curve [4],[27],[31]. Note that the probability scale must be labeled "percentage of time," since the annual series was not used. Duration curves are useful in predicting the availability and variability of sustained flows, but, again, they do not represent the actual sequence of flows.

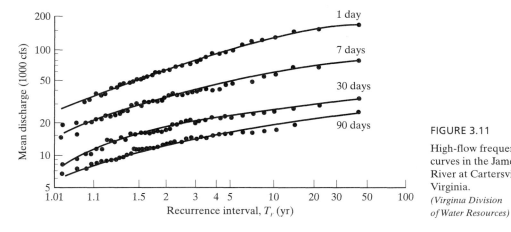

FIGURE 3.11

High-flow frequency
curves in the James
River at Cartersville,
Virginia.
*(Virginia Division
of Water Resources)*

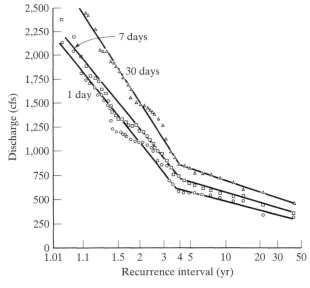

FIGURE 3.12

Low-flow frequency curves in the James River
at Cartersville, Virginia.
(Virginia Division of Water Resources)

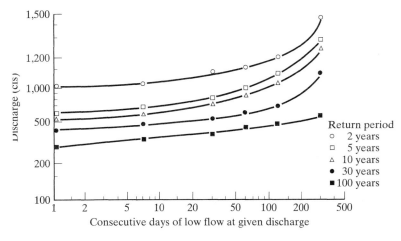

FIGURE 3.13

Low flow duration–frequency
curves in the James River at
Cartersville, Virginia. Drainage
area: 6242 mi².
*(Virginia Division of Water
Resources)*

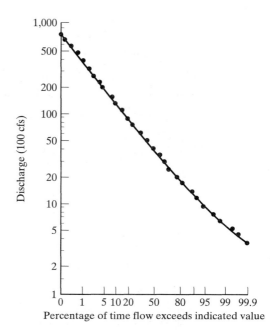

FIGURE 3.14

Flow duration curve for the James River at
Cartersville, Virginia.
(Virginia Division of Water Resources)

3.9   LINEAR REGRESSION AND CORRELATION

Correlation and regression procedures are widely used in hydrology and other sciences [6],[32]. The premise of the methods is that one variable is often conditioned by the value of another, or of several others, or the distribution of one may be conditioned by the value of another. Just as there are probability density functions (PDFs) for evaluating the *marginal* probability of a variable (see Section 3.4), so also are there PDFs for the *conditional* probabilities (also described in Section 3.4) of variables. The concept is illustrated in Fig. 3.15. For two variables, the bivariate density function, $f(y \mid x_1)$, plotted in the vertical on the figure, changes for each value of x. The one shown applies only to variations in y when $x = x_1$. Different distributions might occur for other values of x.

A measure of the degree of linear correlation between two variables x and y is the *linear correlation coefficient*, $\rho_{x, y}$. A value of $\rho_{x, y} = 0$ indicates a lack of linear correlation and $\rho_{x, y} = \pm 1.0$ means perfect correlation. The correlation coefficient is found from:

$$\rho_{x, y} = \frac{\text{cov}(x, y)}{\sigma_x \sigma_y} = \frac{\sigma_{x, y}}{\sigma_x \sigma_y} \tag{3.47}$$

where σ_x and σ_y are the population variances of each variable, respectively (see Eq. 3.10), and $\text{cov}(x, y)$ is the *covariance* shared by the two variables, defined as:

$$\text{cov}(x, y) = \sigma_{x, y} = \int_{-\infty}^{\infty} \int_{-\infty}^{\infty} (x - \mu_x)(y - \mu_y) f(x, y) \, dy \, dx \tag{3.48}$$

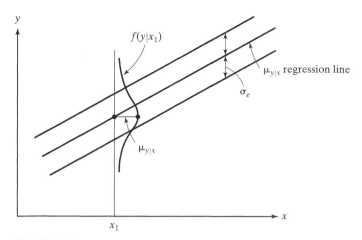

FIGURE 3.15

Bivariate regression with conditional probability function.

The sample correlation coefficient, $r = s_{x,y}/s_x s_y$, is used to estimate $\rho_{x,y}$. The sample covariance is found from the square root of:

$$s_{x,y}^2 = \frac{\Sigma(x_i - \bar{x})(y_i - \bar{y})}{n - 1} \tag{3.49}$$

The regression line shown in Fig. 3.15 is derived to pass through the mean values of the distributions, so that for any given value of x, the mean value of $y \mid x$ (read "y given x") can be estimated by the regression line. The *standard error* of the estimate of $y \mid x$ is depicted by the line drawn through the conditional distributions at a distance of one standard deviation from the mean. If the conditional distributions at all x-values are normal, it can be shown that the mean value, $\mu_{y|x}$, of the conditional distribution is related to the means of x and y, or:

$$\mu_{y|x} = \mu_y + \rho\frac{\sigma_y}{\sigma_x}(x - \mu_x) \tag{3.50}$$

and the variance is:

$$\sigma_{y|x}^2 = \frac{\sigma_e^2}{N}\left[1 + \frac{(x - \mu_x)^2}{\sigma_x^2}\right] \tag{3.51}$$

where:

$$\sigma_e^2 = \sigma_y^2(1 - \rho^2) \tag{3.52}$$

which is the variance of the residuals of the regression. Just as the mean of the distribution requires substitution of the given value of x into Eq. 3.50, so also does the variance, Eq. 3.51. When the value of x in Fig. 3.15 is set equal to $\bar{x}$, the standard error of the mean is:

$$\sigma_{\bar{y}|\bar{x}} = \frac{\sigma_e}{\sqrt{N}} \tag{3.53}$$

Equation 3.50 is linear and expresses the linear dependence between y and x as shown in Fig. 3.14. The mean value of y can be computed for fixed values of x. Also, if the correlation between them is significant, one can predict the values of y with less error than the marginal distribution of y alone. In fact, from Eq. 3.52, the fraction of the original variance explained or accounted by the regression is:

$$\rho^2 = 1 - \frac{\sigma_e^2}{\sigma_y^2} \tag{3.54}$$

It can also be seen from Eq. 3.50 that the slope of the regression line is:

$$\rho \frac{\sigma_y}{\sigma_x} = \frac{\mu_{y|x} - \mu_y}{x - \mu_x} \tag{3.55}$$

or, if x and y are standardized, then ρ itself is the slope, where:

$$\rho = \frac{(\mu_{y|x} - \mu_y)/\sigma_y}{(x - \mu_x)/\sigma_x} \tag{3.56}$$

The bivariate case can be expanded to cover higher-order, multivariate distributions.

Deriving Regression Equations

Regression lines as expressed by Eq. 3.50 and shown in Fig. 3.15 are useful in explaining linear dependence and, where significant correlation exists, in making predictions. For the bivariate case, in general, the procedure is to fit a linear model to a sample of random variables observed in pairs (see Fig. 3.16). The fitting technique is the method of least squares, which minimizes the sum of the residuals squared. Residuals as shown in the figure are the difference, vertically in this instance, from the value of y predicted by the line and the y value observed for the same corresponding value of x. The line to be fitted is:

$$y_i = \alpha + \beta x_i \tag{3.57}$$

The best estimates of α and β are sought. Thus to minimize:

$$\sum (y_i - \hat{y}_i)^2 = \sum [y_i - (\alpha + \beta x_i)]^2 \tag{3.58}$$

where y_i are the observed values and $\hat{y}_i$ are the estimated values from Eq. 3.57, take partial derivatives as follows:

$$\frac{\partial}{\partial \alpha} \left\{ \sum [y_i - (\alpha + \beta x_i)]^2 \right\} \tag{3.59}$$

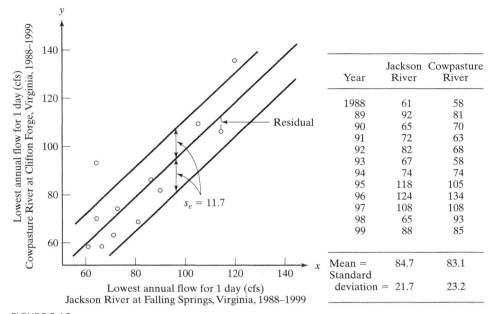

Year	Jackson River	Cowpasture River
1988	61	58
89	92	81
90	65	70
91	72	63
92	82	68
93	67	58
94	74	74
95	118	105
96	124	134
97	108	108
98	65	93
99	88	85
Mean =	84.7	83.1
Standard deviation =	21.7	23.2

FIGURE 3.16

Cross-correlation of low flows. Regression line: $Y = 4.94 + 0.923X$; $r = 0.86$.

$$\frac{\partial}{\partial \beta}\left\{\sum [y_i - (\alpha + \beta x_i)]^2\right\} \tag{3.60}$$

After carrying out the differentiations and summations, two equations result in α and β, called *normal equations*:

$$\sum y_i - n\alpha - \beta \sum x_i = 0 \tag{3.61}$$

$$\sum x_i y_i - \alpha \sum x_1 - \beta \sum x_i^2 = 0 \tag{3.62}$$

Solving Eqs. 3.61 and 3.62 simultaneously yields:

$$\alpha = \frac{\sum y_i}{n} - \frac{\beta \sum x_i}{n} = \bar{y} - \beta \bar{x} \tag{3.63}$$

$$\beta = \frac{\sum x_i y_i - \left(\sum x_i \sum y_i / n\right)}{\sum x_i^2 - [(\sum x_i)^2 / n]} \tag{3.64}$$

Recall that the slope is $\rho(\sigma_y/\sigma_x)$, or as estimated from sample data:

$$\beta = r\frac{s_y}{s_x} \tag{3.65}$$

Also, the unexplained variance in the regression equation is:

$$\sigma_e^2 = \sigma_y^2(1 - \rho^2) \tag{3.66}$$

the square root of which is the standard deviation of residuals (see Fig. 3.15) and is called the *standard error of estimate*. These can be estimated from:

$$s_e^2 = \frac{n-1}{n-2}s_y^2(1 - r^2) \tag{3.67}$$

or

$$s_e^2 = \frac{1}{n-2}\sum(y_i - \hat{y}_i)^2 \tag{3.68}$$

where y_i and $\hat{y}_i$ are as defined previously (see Eq. 3.58).

Many hydrologic variables are linearly related, and after estimating the regression coefficients, prediction of y can be made for any value of x within the range of observed x values. Extrapolation outside the range is often performed but should be done with caution. Equation 3.51 shows that the variance in the estimate of y for a given x value becomes large when x is several standard deviations above or below the mean.

Example 3.12

The lowest annual flows for a 12-yr period on the Jackson and Cowpasture Rivers are tabulated in Fig. 3.16. The stations are upstream of the confluence of the two rivers that form the James River. Find the regression equation and the correlation between low flows.

Solution

1. The basic statistics are $\Sigma x = 1{,}016$; $\Sigma y = 997$; $\Sigma x^2 = 91{,}216$; $\Sigma y^2 = 88{,}777$; and $\Sigma xy = 89{,}209$.

2. For the two-variable regression, α and β are found from Eqs. 3.63 and 3.64.

$$\beta = \frac{89{,}209 - (1{,}016 \times 997/12)}{91{,}216 - [(1{,}016)^2/12)]} = 0.923$$

$$\alpha = \frac{997}{12} - \frac{(0.923)(1{,}016)}{12} = 4.91$$

The regression is $y = 4.91 + 0.923x$.

3. The correlation coefficient from Eq. 3.65 is:

$$r = \frac{(0.923)(21.7)}{23.2} = 0.86$$

4. From Eq. 3.67 the standard error of estimate, s_e, is 11.7, which is plotted as limits around the regression line in Fig. 3.16.

Coefficient of Determination for the Regression

A regression equation replaces (and extends) the data used in its development. Because the equation cannot reproduce all the base data, the process results in the loss of some information. This not only includes loss of information about particular pairs of data, but also about the variability of the data. The variance s_y^2 is a statistical measure of the variability of the measured values of y. The greater the value of s_y^2, the wider the spread of points around the mean. The percentage of information about the variance in y that is retained in, or explained by, the regression equation is called the *coefficient of determination*, C_D. To determine its value, the residuals or departures (differences between actual and estimated y values) have known variance σ_ϵ^2, which represents the unaccounted variance in the regression equation. The explained variance would be the difference, $\sigma_y^2 - \sigma_\epsilon^2$, and the percentage retained (coefficient of determination) is:

$$C_D = \frac{\sigma_y^2 - \sigma_\epsilon^2}{\sigma_y^2} = 1 - \frac{\sigma_\epsilon^2}{\sigma_y^2} \tag{3.69}$$

Comparison with Eq. 3.52 reveals that:

$$C_D = \rho^2 \tag{3.70}$$

Thus the square of the correlation coefficient ρ is the percentage of σ_y^2 explained by the regression. For any sample of data, the coefficient of determination r^2 is estimated as $s_{x,y}^2 / s_x^2 s_y^2$. A large r^2 indicates a good fit of the regression equation to the data because the equation accounts for or is able to explain a large percentage of the variation in the data.

Example 3.13

Determine the coefficient of determination for the regression in Example 3.12.

Solution. From Eq. 3.70, the coefficient of determination, r^2, is 0.7396. Thus, the regression equation adequately explains or "accounts for" about 74 percent of the original information about y contained in the raw data. Twenty-six percent of the information is lost.

The bivariate example can be extended to multiple linear regressions. For example, the linear model in three variables, with y the dependent variable and x_1 and x_2 the independent variables, has the form:

$$y = \alpha + \beta_1 x_1 + \beta_2 x_2 \tag{3.71}$$

The normal equations are:

$$\sum y = \alpha n + \beta_1 \sum x_1 + \beta_2 \sum x_2 \tag{3.72}$$

$$\sum yx_1 = \alpha \sum x_1 + \beta_1 \sum x_1^2 + \beta_2 \sum x_1 x_2 \tag{3.73}$$

$$\sum yx_2 = \alpha \sum x_2 + \beta_1 \sum x_1 x_2 + \beta_2 \sum x_2^2 \tag{3.74}$$

The square of the standard error of estimate is:

$$s_e^2 = \frac{1}{n-3} \sum (y_i - \bar{y}_i)^2 \tag{3.75}$$

where y_i are the observed values and $\bar{y}_i$ are predicted by Eq. 3.71. The *multiple correlation coefficient* is:

$$R = \left(1 - \frac{s_e^2}{s_y^2}\right)^{1/2} \tag{3.76}$$

Linear Transformations in Hydrology

Strong nonlinear bivariate and multivariate correlations are also common in hydrology, and various mathematical models have been used to describe the relations. Parabolic, exponential, hyperbolic, power, and other forms have provided better graphical fits than straight lines. Because of difficulties in the derivation of normal equations using least squares for these models, many can be transformed to linear forms. The most familiar transformation is a linearization of multiplicative nonlinear relations by using logarithms. For example, the equation:

$$y = \alpha x_1^{\beta_1} x_2^{\beta_2} \tag{3.77}$$

becomes linear when logarithms are taken, or:

$$\log y = \log \alpha + \beta_1 \log x_1 + \beta_2 \log x_2 \tag{3.78}$$

The log transformation procedure results in a linear form when the logarithms of one or both sets of measurements are substituted in Eqs. 3.63 and 3.64. For example, if a bivariate parabolic form $Y = aX^b$ is suggested by the data, logarithms allow use of the linear form $\log Y = \log a + b \log X$. The normal equations can be used by redefining $y = \log Y$, $x = \log X$, $\alpha = \log a$, and $\beta = b$, thereby transforming the equation to $y = \alpha + \beta x$. The regression can now be performed on the logarithms, values of α and β are determined, and the estimate of a is found by taking the antilog of α. This transformation is possible for several other nonlinear models, some of which are shown in

TABLE 3.12   Linear Transformations of Nonlinear Forms

Equation	Abscissa	Ordinate	Equation in linear form
$Y = A + BX$	X	Y	$[Y] = A + B[X]$
$Y = Be^{AX}$	X	$\log Y$	$[\log Y] = \log B + A(\log e)[X]$
$Y = AX^B$	$\log X$	$\log Y$	$[\log Y] = \log A + B[\log X]$
$Y = AB^X$	X	$\log Y$	$[\log Y] = \log A + (\log B)[X]$

Note: Variables in brackets are the regression variates.

Table 3.12. The variables x and y must be nonnegative, with values preferably greater than 1.0 to avoid problems with the log transformation.

Example 3.14

Using Table 3.13, find the regression equation and multiple correlation coefficient relating the standard deviation of flow logarithms with the logarithms of DA, the drainage area size, and the number of rainy days per year. X_1 should be set equal to $(1 + \log s)$ to avoid negative values.

Solution

1.   From Eqs. 3.72, 3.73, and 3.74, the parameters are:

$$\alpha = 1.34; \quad \beta_1 = -0.013; \quad \beta_2 = -0.49$$

and the regression equation is:

$$X_1 = 1.34 - 0.013X_2 - 0.49X_3$$

or $$\log s = 0.34 - 0.013 \log(\text{DA}) - 0.49 \log(\text{days})$$

2.   The multiple correlation coefficient from Eqs. 3.75 and 3.76 is $R = 0.56$.

SUMMARY

Statistical methods, especially frequency analyses, have widespread applications in hydrologic science, analysis, and design. This chapter introduces most of the concepts and procedures that are presented in undergraduate courses in hydrology. Principally, this chapter presents the commonly used methods and distribution properties for analyzing frequencies and recurrence intervals of random events observed at a point. Numerous applications of the methods are included in subsequent chapters. Readers seeking more advanced presentations are referred to other standard works [1],[4],[5].

TABLE 3.13   Logarithmic Data for 50 Gauging Stations

$X_1 = 1 + \log s$		$X_2 = \log DA$		$X_3 = \log$ number of rainy days per year			
Station number (1)	X_2 (2)	X_3 (3)	X_1 (4)	Station number (5)	X_2 (6)	X_3 (7)	X_1 (8)
1	1.61	2.11	0.29	33	1.94	1.87	0.20
2	2.89	2.12	0.18	34	2.73	1.36	0.58
3	4.38	2.11	0.17	35	3.63	1.81	0.64
4	3.20	2.04	0.44	36	1.91	1.58	0.37
5	3.92	2.07	0.38	37	2.26	1.48	0.27
6	1.61	2.04	0.37	38	2.97	1.89	0.54
7	3.21	2.09	0.30	39	0.70	1.32	0.63
8	3.65	1.99	0.35	40	0.30	1.54	0.78
9	3.23	2.15	0.16	41	3.38	1.62	0.46
10	4.33	2.08	0.11	42	2.87	2.03	0.44
11	1.60	2.09	0.32	43	2.42	2.26	0.24
12	2.82	2.00	0.34	44	4.53	1.93	−0.03
13	2.40	2.00	0.25	45	3.04	1.78	0.30
14	3.69	2.09	0.43	46	4.13	2.00	0.17
15	2.18	2.19	0.27	47	1.49	2.01	0.14
16	2.09	2.17	0.25	48	5.37	1.95	0.10
17	4.48	1.91	0.52	49	1.36	2.11	0.27
18	4.95	1.95	0.18	50	2.31	2.23	0.18
19	2.21	1.97	0.39				
20	3.41	2.08	0.40	ΣX	147.55	96.24	17.89
21	4.82	1.88	0.25	$\overline{X}$	2.951	1.925	0.358
22	1.78	1.93	0.23	$\Sigma X X_2$	503.7779	285.5627	51.1527
23	4.39	1.74	0.54	$\Sigma X \Sigma X_2/n$	435.4200	284.0042	52.7934
24	3.23	2.01	0.51	$\Sigma x x_2$	68.3579	1.5585	−1.6407
25	3.58	2.04	0.45				
26	1.64	1.78	0.63	$\Sigma X X_3$		187.5912	33.2598
27	4.58	1.76	0.45	$\Sigma X \Sigma X_3/n$		185.2428	34.4347
28	3.26	1.93	0.59	$\Sigma x x_3$	1.5585	2.3484	−1.1749
29	4.29	1.81	0.46				
30	1.23	1.89	0.32	$\Sigma X X_1$			8.1635
31	3.44	1.48	0.96	$\Sigma X \Sigma X_1/n$			6.4010
32	2.11	1.97	0.12	$\Sigma x x_1$	−1.6407	−1.1749	1.7625

Note: $x = X - \overline{X}$.
(After Beard [27])

PROBLEMS

SECTION 3.1: RANDOM VARIABLES AND STATISTICS

3.1 List 10 random variables within the field of hydrology. List 10 random variables from outside the field of hydrology.

3.2 List five random discrete variables within the field of hydrology.

SECTION 3.2: PROBABILITY DISTRIBUTIONS

3.3 For your hometown, sketch how Fig. 3.1 might appear if the data were actually collected. Base your sketch on your recollection; do not attempt to find the data needed.

3.4 Sketch how Fig. 3.1 might appear for Seattle, Washington, and Phoenix, Arizona.

3.5 The distribution of mean annual rainfall at 35 stations in the James River Basin, Virginia, is given in the following summary:

Interval (2-in. groupings)	36 or 37 in.	38 or 39 in.	40 or 41 in.	42 or 43 in.
Number of observations	2	4	7	9

Interval (2-in. groupings)	44 or 45 in.	46 or 47 in.	48 or 49 in.	50 or 51 in.
Number of observations	5	4	2	2

Compute the relative frequencies and plot the frequency distribution and the cumulative distribution. Estimate the probability that the mean annual rainfall (a) will exceed 40 in., (b) will exceed 50 in., and (c) will be between these values.

3.6 A normally distributed random variable has a mean of 4.0 and a standard deviation of 2.0. Use Table B.1 to determine the value of

$$\int_8^\infty f(x)dx$$

3.7 For the function described below, find (a) the number b that will make the function a probability density function, and (b) the probability that a single measurement of x will be less than $\frac{1}{2}$.

$$f(x) = \begin{cases} 0 & \text{for } x < 0 \\ 3x^2/8 & \text{for } 0 \le x \le b \\ 0 & \text{for } x > b \end{cases}$$

SECTION 3.3: DISTRIBUTION STATISTICS

3.8 A given set of data has a symmetric, zero-skew histogram. Determine the frequency and return period of the made. The mode is defined as the value exceeded by half the values.

3.9 The pan-evaporation data (in.) for the month of July at a site in Missouri are

9.7	11.7	11.2	11.3	11.5
11.2	8.8	11.4	11.8	8.9
9.3	9.2	9.3	9.3	10.4
9.8	8.7	11.5	10.9	10.2

Determine the mean, standard deviation, and coefficient of variation. What are the standard errors of these statistics?

3.10 If the mode of a PDF is considerably larger than the median, would the skew most likely be positive or negative?

3.11 The mean July precipitation at a station is half as large as the mode. Sketch the probability density function, label the axes, and state (a) whether the distribution is skewed left or right, and (b) whether the skew is positive or negative.

SECTION 3.4: PROBABILITY APPLICATIONS : FREQUENCY AND RETURN PERIOD

3.12 In the past 60 years, a discharge of 30,000 cfs at a stream-gauging station was equaled or exceeded only three times. Determine the average return period (years) of this value.

3.13 Annual flood records for a 10-year period are given by:

Year	1	2	3	4	5	6	7	8	9	10
Flood	300	700	200	400	1000	900	800	500	100	600

Mean = 550 cfs, median = 550 cfs, standard deviation = 300 cfs. Use an annual series and the definition of frequency in a frequency analysis to determine the magnitude of the 4-year flood. Compare this historical value with the analytical 4-year flood obtained by assuming floods follow a normal distribution.

3.14 A reservoir in the locale of Problem 3.48 will overfill when the annual precipitation exceeds 30 in. Determine the probability that the reservoir will overfill (a) next year, (b) at least once in three successive years, and (c) in each of three successive years.

3.15 The probabilities of events E_1 and E_2 are each 0.3. What is the probability that E_1 or E_2 will occur (a) when the events are independent but not mutually exclusive, and (b) when the probability of E_1 given E_2 is 0.1?

3.16 Events A and B are independent events having marginal probabilities of 0.4 and 0.5, respectively. Determine for a single trial (a) the probability that both A and B will occur simultaneously, and (b) the probability that neither occurs.

3.17 The conditional probability, $P(E_1 \mid E_2)$, of a power failure (given that a flood occurs) is 0.9, and the conditional probability, $P(E_2 \mid E_1)$, of a flood (given that a power failure occurs) is 0.2. If the joint probability, $P(E_1 \text{ and } E_2)$, of a power failure and a flood is 0.1, determine the marginal probabilities, $P(E_1)$ and $P(E_2)$.

3.18 Describe two random events that are (a) mutually exclusive, (b) dependent, (c) both mutually exclusive and dependent, and (d) neither mutually exclusive nor dependent.

3.19 Events A and B are independent and have marginal probabilities of 0.4 and 0.5, respectively. Determine the following for a single trial:

(a) The probability that both A and B occur.

(b) The probability that neither occurs.

(c) The probability that B, but not A, occurs.

3.20 Existing records reveal the following information about events A and B, where $A = $ a long March warm spell and $B = $ an April flood:

Year	1	2	3	4	5	6	7	8	9	10
$A = $ warm March?	No	No	Yes	No	Yes	No	Yes	No	Yes	No
$B = $ April flood?	Yes	No	No	Yes	Yes	Yes	No	Yes	Yes	No

On the basis of the 10-year record, answer the following:

(a) Are variables A and B independent? Prove.

(b) Are variables A and B mutually exclusive? Prove.

(c) Determine the marginal probability of an April flood.

(d) Determine the probability of having a cold March next year.

(e) Determine the probability (one value) of having both a cold March and a flood-free April next year.

(f) If a long March warm spell has just ended today, what is the best estimate of the probability of a flood in April?

3.21 Two dependent events are A = a flood will occur in Omaha next year and B = an ice-jam will form near Omaha in the Missouri River next year. Use your judgment to rank from largest to smallest the following probabilities: $P(A), P(A$ and $B), P(A$ or $B), P(A \mid B)$.

3.22 The probability of having a specified return period, T_r, is defined as:

$$P\genfrac{}{}{0pt}{}{\text{(annual value will be equaled or exceeded}}{\text{exactly once in a period of } t = T_r \text{ years)}} = \left(1 - \frac{1}{T_r}\right)^{t-1}$$

Also,

$$P\genfrac{}{}{0pt}{}{\text{(annual value will be equaled or exceeded}}{\text{exactly } r \text{ times in a period of } n \text{ years)}} = \frac{n!}{r!(n-r)!} P^{n-r}(1-P)^r$$

(a) According to the descriptions in parentheses, the second probability should equal the first when n and r are equal to what values?

(b) Show that both equations result in the same probability for an annual value whose frequency is $33\frac{1}{3}\%$ and the return period is $T_r = t = 3$ years. Discuss.

3.23 What return period must an engineer use in the design of a bridge opening for a 50% risk that flooding will occur at least once in two successive years? Repeat for a risk of 100%.

3.24 A temporary cofferdam is to be built to protect the 5-year construction activity for a major cross-valley dam. If the cofferdam is designed to withstand the 20-year flood, what is the probability that the structure will be overtopped (a) in the first year, (b) in the third year exactly, (c) at least once in the 5-year construction period, and (d) not at all during the 5-year period?

3.25 A 33-year record of peak annual flow rates was subjected to a frequency analysis. The median value is defined as the midvalue in the table of rank-ordered magnitudes. Estimate the following:

(a) The probability that the annual peak will equal or exceed the median value in any single year.

(b) The average return period of the median value.

(c) The probability that the annual peak in 1993 will equal or exceed the median value.

(d) The probability that the peak flow rate next year will be less than the median value.

(e) The probability that the peak flow rate in all of the next 10 successive years will be less than the median value.

(f) The probability that the peak flow rate will equal or exceed the median value at least once in 10 successive years.

(g) The probability that the peak flow rates in both of two consecutive years will equal or exceed the median value.

(h) The probability that, for a 2-year period, the peak flow rate will equal or exceed the median value in the second year but not in the first.

3.26 A temporary flood wall has been constructed to protect several homes in the floodplain. The wall was built to withstand any discharge up to the 20-year flood magnitude. The wall will be removed at the end of the 3-year period after all the homes have been relocated. Determine the probabilities of the following events:

(a) The wall will be overtopped in any year.

(b) The wall will not be overtopped during the relocation operation.

(c) The wall will be overtopped at least once before all the homes are relocated.

(d) The wall will be overtopped exactly once before all the homes are relocated.

(e) The wall will be adequate for the first 2 years and then overtopped in the third year.

3.27 Wave heights and their respective return periods are known for a 40-mi-long reservoir. Owners of a downstream campsite will accept a 25% risk that a protective wall will be overtopped by waves at least once in a 20-year period. Determine the minimum height of the protective wall.

Wave height (ft)	Return period (years)
10.0	100
8.5	50
7.4	30
5.0	10
3.5	5

3.28 Assume that the channel capacity of 12,000 cfs near a private home was equaled or exceeded in 3 of the past 60 years. Find the following:

(a) The frequency of the 12,000-cfs value.

(b) The probability that the home will be flooded next year.

(c) The return period of the 12,000-cfs value.

(d) The probability that the home will not be flooded next year.

(e) The probability of two consecutive, safe years.

(f) The probability of a flood at least once in the next 20 years.

(g) The probability of a flood in the second, but not the first, of two consecutive years.

(h) The 20-year flood risk.

SECTIONS 3.5 AND 3.6: TYPES OF CONTINUOUS PROBABILITY DISTRIBUTION FUNCTIONS

3.29 Assume that the Anniston, Alabama, data in Table 3.1 follow a normal distribution. Use the standard normal distribution in Table B.1 to determine the rain depth that would have a recurrence interval of 100 years. Use the same table to estimate the recurrence interval of the 1988 annual rain of 99 inches.

3.30 Determine the probability that a measurement of a hydrologic variable with a normal probability density function will fall between the mean and one standard deviation above the mean.

3.31 Determine the probability that a measurement of a hydrologic variable with a normal probability density function will fall in the range of the mean ± 3 standard deviations.

3.32 The total annual runoff from a small drainage basin is determined to be approximately normal with a mean of 14.0 in. and a variance of 9.0 in.^{2}. Determine the probability that the annual runoff from the basin will be less than 11.0 in. in all three of the next three consecutive years.

3.33 A normal variable X has a mean of 5.0 and a standard deviation of 1.0. Determine the value of X that has a cumulative probability of 0.330.

3.34 For a standard normal density function, use Table B.1 to determine the value of

$$\int_{\mu-2\sigma}^{\mu+\sigma} f(x)dx$$

3.35 The random variable x represents depth of precipitation in July. Between values of $x = 0$ and $x = 30$, the probability density function has the equation $f(x) = x/40\mu_x$. In the past, the average July precipitation μ_x was 30 in.

(a) Determine the probability that next July's precipitation will not exceed 20 in.

(b) Determine the single probability that the July precipitation will equal or exceed 30 in. in all of five consecutive years.

3.36 The random variable x represents depth of precipitation in July. Between values of $x = 0$ and $x = 30$, the probability density function has the equation $f(x) = x/1200$.

(a) Determine the probability that next July's precipitation will not exceed 20 in.

(b) Determine the probability that next July's precipitation will equal or exceed 30 in.

3.37 Complete the following mathematical statements about the properties of a PDF by inserting in the boxes on the left the correct item number from the right. Assume that X is a series of annual occurrences from a normal distribution.

(a) $\displaystyle\int_{-\infty}^{\infty} f(x)dx = \square$ 1. Zero

(b) $\displaystyle\int_{-\infty}^{m_1} f(x)dx = \square$ 2. Unity

(c) $\displaystyle\int_{\mu}^{u+\square} f(x)dx = 0.34$ 3. Value with 5% chance of exceedance each year

(d) $\displaystyle\int_{m_1}^{m_2} f(x)dx = \square$ 4. 0.68

(e) $\displaystyle\int_{-\infty}^{\square} f(x)dx = 0.5$ 5. Value expected every 50 years on the average

(f) $\displaystyle\int_{\mu-\sigma}^{\mu+\sigma} f(x)dx = \square$ 6. $P(X \le m_1)$

(g) $\displaystyle\int_{\square}^{\infty} f(x)dx = 0.02$ 7. $P(m_1 \le X \le m_2)$

(h) $\displaystyle\int_{\mu}^{\mu} f(x)dx = \square$ 8. $P(m_1 \ge X \ge m_2)$

(i) $1 - \displaystyle\int_{m_1}^{m_2} f(x)dx = \square$ 9. Median

(j) $\displaystyle\int_{-\infty}^{\square} f(x)dx = 0.95$ 10. Standard deviation

3.38 The mean monthly temperature for September at a weather station is found to be normally distributed. The mean is 65.5°F, the variance is 39.3°F², and the record is complete for 63 years. With the aid of Table B.1, find (a) the midrange within which two-thirds of all future mean monthly values are expected to fall, (b) the midrange within which 95% of all future values are expected, (c) the limit below which 80% of all future values are expected, and (d) the values that are expected to be exceeded with a frequency of once in 10 years and once in 100 years. Verify the results by plotting the cumulative distribution on normal probability paper.

SECTION 3.7: FREQUENCY ANALYSIS

3.39 The total annual pumpage (in acre-ft) over the last 30 years from a fully developed irrigation well field was observed as follows:

2450	3300	3400	3650	3800
2650	3150	3100	3500	2850
3050	4300	3300	3300	3150
2100	3300	3650	3150	3550
2900	3250	3000	3400	3750
3900	3600	3150	3600	3000

Form an array of the data and plot the apparent frequencies on normal probability paper. Compute and plot the mean of the data and draw the line of best fit through the mean and as close as possible to the other points by eye. Estimate from the line the standard deviation and the highest value expected to be equaled or exceeded once in 50 years.

3.40 For a 60-year record of precipitation intensities and durations, a 30-min intensity of 2.50 in./hr was equaled or exceeded a total of 85 times. All but 5 of the 60 years experienced one or more 30-min intensities equaling or exceeding the 2.50-in./hr value. Use the Kimball formula to determine the return period of this intensity using (a) a partial-duration series and (b) an annual series.

3.41 Given the following values of peak flow rates for a small stream, determine the return period (years) for a flood of 100 cfs by first using annual peaks for an annual series and then by using all the data in a partial-duration series.

Year	Date	Peak (cfs)	Year	Date	Peak (cfs)
1984	June 1	90	1989	May 11	800
	Aug. 3	300		June 8	700
1985	June 7	60		Sept. 4	90
1986	July 2	80	1990	Aug. 8	400
1987	May 18	100	1991	May 9	30
	June 3	90	1992	Sept. 8	700
1988	July 4	40	1993	May 4	80

3.42 Recorded maximum depths (in.) of precipitation for a 30-min duration at a single station are:

Year	Date	Depth (in.)	Year	Date	Depth (in.)
1963	May 3	2.0	1968	Aug. 8	4.0
	June 3	1.0	1969	May 6	6.0
1964	June 7	1.0		June 8	5.0
1965	July 2	1.0		Sept. 4	1.0
1966	June 1	1.0	1970	May 4	1.0
	Aug. 3	3.0	1971	Sept. 8	5.0
1967	July 4	1.0	1977	May 9	1.0

(a) Determine the return period (years) for a depth of 2.0 in. using the California method with an annual series.

(b) Repeat part (a) using a partial-duration series.

(c) Determine from the partial-duration series the depth of 30-min rain expected to be equaled or exceeded (on the average) once every 8 years.

3.43 For a 60-year record of precipitation intensities and durations, a 30-min intensity of 2.50 in./hr was equaled or exceeded a total of 85 times. All but 5 of the 60 years experienced one or more 30-min intensities equaling or exceeding the 2.50-in./hr value. Use the Weibull formula to determine the return period of this intensity using (a) a partial-duration series and (b) an annual series.

3.44 On which type of plotting paper (probability, log–probability, rectangular coordinate, log–log, semilog, extreme-value, none) would each of the following plot as a straight line?

(a) Normal frequency distribution.

(b) Gumbel frequency distribution.

(c) $Y = 3X + 4$

(d) Lognormal frequency distribution.

(e) Pearson type III with a skew of zero.

(f) $Q = 43A^{0.7}$

(g) Log–Pearson type III with a skew of logarithms equal to zero.

(h) Pearson type III with a skew of 3.0.

3.45 Using the data of Problem 3.60, compute the mean and variance of the discharges, and, using frequency factors from Table 3.7, find the Gumbel estimates of the 50- and 100-year events.

3.46 The total annual runoff from a small drainage basin is determined to be approximately normal with a mean of 14.0 in. and a standard deviation of 3 in. Determine the probability that the annual runoff from the basin will be less than 8.0 in. in the second year only of the next three consecutive years.

3.47 Annual floods for a stream are normally distributed with a mean of 30,000 cfs and a variance of 1×10^6 cfs^2. Determine the average return period T, of a 32,000-cfs flood in the stream.

3.48 The 80-year record of annual precipitation at a midwestern gauge location has a range between 14 in. in 1972 and 42 in. in 2001. The record has a mean of 27.6 in. and a standard deviation of 6.06 in. Assuming a normal distribution, (a) plot the frequency curve on probability paper, (b) determine the probability of a drought worse than the 1972 value, and (c) determine the recurrence interval of the 2001 maximum depth and compare it with the apparent recurrence interval.

3.49 Annual floods for a small river are reported to follow a normal probability distribution. The 2-year flood for the basin has been estimated as 40,000 cfs and the 10-year flood as 52,820 cfs. Using normal frequency factors, determine the magnitude of the 25-year flood.

3.50 Annual floods for a stream have a normal frequency distribution. The 2-year flood is 40,000 cfs and the 10-year flood is 52,820 cfs. Determine the magnitude of the 25-year flood.

3.51 Six years of peak runoff rates are given below. Assume that the floods follow exactly a normal distribution and determine the magnitude of the 50-year peak.

Year	1	2	3	4	5	6
Runoff (cfs)	200	800	500	600	400	500

3.52 The 80-year record of annual precipitation at Lincoln, Nebraska, yields a range of values between 10 and 50 in. with a mean annual value of 25.00 in. and a standard deviation of 5.30 in. Because annual precipitation represents a sum of many random variables (i.e., depth of precipitation for each day of the year), assume that annual precipitation is normally distributed.

(a) In 1936 the precipitation at Lincoln was a mere 14 in. Determine the probability that the annual precipitation will be 14 in. or less next year.

(b) In 1965 Lincoln received 42 in. On the average, this amount would be equaled or exceeded once in how many years?

(c) Compare the theoretical and apparent return periods of the record-high value of 50.00 in.

3.53 Annual floods (cfs) at a particular site on a river follow a zero-skew log–Pearson type III distribution. If the mean of logarithms (base 10) of annual floods is 2.946 and the standard deviation of base-10 logarithms is 1.000, determine the magnitude of the 50-year flood.

3.54 Determine the 50-year peak (cfs) for a log–Pearson type III distribution of annual peaks for a major river if the skew coefficient of logarithms (base 10) is −0.1, the mean of logarithms (base 10) is 3.0, and the standard deviation of base 10 logarithms is 1.0.

3.55 A Pearson type III variable X has a mean of 4.0, a standard deviation of 2.0, and a coefficient of skew of 0. Determine the value (four significant figures) of $\int_0^6 (X)dX$.

3.56 Annual floods (cfs) at a particular site on a river follow a zero-skew log–Pearson type III distribution. If the mean of logarithms (base 10) of annual floods is 1.733 and the standard deviation of base-10 logarithms is 1.420, determine the magnitude of the 100-year flood.

3.57 The following parameters were computed for a stream near Lincoln, Nebraska:

Period of record = 1980–1999, inclusive
Mean annual flood = 7000 cfs
Standard deviation of annual floods = 1000 cfs
Skew coefficient of annual floods = 1.0
Mean of logarithms (base 10) of annual floods = 3.52
Standard deviation of logarithms = 0.50
Coefficient of skew of logarithms = 2.0

Determine the magnitude of the 25-year flood by assuming that the peaks follow a (a) log–Pearson type III distribution, (b) Gumbel distribution, and (c) lognormal distribution.

3.58 The following parameters were computed for a stream:

Period of record = 1970–1994, inclusive

Mean annual flood = 7000 cfs

Standard deviation of annual floods = 1000 cfs

Skew coefficient of annual floods = 2.0

Mean of logarithms (base 10) of annual floods = 3.52

Standard deviation of logarithms = 0.50

Coefficient of skew of logarithms = −2.0

Determine the magnitude of the 25-year flood by assuming that the peaks follow a (a) log–Pearson type III distribution, (b) Gumbel distribution, and (c) lognormal distribution.

3.59 Peak annual discharge rates in the Elkhorn River at Waterloo, Nebraska, yield the following statistics:

Period of record = 1930–1969, inclusive

Mean flood = 16,900 cfs

Standard deviation = 17,600 cfs

Skew of annual floods = 0.8

Mean of logarithms (base 10) = 4.0923

Standard deviation of logarithms = 0.3045

Skew of logarithms = 2.5

(a) Determine the 100-year flood magnitude using the uniform technique adopted by the U.S. Water Resources Council for all federal evaluations.

(b) Determine the 100-year flood magnitude assuming that the floods follow a two-parameter gamma distribution.

3.60 The 20-year record of annual flood peaks on Furr's Run are as follows:

Year	Q (cfs)	Year	Q (cfs)
1951	1060	1961	1350
1952	2820	1962	1140
1953	1970	1963	2100
1954	1760	1964	1090
1955	1650	1965	2890
1956	1140	1966	1100
1957	1020	1967	1840
1958	2260	1968	1710
1959	1650	1969	1630
1960	870	1970	1260

Log the values of peak discharge and compute the mean, standard deviation, and skewness coefficient. Plot the data on lognormal probability paper. With the aid of Table B.2, fit both the lognormal and log–Pearson III distributions. Compare estimates of the 50- and 100-year events.

3.61 Perform a complete frequency analysis on one of the three 33-year records given in the table. Fit a Pearson type III or log–Pearson III and compare with the normal or lognormal of best fit. Plot the data and place control curves around the theoretical curve of best fit using Table 3.9.

Year	Trempeuleau River Dodge, WI (DA = 643 mi^2) Q_{peak} (cfs)	Bow River Banff, Alberta, Canada (DA = 858 mi^2) Q_{peak} (cfs)	James River Scottsville, VA (DA = 4570 mi^2) Q_{peak} (cfs)
1928	3,700	10,200	75,600
1929	1,700	7,590	44,700
1930	3,360	9,280	45,800
1931	1,650	6,610	21,100
1932	3,600	9,850	31,400
1933	11,000	11,000	59,500
1934	2,570	9,490	38,800
1935	4,490	6,940	93,400
1936	7,180	7,720	126,000
1937	1,780	5,210	62,200
1938	3,170	7,770	87,400
1939	6,400	6,270	68,400
1940	3,120	7,220	130,000
1941	2,890	4,450	27,100
1942	5,680	5,850	80,600
1943	5,060	7,380	95,200
1944	2,040	5,590	133,000
1945	8,120	4,450	57,000
1946	4,570	7,210	41,200
1947	5,410	5,880	33,200
1948	4,840	10,320	59,600
1949	1,920	4,290	94,200
1950	3,600	10,080	73,300
1951	4,840	8,570	64,900
1952	6,950	5,460	54,500
1953	4,040	9,180	67,000
1954	5,710	10,120	62,900
1955	10,400	8,680	70,000
1956	17,400	9,060	20,400
1957	713	5,360	64,200
1958	1,140	6,730	44,500
1959	8,000	7,480	29,300
1960	1,480	6,440	64,200

3.62 Compare results of Problem 3.61 with estimates by Gumbel's extreme-value distribution for the 50- and 100-year events.

SECTION 3.8: FLOW DURATION ANALYSIS

3.63 From the data given in the accompanying table of low flows, prepare a set of low-flow frequency curves for the daily, weekly, and monthly durations.

Lowest Mean Discharge (cfs) for the Following Number of Consecutive Days, Maury River Near Buena Vista, VA

Year	1-day	7-day	30-day
1939	100.0	103.0	125.0
1940	167.0	171.0	194.0
1941	22.0	59.4	69.1
1942	101.0	127.0	173.0
1943	86.0	93.9	103.0
1944	62.0	65.9	77.4
1945	78.0	80.7	90.3
1946	76.0	78.6	87.1
1947	97.0	102.0	123.0
1948	154.0	176.0	215.0
1949	136.0	138.0	163.0
1950	113.0	125.0	139.0
1951	95.0	95.3	101.0
1952	115.0	116.0	120.0
1953	85.0	86.1	90.8
1954	68.0	70.0	81.7
1955	83.0	96.1	99.9
1956	64.0	66.3	71.7
1957	62.0	64.1	75.8
1958	88.0	92.6	107.0
1959	76.0	80.9	117.0
1960	83.0	91.7	103.0
1961	99.0	103.0	152.0
1962	90.0	95.0	105.0
1963	60.0	60.6	70.8
1964	51.0	54.1	62.0
1965	64.0	68.7	76.2

3.64 For the 7-day low flows at Buena Vista given in Problem 3.63, attempt to fit a straight-line frequency curve on lognormal or extreme-value probability paper, proceeding as follows: From the original plot of the data, estimate the lowest flow (say, q) at the high recurrence intervals; subtract this flow from all observed flows ($Q - q = Q'$); and replot Q' versus the original recurrence intervals. Repeat if necessary. The best-fitting curve will be a three-parameter frequency distribution.

SECTION 3.9: LINEAR REGRESSION AND CORRELATION

3.65 The square of the linear correlation coefficient is called the proportion of the variance that is "explained by the regression." Describe the meaning of this phrase by evaluating the equations given in the text. What variance is explained, and what does the term *explained* mean?

3.66 Twenty measured pairs of values of normally distributed variables X and Y are analyzed, yielding values of $\overline{X} = 30$, $\overline{Y} = 20$, $s_x = 20$, and $s_y = 0$. Determine the values of a and b and the standard deviation of residuals for a least-squares fit using the linear equation $Y = a + bX$.

3.67 The least-squares estimates of A and B in the bivariate regression equation $y = A + BX$ are $A = 2.0$ and $B = 1.0$, where y is a transformation defined as $\log_{10} Y$. If Y and X are related by $Y = a(b)^x$, determine the values of a and b.

3.68 Given a table of ten values of mean annual floods and corresponding drainage areas for a number of drainage basins, state how linear regression techniques would be used to determine the coefficient and exponent (p and q) in the equation $Q_{2\,33} = pA^q$.

3.69 What choice of transformed variables Y and X would provide a linear transformation for $y = a/(x^3 + b)$? Also, if a regression on these transformed variables yields $Y = 100 + 10X$, determine the corresponding values of a and b. Would the linear transformation be applicable to all possible pairs and values of x and y?

3.70 Observations for the past 10 years of withdrawals and estimated recharge of an artesian aquifer are given in the following table. Find the means, variances, standard deviations, covariance, and correlation coefficient.

Measured discharge (1000 acre-ft)	Estimated recharge (1000 acre-ft)
12.2	12.0
10.4	9.8
10.6	11.0
12.6	13.2
14.2	14.6
13.0	14.0
14.0	14.0
12.0	12.4
10.4	10.4
11.4	11.6

3.71 Fit a regression equation to the data in Problem 3.70, treating discharge as the dependent variable. Compute the standard error of estimate. Estimate the expected discharge when recharge is 13,000 acre-ft. What would be the estimate of discharge if no information were available on recharge? What is the relative improvement provided by the regression estimate?

3.72 From the following observations of variation of the mean annual rainfall with the altitude of the gauge, determine a linear prediction equation for the catchment. How well correlated are rainfall and altitude?

Gauge number	Mean annual rainfall (in.)	Altitude of gauge (1000 ft)
1	22	4.2
2	28	4.4
3	25	4.5
4	31	5.4
5	32	5.6
6	37	5.6
7	36	5.8
8	35	6.0
9	36	6.6
10	46	6.6
11	41	6.8
12	41	7.0

3.73 Estimate the expected rainfall in Problem 3.72 for a gauge to be installed at an altitude of 5500 ft.

3.74 The least-squares estimates of A and B in the bivariate regression equation $Y = A + BX$ are $A = 2.0$ and $B = 3.0$, where Y is a transformation defined as $\log_{10} y$ and X is defined as $\log_{10} x$. If y and x are related by $y = ax^b$, determine the values of a and b.

3.75 Which measure of variation in a regression $Y = a + bX$ is generally larger in magnitude, the standard deviation of Y or the standard deviation of residuals? For what condition would the two values be equal?

3.76 Given a table of values of mean annual floods and corresponding drainage areas for a number of basins in a region, describe how regression analysis could be used to determine the coefficients p and q in the relation $Q_{2.33} = pA^q$.

3.77 The time of rise of flood hydrographs (T_r), defined as the time for a stream to rise from low water to maximum depth following a storm, is related to the stream length (L) and the average slope (S). From the information given below for 11 watersheds in Texas, New Mexico, and Oklahoma, derive a functional relation of the form $T_r = aL^b S^c$.

Watershed number	T_r (min)	L (1000 ft)	S (ft y 1000 ft)
1	150	18.5	7.93
2	90	14.2	19.0
3	60	25.3	12.0
4	60	11.7	13.3
5	100	9.7	11.0
6	75	8.1	15.0
7	90	21.7	16.7
8	30	3.9	146.0
9	30	1.2	20.0
10	45	3.3	64.0
11	50	3.5	33.0

3.78 Repeat the exercise in Problem 3.77 by fitting the relation $T_r = dF^e$, where $F = L/\sqrt{S}$ with L in mi and S in ft/mi. Plot the results on log–log paper.

REFERENCES

[1] D. R. Maidment (ed.), *Handbook of Hydrology*, New York: McGraw-Hill, 1993.

[2] K. C. Patra, *Hydrology and Water Resources Engineering*, Boca Raton, FL: CRC Press, 2000.

[3] A. D. Ward and W. J. Elliot, *Environmental Hydrology*, Boca Raton, FL: Lewis Publishers, 1995.

[4] V. T. Chow, D. R. Maidment, and L. W. Mays, *Applied Hydrology*, New York: McGraw-Hill, 1988.

[5] W. Viessman and G. L. Lewis, *Introduction to Hydrology*, 4th ed., New York: HarperCollins, 1996.

[6] J. R. Benjamin and C. Cornell, *Probability, Statistics and Decision for Civil Engineers*, New York: McGraw-Hill, 1969.

[7] A. M. Mood and F. A. Graybill, *Introduction to the Theory of Statistics*, 2nd ed., New York: McGraw-Hill, 1963.

[8] A. J. Duncan, *Quality Control and Statistics*, Homewood, IL: Richard D. Irwin, Inc., 1959.

[9] P. G. Hoel, *Introduction to Mathematical Statistics*, 3rd ed., New York: Wiley, 1962.

[10] J. Aitchison and J. A. C. Brown, *The Log–Normal Distribution*, New York: Cambridge University Press, 1957.

[11] V. T. Chow, "Statistical and Probability Analysis of Hydrologic Data," in *Handbook of Applied Hydrology*, New York: McGraw-Hill, 1964.

[12] H. A. Foster, "Theoretical Frequency Curves," *Trans. ASCE* **87**, 142–203(1924).

[13] L. R. Beard, *Statistical Methods in Hydrology*, Civil Works Investigations, U.S. Army Corps of Engineers, Sacramento District, 1962.

[14] "A Uniform Technique for Determining Flood Flow Frequencies," Bull. No. 17B, U.S. Geological Survey, 1989.

[15] "New Tables of Percentage Points of the Pearson Type III Distribution," Tech. Release No. 38, Central Technical Unit, U.S. Department of Agriculture, 1968.

[16] M. B. Fiering, *Streamflow Synthesis*, Cambridge, MA: Harvard University Press, 1967.

[17] F. E. Perkins, Simulation Lecture Notes, Summer Institute, "Applied Mathematical Programming in Water Resources," University of Nebraska, 1970.

[18] E. J. Gumbel, "The Return Period of Flood Flows," *Ann. Math. Statist.* **12**(2), 163–190(June 1941).

[19] Ven T. Chow, "The Log-Probability and Its Engineering Application," *Proc. ASCE* **80**, 1–25(Nov. 1954).

[20] "Probability Tables and Other Analysis of Extreme Value Data," Series 22, National Bureau of Standards Applied Mathematics, 1953.

[21] E. J. Gumbel, *Statistics of Extremes*, New York: Columbia University Press, 1958.

[22] E. J. Gumbel, "Statistical Theory of Extreme Values for Some Practical Application," National Bureau of Standards, Applied Mathematical Series 33, 1954.

[23] M. A. Benson, "Plotting Positions and Economics of Engineering Planning," *Proc. ASCE J. Hyd. Div.* **88**, 57–71(Nov. 1962).

[24] I. I. Gringorten, "A Plotting Rule for Extreme Probability Paper," *J. Geophys. Res.* **68**(3), 813–814(Feb. 1963).

[25] D. H. Hoggan, *Computer-Assisted Floodplain Hydrology and Hydraulics*, New York: McGraw-Hill, 1989.

[26] Ven T. Chow, "A General Formula for Hydrologic Frequency Analysis," *Trans. Am. Geophys. Union* **32**, 231–237(1951).

[27] L. R. Beard, *Statistical Methods in Hydrology*, Civil Works Investigations, U.S. Army Corps of Engineers, Sacramento District, 1962.

[28] A. Hazen, *Flood Flows*, New York: Wiley, 1930.

[29] Ven T. Chow, "Statistical and Probability Analysis of Hydrologic Data," Sec. 8–1, in *Handbook of Applied Hydrology* (V. T. Chow, ed.), New York: McGraw-Hill, 1964.

[30] P. Victorov, "Effect of Period of Record on Flood Prediction," *Proc. ASCE J. Hyd. Div.* **97**(Nov. 1971).

[31] R. H. McCuen, *Hydrologic Analysis and Design*, Englewood Cliffs, NJ: Prentice-Hall, 1989.

[32] R. S. Gupta, *Hydrology and Hydraulic Systems*, Prospect Heights, IL: Waveland Press, 1995.

C H A P T E R   4

Precipitation

OBJECTIVES

The purpose of this chapter is to:

- Define precipitation, discuss its forms, and describe its spatial and temporal attributes
- Illustrate techniques for estimating areal precipitation amounts for specific storm events and for maximum precipitation-generating conditions.

Precipitation replenishes surface water bodies, renews soil moisture for plants, and recharges aquifers. Its principal forms are rain and snow. The relative importance of these forms is determined by the climate of the area under consideration. In many parts of the western United States, the extent of the snowpack is a determining factor relative to the amount of water that will be available for the summer growing season. In more humid localities, the timing and distribution of rainfall are of principal concern.

Precipitated water follows the paths shown in Figs. 1.1 and 1.2. Some of it may be intercepted, evaporated, and infiltrated and become surface flow. The actual disposition depends on the amount of rainfall, soil moisture conditions, topography, vegetal cover, soil type, and other factors.

Hydrologic modeling and water resources assessments depend upon a knowledge of the form and amount of precipitation occurring in a region of concern over a time period of interest.

4.1   WATER VAPOR

The fraction of water vapor in the atmosphere is very small compared to quantities of other gases present, but it is exceedingly important to our way of life. Precipitation is derived from this atmospheric water. The moisture content of the air is also a significant factor in local evaporation processes. Thus it is necessary for a hydrologist to be acquainted with ways for evaluating the atmospheric water vapor content and to understand the thermodynamic effects of atmospheric moisture [1].

Under most conditions of practical interest (modest ranges of pressure and temperature, provided that the condensation point is excluded), water vapor essentially obeys the gas laws. Atmospheric moisture is derived from evaporation and transpiration, the principal source being evaporation from the oceans. Precipitation over the United States comes largely from oceanic evaporation, the water vapor being transpirated over the continent by the primary atmospheric circulation system.

Measures of water vapor or atmospheric humidity are related basically to conditions of evaporation and condensation occurring over a level surface of pure water. Consider a closed system containing approximately equal volumes of water and air maintained at the same temperature. If the initial condition of the air is dry, evaporation takes place and the quantity of water vapor in the air increases. A measurement of pressure in the airspace will reveal that as evaporation proceeds, pressure in the airspace increases because of an increase in partial pressure of the water vapor (vapor pressure). Evaporation continues until vapor pressure of the overlying air equals the surface vapor pressure [a measure of the excess of water molecules leaving (evaporating from) the water surface over those returning]. At this point, evaporation ceases, and if the temperatures of the air space and water are equal, the airspace is said to be *saturated*. If the container had been open instead of closed, the equilibrium would not have been reached, and all the water would eventually have evaporated. Some commonly used measures of atmospheric moisture or humidity are vapor pressure, absolute humidity, specific humidity, mixing ratio, relative humidity, and dew point temperature.

Amount of Precipitable Water

Estimates of the amount of precipitation that might occur over a given region with favorable conditions are often useful. These may be obtained by calculating the amount of water contained in a column of atmosphere extending up from the earth's surface. This quantity is known as the *precipitable water W*, although it cannot all be removed from the atmosphere by natural processes. Precipitable water is usually expressed in centimeters or inches.

An equation for computing the amount of precipitable water in the atmosphere can be derived as follows. Consider a column of air having a square base 1 cm on a side. The total water mass contained in this column between elevation zero and some height z would be:

$$W = \int_0^z \rho_w dz \qquad (4.1)$$

where ρ_w = the absolute humidity and W is the depth of precipitable water in centimeters. The integral can be evaluated graphically or by dividing the atmosphere into layers of approximately uniform specific humidities, solving for these individually, and then summing (for additional information on this topic, see Ref. 2).

Geographic and Temporal Variations

The quantity of atmospheric water vapor varies with location and time. These variations may be attributed mainly to temperature and source of supply considerations. The greatest concentrations can be found near the ocean surface in the tropics, the concentrations generally decreasing with latitude, altitude, and distance inland from coastal areas.

About half the atmospheric moisture can be found within the first mile above the earth's surface. This is because the vertical transport of vapor is mainly through convective action, which is slight at higher altitudes. It is also of interest that there is not necessarily any relation between the amount of atmospheric water vapor over a region and the resulting precipitation. The amount of water vapor contained over dry areas of the Southwest, for example, at times exceeds that over considerably more humid northern regions, even though the latter areas experience precipitation while the former do not.

4.2   PRECIPITATION

Precipitation is the primary input vector of the hydrologic cycle. Its forms are rain, snow, and hail and variations of these such as drizzle and sleet. Precipitation is derived from atmospheric water, its form and quantity thus being influenced by the action of other climatic factors such as wind, temperature, and atmospheric pressure. Atmospheric moisture is a necessary but not sufficient condition for precipitation. Continental air masses are usually very dry so that most precipitation is derived from moist maritime air that originates over the oceans. In North America about 50 percent of the evaporated water is taken up by continental air and moves back again to the sea.

Formation of Precipitation

Two processes are considered to be capable of supporting the growth of droplets of sufficient mass (droplets from about 500 to 4000 μm in diameter) to overcome air resistance and consequently fall to the earth as precipitation. These are known as the *ice crystal process* and the *coalescence process*.

The coalescence process is one by which the small cloud droplets increase their size due to contact with other droplets through collision. Water droplets may be considered as falling bodies that are subjected to both gravitational and air resistance effects. Fall velocities at equilibrium (terminal velocities) are proportional to the square of the radius of the droplet; thus the larger droplets will descend more quickly than the smaller ones. As a result, smaller droplets are often overtaken by larger droplets, and the resulting collisions tend to unite the drops, producing increasingly larger particles. Very large drops (order of 7 mm in diameter) break up into small droplets that repeat the coalescence process and produce somewhat of a chain effect. In this manner, sufficiently large raindrops may be produced to generate significant precipitation. This process is considered to be particularly important in tropical regions or in warm clouds.

An important type of growth is known to occur if ice crystals and water droplets are found to exist together at subfreezing temperatures down to about $-40°C$. Under these conditions, certain particles of clay minerals and organic and ordinary ocean salts serve as freezing nuclei so that ice crystals are formed. The vapor pressure under these conditions is higher over the water droplets than over the ice crystals, and thus condensation occurs on the surface of the crystals. The ice crystals grow in size, and uneven particle size distributions develop, which further favor growth through contact with other particles. This is considered to be a very important precipitation-producing mechanism.

The artificial inducement of precipitation has been studied extensively, and these studies are continuing. It has been demonstrated that condensation nuclei supplied to clouds can induce precipitation. The ability of humans to ensure the production of precipitation or to control its geographic location or timing has not yet been attained, however.

Many legal as well as technological problems are associated with the prospects of "rain-making" processes. Of interest here is the impact on hydrologic estimates that uncontrolled or only partially controlled artificial precipitation might have. Many naturally occurring hydrologic variables are considered as statistical variates that are either randomly distributed or distributed with a random component. If the distribution or time series of the variable can be modeled, an inference as to the frequency of occurrence of significant hydrologic events of a given magnitude (such as precipitation) can be made. If, however, artificial controls are used and if the effects of these cannot be reliably predicted, frequency analyses may prove to be totally unreliable tools.

Precipitation Types

Dynamic or adiabatic cooling is the primary cause of condensation and is responsible for most rainfall. Thus it can be seen that vertical transport of air masses is a requirement for precipitation. Precipitation may be classified according to the conditions that generate vertical air motion. In this respect, the three major categories of precipitation type are *convective, orographic,* and *cyclonic.*

Convective Precipitation   Convective precipitation is typical of the tropics and is brought about by heating of the air at the interface with the ground. This heated air expands with a resultant reduction in weight. During this period, increasing quantities of water vapor are taken up; the warm moisture-laden air becomes unstable; and pronounced vertical currents are developed. Dynamic cooling takes place, causing condensation and precipitation. Convective precipitation may be in the form of light showers or storms of extremely high intensity (thunderstorms are a typical example).

Orographic Precipitation   Orographic precipitation results from the mechanical lifting of moist horizontal air currents over natural barriers such as mountain ranges. This type of precipitation is very common on the West Coast of the United States where moisture-laden air from the Pacific Ocean is intercepted by coastal hills and mountains. Factors that are important in this process include land elevation, local slope, orientation of land slope, and distance from the moisture source.

In dealing with orographic precipitation, it is common to divide the region under study into zones for which influences aside from elevation are believed to be reasonably constant. For each of these zones, a relation between rainfall and elevation is developed for use in producing isohyetal maps (see Section 4.5).

Cyclonic Precipitation   Cyclonic precipitation is associated with the movement of air masses from high-pressure regions to low-pressure regions. These pressure differences are created by the unequal heating of the earth's surface.

Cyclonic precipitation may be classified as frontal or nonfrontal. Any barometric low can produce nonfrontal precipitation as air is lifted through horizontal convergence of the inflow into a low-pressure area. Frontal precipitation results from the lifting of warm air over cold air at the contact zone between air masses having different characteristics. If the air masses are moving so that warm air replaces colder air, the front is known as a *warm front*; if, on the other hand, cold air displaces warm air, the front is said to be *cold*. If the front is not in motion, it is said to be a *stationary front*. Figure 4.1 illustrates a vertical section through a frontal surface.

Thunderstorms

Many areas of the United States are subjected to severe convective storms, which are generally identified as thunderstorms because of their electrical nature. These storms, although usually very local in nature, are often productive of very intense rainfalls that are highly significant when local and urban drainage works are considered.

Thunderstorm cells develop from vertical air movements associated with intense surface heating or orographic effects. There are three primary stages in the life history of a thunderstorm. These are the *cumulus stage*, the *mature stage*, and the *dissipating stage*. Figure 4.2 illustrates each of these stages.

All thunderstorms begin as cumulus clouds, although few such clouds ever reach the stage of development needed to produce such a storm. The cumulus stage is characterized by strong updrafts that often reach altitudes of over 25,000 ft. Vertical wind speeds at upper levels are often as great as 35 mph. As indicated in Fig. 4.2a, there is considerable horizontal inflow of air (entrainment) during the cumulus stage. This is an important element in the development of the storm, as additional moisture is provided. Air temperatures inside the cell are greater than those outside, as indicated by

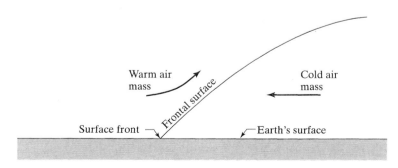

FIGURE 4.1

Vertical cross section through a frontal surface.

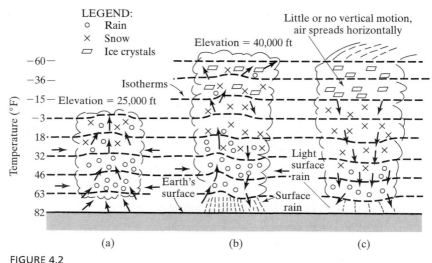

FIGURE 4.2

Cumulus, mature, and dissipating stages of a thunderstorm cell.
(Department of the Army)

the convexity of the isotherms viewed from above. The number and size of the water droplets increase as the stage progresses. The duration of the cumulus stage is approximately 10–15 min.

The strong updrafts and entrainment support increased condensation and the development of water droplets and ice crystals. Finally, when the particles increase in size and number so that surface precipitation occurs, the storm is said to be in the mature stage. In this stage strong downdrafts are created as falling rain and ice crystals cool the air below. Updraft velocities at the higher altitudes reach up to 70 mph in the early periods of the mature stage. Downdraft speeds of over 20 mph are usually above about 5,000 ft in elevation. At lower levels, frictional resistance tends to decrease the downdraft velocity. Gusty surface winds move outward from the region of rainfall. Heavy precipitation is often derived during this preiod, which is usually on the order of 15–30 min.

In the final or dissipating stage, the downdraft becomes predominant until all the air within the cell is descending and being dynamically heated. Since the updraft ceases, the mechanism for condensation ends and precipitation tails off and ends.

Precipitation Data

Considerable data on precipitation are available in publications of the National Weather Service [4],[5]. Other sources include various state and federal agencies engaged in water resources work. For regional studies it is recommended that all possible data be compiled; often the establishment of a gauging network will be necessary.

Precipitation Variability

Precipitation varies geographically, temporally, and seasonally (see Figs. 4.3 and 4.4). It should be understood that both regional and temporal variations in precipitation are

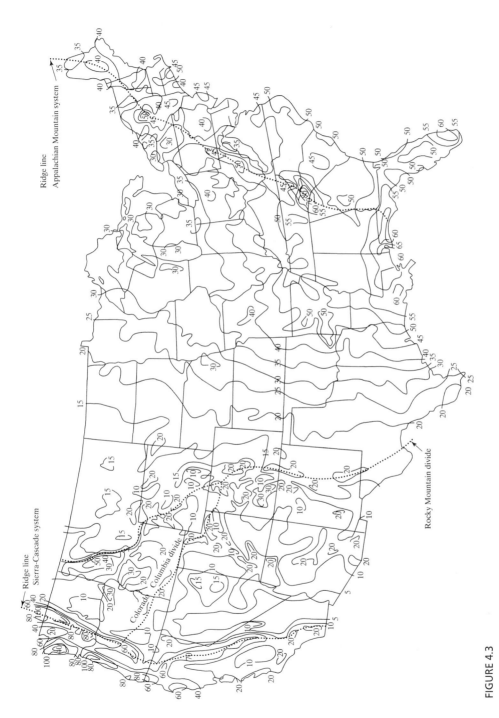

FIGURE 4.3

Mean annual precipitation in inches.

(*U.S. Department of Agriculture, Soil Conservation Service*)

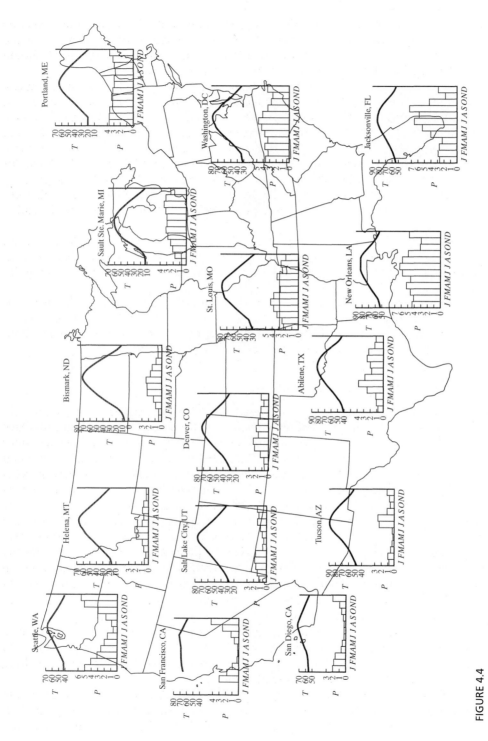

FIGURE 4.4

Precipitation and temperature distributions: *T*, mean monthly temperature (°F); *P*, mean monthly precipitation (in.).
(U.S. Department of Agriculture, Soil Conservation Service)

important in water resources planning and hydrologic studies. For example, it may be very important to know that the cycle of minimum precipitation coincides with the peak growing season in a particular area, or that the period of heaviest rainfall should be avoided in scheduling certain construction activities.

Precipitation amounts sometimes vary considerably within short distances. Records have shown differences of 20 percent or more in the catch of rain gauges less than 20 ft apart. Precipitation is usually measured with a rain gauge placed in the open so that no obstacle projects within the inverted conical surface having the top of the gauge as its apex and a slope of 45°. The catch of a gauge is influenced by the wind, which usually causes low readings. Various devices such as Nipher and Alter shields have been designed to minimize this error in measurement. Precipitation gauges may be of the recording or nonrecording type. The former are required if the time distribution of precipitation is to be known. Information about the features of gauges is readily available [3].

Because precipitation varies spatially, it is usually necessary to use the data from several gauges to estimate the average precipitation for an area and to evaluate its reliability. This is especially important in forested areas where the variation tends to be large.

Time variations in rainfall intensity are extremely important in the rainfall-runoff process, particularly in urban areas (see Fig. 4.5a). The areal distribution is also significant and highly correlated with the time history of outflow (see Fig. 4.5b). These considerations are discussed in greater detail in following chapters.

4.3   DISTRIBUTION OF PRECIPITATION

Total precipitation is distributed in numerous ways. That intercepted by vegetation and trees may be equivalent to the total precipitation input for relatively small storms. Once interception storage is filled, raindrops begin falling from leaves and grass, where water stored on these surfaces eventually becomes depleted through evaporation. Precipitation that reaches the ground may take several paths. Some water will fill depressions and eventually evaporate; some will infiltrate the soil. Part of the infiltrated water may strike relatively impervious strata near the soil surface and flow approximately parallel to it as interflow until an outlet is reached. Other portions may replenish soil moisture in the upper soil zone, and some infiltrated water may reach the groundwater reservoir that sustains dry weather streamflow. The component of the precipitation input that exceeds the local infiltration rate will develop a film of water on the surface (surface detention) until overland flow commences. Detention depths varying from $\frac{1}{8}$ to $1\frac{1}{2}$ in. for various conditions of slope and surface type have been reported [3]. Overland flow ultimately reaches defined channels and becomes streamflow.

Figure 4.6 illustrates in a general way the disposition of a uniform storm input to a natural drainage basin. Although such an input is not to be expected in nature, the indicated relations are representative of actual conditions. Modifications resulting from nonuniform storms will be discussed as they arise.

In Fig. 4.6a note that the storm input is distributed uniformly over time t_a at a rate equal to i (dimensionally equal to LT^{-1}). This input is dissected into components i_1 through i_4, the sum of which is equal to i at any time t. Figure 4.6b illustrates the

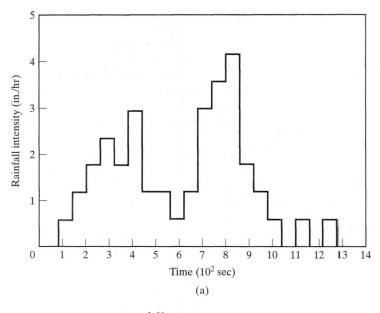

(a)

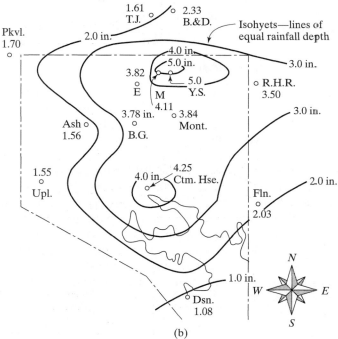

(b)

FIGURE 4.5

(a) Rainfall distribution in a convective storm June 1960, Baltimore, Maryland.
(b) Isohyetal pattern, storm of September 10, 1957, Baltimore, Maryland.
○, recording rain gauge.

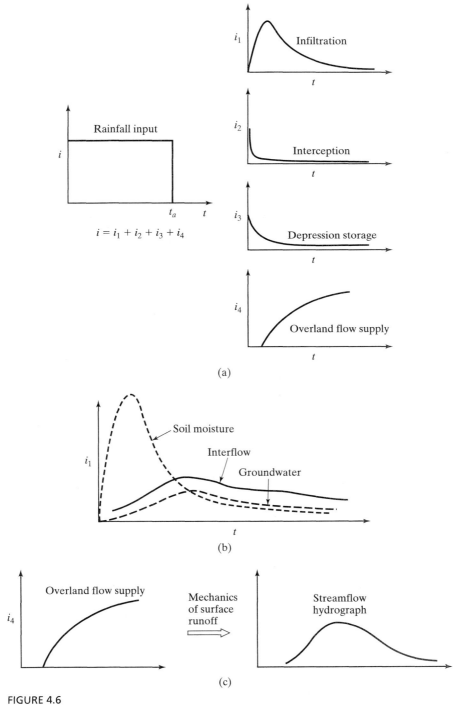

FIGURE 4.6

The runoff process: (a) disposition of precipitation, (b) components of infiltration, and (c) disposition of overland flow supply.

manner in which infiltrated water is further subdivided into interflow, groundwater, and soil moisture. Figure 4.6c shows the transition from overland flow supply into streamflow. The mechanics of these processes will be treated in detail in later sections. The nature of the curves presented depicts the general runoff process. It should be realized, however, that actual graphs of infiltration and/or other factors versus time might appear quite different in form and relative magnitude when compared with these illustrations because of the effects of nonuniform storm patterns, antecedent conditions, and other factors.

The rate and areal distribution of runoff from a drainage basin are determined by a combination of physiographic and climatic factors. Important climatic factors include the form of precipitation (rain, snow, hail), the type of precipitation (convective, orographic, cyclonic), the quantity and time distribution of the precipitation, the character of the regional vegetative cover, prevailing evapotranspiration characteristics, and the status of the soil moisture reservoir. Physiographic factors of significance include geometric properties of the drainage basin, land-use characteristics, soil type, geologic structure, and characteristics of drainage channels (geometry, slope, roughness, and storage capacity).

Large drainage basins often react differently from smaller ones when subjected to a precipitation input. This can be explained in part by such factors as geologic age, relative impact of land-use practices, size differential, variations in storage characteristics, and other causes. Chow defines a small watershed as a drainage basin whose characteristics do not filter out (1) fluctuations characteristic of high-intensity, short-duration storms; or (2) the effects of land management practices [6]. On this basis, small basins may vary from less than an acre up to 100 mi^2. A large basin is one in which channel storage effectively filters out the high frequencies of imposed precipitation and effects of land-use practices.

4.4 POINT PRECIPITATION

Precipitation events are recorded by gauges at specific locations. The resulting data permit determination of the frequency and character of precipitation events in the vicinity of the site. Point precipitation data are used collectively to estimate areal variability of rain and snow and are also used individually for developing design storm characteristics for small urban or other watersheds.

Point rainfall data are used to derive intensity–duration–frequency curves such as those shown in Fig. 4.7. Such curves are used in the rational method for urban storm drainage design (Chapter 12); their construction is discussed in Chapter 7. In applying the rational method, a rainfall intensity is used which represents the average intensity of a storm of given frequency for a selected duration. The frequency chosen should reflect the economics of flood damage reduction. Frequencies of up to 100 years are commonly used where residential areas are to be protected. For higher-value districts and critical facilities, up to 500 years or higher return periods are often selected. Local conditions and practice normally dictate the selection of these design criteria. (Executive Order 11988, Floodplain Management, 1977).

It is occasionally necessary to estimate point rainfall at a given location from recorded values at surrounding sites. This can be done to complete missing records or

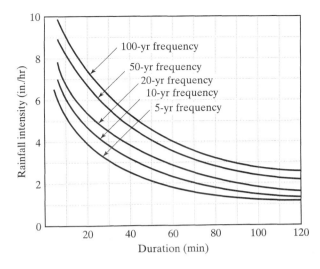

FIGURE 4.7

Typical intensity–duration–frequency curves for Baltimore, Maryland, and vicinity.

to determine a representative precipitation to be used at the point of interest. The National Weather Service has developed a procedure for this which has been verified on both theoretical and empirical bases [7].

Consider that rainfall is to be calculated for point A in Fig. 4.8. Establish a set of axes running through A and determine the absolute coordinates of the nearest surrounding points B, C, D, E, and F. These are recorded in columns 3 and 4 of Table 4.1. The estimated precipitation at A is determined as a weighted average of the other five points. The weights are reciprocals of the sums of the squares of ΔX and ΔY; that is, $D^2 = \Delta X^2 + \Delta Y^2$, and $W = 1/D^2$. The estimated rainfall at the point of interest is given

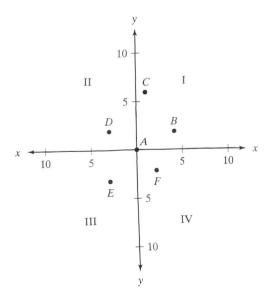

FIGURE 4.8

Four quadrants surrounding precipitation station A.

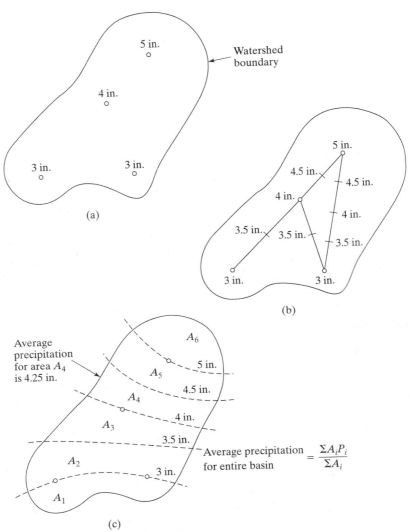

FIGURE 4.11

Construction of an isohyetal map: (a) Locate rain gauges and plot values; (b) interpolate between gauges; and (c) plot isohyets.

a suitable map and to record the rainfall amounts (Fig. 4.11). Next, an interpolation between gauges is performed and rainfall amounts at selected increments are plotted. Identical depths from each interpolation are then connected to form isohyets (lines of equal rainfall depth). The areal average is the weighted average of depths between isohyets, that is, the mean value between the isohyets. The isohyetal method is the most accurate approach for determining average precipitation over an area, but its proper use requires a skilled analyst and careful attention to topographic and other factors that impact on areal variability. Figure 4.12 illustrates the representation of a major storm event in North Carolina by an isohyetal map.

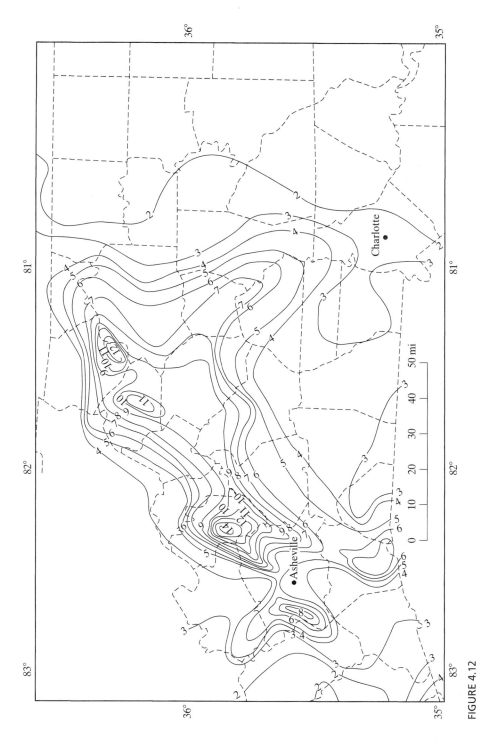

FIGURE 4.12

Map of Asheville-Statesville, North Carolina, area showing the precipitation that caused the flood of November 1977. Total precipitation is given in inches for the period from Friday, November 4 at 7 a.m. to Monday, November 7 at 7 a.m. *(National Weather Service)*

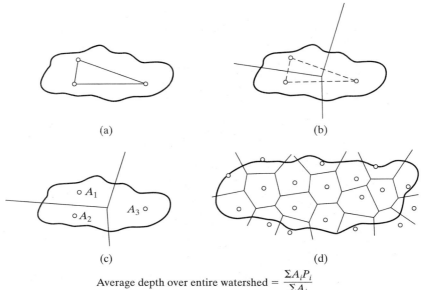

(a)   (b)

(c)   (d)

$$\text{Average depth over entire watershed} = \frac{\Sigma A_i P_i}{\Sigma A_i}$$

FIGURE 4.13

Construction of a Thiessen diagram: (a) Connect rain gauge locations; (b) draw perpendicular bisectors; and (c) calculate Thiessen weights (A_1, A_2, A_3). (d) A completed network.

Thiessen Method

Another method of calculating areal rainfall averages is the Thiessen method. In this procedure the area is subdivided into polygonal subareas using rain gauges as centers. The subareas are used as weights in estimating the watershed average depth. Thiessen diagrams are constructed as shown in Fig. 4.13. This procedure is not suitable for mountainous areas because of orographic influences. The Thiessen network is fixed for a given gauge configuration, and polygons must be reconstructed if any gauges are relocated.

Accuracy

Irrespective of the method used for estimating areal precipitation, the location of the gauge used in deriving the estimate relative to the point of application of the estimate must be taken into consideration. In mountainous localities, vertical distances may be more important than horizontal ones. For gentle landscapes, horizontal spacings are the most important. When a precipitation gauging network is to be developed, both spacing and arrangement of gauges must be considered.

Example 4.1

Given the drainage area of Fig. 4.14 and the rainfall data displayed in column 3 of Table 4.2, calculate the average rainfall over the area using (a) the arithmetic mean and (b) the Thiessen polygon weighting system.

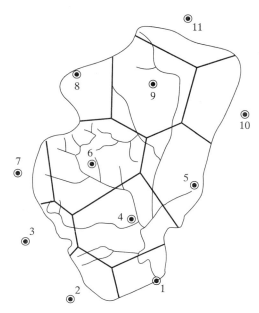

FIGURE 4.14

Thiessen diagram for Example 4.1.

Solution.

1. Identify those gauges falling within the area boundary. They include gauges 1, 4 through 6, 8, and 9. Averaging the values for these six gauges yields an estimated mean areal rainfall of 3.20 inches.

2. Following the Thiessen method as described in Section 4.5, construct polygons using triangles to connect gauge points. These polygons are shown in Fig. 4.14. Calculate the percent of the total area associated with each gauge

TABLE 4.2   Data and Thiessen Polygon Calculation for Example 4.1.

(1)	(2)	(3)	(4)
Gauge no.	% Area	Precip. (in.)	(2) × (3)
1	5	1.56	0.08
2	4	2.95	0.12
3	3	3.44	0.10
4	15	2.91	0.44
5	11	4.17	0.46
6	19	4.21	0.80
7	4	2.7	0.11
8	7	2.45	0.17
9	21	3.88	0.81
10	6	3.98	0.24
11	5	2.51	0.13
Total	100		3.45

and record as in column 2 of Table 4.2. The Thiessen weighted average is obtained by multiplying the values in column 2 by the values in column 3. The Thiessen average is computed as 3.45 inches of rainfall. The use of a spreadsheet (Table 4.2) facilitates computations and aids in organizing data.

4.6   PROBABLE MAXIMUM PRECIPITATION

The probable maximum precipitation (PMP) is the critical depth-duration-area rainfall relation for a given area and season which would result from a storm containing the most critical meteorological conditions considered probable [9]. Such storm events are used in flood flow estimates by the U.S. Corps of Engineers and other water resources agencies. The critical meteorological conditions are based on analyses of air–mass properties (effective precipitable water, depth of inflow layer, wind, temperature, and other factors), synoptic situations during recorded storms in the region, topography, season, and location of the area. The rainfall derived is termed *probable maximum precipitation* since it is subject to limitations of meteorological theory and data and is based on the most effective combination of factors controlling rainfall intensity [9]. An earlier designation of *maximum possible precipitation* is synonymous.

The seasonal variation of PMP is important in the design and operation of multipurpose structures and in flooding considerations that may occur in combination with snowmelt. In both of these cases, annual probable maximums might be less important than seasonal maximums. Figures 4.15 and 4.16 display 24-hr PMP for the eastern half of the United States for 200-mi^2 watersheds during the month of August (similar figures are available from the National Weather Service).

4.7   GROSS AND NET PRECIPITATION

The net (excess) precipitation that contributes directly to surface runoff is equivalent to the gross precipitation minus losses to interception, storm period evaporation, depression storage, and infiltration. The relation between excess precipitation P_e and gross precipitation P is thus:

$$P_e = P - \Sigma \text{ losses} \qquad (4.2)$$

where the losses include all deductions from the gross storm input.

The paths that water precipitated over an area may take can be represented by flow diagrams of the type given in Fig. 1.2 and by equations of the form of Eq. 4.2. Models such as these are the basis for most hydrologic investigations, and much of the content of this book is devoted to the conceptualization of individual components of the various hydrologic processes and to synthesizing these components into holistic representations of hydrologic events.

4.8   PRECIPITATION FREQUENCY ANALYSIS

Point precipitation data described in Section 4.4 are subjected to frequency analyses to develop frequency relationships among rainfall depth, intensity, and storm duration.

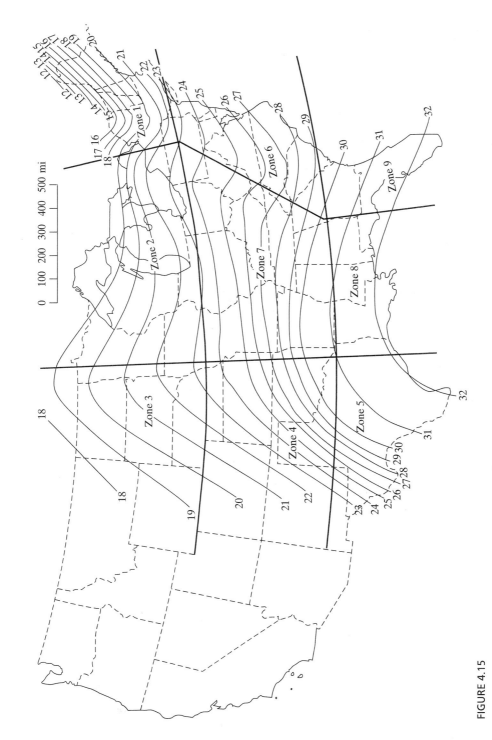

FIGURE 4.15

Probable maximum precipitation (in inches) for 200 mi² in 24 hr.

(*U.S. Department of Commerce, National Weather Service*)

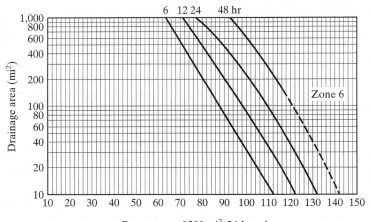

FIGURE 4.16

Seasonal variation, depth-area-duration relations. Percentage to be applied to 200-mi^2, 24-hr probable maximum precipitation values for August in Zone 6. *(U.S. Department of Commerce, National Weather Service)*

These relationships, known as intensity–duration–frequency (IDF) curves, are used in designing storm water management facilities and floodway reservations. Such designs are based on estimates of worst-case scenarios of rainfall intensity and duration during a given interval of time. This often requires analyzing several storms of different durations to find the most critical event for the selected design frequency. Typical recurrence intervals for designing storm water structures are given in Chapter 13.

The IDF curves for Baltimore, Maryland, shown in Fig. 4.7 are representative of those encountered in practice. Similar curves for other locations are available in the literature [10]–[13]. IDF curves are derived from analysis of rainfall data at a specific point; however, development of such curves may also involve hourly data from more than one recording gauge in the vicinity. Curves such as these are generally considered to be applicable for surface areas up to about five square miles. For larger drainage areas, the point data tend to overestimate the areal average rainfall, and adjustments (see Chapter 13) are necessary to determine areal depths for any desired frequency of occurrence.

The correct interpretation of any point value obtained from Fig. 4.7 is that on the average, for any given time duration, storms having a uniform intensity (i) for that duration would have a recurrence interval equal to the corresponding curve value. For example, in any time duration of 90 minutes, Baltimore could experience a peak 2.0-in./hr storm once every 20 years. The 20-yr, 90-min design storm for Baltimore would have a depth of $D = 3.0$ in., where D = rainfall intensity multiplied by the storm duration. A 20-yr, 30-min design storm would have an intensity of 4.6 in./hr but result in a depth of only 2.3 in. Although the latter storm produces less depth, its higher intensity could be the governing factor in determining the size of drainage works. The probability of occurrence of both storms would be the same, however.

Development of IDF Curves

Point rainfall data from weighing or tipping-bucket rain gauges provide the basis for calculating IDF curves for a given location. Rainfall depths occurring over selected time increments are compiled and arrayed by duration and storm. For any specified duration, the rainfall depths that occurred are ranked in decreasing order of magnitude and are assigned a rank order number m. The commonly used formula for computing plotting positions is:

$$P = m/(n + 1) \qquad (4.3)$$

where n is the number of observations and P is the estimated probability of values being greater than or equal to the ranked value. Note that where droughts are being studied, the data would be ranked from the lowest to the highest, and P would be the estimated probability of values being less than or equal to the ranked value.

Extreme hydrologic events are commonly estimated using annual series or partial-duration series (Chapter 3) of historical data. An annual series is comprised of a single extreme event (this may be either a maximum or a minimum) for each year of record. A partial-duration series is comprised of all events exceeding a selected base value for each year of record. Such a series is used in cases where more than one event of significance occurs per year. In performing frequency analyses on historical data, at least 10 years of record should be used. The following example illustrates the IDF procedure for both an annual and a partial-duration series for an 11-year record.

Example 4.2

Perform a frequency analysis of the 30-min Baltimore rainfall data in Table 4.3 as an annual and a partial-duration series and plot the results.

Solution.   In Table 4.3 the maximum rainfall depths that occurred for any 30-min period during excessive rainfalls at Baltimore, Maryland, 1945–1954, are shown in the order of occurrence. The 65 observations represent a complete series. The 11 maximum annual events are underlined and represent the annual series. The greatest 11 events throughout the record are identified by an asterisk and represent the partial-duration series.

The solution is shown in Table 4.4, and the data are plotted in Fig. 4.17. Note that the larger numbers occur in both series, and hence recurrence intervals for the less frequent events are the same.

The preceding example leads to consideration of the frequency analysis of rainfall depth or intensity for various durations of rainfall. Design problems often require the estimation of expected intensities for a critical time period. Frequency analysis of the rainfall record for periods other than the 30-min duration—for example, the maximum 5-, 10-, 20-, and 60-min occurrences—would yield a family of curves similar to those of Fig. 4.18. The usual method of presenting these data is to convert depth in inches to an intensity in in./hr and to summarize the data in intensity–duration–frequency curves as shown in Fig. 4.7. These curves are typical of the point analysis of rainfall data. It should be emphasized that the frequency curves join occurrences that are not necessarily from the same storm; that is, they do not represent a sequence of intensities during a single storm but only the average intensity expected for a specific duration.

TABLE 4.3 Maximum 30-min Rainfall Depths, Baltimore, MD, 1945–1954

Year	Storm number	RF depth (in.)	Year	Storm number	RF depth (in.)	Year	Storm number	RF depth (in.)
1945	1	0.38	1948	1	1.33*	1953	1	0.40
	2	0.47		2	0.65		2	0.45
	3	0.39		3	0.47		3	0.53
	4	0.76		4	0.84		4	2.50*
	5	0.56		5	0.68		5	1.03
	6	0.35		6	0.63		6	0.75
	7	0.43		7	0.47		7	0.70
	8	0.40	1949	1	0.52		8	1.00*
	9	0.36		2	0.49	1954	1	0.42
1946	1	0.62	1950	1	0.55		2	0.70
	2	0.55		2	0.63		3	0.85
	3	0.88		3	0.69		4	0.60
	4	0.47		4	1.27*			
	5	0.36		5	1.10*	1955	1	0.70
	6	1.15*	1951	1	0.88		2	0.95
	7	0.75		2	0.97		3	1.02
	8	1.53*		3	0.59		4	0.50
	9	0.51		4	0.46		5	0.65
1947	1	0.88		5	0.50		6	0.55
	2	2.04*		6	0.55		7	0.52
	3	0.76	1952	1	0.47		8	0.45
	4	0.97		2	1.20*		9	0.54
	5	0.71		3	0.93		10	0.60
	6	1.07*		4	0.70		11	0.80
	7	0.94		5	0.57		12	0.95
	8	1.20*		6	0.46			
				7	0.48			
				8	1.30*			

Note: Underlined items are the annual series. Asterisks identify the partial-duration series.

IDF Data Sources

The principal source of precipitation data is the National Weather Service (NWS). The agency has six regional offices and is part of the National Oceanic and Atmospheric Administration (NOAA). Precipitation measurements are made daily at over 20,000 U.S. locations. State data are compiled by satellite and other means in the Weather Service Forecast Office (WSFO). Results of IDF analyses for most major urban locations have been compiled [13]. For additional information and Web addresses, see also Chapter 2).

Maximum average rainfall depths have been published by the U.S. Weather Bureau [13] for durations between 30 min and 24 hr and for recurrence intervals

TABLE 4.4 Annual and Partial Series Rainfall Depths

Order	Depth (in.)		Recurrence interval, $(n + 1)/m$
	Annual series	Partial series	
1	2.50	2.50	12
2	2.04	2.04	6
3	1.53	1.53	4
4	1.33	1.33	3
5	1.30	1.30	2.4
6	1.27	1.27	2
7	1.02	1.20	1.7
8	0.97	1.20	1.5
9	0.85	1.15	1.3
10	0.76	1.10	1.2
11	0.52	1.07	1.1

between 1 and 100 years. Depth–duration–frequency curves can be constructed for any location by plotting successive values from the various rainfall maps, preferably on logarithmic paper to facilitate fitting flatter curves. Correction factors are given to permit estimates of depths for durations less than 30 min.

Formulas for IDF Curves

Regression analysis can be used to fit intensity–duration–frequency curves, and the constants can be interpreted as regional characteristics. Many formulas have been used to fit these curves, but most of them are in a form with intensity (i) inversely proportional to duration (t). Steel [14] has used a model of the form $i = A/(t + B)$ to fit rainfall

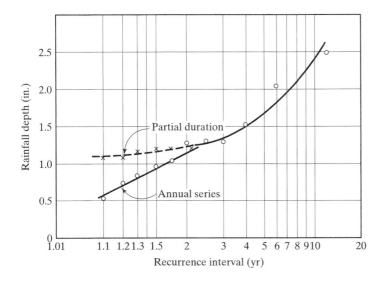

FIGURE 4.17

Difference in annual and partial-duration series, 11-year record of maximum 30-min durations, Baltimore, Maryland.

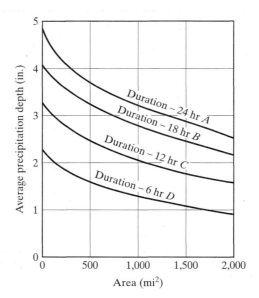

FIGURE 4.18

Depth–area–duration curves for
24-hr storm.

data throughout the United States. The constants A and B therefore serve as characteristic features of both the rainfall region and the frequency of occurrence in
each area.

Example 4.3

Fit the following rainfall data to determine the 10-year intensity–duration–frequency
curve.

t = duration (min)	5	10	15	30	60	120
i = intensity (in./hr)	7.1	5.9	5.1	3.8	2.3	1.4
$1/i$	0.14	0.17	0.20	0.26	0.43	0.71

Solution

1. A model of the form $i = A/(t + B)$ can be expressed in linear form as
    $1/i = t/A + B/A$.
2. The regression of $1/i$ versus t yields $1/i = 0.005t + 0.12$, from which
    $A = 200$ and $B = 24$.
3. Thus the rainfall formula is $i = 200/(t + 24)$. The correlation coefficient is
    -0.997.

Example 4.4

Table 4.5 catalogs data for numerous storms having a range of intensities and durations. The entire record spans 40 years. By interpolating the values in the table,
estimate the time versus intensity values for the 5-year storm.

TABLE 4.5 Intense Rainfalls for 40-Year Record

Row No.	Storm duration (min)	(2)	(3)	(4)	(5)	(6)	(7)	(8)	(9)	(10)	(11)	(12)	(13)
		Row immediately below indicates rainfall intensity in in./hr											
		1.0	1.25	1.5	1.75	2.0	2.5	3.0	4.0	5.0	6.0	7.0	8.0
(1)	5							117	46	19	13	5	2
(2)	10					119	76	46	13	8	3	2	1
(3)	15				99	84	43	19	9	3	1	1	
(4)	20			96	62	42	16	12	5	2	1		
(5)	30	97	69	53	28	23	9	5	2	1			
(6)	40	68	51	26	13	12	4	3	1				
(7)	50	49	27	19	10	9	3	2					
(8)	60	38	17	13	6	4	3	1					
(9)	80	20	12	4	1	2							
(10)	100	12	3	1									
(11)	120	7	1										

Values under columns 2–13 in rows 1–11 are the number of storms of the duration shown in the left column for the intensity given below the column number.

Solution. The 5-year storm would be equaled or exceeded $40/5 = 8$ times in 40 years. Using this as the reference value, interpolations by row and column can be made. For example, for row 1, the interpolation would be as follows:

$$7 - [(8 - 5)/(13 - 5)] \times 1 = 7 - (3/8) = 6.63$$

For column 11, the interpolation would be as follows:

$$10 - [(8 - 3)/(13 - 3)] \times 5 = 10 - (5/10) \times 5 = 7.5$$

Similar operations made on all rows and columns yield the values shown below.

Row Interpolation

Time	5	10	15	20	30	40	50	60	80	100
Rainfall intensity	6.63	5.00	4.17	3.57	2.63	2.25	2.08	1.83	1.38	1.11

Column Interpolation

Rainfall intensity	6.00	5.00	4.00	3.00	2.50	2.00	1.75	1.50	1.25	1.00
Time	7.50	10.00	16.25	25.71	32.00	52.00	55.00	71.11	88.89	116.00

Plotting these data, the IDF curve for the 5-year storm can be obtained as shown in Fig. 4.19.

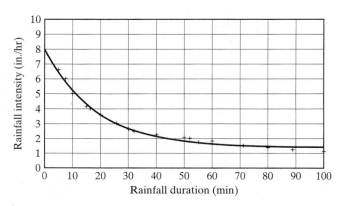

FIGURE 4.19

Intensity–duration–frequency curve, 5-year storm

Example 4.5

You are given average annual values of precipitation (inches) for the period from 1889 to 1997 for Tallahassee, Florida (see Table 4.6). Arrange the data in order of magnitude from highest to lowest and calculate the exceedance probability for each annual value. Use Eq. 4.3 and determine the values as percentages. Plot the exceedance probabilities versus average annual precipitation.

Solution. The data are arranged by descending order of magnitude in columns 2, 5, 8, 11, 14, and 27. The probabilities are calculated on the spreadsheet using Eq. 4.3 and multiplying by 100 to convert to percent. Figure 4.20 is a plot relating exceedance probability to average annual precipitation.

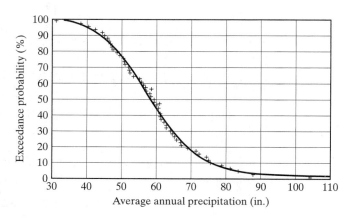

FIGURE 4.20

Exceedance probabilities for average annual precipitation, Tallahassee, Florida

TABLE 4.6 Tallahassee, FL, Average Annual Precipitation 1889–1997*.

1	2	3	4	5	6	7	8	9	10	11	12	13	14	15	16	17	18
1	104.2	0.9	19	71.4	17.3	37	62.8	33.6	55	59.3	50.0	73	52.2	66.4	91	47.0	82.7
2	89.9	1.8	20	69.5	18.2	38	62.6	34.5	56	58.5	50.9	74	52.1	67.3	92	47.0	83.6
3	87.8	2.7	21	69.0	19.1	39	62.1	35.5	57	58.3	51.8	75	52.0	68.2	93	46.5	84.5
4	85.9	3.6	22	67.8	20.0	40	61.9	36.4	58	58.3	52.7	76	51.9	69.1	94	46.2	85.4
5	83.5	4.5	23	67.1	20.9	41	61.3	37.3	59	58.0	53.6	77	51.8	70.0	95	46.1	86.4
6	81.8	5.5	24	67.0	21.8	42	61.3	38.2	60	57.2	54.5	78	50.8	70.9	96	45.7	87.3
7	81.2	6.4	25	66.9	22.7	43	61.1	39.1	61	56.9	55.4	79	50.7	71.8	97	45.7	88.2
8	80.5	7.3	26	66.1	23.6	44	61.1	40.0	62	56.8	56.4	80	50.7	72.7	98	45.0	89.1
9	78.8	8.2	27	65.8	24.5	45	61.0	40.9	63	56.7	57.3	81	50.6	73.6	99	44.9	90.0
10	77.0	9.1	28	65.6	25.5	46	60.9	41.8	64	56.6	58.2	82	50.3	74.5	100	44.7	90.9
11	75.6	10.0	29	65.1	26.4	47	60.8	42.7	65	56.2	59.1	83	50.3	75.4	101	44.2	91.8
12	75.5	10.9	30	64.8	27.3	48	60.6	43.6	66	56.1	60.0	84	50.0	76.4	102	44.1	92.7
13	75.1	11.8	31	64.6	28.2	49	60.5	44.5	67	55.6	60.9	85	49.3	77.3	103	42.2	93.6
14	74.5	12.7	32	64.3	29.1	50	60.4	45.5	68	55.1	61.8	86	48.5	78.2	104	40.8	94.5
15	74.4	13.6	33	64.0	30.0	51	60.2	46.4	69	55.0	62.7	87	48.4	79.1	105	40.3	95.4
16	74.3	14.5	34	63.6	30.9	52	60.0	47.3	70	53.9	63.6	88	47.3	80.0	106	39.2	96.4
17	72.3	15.5	35	62.9	31.8	53	59.4	48.2	71	53.3	64.5	89	47.3	80.9	107	38.1	97.3
18	71.8	16.4	36	62.9	32.7	54	59.3	49.1	72	53.0	65.4	90	47.1	81.8	108	38.0	98.2
															109	31.0	99.1

*Data are rank ordered from highest to lowest.
Columns 1, 4, 7, 10, 13, and 16 are the rank-ordered numbers for the rainfall data.
Columns 2, 5, 8, 11, 14, and 17 are the rank-ordered rainfall data for the 109-year record.
Columns 3, 6, 9, 12, 15, and 18 are the calculated probabilities that the rainfall value will be equaled or exceeded (values shown are in percent).

4.9   SNOW ACCUMULATION

Information on the areal distribution of snow cover is required for water supply fore-casting, flood potential investigations, and other hydrologic analyses. Observations of snow cover are generally obtained by ground and air reconnaissance and photography. Between snow surveys, approximations of the extent of the snow cover are based on available hydrometeorological data. Snow cover depletion patterns within a basin are usually somewhat uniform from year to year; thus snow cover indexes can often be developed from data gathered at a few representative stations.

Estimating the areal distribution of snowfall poses some problems. Taking arithmetic averages or using Thiessen polygons does not commonly provide reliable estimates of areal snow distribution from point gaugings. This is because orographic and topographic effects are often pronounced, and gauging networks frequently are not dense enough to permit the straightforward use of normal averaging techniques. However, regional orographic effects are relatively constant from year to year and storm to storm for tracts that are small when compared with the areal extent of gen-eral storms occurring in the region [15]. This circumstance permits many useful approaches in estimating the areal snow distribution once the basic pattern has been found for a region.

One method used to estimate basin precipitation from point observations assumes that the ratio of station precipitation to basin precipitation is approximately constant for a storm or storms. This can be stated as [15]:

$$\frac{P_b}{P_a} = \frac{N_b}{N_a} \tag{4.4}$$

or

$$P_b = \frac{P_a N_b}{N_a} \tag{4.5}$$

where P_b = the basin precipitation
P_a = the observed precipitation at a point or group of stations
N_b = the annual precipitation for the basin
N_a = the normal annual precipitation for the control station or stations

The normal annual precipitation is determined from a map (carefully prepared if it is to be representative) displaying the mean annual isohyets for the region. The precipi-tation is determined by using a planimeter on the areas between the isohyets. If the number of stations used and their distribution adequately depict the basin, Eq. 4.5 can provide a good approximation. For stations not uniformly distributed, weighting coef-ficients based on the percentage of the basin area portrayed by a gauge are sometimes used in determining N_a for the group.

Another system used in estimating areal snowfall is the isopercental method. In this approach, the storm or annual station precipitation is expressed as a percentage of the normal annual total. Isopercental lines are drawn and can be superimposed on a normal annual precipitation (NAP) map to produce new isohyets representing the storm of interest. A NAP map indicates the general nature of the basin's topographic effects, while the isopercental map shows the deviations from this pattern. The advantage of this method over preparing an isohyetal map directly is that relatively consistent storm pattern features of the NAP map can be taken into consideration as well as observed individual storm variations.

Snow Measurements and Surveys

Snow measurements are obtained through the use of standard and recording rain gauges, seasonal storage precipitation gauges, snow boards, and snow stakes. Rain gauges are usually equipped with shields to reduce the effect of wind. Snow boards are about 16 in. square, laid on the snow so that new snowfall which accumulates between observation periods will be found above them. Care must be taken to assure that adverse wind effects or other conditions do not produce an erroneous sample at the gauging location. Snow stakes are calibrated wooden posts driven into the ground for periodic observation of the snow depth or inserted into the snowpack to determine its depth.

Direct measurements of snow depth at a single station are generally not very useful in making estimates of the distributon over large areas, since the measured depth may be highly unrepresentative because of drifting or blowing. To circumvent this problem, snow-surveying procedures have been developed. Such surveys provide information on the snow depth, water equivalent, density, and quality at various points along a snow course. All these measures are of direct use to a hydrologist.

The water equivalent is the depth of water that would weigh the same amount as that of the sample. In this way snow can be described in terms of inches of water. Density is the percentage of snow volume that would be occupied by its water equivalent. The quality of the snow relates to the ice content of the snowpack and is expressed as a decimal fraction. It is the ratio of the weight of the ice content to the total weight. Snow quality is usually about 0.95 except during periods of rapid melt, when it may drop to 0.70–0.80 or less. The thermal quality of snow, Q_t, is the ratio of heat required to produce a particular amount of water from the snow, to the quantity of heat needed to produce the same amount of melt from pure ice at 32°F. Values of Q_t may exceed 100 percent at subfreezing temperatures. The density of dry snow is approximately 10 percent but there is considerable variability between samples. With aging, the density of snow increases to values on the order of 50 percent or greater.

A snow course includes a series of sampling locations, normally not fewer than 10 in number [6]. The various stations are spaced about 50–100 ft apart in a geometric pattern designed in advance. Points are permanently marked so that the same locations will be surveyed each year—this is very important if snow course memoranda are to be correlated with areal snowcover and depth, expected runoff potential, or other significant factors. Survey data are obtained directly by foresters and others, by aerial photographs and observations, and by automatic recording stations that telemeter information to a central processing location.

In the western United States the Soil Conservation Service coordinates many snow surveys. Various states, federal agencies, and private enterprises are also engaged in this type of activity. Sources of snow survey data are summarized in Ref. 6.

Basin-Wide Water Equivalent

In the final analysis, it is the water equivalent of the snowpack that determines runoff. Basin water equivalent may be given as an index or reported in a quantitative manner such as inches or centimeters of depth for the watershed under investigation.

The customary procedure for determining the basin water equivalent is to take observed data from snow course stations and to provide an index of basin conditions. Various procedures employ averages, weighted averages, and other approaches to accomplish this [15]. The important point to remember is that the usefulness of any index is based on how well it represents the overall basin conditions, not on how favorably it describes a particular point value. Indexes do not actually provide a quantitative evaluation of the property they cover. Instead, they give relative changes in the factor. By introducing additional data, however, an index can be used in a prediction equation. For example, if the basin water equivalent can be estimated by subtracting the runoff and loss components from the precipitation input, the index can be correlated with actual basin water equivalent in a quantitative manner.

SUMMARY

Precipitation is the source of fresh water replenishment for the planet Earth. Too much or too little can mean the difference between prosperity and disaster. In between these extremes are the normal precipitation events that are experienced with a frequency and intensity related mainly to geographic position and topographic features.

After reading this chapter you should understand that both the timing and amount of precipitation occurring over an area are important and that there is considerable geographic variability in precipitation. You should be able to estimate areal precipitation amounts from gauge data and conceptualize simple hydrologic process models. It should be recognized that average values of precipitation for a region shed some light on the quantity of water that might be made available for various uses, while a knowledge of the time-distribution and time-disposition of precipitation are requisites for developing management plans for periods of excess and shortage.

PROBLEMS

4.1 Rain gauge X was out of operation for a month during which there was a storm. The rainfall amounts at three adjacent stations A, B, and C were 4.2, 3.7, and 4.9 in., respectively. The average annual precipitation amounts for the gauges are $X = 36.5$, $A = 42.1$, $B = 37.1$, and $C = 39.8$ in. The delta x and delta y values respectively for each station are $X = 0, 0$; $A = 3, 7$; $B = 4, 6$; and $C = 5, 9$. Using a weighted average, estimate the amount of rainfall for gauge X.

4.2 Compute the rainfall for gauge X in Problem 4.1 if the storm readings at A, B, and C were 3.7, 4.1, and 4.8 in., respectively.

4.3 Compute the mean annual precipitation for the watershed in the following figure using the arithmetic mean, the Thiessen polygon method, and the isohyetal method. The gauge readings for gauges A–K, respectively, are: 29.79, 34.97, 25.6, 24.27, 24.6, 42.61, 42.35, 15.51, 39.99, 43.04, and 28.41.

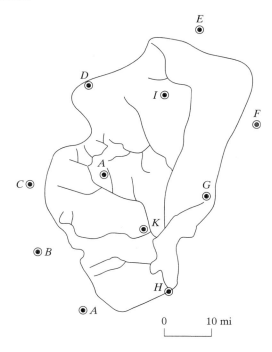

4.4 Compute the mean annual precipitation for the watershed in the figure for Problem 4.3 using the arithmetic mean and the Thiessen polygon method. The gauge readings for gauges A–K, respectively, are: 28.1, 33.7, 25.6, 23.9, 24.6, 40.7, 41.3, 37.2, 38.7, 41.1, and 29.3.

4.5 The chart from a rain gauge shown in the sketch represents a record that you must interpret. Find the average rainfall intensity (rate) between 6 a.m. and noon on August 10. Also find the total precipitation on August 10 and August 11.

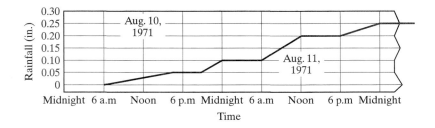

4.6 Refer to the chart of Problem 4.5. Calculate the rainfall intensity for the period between 6 a.m. and noon on August 11. Would you consider this to be a period of intense rainfall?

4.7 Use the map of Fig. 4.5 and from it construct a set of Thiessen polygons. Using these, estimate the mean rainfall for the region.

4.8 A mean draft of 100 mgd is produced from a drainage area of 200 mi^2. At the flow line the reservoir is estimated to cover 4000 acres. The annual rainfall is 37 in., the mean annual runoff is 10 in., and the mean annual lake evaporation is 30 in. Find the net gain or loss in storage. Compute the volume of water evaporated. How significant is this amount?

4.9 A mean draft of 380,000 m³/day is produced from a drainage area of 330 km². At the flow line, the reservoir is estimated to cover about 1600 hectares. The annual rainfall is 96.5 cm, the mean annual runoff is 22.8 cm, and the mean annual lake evaporation is 77.1 cm. Find the net gain or loss in storage and compute the volume of water evaporated. Calculate volumes in cubic meters.

4.10 Drainage areas within each of the isohyetal lines for a storm are tabulated for a watershed. Use the isohyetal method to determine the average precipitation depth within the basin for the storm. Make a conceptual sketch.

Isohyetal interval (in.)	Area (acres)
0–2	2700
2–4	1900
4–6	1000
6–8	0

4.11 Rework Problem 4.10 if the values in column 2 of the table are 2,500, 2,100, 1,200, and 300, respectively.

4.12 Discuss how you would go about collecting data for analysis of the water budget of a region. What agencies would you contact? What other sources of information would you seek out?

4.13 For an area of your choice, plot the mean monthly precipitation versus time. Explain how this fits the pattern of seasonal water uses for the area. Will the form of precipitation be an important consideration?

4.14 The following table summarizes the number of occurrences of intensities of various durations for a 34-year record of rainfall. Maximum intensities for the given durations were determined for all excessive storms and a count made of the exceedances. Interpolate for the average number of exceedances expected on a 5-year frequency and plot the 5-year intensity–duration–frequency curve.

Duration (min)	Intensity (in./hr) 1.0	2.0	3.0	4.0	5.0	6.0	7.0
5			73	48	21	9	2
10		68	51	26	11	3	1
15	72	35	23	11	5	1	
30	29	17	7	3	1		
60	15	6	2	1			
120	8	1					

Number of times stated intensities were equaled or exceeded

4.15 Fit the formula $i = A/(t + B)$ to the data derived in Problem 4.14 for the 5-year intensity–duration–frequency curve.

4.16 Recorded maximum depths (in.) of precipitation for a 30-min duration at a single station are:

Year	Date	Depth (in.)
1963	May 3	2.0
	June 3	1.0
1964	June 7	1.0
1965	June 2	1.0
1966	June 1	1.0
	Aug. 3	3.0
1967	July 4	1.0
1968	Aug. 8	4.0
1969	May 6	6.0
	June 8	5.0
	Sept. 4	1.0
1970	May 4	1.0
1971	Sept. 8	5.0
1977	May 9	1.0

(a) Determine the return period (years) for a depth of 2.0 in. using the California method with an annual series.

(b) Repeat Part (a) using a partial-duration series.

(c) Determine from the partial-duration series the depth of 30-min rain expected to be equaled or exceeded (on the average) once every 8 years.

4.17 For a 60-year record of precipitation intensities and durations, a 30-min intensity of 2.50 in./hr was equaled or exceeded a total of 85 times. All but 5 of the 60 years experienced one or more 30-min intensities equaling or exceeding the 2.50-in./hr value. Use the Weibull formula to determine the return period of this intensity using (a) a partial series and (b) an annual series.

REFERENCES

[1] Tennessee Valley Authority, "Heat and Mass Transfer Between a Water Surface and the Atmosphere," Lab. Rep. No. 14, TVA Engineering Lab, Norris, TN, Apr. 1972.

[2] A. L. Shands, "Mean Precipitable Water in the United States," U.S. Weather Bureau, Tech. Paper No. 10, 1949.

[3] R. K. Linsley, Jr., M. A. Kohler, and J. L. H. Paulhus, *Applied Hydrology*, New York: McGraw-Hill, 1949.

[4] D. W. Miller, J. J. Geraghty, and R. S. Collins, *Water Atlas of the United States*, Port Washington, NY: Water Information Center, 1963.

[5] U.S. Weather Bureau, Tech. Papers 1–33. Washington, D.C.: U.S. Government Printing Office.

[6] Ven Te Chow (ed.), *Handbook of Applied Hydrology*, New York: McGraw-Hill, 1964.

[7] Staff, Hydrologic Research Laboratory, "National Weather Service River Forecast System Forecast Procedures," NOAA Tech. Mem. NWS HYDRO 14, National Weather Service, Silver Spring, MD, Dec. 1972.

[8] V. Mockus, Sec. 4, in *SCS National Engineering Handbook on Hydrology*, Washington, D.C.: Soil Conservation Service, Aug. 1972.

[9] J. T. Riedel, J. F. Appleby, and R. W. Schloemer, "Seasonal Variation of the Probable Maximum Precipitation East of the 105th Meridian for Areas from 10 to 1000 Square Miles and Durations of 6, 12, 24, and 48 Hours," Hydrometeorological Rept. No. 33, U.S. Weather Bureau, Washington, D.C., 1967.

[10] U.S. Weather Bureau, "Maximum Recorded United States Rainfall," Technical Paper No. 2, U.S. Department of Commerce, Washington, D.C., 1947.

[11] U.S. Department of Commerce, NOAA and U.S. Army Corps of Engineers, "Probable Maximum Precipitation Estimates, United States East of the 105th Meridian," Hydrometeorological Report No. 51, Washington, D.C., 1978.

[12] U.S. Department of Commerce, NOAA, "Probable Maximum Precipitation Estimates, United States between the Continental Divide and the 103rd Meridian," Hydrometeorological Report No. 55, Washington, D.C., 1983.

[13] U.S. Weather Bureau, "Rainfall Frequency Atlas of the United States," Technical Paper No. 40, U.S. Department of Commerce, Washington, D.C., 1961.

[14] E.W. Steel, *Water Supply and Sewerage*, 4th ed., New York: McGraw-Hill, 1960.

[15] U.S. Army Corps of Engineers, "Snow Hydrology," NTIS PB 151 660, North Pacific Division, Portland, OR, June 1956.

C H A P T E R   5

Interception and Depression Storage

OBJECTIVES

The purpose of this chapter is to:

- Define interception and depression storage
- Explain how these mechanisms affect the quantity of precipitated water available for groundwater replenishment and surface flow
- Present some methods for estimating the quantities of water intercepted and stored in depressions during precipitation events.

Figure 1.2 indicates the paths that precipitated water may take as it reaches the earth. The first encounters are with intercepting surfaces such as trees, plants, grass, and structures. Water in excess of interception capacity then begins to fill surface depressions. A film of water also builds up over the ground surface. This is known as *surface detention*. Once this film is of sufficient depth, surface flow toward defined channels commences, providing that the rate at which water seeps into the ground is less than the rate of surface supply. This chapter deals with the first two mechanisms by which the gross precipitation input becomes transformed into net precipitation.

5.1   INTERCEPTION

Part of the storm precipitation that occurs is intercepted by vegetation and other forms of cover on the drainage area. *Interception* can be defined as that segment of the gross precipitation input which wets and adheres to aboveground objects until it is returned to the atmosphere through evaporation. Precipitation striking vegetation may be retained on leaves or blades of grass, flow down the stems of plants and become stemflow, or fall off the leaves to become part of the throughfall. The modifying effect that a forest canopy can have on rainfall intensity at the ground (the throughfall) can be put to practical use in watershed management schemes.

The amount of water intercepted is a function of (1) the storm character, (2) the species, age, and density of prevailing plants and trees, and (3) the season of the year. Usually about 10–20 percent of the precipitation that falls during the growing season is intercepted and returned to the hydrologic cycle by evaporation. Water losses by interception are especially pronounced under dense closed forest stands—as much as 25 percent of the total annual precipitation. Schomaker has reported that the average annual interception loss by Douglas fir stands in western Oregon and Washington is about 24 percent [1]. A 10-year-old loblolly pine plantation in the South showed losses on a yearly basis of approximately 14 percent, while Ponderosa pine forests in California were found to intercept about 12 percent of the annual precipitation. Mean interception losses of approximately 13 percent of gross summer rainfall were reported for hardwood stands in the White Mountains of New Hampshire. Additional information given in Table 5.1 includes some data on interception measurements obtained in Maine from a mature spruce-fir stand, a moderately well-stocked white and gray birch stand, and an improved pasture [1].

Lull indicates that oak or aspen leaves may retain as much as 100 drops of water [2]. For a well-developed tree, interception storage on the order of 0.06 in. of precipitation could therefore be expected on the basis of an average retention of about 20 drops per leaf. For light showers (where gross precipitation $P < 0.01$ in.), 100 percent interception might occur, whereas for showers where $P > 0.04$ in., losses in the range of 10–40 percent are realistic [3].

Figure 5.1 illustrates the general time distribution pattern of interception loss intensity. Most interception loss develops during the initial storm period and the rate of interception rapidly approaches zero thereafter [1]–[6]. Potential storm interception losses can be estimated by using [2],[3],[6]:

$$L_i = S + KEt \tag{5.1}$$

where L_i = the volume of water intercepted (in.)

S = the interception storage that will be retained on the foliage against the forces of wind and gravity (usually varies between 0.01 and 0.05 in.)

K = the ratio of surface area of intercepting leaves to horizontal projection of this area

E = the amount of water evaporated per hour during the precipitation period (in.)

t = time (hr)

Equation 5.1 is based on the assumption that rainfall is sufficient to fully satisfy the storage term S. The following equation was designed to account for the rainfall amount [7]–[9]:

$$L_i = S(1 - e^{-P/S}) + KEt \tag{5.2}$$

TABLE 5.1   Weekly Average Precipitation Catch of Standard U.S. Weather
Bureau–Type Rain Gauges Located in a Spruce-Fir Stand, a Hardwood
Stand, and a Pasture During the Winter of 1965–1966

Measuring date[a]	Weekly average precipitation catch (in. of equivalent rain)			Percent interception by forest cover	
	Spruce-fir	Birch	Pasture	Spruce-fir	Birch
11/9/65	0.24	0.33	0.39	38	15
11/16/65	1.01	1.25	1.45	30	14
11/23/65	1.01	1.23	1.36	26	10
12/10/65[b]	1.41	1.65	1.79	21	8
12/17/65	0.55	0.81	0.87	37	7
12/30/65	0.66	0.95	1.08	39	12
1/4/66	0.20	0.25	0.26	23	4
1/12/66	0.36	0.55	0.61	41	10
1/18/66	Trace	Trace	Trace	—	—
1/25/66	0.25	0.58	0.59	58	2
2/1/66	1.38	1.91	1.96	30	3
2/8/66	0.05	0.07	0.06	17	16
2/11/66	0.29[c]	0.02	Trace	—	—
2/15/66	0.76	0.81	0.98	22	17
2/21/66	0.17	0.22	0.22	23	0
3/2/66	0.86	1.23	1.45	41	16
3/7/66	0.76	0.84	0.97	22	13
3/15/66	0	0	0	—	—
3/29/66	0.73	1.13	1.27	43	11
Total	10.69	13.83	15.31	30.2	9.5

[a] The period between measuring dates is 7 days, except when precipitation occurred on the seventh day. In this event, measurement was postponed until precipitation ceased.
[b] Measurements were delayed until a method was devised to melt frozen precipitation on the site.
[c] This measurement in the spruce stand was the result of foliage drip during a thaw from previously intercepted snow.
Source: After C. E. Schomaker, "The Effect of Forest and Pasture on the Disposition of Precipitation," Maine Farm Res. (July 1966).

where P = rainfall and e is the base of natural logarithms. Note in Eqs. 5.1 and 5.2 that the storm time duration t is given in hours, while L_i, S, and E are commonly measured in inches or millimeters.

It is important to recognize that forms of vegetation other than trees can also intercept large quantities of water. Grasses, crops, and shrubs often have leaf-area to ground-area ratios that are similar to those for forests. Table 5.2 summarizes some observations that have been made on crops during growing seasons and on a variety of grasses. Intercepted amounts are about the same as those for forests, but since some of these types of vegetation exist only until harvest, their annual impact on interception is generally less than that of forested areas.

Precipitation type, rainfall intensity and duration, wind, and atmospheric conditions affecting evaporation are factors that serve to determine interception losses.

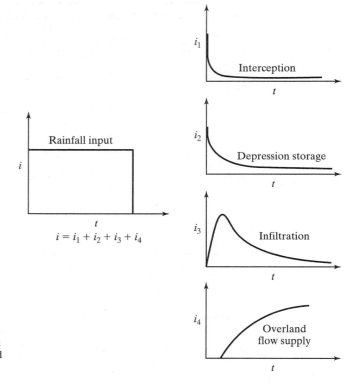

FIGURE 5.1

Disposition of rainfall input in terms of interception, depression storage, infiltration, and overland flow.

Snow interception, while highly visible, usually is not a major loss since much of the intercepted snowfall is eventually transmitted to the ground by wind action and melt. Interception during rainfall events is commonly greater than for snowfall events. In both cases, wind velocity is an important factor.

The importance of interception in hydrologic modeling is tied to the purpose of the model. Estimates of loss to gross precipitation through interception can be significant in annual or long-term models, but for heavy rainfalls during individual storm events, accounting for interception may be unnecessary. It is important for the modeler to assess carefully both the time frame of the model and the volume of precipitation with which one must deal.

Equations 5.1 and 5.2 can be used to estimate total interception losses, but for detailed analyses of individual storms, it is necessary to deal with the areal variability of such losses. General equations for estimating such losses are not available, however. Most research has been related to particular species or experimental plots strongly associated with a given locality. In addition, the loss function varies with the storm's character. If adequate experimental data are available, the nature of the variance of interception versus time might be inferred. Otherwise, common practice is to deduct the estimated volume entirely from the initial period of the storm (called *initial abstraction*).

TABLE 5.2   Observed Percentages of Interception by Various Crops and Grasses[a]

Vegetation type	Intercepted (%)	Comments
Crops		
Alfalfa	36	
Corn	16	
Soybeans	15	
Oats	7	
Grasses[b]		
Little bluestem	50–60	
Big bluestem	57	
Tall panic grass	57	Water applied at rate of $\frac{1}{2}$ in. in 30 min
Bindweed	17	
Buffalo grass	31	
Blue grass	17	Prior to harvest
Mixed species	26	
Natural grasses	14–19	

[a] Values rounded to nearest percent. Data for table were obtained from Refs. 2, 4, and 5.
[b] Grass heights vary up to about 36 in.

Example 5.1

Using the following equations developed by Horton [6] for interception by ash and oak trees, estimate the interception loss beneath these trees for a storm having a total precipitation of 1.5 in.

Solution

1.   For ash trees:

$$L_i = 0.015 + 0.23P$$
$$= 0.015 + 0.23(1.5) = 0.36 \text{ in.}$$

2.   For oak trees:

$$L_i = 0.03 + 0.22P$$
$$= 0.03 + 0.22(1.5) = 0.36 \text{ in.}$$

5.2   THROUGHFALL

A number of relationships for estimating throughfall for a variety of forest types have been developed [9]–[12]. Determining factors for throughfall quantities include canopy coverage, total leaf area, number and type of layers of vegetation, wind velocity, and rainfall intensity. The areal variability of these factors results in little or no throughfall in some locations and considerable throughfall in others. In general, prediction equations for throughfall must include measures of canopy surface area and cover as prime variables. An example of a throughfall relationship for an eastern United States hardwood forest follows [12].

For the growing season:

$$T_h = 0.901P - 0.031n \tag{5.3}$$

For the dormant season:

$$T_h = 0.914P - 0.015n \tag{5.4}$$

where T_h = throughfall (in.)
$\quad\quad\quad P$ = total precipitation (in.)
$\quad\quad\quad n$ = number of storms

5.3   DEPRESSION STORAGE

Precipitation that reaches the ground may infiltrate, flow over the surface, or become trapped in numerous small depressions from which the only escape is evaporation or infiltration. The nature of depressions, as well as their size, is largely a function of the original land form and local land-use practices. Because of extreme variability in the nature of depressions and the paucity of sufficient measurements, no generalized relation with enough specified parameters for all cases is feasible. A rational model can, however, be suggested.

Figure 5.1 illustrates the disposition of a precipitation input. A study of it shows that the rate at which depression storage is filled rapidly declines after the initiation of a precipitation event. Ultimately, the amount of precipitation going into depression storage will approach zero, given that there is a large enough volume of precipitation to exceed other losses to surface storage such as infiltration and evaporation. Ultimately, all the water stored in depressions will either evaporate or seep into the ground. Finally, it should be understood that the geometry of a land surface is usually complex and thus depressions vary widely in size, degree of interconnection, and contributing drainage area. In general, depressions may be looked upon as miniature reservoirs and as such they are subject to similar analytical techniques.

According to Linsley et al. [13] the volume of water stored by surface depressions at any given time can be approximated using:

$$V = S_d(1 - e^{-kP_e}) \tag{5.5}$$

where V = the volume actually in storage at some time of interest
$\quad\quad\quad S_d$ = the maximum storage capacity of the depressions
$\quad\quad\quad P_e$ = the rainfall excess (gross rainfall minus evaporation, interception, and infiltration)
$\quad\quad\quad k$ = a constant equivalent to $1/S_d$

The value of the constant can be determined by considering that if $P_e \approx 0$, essentially all the water will fill depressions and dV/dP_e will equal one. This requires that $k = 1/S_d$. Estimates of S_d may be secured by making sample field measurements of the area under study. Combining such data with estimates of P_e permits a determination of V. The manner in which V varies with time must still be estimated if depression storage losses are to be abstracted from the gross rainfall input.

One assumption regarding dV/dt is that all depressions must be full before overland flow supply begins. Actually, this would not agree with reality unless the locations of depressions were graded with the largest ones occurring downstream. If the depression storage were abstracted in this manner, the total volume would be deducted from the initial storm period such as shown by the shaded area in Fig. 5.2. Such postulates have been used with satisfactory results under special circumstances [14].

Depression storage intensity can also be estimated using Eq. 5.5. If the overland flow supply rate σ plus depression storage intensity equals $i - f$, where i is the rainfall intensity reaching the ground and f is the infiltration rate, then the ratio of overland flow supply to overland flow plus depression storage supply can be proved equal to:

$$\frac{\sigma}{i - f} = 1 - e^{-kP_e} \tag{5.6}$$

This expression can be derived by adjudging:

$$\frac{\sigma}{i - f} = \frac{i - f - v}{i - f} \tag{5.7}$$

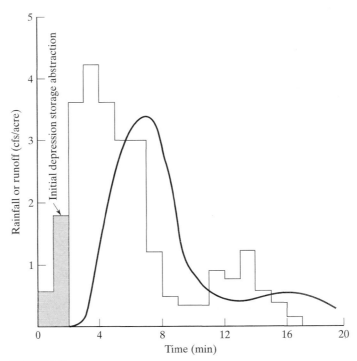

FIGURE 5.2

Simple depression storage abstraction scheme.

and noting that v is equal to the derivative of Eq. 5.5 with respect to time. Then:

$$v = \frac{d}{dt}S_d(1 - e^{-kP_e}) \tag{5.8}$$

$$v = (S_d k e^{-kP_e})\frac{dP_e}{dt} \tag{5.9}$$

It was shown that $k = 1/S_d$ so that:

$$v = e^{-kP_e}\frac{dP_e}{dt} \tag{5.10}$$

The excess precipitation P_e equals the gross rainfall minus infiltrated water, and since the derivative with respect to time can be replaced by the equivalent intensity $(i - f)$, the intensity of depression storage becomes:

$$v = e^{-kP_e}(i - f) \tag{5.11}$$

Inserting in Eq. 5.7, we obtain:

$$\frac{\sigma}{i - f} = \frac{(i - f) - (i - f)e^{-kP_e}}{i - f} \tag{5.12}$$

and

$$\frac{\sigma}{i - f} = \frac{(i - f)(1 - e^{-kP_e})}{i - f} \tag{5.13}$$

$$\frac{\sigma}{i - f} = 1 - e^{-kP_e} \tag{5.14}$$

Figure 5.3 illustrates a plot of this function versus the mass overland flow and depression storage supply $(P - F)$, where F is the accumulated mass infiltration [15] and P is the gross precipitation. In the plot mean depths of 0.25 in. for turf and 0.0625 in. for pavements were assumed. Maximum depths were 0.50 and 0.125 in., respectively.

The figure also depicts the effect on estimated overland flow supply rate, which is derived from the choice of the depression storage model. Three models are shown in the figure; the first one assumes that all depressions are full before overland flow begins. For a turf area having depressions with a mean depth of 0.25 in., the figure shows that for $P - F$ values less than 0.25 in., there is no overland flow supply, while for $P - F$ values greater than 0.25 in., the overland flow supply is equal to $i - f$.

For the exponential model (Model 2), σ always will be greater than zero. Tholin and Kiefer have recommended that a relation between the three models previously mentioned is likely more representative of fully developed urban areas [15]. A cumulative normal probability curve was selected for this representation and is also described in Fig. 5.3 (Model 3).

Figure 5.3 shows that as rainfall accumulation increases, in the limit, the ratio of $\sigma/(i - f)$ increases to a maximum value of 1. It is important to understand that the overland flow supply rate determined using Eq. 5.6 represents the amount of the gross precipitation that can be delivered overland after infiltration and depression storage losses have been deducted. The overland flow eventually becomes the runoff that produces a streamflow hydrograph (see Fig. 1.2 and Chapter 9).

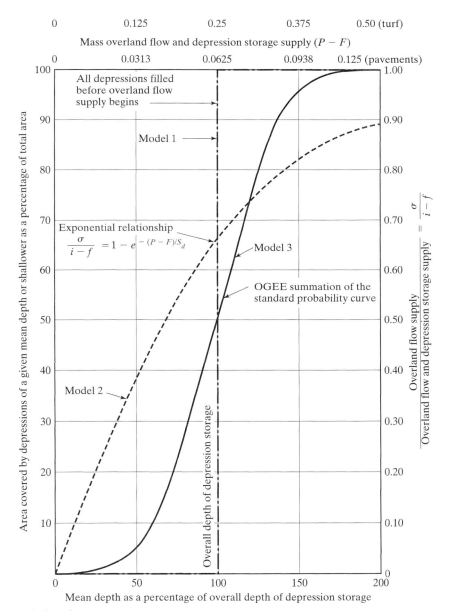

FIGURE 5.3

Depth distribution curve of depression storage. Enter graph from top, read down to selected curve, and project right or left as desired.

(After Tholin and Kiefer [15])

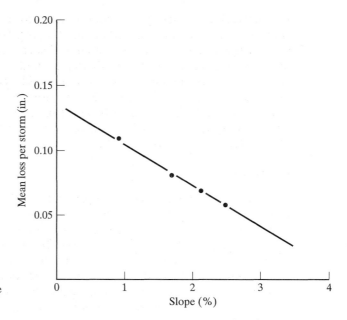

FIGURE 5.4

Depression storage loss versus
slope for four impervious drainage
areas *(from Viessman [14]).*

   Depression storage deductions are usually made from the first part of the storm
as illustrated in Fig. 5.2. The amount to be deducted is a function of topography, ground
cover, and extent and type of land development. During major storms, this loss is often
considered to be negligible. Some guidelines for estimating depression storage losses
have been developed based on studies of experimental and other watersheds. Values

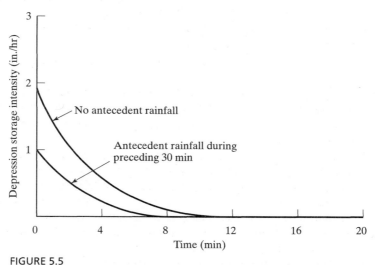

FIGURE 5.5

Depression storage intensity versus time for an impervious area *(after Turner [17]).*

for depression storage losses from intense storms reported by Hicks are 0.20 in. for sand, 0.15 in. for loam, and 0.10 in. for clay [16]. Tholin and Kiefer have used values of 0.25 in. in pervious urban areas and 0.0625 in. for pavements [15]. Studies of four small impervious drainage areas by Viessman yielded the information shown in Fig. 5.4, where mean depression storage loss is highly correlated with slope. This is easily understood, since a given depression will hold its maximum volume if horizontally oriented. Using very limited data from a small, paved-street section, Turner devised the curves shown in Fig. 5.5 [17]. Other sources of data related to surface storage are available in the literature [2],[18],[19].

SUMMARY

Accounting for the disposition of precipitation is an important part of the hydrologic modeling process. Two abstractions from the precipitation input, interception, and depression storage were covered in this chapter.

Interception losses during the course of a year may be substantial, but during intense storms, they may be sufficiently small to neglect. Precipitation type, rainfall intensity and duration, wind, and atmospheric conditions affecting evaporation are factors that serve to determine interception losses for a given forest stand or ground cover configuration. Interception during rainfall events is commonly greater than for snowfall events.

Depression storage deductions occur early in a storm sequence, and they are a function of topography, ground cover, and extent and type of land development. During major storms, this loss is often considered to be negligible.

PROBLEMS

5.1 Using the precipitation input of Fig. 5.2, estimate the volume of depression storage for a 3-acre paved drainage area. State the volume in cubic feet and cubic meters. Convert it to equivalent depth over the area in inches and centimeters.

5.2 Estimate the percentage of the total volume of rainfall that is indicated as depression storage in Fig. 5.2.

5.3 Using the average annual precipitation for your state, estimate the annual interception loss.

5.4 Using Eqs. 5.3 and 5.4, estimate the throughfall in inches for 28 in. of rainfall during the growing season (21 events) and 17 in. of rainfall during the dormant season (13 events).

5.5 Rework Problem 5.4 for a rainfall of 21 in. in the growing season (18 events) and 22 in. in the dormant season (16 events).

5.6 Using Horton's equations given in Example 5.1, estimate the interception losses by ash and oak trees for a storm having a total precipitation of 1.33 in.

5.7 Refer to Fig. 5.3 and estimate the ratio of overland flow supply to overland flow and depression storage supply if the area is turf, the OGEE summation curve is the model, and the mean depth of depression storage is (a) 75% and (b) 125%.

5.8 Explain how a relationship such as that given in Fig. 5.3 could be used in a simulation model of the rainfall-runoff process.

5.9 Using Fig. 5.4, estimate the percentage of rainfall that would be lost to depression storage for a 10-acre parking lot having a mean slope of 1%. Repeat for a slope of 3%. Using the total rainfall volume determined in Problem 5.2, estimate the equivalent depth over the area of the depression storage loss for both slopes. State depths in millimeters and inches.

REFERENCES

[1] C. E. Schomaker, "The Effect of Forest and Pasture on the Disposition of Precipitation," *Maine Farm Res.* (July 1966).

[2] Ven Te Chow (ed.), *Handbook of Applied Hydrology*, New York: McGraw-Hill, 1964, p 6–24.

[3] Joseph Kittredge, *Forest Influences*, New York: McGraw-Hill, 1948.

[4] O. R. Clark, "Interception of Rainfall by Herbaceous Vegetation," *Science* **86**(2243), 591–592(1937).

[5] J. S. Beard, "Results of the Mountain Home Rainfall Interception and Infiltration Project on Black Wattle, 1953–1954," *J. S. Afr. Forestry Assoc.* **27**, 72–85(1956).

[6] R. E. Horton, "Rainfall Interception," *Monthly Weather Rev.* **47**, 603–623(1919).

[7] R. A. Meriam, "A Note on the Interception Loss Equation," *J. Geophys. Res.* **65**, 3850–3851(1960).

[8] D. M. Gray (ed.), *Handbook on the Principles of Hydrology*, National Research Council, Canada, Port Washington: Water Information Center, Inc., 1973.

[9] K. N. Brooks, P. F. Folliott, H. M. Gregersen, and J. L. Thames, *Hydrology and the Management of Watersheds*, Ames, IA: Iowa State University Press/Ames, 1991.

[10] G. J. Blake, "The Interception Process," in *Prediction in Catchment Hydrology*, National Symposium on Hydrology, T. G. Chapman and F. X. Dunin, eds., Melbourne Aust. Acad. Sci., 59–81, 1975.

[11] F. A. Roth, II and M. Chang, "Throughfall in Planted Stands of Fourth Southern Pines Species in East Texas," *Water Resources Bulletin* **17**, 880–885(1981).

[12] J. D. Helvey and J. H. Patric, "Canopy and Litter Interception by Hardwoods of Eastern United States," *Water Resour. Res.* **1**, 193–206(1965).

[13] R. K. Linsley, Jr., M. A. Kohler, and J. L. H. Paulhus, *Applied Hydrology*, New York: McGraw-Hill, 1949.

[14] Warren Viessman, Jr., "A Linear Model for Synthesizing Hydrographs for Small Drainage Areas," paper presented at the Forty-eighth Annual Meeting of the American Geophysical Union, Washington, D.C., Apr. 1967.

[15] A. L. Tholin and C. J. Kiefer, "The Hydrology of Urban Runoff," *Trans. ASCE* **125**, 1308–1379(1960).

[16] W. I. Hicks, "A Method of Computing Urban Runoff," *Trans. ASCE* **109**, 1217–1253(1944).

[17] L. B. Turner, "Abstraction of Depression Storage from Storms on Small Impervious Areas," Master's thesis, University of Maine, Orono, ME, Aug. 1967.

[18] R. E. Horton, *Surface Runoff Phenomena: Part I, Analysis of the Hydrograph*, Horton Hydrol. Lab. Pub. 101, Ann Arbor, MI: Edwards Bros., 1935.

[19] L. F. Huggins and E. J. Monke, "The Mathematical Simulation of the Hydrology of Small Watersheds," Tech. Rept. 1, Purdue University Water Resources Center, Lafayette, IN, Aug. 1966.

C H A P T E R   6

Evaporation and Transpiration

OBJECTIVES

The purpose of this chapter is to:

- ■ Define evaporation and transpiration
- ■ Describe methods for estimating the quantities of water that are evaporated and transpired
- ■ Indicate the importance of evapotranspiration in hydrologic modeling.

Evaporation is the process by which water is transferred from the land and water masses of the earth to the atmosphere. *Transpiration* is the evaporation counterpart for plants. It is the process by which soil moisture taken up by vegetation is eventually evaporated as it exits at plant pores. Evaporation and transpiration combined (*evapotranspiration*) generally constitute the largest component of losses in rainfall-runoff sequences. Accordingly, good estimates of evapotranspiration are a requisite for hydrologic modeling.

On the average, about 40,000 billion gallons per day (bgd) of water moves across the coterminous United States in the form of water vapor [1]. Of this amount, approximately 10 percent is precipitated. The remainder continues to move in atmospheric suspension. Of the precipitated amount (about 4,200 bgd), about two-thirds (2,750 bgd) is evaporated from wet surfaces or transpired from vegetation (see also Fig. 1.3). Evaporation is particularly significant over large bodies of water such as lakes, reservoirs, and the ocean. And estimates of evaporation are critical elements in the design and operation of reservoirs. In the cool, humid northeastern United States, annual evaporation amounts range from about 20–30 inches, while in the warm, dry Southwest, annual figures are on the order of 80 or more inches per year (Fig. 6.1). Transpiration is an important component of the water budget of heavily vegetated areas and is of particular concern to the producers of agricultural products.

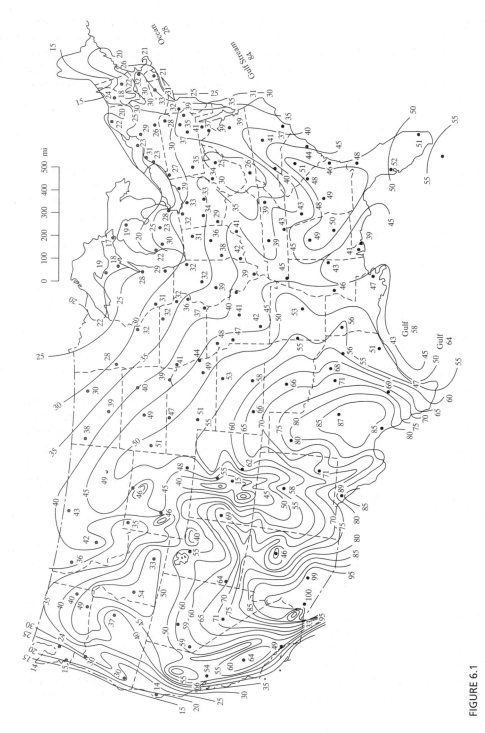

FIGURE 6.1

Mean annual evaporation from shallow lakes and reservoirs, in inches.

NOTE: Evaporation from large, deep lakes and reservoirs, particularly in arid regions, will be substantially less in spring and summer, greater in fall and winter, and less for the year than the values shown here. Evaporation from the surfaces of soil and vegetation immediately after rains or irrigation will begin at greater rates and diminish rapidly with the supply of available moisture. Significant local differences in topography and climate in mountainous regions cause large local differences in evaporation not adequately shown here, particularly in the western states.

(*U.S. Department of Agriculture, Soil Conservation Service*)

Evaporation and transpiration rates depend upon temperature, vapor pressure, wind velocity, and the nature of the evaporating surface. Fig. 6.1 gives mean annual evaporation for shallow lakes and reservoirs, Table 6.1 gives the adjusted mean monthly Class A pan evaporation for 40 selected stations in the United States, and Fig. 6.2 shows the mean monthly percent of annual evaporation for the stations shown in Table 6.1 [2], [3].

6.1   EVAPORATION

Because there is a continuous exchange of water molecules between an evaporating surface and its overlying atmosphere, it is common in hydrologic practice to define evaporation as the *net rate of vapor transfer*. It is a function of solar radiation, differences in vapor pressure between a water surface and the overlying air, temperature, wind, atmospheric pressure, and the quality of evaporating water. Conversion of snow or ice into water vapor is in reality sublimation rather than evaporation, since water molecules do not pass through a liquid phase. Otherwise, the effects of these two processes are the same.

Evaporation from a particular surface is directly related to the opportunity for evaporation (availability of water) provided by that surface. For open bodies of water, evaporation opportunity is 100 percent, while for soils it varies from a high of 100 percent when the soil is highly saturated—for example, during storm periods—to essentially zero at stages of very low moisture content. Other types of surface provide diverse degrees of evaporation opportunity and, except in rare cases, these will almost always vary widely with time.

Direct measurements of evaporation are not easily obtained for large bodies of water because of the extensive surfaces involved. In fact, of all variables included in the general hydrologic equation, surface runoff is the only one that readily permits direct evaluation, since it is confined within well-defined geometric boundaries that permit determination of both rate and cross-sectional area of flow. The choice of method used to determine evaporation depends on the required accuracy of results and the type of instrumentation available. Accuracy is related to the varying degree of reliability with which the method's parameters can be determined.

6.2   ESTIMATING EVAPORATION

The methods applicable to estimating evaporation are the water budget, the energy budget, mass transfer techniques, and the use of pans. Usually, instrumentation for energy budget and mass transfer methods is quite expensive and the cost to maintain observations is substantial. For these reasons, the water budget method and use of evaporation pans are more common. The pan method is the least expensive and will frequently provide good estimates of annual evaporation. Any approach selected is dependent, however, on the degree of accuracy required. As our ability to evaluate the terms in the water budget and energy budget improves, so will the resulting estimates of evaporation.

TABLE 6.1 Adjusted Mean Monthly Class A Pan Evaporation for Selected Stations 1956–1970

Station name	Map ID*	State index no.**	Station index no.**	Percent of Annual Evaporation														Annual inches
				Jan	Feb	Mar	Apr	May	Jun	Jul	Aug	Sep	Oct	Nov	Dec	May thru Oct	Nov thru Apr	
Fairhope 2NE, Ala.	1	1	2813	3.7	4.8	7.8	9.8	12.5	12.5	12.3	11.1	9.3	7.6	4.8	3.8	65	35	50.97
Bartlett Dam, Ariz.	2	2	0632	3.5	4.0	6.1	8.7	12.0	13.8	13.7	11.6	10.1	7.9	4.9	3.9	69	31	121.3
Bacus Ranch, Calif.	3	4	418	3.0	3.5	6.6	8.7	11.5	14.0	14.5	14.7	10.0	7.1	3.6	2.7	72	28	120.56
Sacramento, Calif. (Met)	4	4	7630	1.8	3.1	5.4	8.4	11.9	15.4	16.2	14.5	11.0	7.2	3.3	1.8	76	24	69.70
Wagon Wheel Gap, Colo.	5	5	8742				14.0	16.0	14.1	12.0	10.7	7.1				74	26	50.95
Hartford, Conn. (Met)	6	6	3456	2.6	3.1	5.8	10.1	13.3	14.3	15.1	13.7	9.0	6.4	4.0	2.5	72	28	42.52
Tamiami Trail, Fla.	7	8	8780	5.3	5.9	8.4	10.4	10.9	10.2	10.6	10.1	8.8	8.2	6.0	5.2	59	41	56.48
Experiment, Ga.	8	9	3271	4.1	4.5	7.3	10.0	12.3	12.6	12.4	11.4	9.3	6.7	5.1	4.2	65	35	64.65
Moscow, U of I, Idaho	9	10	6152				6.8	12.0	14.1	19.3	17.7	11.6	6.0			81	19	45.25
Pocatello, Idaho	10	10	7211	1.6	2.3	5.8	8.1	11.9	14.5	19.1	15.1	10.5	6.5	2.9	1.7	78	22	60.98
Ames, Iowa	11	13	205				10.0	14.6	15.8	15.5	13.3	9.3	7.6	3.4		76	24	50.10
Toronto Dam, Kans.	12	14	8191	2.3	3.4	6.6	10.3	12.6	12.5	15.0	14.3	9.5	7.6	4.1	1.7	72	28	61.19
Tribune, Kans.	13	14	8235				9.0	11.8	13.9	15.7	13.9	9.9				73	27	92.98
Madisonville, Ky.	14	15	5067				11.1	13.1	13.9	14.6	13.2	9.6	7.8			72	28	55.26
Urbana, Ill.	15	11	8750				8.6	13.3	15.0	15.2	13.6	10.3	7.3	3.8		75	25	49.46
Woodworth State Forest, La.	16	16	9865	3.4	4.4	7.3	9.4	12.1	13.1	13.0	12.5	9.2	7.7	4.5	3.4	68	32	48.86
Caribou, Maine (Met)	17	17	1175	1.8	2.4	5.0	8.3	15.4	16.0	16.4	13.9	9.0	6.5	3.2	2.1	77	23	22.25
Rochester, Mass.	18	19	6938				8.1	13.0	15.0	14.6	13.0	8.7	5.4			70	30	35.71
East Lansing Hort. Farm, Mich.	19	20	2395				9.4	13.7	15.3	16.2	14.0	9.6	6.4	2.3		75	25	44.53
Scott, Miss.	20	22	7886	3.0	3.4	6.8	9.6	12.9	13.8	13.4	11.9	9.2	7.0	4.3	3.1	68	32	60.99
Weldon Springs																		

Station																		
Farm, Mo.	21	23	8805				9.5	11.9	13.7	14.5	13.5	10.5	7.5	4.0		72	28	48.08
Bozeman Agric. Col., Mont.	22	24	1044				7.8	12.6	13.9	19.0	16.6	10.3	5.9			78	22	47.06
Medicine Creek Dam, Nebr.	23	25	5388	3.1	3.7		9.9	12.4	14.2	15.5	14.4	10.5	7.5			74	26	70.60
Boulder City, Nev.	24	26	1071		3.7	6.4	8.9	12.4	14.3	14.8	12.9	9.9	6.9	3.8	2.8	71	29	109.73
Topaz Lake, Nev.	25	26	8186				8.4	11.8	13.6	15.6	14.5	10.9	7.2	3.3		74	26	82.07
Elephant Butte Dam, N. Mex.	26	29	2848	2.9	4.3	7.5	11.1	13.7	14.8	12.5	10.6	8.5	6.8	4.2	2.8	67	33	116.86
El Vado Dam, N. Mex.	27	29	2837			9.9	10.4	15.1	14.4	14.5	11.5	9.3	6.1			71	29	57.91
Aurora Research Farm, N.Y.	28	30	331				12.5	15.4	16.7	14.3	10.1	6.8				76	24	41.08
Chapel Hill, N.C.	29	31	1677	3.1	4.7	7.8	10.5	12.3	12.6	13.2	11.8	9.3	6.9	4.7	3.2	66	34	52.89
Wooster Exp. Sta., Ohio	30	33	9312				9.1	12.6	15.1	15.5	13.7	9.9	7.1			74	26	46.12
Canton Dam, Okla.	31	34	1445	2.6	4.0	6.8	9.9	11.5	12.5	14.2	13.6	9.3	7.5	4.6	3.4	69	31	77.51
Detroit Power House, Oreg.	32	35	2292	.4	2.2	4.4	6.4	11.8	15.7	21.8	17.9	11.0	5.2	2.4	1.1	83	17	39.74
Redfield, S. Dak.	33	39	7052		3.7		9.6	13.3	14.5	16.9	15.9	11.0	7.2			79	21	51.83
Neptune, Tenn.	34	40	6454	2.4	3.7	6.8	10.5	12.0	13.8	14.0	12.5	9.3	7.1	4.2	3.5	69	31	46.47
Grapevine, Tex.	35	41	3691	3.1	4.0	7.2	8.7	10.3	12.4	14.5	13.9	9.8	7.4	4.9	3.9	68	32	84.81
Welasco, Tex.	36	41	9588	4.1	4.8	7.3	9.3	10.7	11.3	13.2	12.8	9.4	7.3	5.4	4.2	65	35	85.70
Ysletta, Tex.	37	41	9966	3.6	4.9	7.7	13.3	13.9	12.9	10.1	8.8	6.6	4.3	3.1		65	35	108.76
Utah Lake, Utah	38	42	8973			5.7	9.1	13.3	15.4	17.7	15.3	10.7	6.6			79	21	56.12
Templeau Dam, Wis.	39	47	8589				14.3	15.8	16.5	13.6	9.6	8.2				78	22	39.29
Heart Mountain, Wyo.	40	48	4411				6.9	13.5	13.9	16.3	14.8	9.5	6.4			74	26	49.36

*Plot identification number for Fig. 6.2.

**NOAA-EDIS *Climatological Data*

Source: Reference 2

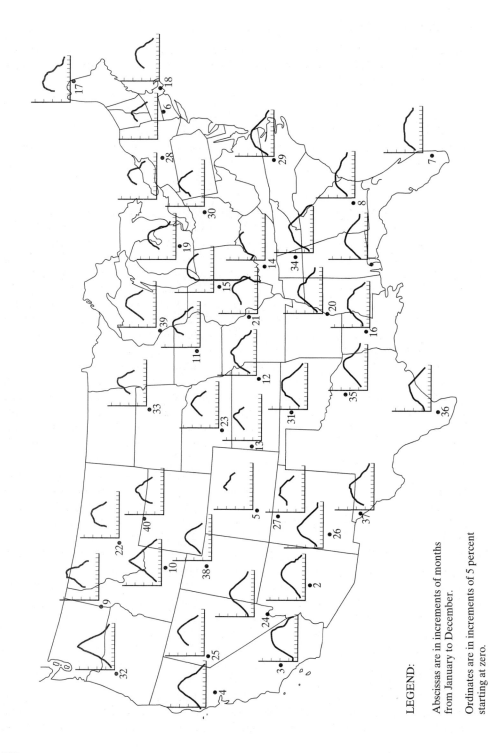

LEGEND:

Abscissas are in increments of months
from January to December.

Ordinates are in increments of 5 percent
starting at zero.

FIGURE 6.2

Mean monthly percent of annual precipitation for the 40 stations shown in Table 6.1 [2].

Water Budget Calculations

The water budget method for determining evaporation is a very simple procedure, but it seldom produces reliable results. In this method, reservoir (lake, pool) evaporation E_s can be computed by rearranging Eq. 1.1:

$$E_s = P + R_1 - R_2 + R_g - T_s - I - \Delta S_s \qquad (6.1)$$

It is useful to deal with the net transfer of seepage through the ground, $R_g - I = O_s$, and consider that the transpiration term T_s equals zero. With these few modifications, Eq. 6.1 becomes:

$$E_s = P + R_1 - R_2 + O_s - \Delta S_s \qquad (6.2)$$

All the terms are in volume units for a time period of interest, and Δt should be at least a week. In general, however, the method would more likely be used to estimate monthly or annual evaporation from a particular reservoir. Note that all errors in measuring inflow, precipitation, net seepage, and change in storage are reflected in the final estimate of evaporation. Precipitation, runoff, and changes in storage can often be determined within reasonable limits of accuracy, but evaluation of the net seepage O_s is frequently subject to appreciable errors; if the magnitude of O_s is on the order of E_s, very large errors are possible. Seepage estimates usually come from measurements of groundwater levels and/or soil permeability. The water budget is usable on a continuous basis if a stage-seepage relation for the lake can be established.

In cases where the water budget for a lake is defined by only two unknowns, net seepage and evaporation, these losses can be separated and evaluated in a relatively simple manner by assuming that evaporation is proportional to the product $u(e_0 - e_a)$. This is the mass transfer product described later in the section on mass transfer techniques. The variables are wind velocity u, saturation vapor pressure e_0 (related to water surface temperature), and vapor pressure of the air e_a. When the product $u(e_0 - e_a)$ is zero, evaporation may be neglected.

Periods of no surface inflow or outflow are desirable for net seepage determination, since during such intervals the only losses are evaporation and seepage. Under these conditions, whenever the mass transfer product is equal to zero, the change in water elevation is considered equivalent to the net seepage loss. Normally, a daily plot of change in elevation versus $u(e_0 - e_a)$ is obtained and a best-fitting line constructed. The intercept of this line on the change in stage axis is the net seepage rate. Values of net seepage estimated in this manner can be used on a long-term basis in the water budget equation if the net seepage does not change appreciably over extended periods. Unfortunately, this condition is rarely representative, since net seepage is a function of reservoir stage and season of the year in many cases. Unless these effects can be calculated, net seepage values determined from limited data have little utility.

Good estimates of evaporation using the water budget equation have been obtained, as exemplified by research conducted on Lake Hefner in Oklahoma [4]. Under optimal conditions, the order of accuracy of the method is about 10 percent.

Example 1.1 illustrated the use of the water budget for estimating basin evapotranspiration. For such estimates, reliable values can be expected if the period of time chosen is 1 year or longer. Short-period values may also be obtained if observations are adequate. Mean annual evapotranspiration is successfully judged by using long-time averages of precipitation and surface flows and credible information on the fluctuation of storage. Adequate short-period estimates are also possible if variables in the budget equation can be satisfactorily quantified on a short-term basis.

Energy Budget Method

The energy budget method illustrates an application of the continuity equation written in terms of energy. It has been employed to compute the evaporation from oceans and lakes, that is, for Lake Hefner in Oklahoma and at Elephant Butte Reservoir in New Mexico [4],[5]. The equation accounts for incoming and outgoing energy balanced by the amount of energy stored in the system. The accuracy of estimates of evaporation using the energy budget is highly dependent on the reliability and preciseness of measurement data. Under good conditions, average errors of perhaps 10 percent for summer periods and 20 percent for winter months can be expected.

The energy budget equation for a lake may be written as:

$$Q_0 = Q_s - Q_r + Q_a - Q_{ar} + Q_v - Q_{bs} - Q_e - Q_h - Q_w \qquad (6.3)$$

where Q_0 = increase in stored energy by the water
   Q_s = solar radiation incident at the water surface
   Q_r = reflected solar radiation
   Q_a = incoming long-wave radiation from the atmosphere
   Q_v = net energy advected (net energy content of incoming and outgoing water) into the water body
   Q_{ar} = reflected long-wave radiation
   Q_{bs} = long-wave radiation emitted by the water
   Q_e = energy used in evaporation
   Q_h = energy conducted from water mass as sensible heat
   Q_w = energy advected by evaporated water

All the terms are in calories per square centimeter per day (cal/cm^2-day). Heating brought about by chemical changes and biological processes is neglected as is the energy transfer that occurs at the water-ground interface. The transformation of kinetic energy into thermal energy is also excluded. These factors are usually very small, in a quantitative sense, when compared with other terms in the budget if large reservoirs are considered. As a result, their omission has little effect on the reliability of results.

During winter months when ice cover is partial or complete, the energy budget only occasionally yields adequate results because it is difficult to measure reflected solar radiation, ice-surface temperature, and the areal extent of the ice cover. Daily evaporation estimates based on the energy budget are not feasible in most cases because reliable determination of changes in stored energy for such short periods is impractical. Periods of a week or longer are more likely to provide satisfactory measurements.

In using the energy budget approach, it has been demonstrated that the required accuracy of measurement is not the same for all variables [5]. For example, errors in measurement of incoming long-wave radiation as small as 2 percent can introduce errors of 3 to 15 percent in estimates of monthly evaporation, while errors on the order of 10 percent in measurements of reflected solar energy may cause errors of only 1 to 5 percent in calculated monthly evaporation.

To permit the determination of evaporation by Eq. 6.3, it is common to use the following relation:

$$B = \frac{Q_h}{Q_e} \tag{6.4}$$

where B is known as Bowen's ratio [6], and:

$$Q_w = \frac{c_p Q_e (T_e - T_b)}{L} \tag{6.5}$$

where c_p = the specific heat of water (cal/g-°C)
T_e = the temperature of evaporated water (°C)
T_b = the temperature of an arbitrary datum usually taken as 0°C
L = the latent heat of vaporization (cal/g)

Introducing these expressions in Eq. 6.3 and solving for Q_e, we obtain:

$$Q_e = \frac{Q_s - Q_r + Q_a - Q_{ar} - Q_{bs} - Q_0 + Q_v}{1 + B + c_p(T_e - T_b)/L} \tag{6.6}$$

To determine the depth of water evaporated per unit time, the following expression may be used:

$$E = \frac{Q_e}{\rho L} \tag{6.7}$$

where E = evaporation (cm³/cm²-day)
ρ = the mass density of evaporated water (g/cm³)

The energy budget equation thus becomes:

$$E = \frac{Q_s - Q_r + Q_a - Q_{ar} - Q_{bs} - Q_0 + Q_v}{\rho[L(1 + B) + c_p(T_e - T_b)]} \tag{6.8}$$

The Bowen ratio can be computed using:

$$B = 0.61 \frac{p}{1000} \frac{(T_0 - T_a)}{(e_0 - e_a)} \tag{6.9}$$

where p = the atmospheric pressure (mb)
T_0 = the water surface temperature (°C)
T_a = the air temperature (°C)

e_0 = the saturation vapor pressure at the water surface temperature (mb)

e_a = the vapor pressure of the air (mb)

This expression circumvents the problem of evaluating the sensible heat term, which does not lend itself to direct measurement.

Mass Transfer Techniques

Mass transfer equations are based primarily on the concept of the turbulent transfer of water vapor (by eddy motion) from an evaporating surface to the atmosphere. Many equations, both theoretical and empirical, have been developed. Most are similar in form to a relation between evaporation and vapor pressure first recognized by Dalton [7]:

$$E = \kappa(e_0 - e_a) \tag{6.10}$$

where E = direct evaporation

κ = a coefficient dependent on the wind velocity, atmospheric pressure, and other factors

e_0, e_a = the saturation vapor pressure at the water surface temperature and the vapor pressure of air, respectively

Theoretical mass transfer equations are based on the concepts of discontinuous and continuous mixing at the air-liquid interface.

Empirical approaches often require exacting and costly instrumentation and observations, so their general utility is limited. The complexity of the equations varies from simple expressions such as Eq. 6.10 to complex relations like Sutton's equation [8] for a circular lake of radius r:

$$E = \frac{0.623}{p} G' \rho u^{(2-n)/(2+n)} r^{(4+n)/(2+n)} (e_0 - e_a) \tag{6.11}$$

where E = evaporation (cm/day)

ρ = mass density of the air (g/cm^3)

u = average wind velocity (cm/sec)

r = radius of circular lake (cm)

p = atmospheric pressure (mb)

e_0, e_a = as previously defined

n = an empirical constant

G' = a complex function

A commonly used empirical equation has been developed by Meyer [9]. This equation takes the form:

$$E = C(e_0 - e_a)\left(1 + \frac{W}{10}\right) \tag{6.12}$$

where E = the daily evaporation in inches of depth

e_0, e_a = as previously defined but in units of in. Hg

W = the wind velocity in mph measured about 25 ft above the water surface

C = pan empirical coefficient

For daily data on an ordinary lake, C is about 0.36. For wet soil surfaces, small puddles, and shallow pans, the value of C is approximately 0.50.

Another mass transfer equation used to estimate the rate of evaporation is one developed by Dunne [10],[11]. It takes the following form:

$$E = (0.013 + 0.00016u_2)e_a[(100 - R_h)/100] \tag{6.13}$$

where E is the evaporation rate in cm/day, u_2 is the wind velocity measured at 2 m above the surface in km/day, e_a is defined as before but in millibars, and R_h is the relative humidity given in percent.

Example 6.1

Using the Meyer and Dunne equations, find the daily evaporation rate for a lake given that the mean value for air temperature was 87°F, the mean value for water temperature was 63°F, the average wind speed was 10 mph, and the relative humidity was 20%. Refer to Appendix Table A.2 for vapor pressure values.

Solution.

1.   Interpolating from Table A.2, we find that:

$$e_0 = 0.58 \text{ in. Hg}$$

$$e_a = 1.29 \times 0.20 = 0.26 \text{ in. Hg} = 8.75 \text{ mb}$$

2.   Using Eq. 6.12, and assuming that $C = 0.36$, we obtain:

$$E = 0.36(0.58 - 0.26)[1 + (10/10)]$$
$$= 0.36 \times 0.32 \times 2$$
$$= 0.23 \text{ in./day}$$

3.   Using Eq. 6.13, after converting wind speed to metric units, we obtain:

$$E = [0.013 + (0.00016 \times 386)] \times 8.75 \times [(100 - 20)/100]$$
$$= 0.075 \times 8.75 \times 0.8$$
$$= 0.527 \text{ cm/day, or } 0.21 \text{ in./day}$$

The two estimates are comparable.

Example 6.2

Consider that the lake of Example 6.1 has a surface area of 1.5 mi.^{2} (a) If the average annual evaporation rate is estimated to be $\frac{2}{3}$ of the average daily rate calculated in

Example 6.1, what volume in million gallons per day (mgd) and cubic meters per day would be lost to evaporation? (b) If the average water use of an urban community is 180 gallons per capita per day (gpcd), how large a community would the daily evaporation rate sustain?

Solution.

1. The surface area of the lake in square feet would be:

$$52,800 \times 5,280 \times 1.5 = 41,817,760 \text{ ft}^2$$
$$41,817,760 \times 0.21 \times 0.67 = 731,811 \text{ ft}^3/\text{day}$$

Converting to mgd:

$$731,811 \times 7.48 = 5,473,946 \text{ mgd}$$

For 1 year the total would be:

$$5,473,946 \times 365 = 1,997,990 \text{ mgy}$$

Converting to cubic meters:

$$731,811 \times 0.0283 = 20,710 \text{ m}^3/\text{day}$$
$$20,710 \times 365 = 7,559,242 \text{ m}^3/\text{year}$$

2. The size of the urban community that could have been supported by the daily evaporation loss would be:

$$5,473,946/180 = 30,411 \text{ people}$$

Investigations of the utility of mass transfer equations conducted at Lake Hefner indicated that a simple equation using wind speed and vapor pressure differences yielded results that were as good as any that were tested [4]. The equation was:

$$E = Nu(e_0 - e_a) \tag{6.14}$$

where E = the evaporation (cm/day)
N = a coefficient
u = the wind velocity at 2 m above the water surface (m/sec)
e_0, e_a = as previously defined (mb)

The value of N can be determined by comparative studies between mass transfer and energy budget methods. This is the preferred approach. If such an evaluation is not available, it can be approximated using:

$$N = \frac{0.0291}{A^{0.05}} \tag{6.15}$$

where A is the surface area of the water in square meters. For values of A less than about $4 \times 10^6 \text{ m}^2$, variations in wind exposure may become important and Eq. 6.15

should be used with caution. When N is based on comparative studies using the energy budget, average errors in evaporation estimates of about 15 percent can be expected, while errors of roughly 30 percent would likely be obtained if Eq. 6.15 were employed.

Use of Evaporation Pans

The most widely used method of finding reservoir evaporation is by means of evaporation pans [12]–[14]. The standard National Weather Bureau Class A pan, built of unpainted galvanized iron, is currently the most popular (see Fig. 2.3). It is 4 ft in diameter, 10 in. deep, and mounted 12 in. above the ground on a wooden frame. Relations developed between pan and actual evaporation from large bodies of water such as lakes indicate multiplying the former by a factor of 0.70–0.75 (pan coefficient) gives the equivalent lake evaporation. Ratios of annual reservoir evaporation to pan evaporation are consistent from year to year and region to region, while monthly ratios often show considerable variation.

Estimates of reservoir evaporation based on short-period pan observations (less than 1 year) may be seriously in error. The use of a pan coefficient to estimate evaporation from an ungauged location should reflect the geographic variability in heat transfer through the sides of the Class A pan. For lakes subjected to significant amounts of advected energy, local pan–lake relations should be established.

Available data indicate that the annual ratio of lake evaporation to Class A pan evaporation is essentially 0.70, provided that net advection is balanced by the change in energy stored, conduction through the pan is negligible, and the pan is located so that its exposure conditions are representative of the body of water being considered. Considerable care must be taken, however, in installing and using evaporation pans. Pans may be sunken, floating, or set above the ground surface. Sunken pans tend to have fewer boundary heat transfer problems but are more difficult to gauge and they are more prone to collect trash. While floating pan observations more closely approximate lake evaporation than shore installations, such pans are not without problems. Commonly there are appreciable boundary effects, and splashing often negates the validity of observations. Pans located above ground exhibit heat exchange problems related to side walls and the bottom, but these difficulties may be largely overcome by the use of insulation. Pans installed above ground are easily erected, gauged, and maintained. Such installations are the most common, characterized by the Class A pan.

An equation for daily Class A pan evaporation, E_a, assuming that air and water temperatures are equal, is:

$$E_a = (e_0 - e_a)^{0.88}(0.42 + 0.0029u_p) \qquad (6.16)$$

where the daily pan evaporation is given in millimeters per day, and u_p is the wind movement 150 mm above the rim of the pan in kilometers per day. The vapor pressure difference term $(e_0 - e_a)$ can be determined using the following equation [14]:

$$e_0 - e_a = 33.86[(0.00738T_a + 0.8072)^8 - (0.00738T_d + 0.8072)^8] \qquad (6.17)$$

where vapor pressures are measured in millibars and the dew point temperature T_d and air temperature T_a are measured in degrees Celsius. Note that this equation is valid as long as T_d equals or exceeds $-27°C$.

Penman, through a simultaneous solution of an aerodynamic equation and an energy balance equation, derived the following equation for daily evaporation E [15]:

$$E = \frac{1}{\Delta + \gamma} (Q_n \Delta + \gamma E_a) \tag{6.18}$$

where Δ is the slope of the saturation vapor pressure versus temperature curve at the air temperature T_a; E_a is the pan evaporation given by Eq. 6.16; Q_n is the net radiant energy expressed in the same units as those of E; and γ is the $[0.61p/1000]$ term in Bowen's ratio (Eq. 6.9) where p is the atmospheric pressure in millibars.

Equation 6.18 can be used to estimate lake evaporation after introducing an appropriate pan coefficient. For practical purposes, Class A pan evaporation is estimated to be approximately 0.70 of pan evaporation provided that (1) any net advection into the lake is balanced by the change in energy storage, (2) the net transfer of sensible heat through the pan is negligible, and (3) the pan exposure is representative [15],[16]. If these conditions are met, annual lake evaporation can be estimated from the following equation:

$$E_L = 0.70 \left(\frac{Q_n \Delta + E_a \gamma}{\Delta + \gamma} \right) \tag{6.19}$$

where E_L is the average daily lake evaporation in units of length, and the other terms are as previously defined. By making allowances for advected energy and heat transfer through the pan, Kohler and coworkers have developed the following expression for calculating annual lake evaporation in inches per day:

$$E_L = 0.70[E_p + 0.00051 P \alpha_p (0.37 + 0.0041 u_p)(T_0 - T_a)^{0.88}] \tag{6.20}$$

where α_p is the proportion of advected energy (Class A pan) used for evaporation, T_a is the air temperature in degrees Fahrenheit, u_p is the wind velocity in mi/day, and P is the atmospheric pressure in inches of mercury. Equation 6.20 assumes that any energy advected into the lake is balanced by a change in energy storage and that the pan exposure is representative [12]. A graphical solution of Eq. 6.20 is given in Fig. 6.3; values of α_p can be obtained using Fig. 6.4. An approximation for estimating α_p for use in computer simulations is the following:

$$\alpha_p = 0.13 + 0.0065 T_0 - (6.0 \times 10^{-8} T_0^3) + 0.016 u_p^{0.36} \tag{6.21}$$

where T_0, the outerface temperature of the pan, is in degrees Fahrenheit and the wind velocity is in miles per day [14]. Equation 6.20 is considered to be reliable where water temperature data are available along with appropriate Class A pan observations [12].

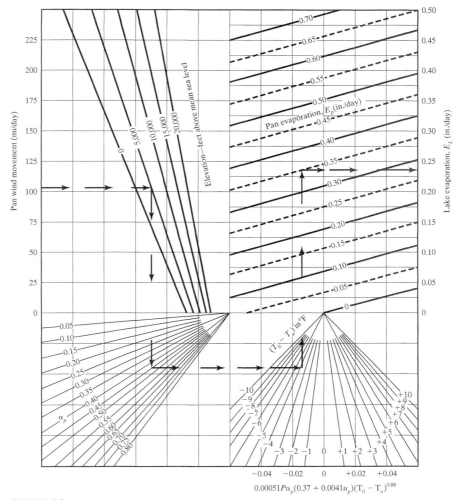

FIGURE 6.3

Graphical solution of Eq. 6.20.

(U.S. Weather Bureau Res. Paper No. 38)

6.3   EVAPORATION CONTROL

Evaporation losses can be greatly significant at any location. Consequently, the concept of evaporation reduction is receiving widespread attention. Evaporation losses from soils can be controlled by employing various types of mulch or by chemical alteration. They may be reduced from open waters by (1) storing water in covered reservoirs, (2) making increased use of underground storage, (3) controlling aquatic growths, (4) building storage reservoirs with minimal surface area, (5) using chemicals, and (6) conveying in closed conduits rather than open channels. Some of these approaches may be

$$\alpha_p = \frac{[(e_0^* - e_a)^{0.88} - (e_0 - e_a)^{0.88}](0.37 + 0.0041u_p)}{[(e_0^* - e_a)^{0.88} - (e_0 - e_a)^{0.88}](0.37 + 0.0041u_p) + (7.6 \times 10^{-11})[(K_0^*)^4 - K_0^4] + 0.000367P[(T_0^* - T_a)^{0.88} - (T_0 - T_a)^{0.88}](0.37 + 0.0041T)}$$

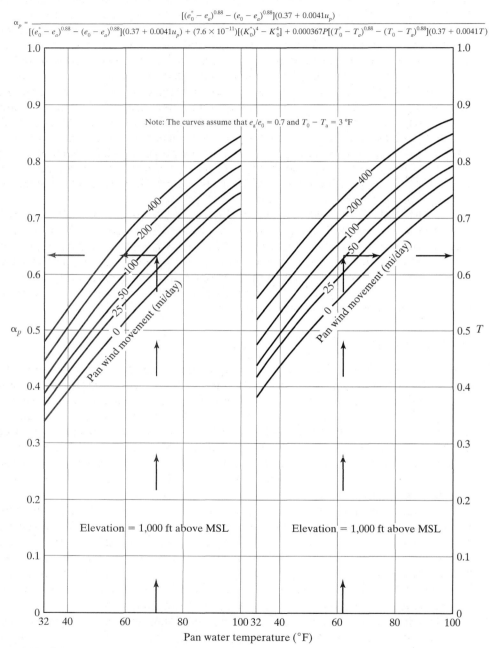

FIGURE 6.4

Proportion of advected energy (into a Class A pan) used for evaporation.

(U.S. Weather Bureau Res. Paper No. 38)

impractical (covering large reservoirs) or uneconomical (large-scale vegetation control). All have potential advantages, however, under the proper circumstances.

The first four approaches need no explanation of the mechanism expected to control evaporation. The fifth method, chemical means, requires further comment. Research has shown that certain types of organic compounds such as hexadecanol and octadecanol form monomolecular films that are effective as evaporation inhibitors [5], [17]. Studies by the Bureau of Reclamation indicate that evaporation may be suppressed by as much as 64 percent with hexadecanol films in 4-ft-diameter pans under controlled conditions. Actual reductions on large bodies of water would be significantly less than this, however, due to problems in maintaining films against wind and wave action. Evaporation reduction in the range of 22–35 percent has been observed for some studies on small lakes of roughly 100 acres in size with reductions of 9–14 percent reported for Lake Hefner in Oklahoma (2500 acres) [18]. Wind was a major problem at Lake Hefner, however.

In Australia, extended tests on medium-sized lakes (less than 2500 acres) have indicated savings of 30–50 percent, although adverse winds were generally not encountered [18]. Although considerable research and development work remains, the use of monomolecular films to control evaporation appears promising. Franzini has indicated that the cost per acre-foot of water saved by evaporation suppression is, in fact, competitive with various alternate means of increasing local water supplies, and these costs will likely decrease with advances in research [17].

6.4   TRANSPIRATION

Root systems of plants absorb water in varying quantities. Most of this water is transmitted through the plant and escapes through pores in the leaf system. This is known as *stomatal transpiration*. Plants also lose water by other mechanisms, but usually this is negligible compared with that lost through the microscopic leaf apertures. Transpiration is basically a process by which water is evaporated from the airspaces in plant leaves. Therefore, it is controlled essentially by the same factors that dominate evaporation, namely, solar radiation, temperature, wind velocity, and vapor pressure gradients. In addition, transpiration is affected to some degree by the character of the plant and plant density.

Soil moisture content, when reduced to the wilting point (stage at which plants wilt and do not recover in a humid atmosphere), also affects transpiration. The effects of decreased soil moisture above the wilting point are not clearly established and are somewhat controversial. Nevertheless, it appears that as long as soil moisture lies between the limits of the wilting point and field capacity (the amount of water retained in a soil against gravity after percolation ceases), transpiration is not materially affected. Saturated soils can sometimes adversely affect plant life.

Diffusion of water vapor from plant leaves to the atmosphere is proportional to the vapor pressure gradient at the leaf-atmosphere interface. Upon absorbing solar radiation, leaves tend to become warmer than the surrounding air (often by as much as 5–10°F). The amount of water vapor held by the air at the leaf-air interface thus increases; more rapid water losses are favored; and transpiration follows a diurnal cycle, which is approximately that of light intensity. It has also been demonstrated that transpiration and the rate of plant growth are related. Below a temperature of about 40°F the amount of water transpired is considered negligible.

Different species and types of plants often display considerably different demands on soil moisture even if the same environmental conditions prevail. For example, an oak tree may transpire as much as 170 quarts of water a day, whereas a corn plant will transpire only about 2 quarts. The area covered by the two root systems is, of course, significantly different. Various species also indicate different patterns in seasonal demands for water. Agricultural products obviously have their periods of greatest transpiration at the peak of the growing season.

Precise values for quantities of water transpired are difficult to acquire, since many variables are active in the process and these range widely from one region to another. Available estimates should be used with caution, and the conditions under which they were obtained should be determined before applying the data. Adequate relations between climatic factors and transpiration become prerequisites if data derived in one climatic region are to have general utility.

Transpiration may be measured in the laboratory by using tanks wherein evaporation is eliminated and water losses are found by weighing. Coefficients must be derived before such data can be applied to field conditions, and even then the observations usually provide little more than an index to field water use. Large-scale field measurements of transpiration are virtually impossible under prevailing field conditions so it is common to find measures of consumptive use (combined evaporation plus transpiration) more widely adopted and of greater value to the practicing hydrologist. Most field observations are made by using lysimeters (grass or crop-covered containers for which a water budget is maintained). Table 6.2 gives some values of consumptive use for several crops in the Montrose area of Colorado [20]. These values are presented only to indicate their order of magnitude in this area during the growing or irrigation season. More complete information on consumptive use by various crops can be found elsewhere [19]–[24].

For many small local projects it is not possible to carry out detailed field studies to determine the consumptive use of crops. In such cases it is common to use either the Blaney–Criddle or Penman method for estimating seasonal consumptive use [23],[24]. The Blaney–Criddle method is briefly described here.

TABLE 6.2   Consumptive Use for Crops in the Montrose, CO, Area During the Irrigation or Growing Season

Crop	Consumptive use (in.)
Alfalfa	26.5
Corn	19.7
Small grain	14.9
Grass hay	23.3
Natural vegetation	37.3

Source: H. F. Blaney, "Water and Our Crops," in Water, The Yearbook of Agriculture, *Washington, D.C.: U.S. Department of Agriculture, 1955.*

The seasonal consumptive use for a particular crop can be computed using the relation:

$$U = k_s B \qquad (6.22)$$

where U = the consumptive use of water during the growing season (in.)

       k_s = a seasonal consumptive use coefficient applicable to a particular crop, empirically derived (Table 6.3)

       B = the summation of monthly consumptive use factors for a given season

TABLE 6.3   Seasonal Consumptive Use Crop Coefficients (k_s) for Irrigated Crops, for Use in Equation 6.22

Crop	Length of normal growing season or period[a]	Consumptive use coefficient k_s[b]	Maximum monthly k[c]
Alfalfa	Between frosts	0.80–0.90	0.95–1.25
Bananas	Full year	0.80–1.00	—
Beans	3 months	0.60–0.70	0.75–0.85
Cocoa	Full year	0.70–0.80	—
Coffee	Full year	0.70–0.80	—
Corn (maize)	4 months	0.75–0.85	0.80–1.20
Cotton	7 months	0.60–0.70	0.75–1.10
Dates	Full year	0.65–0.80	—
Flax	7–8 months	0.70–0.80	—
Grains, small	3 months	0.75–0.85	0.85–1.00
Grain, sorghums	4–5 months	0.70–0.80	0.85–1.10
Oilseeds	3–5 months	0.65–0.75	—
Orchard crops:			
Avocado	Full year	0.50–0.55	—
Grapefruit	Full year	0.55–0.65	—
Orange and lemon	Full year	0.45–0.55	0.65–0.75[d]
Walnuts	Between frosts	0.60–0.70	—
Deciduous	Between frosts	0.60–0.70	0.70–0.95
Pasture crops:			
Grass	Between frosts	0.75–0.85	0.85–1.15
Ladino white clover	Between frosts	0.80–0.85	—
Potatoes	3–5 months	0.65–0.75	0.85–1.00
Rice	3–5 months	1.00–1.10	1.10–1.30
Soybeans	140 days	0.65–0.70	—
Sugar beets	6 months	0.65–0.75	0.85–1.00
Sugarcane	Full year	0.80–0.90	—
Tobacco	4 months	0.70–0.80	—
Tomatoes	4 months	0.65–0.70	—
Truck crops, small	2–4 months	0.60–0.70	—
Vineyard	5–7 months	0.50–0.60	—

[a] Length of season depends largely on variety and time of year when the crop is grown. Annual crops grown during the winter period may take much longer than if grown in the summertime.
[b] The lower values of k_s for use in the Blaney–Criddle formula, $U = k_s B$, are for more humid areas and the higher values are for more arid climates.
[c] Dependent on mean monthly temperature and crop growth stage.
[d] Given by Criddle as "citrus orchard."

Source: From Irrigation Water Requirements, *Technical Release No. 21, Soil Conservation Service, USDA, September 1970.*

TABLE 6.4 Daytime Hours Coefficient (p) for Use in Equation 6.23

Latitude North South	Jan Jul	Feb Aug	Mar Sep	Apr Oct	May Nov	Jun Dec	Jul Jan	Aug Feb	Sep Mar	Oct Apr	Nov May	Dec Jun
60°	0.15	0.20	0.26	0.32	0.38	0.41	0.40	0.34	0.28	0.22	0.17	0.13
50°	0.19	0.23	0.27	0.31	0.34	0.36	0.35	0.32	0.28	0.24	0.20	0.18
40°	0.22	0.24	0.27	0.30	0.32	0.34	0.33	0.31	0.28	0.25	0.22	0.21
30°	0.24	0.25	0.27	0.29	0.31	0.32	0.31	0.30	0.28	0.26	0.24	0.23
20°	0.25	0.26	0.27	0.28	0.29	0.30	0.30	0.29	0.28	0.26	0.25	0.25
10°	0.26	0.27	0.27	0.28	0.28	0.29	0.29	0.28	0.28	0.27	0.26	0.26
0°	0.27	0.27	0.27	0.27	0.27	0.27	0.27	0.27	0.27	0.27	0.27	0.27

Note: Values for p are determined by dividing the mean daily daytime hours for a specified month by the total daytime hours in a year and then multiplying the ratio by 100.

The term B can be expressed as:

$$B = \Sigma\left(\frac{tp}{100}\right) \tag{6.23}$$

where t = the mean monthly temperature (°F)

p = the monthly daytime hours given as percentage of the year (Table 6.4)

If monthly values for the consumptive use coefficient k are available, monthly consumptive use can be found by using:

$$u = \frac{ktp}{100} \tag{6.24}$$

where u is the monthly consumptive use (in.) and the other terms are as previously defined. Selected values of p and k are available in the literature [21],[23]. An example illustrates the use of this equation.

Example 6.3

Determine the monthly consumptive use of an alfalfa crop grown in southern California for the month of July if the average monthly temperature is 72°F, the average value of daytime hours in percentage of the year is 9.88, and the mean monthly consumptive use coefficient for alfalfa is 0.85.

Solution. Using Eq. 6.24 we find that:

$$u = \frac{ktp}{100}$$

$$= 0.85 \times 72 \times \frac{9.88}{100}$$

$$= 6.05 \text{ in. of water}$$

Example 6.4

Determine the seasonal consumptive use of a tomato crop grown in New Jersey if the mean monthly temperatures for May, June, July, and August are 61.6, 70.3, 75.1, and 73.4°F, respectively, and the percent daylight hours for the given months are 10.02, 10.08, 10.22, and 9.54 as percent of the year, respectively.

Solution

1. From Table 6.3, the growing season for tomatoes is 4 months and the range of the consumptive use coefficient is 0.65 to 0.70. Since New Jersey is a humid area, choose the lower value of $k_s = 0.65$.
2. The term B is calculated using Eq. 6.23 as:

$$B = (61.6 \times 10.02/100) + (70.3 \times 10.08/100) + (75.1 \times 10.22/100)$$
$$+ (73.4 \times 9.54/100) = 27.9$$

3. Seasonal consumptive use is determined using Eq. 6.22:

$$U = k_s B$$
$$U = 0.65 \times 27.9$$
$$= 18.1 \text{ in. of water for the 4-month growing season.}$$

Note that the total amount of water to be applied to an irrigated area must include consumptive use plus conveyance and other losses. Thus the amount of water allocated at the source may have to be considerably more than the consumptive use expectation.

6.5   TRANSPIRATION CONTROL

Water conservation through transpiration reduction is being seriously studied, and certain preventative practices are presently in use. Methods of control include the use of chemicals to inhibit water consumption (analogous to the use of films to control surface evaporation except that chemicals are applied in the root zone), harvesting of plants, improved irrigation practices, and actual removal or destruction of certain vegetative types [25].

In arid regions of the Southwest, certain plants known as *phreatophytes* (plants capable of tapping the water table or capillary fringe) transpire enormous quantities of water each year without providing any particular apparent benefit (although this statement is open to question). Many of these plants, such as the salt cedar, grow in stream channels and tend to create flood control problems by restricting channels in addition to using valuable underground water supplies. In New Mexico there have been as many as 43,000 acres of salt cedar along the Pecos River alone. Control of these phreatophytes could result in estimated savings of over 200,000 acre-ft of water in a critically water-short region of the United States [26]. Conservation through transpiration control may be important, but the ecologic consequences of such control practices should be given careful consideration.

6.6   EVAPOTRANSPIRATION

In most cases of practical interest to the hydrologist, only total evaporation from an area—combined evaporation plus transpiration (consumptive use)—is of real interest. Various methods for determining evapotranspiration have been proposed, but there is no one system generally acceptable under all conditions. Basically, there are three major approaches:

1. Theoretical, based on physics of the process
2. Analytical, based on energy or water budgets
3. Empirical

The water budget method was illustrated in Example 1.1. Its adequacy is dependent on the accuracy with which the several terms in the budget equation can be evaluated. The energy budget can also be used to calculate field evapotranspiration in a manner similar to that described previously for lakes. For this application, however, the soil's thermal properties must be known, and temperature and vapor pressure gradients measured at two levels above the ground are needed in Bowen's ratio. For field plots the amount of energy advected usually can be neglected.

Mass transfer equations of the form previously discussed can also be used to estimate evapotranspiration. The Thornthwaite–Holzman equation is a good example of a mass transfer equation that has often been employed for this purpose. However, Linsley and coworkers indicate that there is some question as to the adequate verification of this model to estimate evapotranspiration losses [27]. The equation is expressed as:

$$E = \frac{833\kappa^2(e_1 - e_2)(V_2 - V_1)}{(T + 459.4)\log_e (z_2/z_1)^2} \tag{6.25}$$

where
$$E = \text{evaporation (in./hr)}$$
$$\kappa = \text{von Kármán's constant (0.4)}$$
$$e_1, e_2 = \text{vapor pressures (in. Hg)}$$
$$V_1, V_2 = \text{wind speeds (mph)}$$
$$T = \text{the mean temperature (°F) of the layer between the lower level } z_1 \text{ and the upper level } z_2$$

It is assumed in Eq. 6.25 that the atmosphere is adiabatic and the wind speed and moisture are distributed logarithmically in a vertical direction. In view of the small differences between wind and vapor pressure to be expected at two levels so closely spaced, and since these gradients are directly related to the sought-after evaporation, highly exacting instrumentation is required to get reliable results.

Potential Evapotranspiration

Thornthwaite defined potential evapotranspiration as "the water loss which will occur if at no time there is a deficiency of water in the soil for the use of vegetation." In a practical sense, however, most investigators have assumed that potential evapotranspiration is equal to lake evaporation as determined from National Weather Service Class A pan records. This is not theoretically correct because the albedo (amount of incoming radiation reflected to the atmosphere) of vegetated areas and soils ranges as high as 45 percent [28]. As a result, potential evapotranspiration should be somewhat less than free water surface evaporation. Errors in estimating free water evapotranspiration from pan records are such, however, as to make an adjustment for potential evapotranspiration of questionable value.

An equation for estimating potential evapotranspiration developed by the Agricultural Research Service (ARS) illustrates efforts to include vegetal characteristics and soil moisture in such a calculation. The evapotranspiration potential for any given day is determined as follows [29]:

$$\text{ET} = \text{GI} \times k \times E_p \times \left(\frac{S - \text{SA}}{\text{AWC}} \right)^x \tag{6.26}$$

where
ET = evapotranspiration potential (in./day)
GI = growth index of crop in percentage of maturity
k = ratio of GI to pan evaporation, usually 1.0–1.2 for short grasses, 1.2–1.6 for crops up to shoulder height, and 1.6–2.0 for forest
E_p = pan evaporation (in./day)
S = total porosity
SA = available porosity (unfilled by water)
AWC = porosity drainable only by evapotranspiration
x = AWC/G (G = moisture freely drained by gravity)

The GI curves have been developed by expressing experimental data on daily evapotranspiration for several crops (Fig. 6.5) as a percentage of the annual maximal daily rate (Fig. 6.6). Equation 6.26 is used by the Agricultural Research Service in its USDAHL-74 model of watershed hydrology in combination with GI curves to calculate daily evapotranspiration. Representative values for S, G, and AWC are given in Table 6.5.

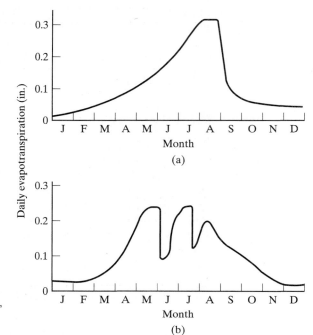

FIGURE 6.5

Average daily consumption of water: (a) for year 1953 by corn, followed by winter wheat under irrigation; (b) for year 1955, with irrigated first-year meadow of alfalfa, red clover, and timothy. Both measurements taken on lysimeter Y 102 C at the Soil and Water Conservation Research Station, Coshocton, Ohio.

(After Holtan et al [29])

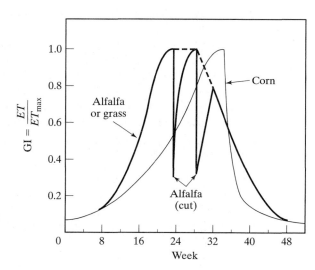

FIGURE 6.6

Growth index $GI = ET/ET_{max}$ from lysimeter records, irrigated corn, and hay for 1955, from Coshocton, Ohio.

(After Holtan et al [29])

TABLE 6.5   Hydrologic Capacities of Soil Texture Classes

Texture class	S^a (%)	G^b (%)	AWC^c (%)	x AWC/G
Coarse sand	24.4	17.7	6.7	0.38
Coarse sandy loam	24.5	15.8	8.7	0.55
Sand	32.3	19.0	13.3	0.70
Loamy sand	37.0	26.9	10.1	0.38
Loamy fine sand	32.6	27.2	5.4	0.20
Sandy loam	30.9	18.6	12.3	0.66
Fine sandy loam	36.6	23.5	13.1	0.56
Very fine sandy loam	32.7	21.0	11.7	0.56
Loam	30.0	14.4	15.6	1.08
Silt loam	31.3	11.4	19.9	1.74
Sandy clay loam	25.3	13.4	11.9	0.89
Clay loam	25.7	13.0	12.7	0.98
Silty clay loam	23.3	8.4	14.9	1.77
Sandy clay	19.4	11.6	7.8	0.67
Silty clay	21.4	9.1	12.3	1.34
Clay	18.8	7.3	11.5	1.58

$^a S$ = total porosity − 15 bar moisture %.
$^b G$ = total porosity − 0.3 bar moisture %.
^{c} AWC = $S - G$.

Source: Adapted from C. B. England, "Land Capability: A Hydrologic Response Unit in Agricultural Watersheds," U.S. Department of Agriculture, ARS 41–172, Sept. 1970 (after H. N. Holtan et al [29]).

6.7   ESTIMATING EVAPOTRANSPIRATION

Transpiration is an important component in the hydrologic budget of vegetated areas, but it is a difficult quantity to measure because of its dependence on phytological variables. It is a function of the number and types of plants, soil moisture and soil type, season, temperature, and average annual precipitation. As noted previously, evaporation and transpiration are commonly estimated in their combined evapotranspiration form.

If the precipitation and net runoff for an area are known and estimates of groundwater flow and storage can be made, rough estimates of ET can be made using the basic hydrologic equation, Eq. 1.1. A more sophisticated approach developed by Penman follows [15]. It is representative of the methods most often used.

The Penman Method

Both the energy budget and mass transport methods for estimating evapotranspiration (ET) have limitations due to the difficulties encountered in estimating parameters and in making other required assumptions. To circumvent some of these problems, Penman developed a method to combine the mass transport and energy budget theories. This widely used method is one of the more reliable approaches to estimating ET rates using climatic data [13],[15],[23],[30].

The Penman equation is of the form of Eq. 5.18; it is theoretically based and shows that ET is directly related to the quantity of radiative energy gained by the exposed surface. In its simplified form, the Penman equation is [15]:

$$ET = \frac{\Delta H + 0.27E}{\Delta + 0.27} \tag{6.27}$$

where Δ = the slope of the saturated vapor pressure curve of air at absolute temperature (mm Hg/°F)

     H = the daily heat budget at the surface (estimate of net radiation) (mm/day)

     E = daily evaporation (mm)

     ET = the evapotranspiration or consumptive use for a given period (mm/day)

The variables E and H are calculated using the following equations:

$$E = 0.35(e_a - e_d)(1 + 0.0098u_2) \tag{6.28}$$

where e_a = the saturation vapor pressure at mean air temperature (mm Hg)

     e_d = the saturation vapor pressure at mean dew point (actual vapor pressure in the air) (mm Hg)

     u_2 = the mean wind speed at 2 m above the ground (mi/day)

The equation used to determine the daily heat budget at the surface, H, is:

$$H = R(1 - r)(0.18 + 0.55S) - B(0.56 - 0.092e_d^{0.5})(0.10 + 0.90S) \tag{6.29}$$

where R = the mean monthly extraterrestrial radiation (mm H_2O evaporated per day)

     r = the estimated percentage of reflecting surface

     B = a temperature-dependent coefficient

     S = the estimated ratio of actual duration of bright sunshine to maximum possible duration of bright sunshine.

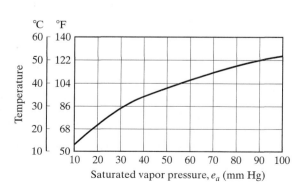

FIGURE 6.7

Relation between temperature and saturated vapor pressure.

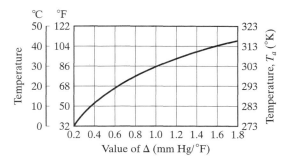

FIGURE 6.8

Temperature versus Δ relation for use with the Penman equation.

(After Criddle [23])

The empirical reflective coefficient r is a function of the time of year, calmness of the water surface, wind velocity, and water quality. Typical ranges for r are 0.05 to 0.12 [31]. Values of e_a and Δ can be obtained from Figs. 6.7 and 6.8; those for R and B can be obtained from Tables 6.6 and 6.7. The use of Penman's equation requires a knowledge of vapor pressures, sunshine duration, net radiation, wind speed, and mean temperature. Unfortunately, regular measurements of these parameters are often unavailable at sites of concern and they must be estimated. Another complication is making a reduction in the value of ET when the calculations are for vegetated surfaces. While results of experiments to quantify reduction factors have not completely resolved the problem, there is evidence that the annual reduction factor is close to unity [32]–[34]. Thus, unless there is evidence to support another value, it appears that using a value of 1 for the reduction coefficient may give satisfactory results for surfaces having varied vegetal covers. Accordingly, any estimate of free water evaporation could be used to estimate ET, providing it is modified by an appropriate reduction coefficient.

TABLE 6.6 Tabulated Values of R, Mean Monthly Intensity of Solar Radiation on a Horizontal Surface,[a] for Use in the Penman Equation

	Latitude (deg)	J	F	M	A	M	J	J	A	S	O	N	D
North	60	1.3	3.5	6.8	11.1	14.6	16.5	15.7	12.7	8.5	4.7	1.9	0.9
	50	3.6	5.9	9.1	12.7	15.4	16.7	16.1	13.9	10.5	7.1	4.3	3.0
	40	6.0	8.3	11.0	13.9	15.9	16.7	16.3	14.8	12.2	9.3	6.7	5.5
	30	8.5	10.5	12.7	14.8	16.0	16.5	16.2	15.3	13.5	11.3	9.1	7.9
	20	10.8	12.3	13.9	15.2	15.7	15.8	15.7	15.3	14.4	12.9	11.2	10.3
	10	12.8	13.9	14.8	15.2	15.0	14.8	14.8	15.0	14.9	14.1	13.1	12.4
	0	14.5	15.0	15.2	14.7	13.9	13.4	13.5	14.2	14.9	15.0	14.6	14.3
South	10	15.8	15.7	15.1	13.8	12.4	11.6	11.9	13.0	14.4	15.3	15.7	15.8
	20	16.8	16.0	14.6	12.5	10.7	9.6	10.0	11.5	13.5	15.3	16.4	16.9
	30	17.3	15.8	13.6	10.8	8.7	7.4	7.8	9.6	12.1	14.8	16.7	17.6
	40	17.3	15.2	12.2	8.8	6.4	5.1	5.6	7.5	10.5	13.8	16.5	17.8
	50	17.1	14.1	10.5	6.6	4.1	2.8	3.3	5.2	8.5	12.5	16.0	17.8
	60	16.6	12.7	8.4	4.3	1.9	0.8	1.2	2.9	6.2	10.7	15.2	17.5

[a] Measured in mm H_2O evaporated per day.

Source: After Criddle [23].

TABLE 6.7 Values of Temperature-Dependent Coefficient B for Use in the Penman Equation

T_a (°K)	B (mm H_2O/day)	T_a (°F)	B (mm H_2O/day)
270	10.73	35	11.48
275	11.51	40	11.96
280	12.40	45	12.45
285	13.20	50	12.94
290	14.26	55	13.45
295	15.30	60	13.96
300	16.34	65	14.52
305	17.46	70	15.10
310	18.60	75	15.65
315	19.85	80	16.25
320	21.15	85	16.85
325	22.50	90	17.46
		95	18.10
		100	18.80

Note: $B = \sigma T_a^4$ where σ is the Boltzmann constant, 2.01×10^{-9} mm/day.

Source: After Criddle [23].

Example 6.5

Using the Penman method, Eqs. 6.27 to 6.29, estimate ET, given the following data: temperature at water surface = 22°C, temperature of air = 33°C, relative humidity = 45%, wind velocity = 1.5 mph (36 mi/day). The month is June at latitude 33° north, r is given as 0.07, and S is found to be 0.70.

Solution.

1. Given the data for temperature, the values of e_a and e_d can be determined. Using Fig. 6.7 or Appendix Table A.2, the saturated vapor pressures are found to be 20.02 and 38.04 mm Hg, respectively. Thus $e_a = 38.04$. For a relative humidity of 40%, $e_d = 38.04 \times 0.45 = 17.12$. Then using Eq. 6.28:

$$E = 0.35(38.04 - 17.12)[1 + (0.0098 \times 36)]$$

$$E = 9.88 \text{ mm/day}$$

2. The value of Δ is found using Fig. 6.8 for the given latitude and month, and R is obtained from Table 6.6. The value of B is obtained from Table 6.7 for a temperature of 33°C. The values found are $\Delta = 1.2$, $R = 16.56$, and $B = 17.69$. Then using Eq. 6.29:

$$H = 16.56(1 - 0.07)[0.18 + (0.55 \times 0.70)] - 17.69[0.56 \\ - (0.092 \times 17.12^{0.5})][0.10 + (0.90 \times 0.70)]$$

$$H = 6.38 \text{ mm/day}$$

3.   Using Eq. 6.27:

$$ET = [(1.2 \times 6.38) + (0.27 \times 9.88)]/(1.2 + 0.27)$$

$$ET = 7.02 \text{ mm/day}$$

Thus the estimated evapotranspiration is 7.02 mm/day.

Simulating Evapotranspiration

The volume of water evaporated and transpired from a watershed can be substantial over weeks, months, and years. For storm-event modeling, ET can often be neglected, but where water budgets are required to make water supply estimates, the ET component must be incorporated. The ET models described in this chapter are typical of those used in models designed to simulate the fate of precipitated water over time (refer to the flow chart of Fig. 1.2). References on this subject abound in the literature [28],[29],[36],[37].

SUMMARY

Figure 1.1 and Table 6.1 show the overall importance of ET in the hydrologic budget. In many regions of the world, annual ET significantly exceeds annual precipitation. As a result, plans for water resources development and use must fully recognize the impact of ET losses on the gross precipitation supply. Such estimates are especially important in regions where irrigated agriculture is practiced.

A number of approaches to estimating ET have been developed. They generally fall into the following classes: theoretical, based on the physics of the process; analytical, based on energy or water budgets; and empirical, based on observations. Equations used in making ET calculations are usually of the type illustrated by Eqs. 6.1, 6.8, 6.10, 6.19, 6.22, and 6.26.

PROBLEMS

6.1 An 8000-mi^2 watershed received 20 in. of precipitation in a 1-year period. The annual streamflow was recorded as 5000 cfs. Roughly estimate the combined amounts of water evaporated and transpired. Qualify your answer.

6.2 Find the daily evaporation from a lake during which the following data were obtained: air temperature 90°F, water temperature 60°F, wind speed 20 mph, and relative humidity 30 percent.

6.3 Find the monthly consumptive use of an alfalfa crop when the mean temperature is 70°F, the average percentage of daytime hours for the year is 10, and the monthly consumptive use coefficient is 0.87.

6.4 During a given month a lake having a surface area of 350 acres has an inflow of 20 cfs, an outflow of 18 cfs, and a total seepage loss of 1 in. The total monthly precipitation is 1.5 in. and the evaporation loss is 4.0 in. Estimate the change in storage.

6.5 What are two methods that might be used to reduce evaporation from a small pond?

6.6 Compute the daily evaporation from a Class A pan if the amounts of water required to bring the level to the fixed point are as follows:

Day	1	2	3	4	5
Rainfall (in.)	0	0.65	0.12	0	0.01
Water added (in.)	0.29	0.55	0.07	0.28	0.10
Evaporation					

6.7 For Problem 6.6, the pan coefficient is 0.70. What is the lake evaporation (in inches) for the 5-day period for a lake with a 250-acre surface area?

6.8 The pan coefficient for a Class A evaporation pan located near a lake is 0.7. A total of 0.50 in. of rain fell during a given day. Determine the depth of evaporation from the lake during the same day if 0.3 in. of water had to be added to the pan at the end of the day in order to restore the water level to its original value at the beginning of the day.

6.9 A 2500-mi^2 drainage basin receives 25 in./yr rainfall. The discharge of the river at the basin outlet is measured at an average of 650 cfs. Assuming that the change in storage for the system is essentially zero, estimate the ET losses for the area in inches and centimeters for the year. State your assumptions.

6.10 Determine the daily evaporation from a lake for a day during which the following mean values were obtained: air temperature 78°F, water temperature 62°F, wind speed 8 mph, and relative humidity 45%.

6.11 Using the Meyer and Dunne equations, find the daily evaporation rate for a lake given that the mean value for air temperature was 80°F, the mean value for water temperature was 60°F, the average wind speed was 10 mph, and the relative humidity was 25%. Refer to Appendix Table A.2 for vapor pressure values.

6.12 Consider that the lake of Problem 6.11 has a surface area of 1.3 mi^2. (a) If the average annual evaporation rate is estimated to be 0.7 of the average daily rate calculated in Problem 6.11, what volume in million gallons per day (mgd) and cubic meters per day would be lost to evaporation? (b) If the average water use of an urban community is 180 gallons per capita per day (gpcd), how large a community would the daily evaporation rate sustain?

6.13 Determine the seasonal consumptive use of truck crops grown in Pennsylvania if the mean monthly temperatures for May, June, July, and August are 62, 71, 76, and 75°F, respectively, and the percent daylight hours for the given months are 10.02, 10.1, 10.3, and 9.6 as percent of the year, respectively.

6.14 Using the Penman method, Eqs. 6.27 to 6.29, estimate ET, given the following data: temperature at water surface = 20°C, temperature of air = 32°C, relative humidity = 45%, and wind velocity = 3 mph. The month is June at latitude 30° north, r is given as 0.08, and S is found to be 0.73.

6.15 Using the Penman method, Eqs. 6.27 to 6.29, estimate ET, given the following data: temperature at water surface = 20°C, temperature of air = 30°C, relative humidity = 40%, and wind velocity = 2 mph (48 mi/day). The month is June at latitude 30° north, r is given as 0.07, and S is found to be 0.75.

REFERENCES

[1] Water Resources Council, *The Nation's Water Resources: 1975–2000,* U.S. Govt. Print. Off., Washington, D.C., 1978.

[2] U.S. Department of Commerce, National Oceanic and Atmospheric Administration, National Weather Service, NOAA Technical Report 33, "Evaporation Atlas of the Contiguous 48 United States," Washington, D.C., June 1982.

[3] U.S. Department of Commerce, National Oceanic and Atmospheric Administration, National Weather Service, NOAA Technical Report NWS 34, "Mean, Monthly, Seasonal, and Annual Pan Evaporation for the United States," Washington, D.C., June 1982.

[4] "Water-Loss Investigations," Vol. 1, Lake Hefner Studies, U.S. Geologic Survey Professional Paper No. 269 (1954). (Reprint of U.S. Geological Survey Circ. 229, 1952.)

[5] N. N. Gunaji, "Evaporation Investigations at Elephant Butte Reservoir, New Mexico," *Int. Assoc. Sci. Hydrol. Pub.* **18**, 308–325(1968).

[6] I. S. Bowen, "The Ratio of Heat Losses by Conduction and by Evaporation from Any Water Surface," *Phys. Rev.* **27**, 779–787(1926).

[7] E. R. Anderson, L. J. Anderson, and J. J. Marciano, "A Review of Evaporation Theory and Development of Instrumentation," Lake Mead Water Loss Investigation; Interim Report, Navy Electronics Lab. Rept. No. 159 (Feb. 1950).

[8] O. G. Sutton, "The Application to Micrometeorology of the Theory of Turbulent Flow over Rough Surfaces," *R. Meteor. Soc. Q. J.* **75**(No. 236), 335–350(Oct. 1949).

[9] A. F. Meyer, "Evaporation from Lakes and Reservoirs," Minnesota Resources Commission, St. Paul, June 1944.

[10] T. Dunne and L. B. Leopold, *Water in Environmental Planning,* San Francisco: Freeman and Co., 1978.

[11] V. M. Ponce, *Engineering Hydrology: Principles and Practices*, Englewood Cliffs, NJ: Prentice Hall, 1989.

[12] M. A. Kohler, T. J. Nordenson, and W. E. Fox, "Evaporation from Pans and Lakes," U.S. Department of Commerce, Weather Bureau, Res. Paper No. 38, Washington, D.C., 1955.

[13] H. T. Haan, H. P. Johnson, and D. L. Brakensiek (eds.), *Hydrologic Modeling of Small Watersheds*, ASAE Monograph No. 5, St. Joseph, MI: American Society of Agricultural Engineers, 1982.

[14] R. K. Linsley, M. A. Kohler, and J. L. H. Paulhus, *Hydrology for Engineers*, 3rd ed., New York: McGraw-Hill, 1982.

[15] H. L. Penman, "Natural Evaporation from Open Water, Bare Soil, and Grass," *Proc. R. Soc. London Ser. A* **193**(1032), 120–145 (Apr. 1948).

[16] F. G. Millar, "Evaporation from Free Water Surfaces," Canada Department of Transport, Division of Meteorological Services, *Can. Meteor. Mem.* Vol. 1, No. 2, 1937.

[17] J. B. Franzini, "Evaporation Suppression Research," *Water and Sewage Works* (May 1961).

[18] Victor K. La Mer, "The Case for Evaporation Suppression," *Chem. Eng.* (June 10, 1963).

[19] D. R. Maidment (ed.), *Handbook of Hydrology*, New York: McGraw-Hill, 1993.

[20] H. F. Blaney, "Water and Our Crops," in *Water, the Yearbook of Agriculture*, Washington, D.C.: U.S. Department of Agriculture, 1955.

[21] H. F. Blaney, "Monthly Consumptive Use Requirements for Irrigated Crops," *Proc. ASCE, J. Irrigation Drainage Div.* **85**(IR1), 1–12(Mar. 1959).

[22] Ven Te Chow (ed.), *Handbook of Applied Hydrology*, New York: McGraw-Hill, 1964.

[23] W. D. Criddle, "Methods of Computing Consumptive Use of Water," *Proc. ASCE J. Irrigation Drainage Div.* **84**(IR1), 1–27(Jan. 1958).

[24] H. L. Penman, "Estimating Evaporation," *Trans. Am. Geophys. Union* **37**(1), 43–50(1956).

[25] N. J. Roberts, "Reduction of Transpiration," *J. Geophy. Res.* **66**(10), 3309–3312(Oct. 1961).

[26] E. H. Hughes, "Research on Control of Phreatophytes," Proc. Ninth Annual Water Conference, New Mexico State University, Las Cruces, March 1964.

[27] R. K. Linsley, Jr., M. A. Kohler, and J. L. H. Paulhus, *Hydrology for Engineers*, New York: McGraw-Hill, 1958.

[28] Staff, Hydrologic Research Laboratory, "National Weather Service River Forecast System Forecast Procedures," NWS HYDRO 14, U.S. Department of Commerce, Washington, D.C., Dec. 1972.

[29] H. N. Holtan, G. J. Stiltner, W. H. Henson, and N. C. Lopez, "USDAHL-74 Revised Model of Watershed Hydrology," ARS Tech. Bull. No. 1518, U.S. Department of Agriculture, Washington, D.C., 1975.

[30] N. J. Rosenberg, H. E. Hart, and K. W. Brown, "Evapotranspiration—Review of Research," MP 20, Agricultural Experiment Station, University of Nebraska, Lincoln, 1968.

[31] R. H. McCuen, *Hydrologic Analysis and Design*, Englewood Cliffs, NJ: Prentice Hall, 1989.

[32] W. O. Pruitt and F. J. Lourence, "Correlation of Climatological Data with Water Requirements of Crops," University of California Water Science Eng. Paper 9001, Davis, June 1968.

[33] C. H. M. Van Bavel, "Potential Evaporation: The Combination Concept and Its Experimental Verification," *Water Resources Res.* **2**, 455–467(1966).

[34] H. F. Blaney, "Discussion of Paper by H. L. Penman, 'Estimating Evaporation,' " *Trans. Am. Geophys. Union* **37**, 46–48(Feb. 1956).

[35] N. H. Crawford and R. K. Linsley, Jr., "Digital Simulation in Hydrology: Stanford Watershed Model IV," Tech. Rept. 39, Department of Civil Engineering, Stanford University, July 1966.

[36] W. C. Huber, J. P. Heaney, S. J. Nix, R. E. Dickinson, and D. J. Polmann, *Storm Water Management Model User's Manual, Version III*, EPA-600/2-84-109a (NTIS PB84–198423). Cincinnati, OH: Environmental Protection Agency, Nov. 1981.

[37] L. A. Roesner, R. P. Shubinski, and J. A. Aldrich, *Storm Water Management Model User's Manual, Version III: Addendum I, Extran*, EPA-600/2-84-109b (NTIS PB84–198431). Cincinnati, OH: Environmental Protection Agency, Nov. 1981.

C H A P T E R   7

Infiltration

OBJECTIVES

The purpose of this chapter is to:

- Define infiltration
- Indicate the role infiltration plays in affecting runoff quantities and in replenishing soil moisture and groundwater storage
- Review infiltration models and illustrate their use.

Infiltration is that process by which precipitation moves downward through the surface of the earth and replenishes soil moisture, recharges aquifers, and ultimately supports streamflows during dry periods. Along with interception, depression storage, and storm period evaporation, it determines the availability, if any, of the precipitation input for generating overland flows (Fig. 1.2). Furthermore, infiltration rates influence the timing of overland flow inputs to channelized systems. Accordingly, infiltration is an important component of any hydrologic model.

The rate f at which infiltration occurs is influenced by such factors as the type and extent of vegetal cover, the condition of the surface crust, temperature, rainfall intensity, physical properties of the soil, and water quality.

The rate at which water is transmitted through the surface layer is highly dependent on the condition of the surface. For example, inwash of fine materials may seal the surface so that infiltration rates are low even when the underlying soils are highly permeable. After water crosses the surface interface, its rate of downward movement is controlled by the transmission characteristics of the underlying soil profile. The volume of storage available below ground is also a factor affecting infiltration rates.

Considerable research on infiltration has taken place, but considering the infinite combinations of soil and other factors existing in nature, no perfectly quantified general relation exists.

7.1   MEASURING INFILTRATION

Commonly used methods for determining infiltration capacity are hydrograph analyses and infiltrometer studies. Infiltrometers are usually classified as rainfall simulators or flooding devices. In the former, artificial rainfall is simulated over a small test plot and the infiltration calculated from observations of rainfall and runoff, with consideration given to depression storage and surface detention [1]. Flooding infiltrometers are usually rings or tubes inserted in the ground. Water is applied and maintained at a constant level and observations made of the rate of replenishment required.

Estimates of infiltration based on hydrograph analyses have the advantage over infiltrometers of relating more directly to prevailing conditions of precipitation and field. However, they are no better than the precision with which rainfall and runoff are measured. Of particular importance in such studies is the areal variability of rainfall. Several methods have been developed and are in use. Reference 1 gives a good description of these methods.

7.2   CALCULATING INFILTRATION

Infiltration calculations vary in sophistication from the application of reported average rates for specific soil types and vegetal covers to the use of differential equations governing the flow of water in unsaturated porous media. For small urban areas that respond rapidly to storm input, more precise methods are sometimes warranted. On large watersheds subject to peak flow production from prolonged storms, average or representative values may be adequate.

The infiltration process is complicated at best. Even under ideal conditions (uniform soil properties and known fluid properties), conditions rarely encountered in practice, the process is difficult to characterize. Accordingly, there has been considerable study of the infiltration process. Most of these efforts have related to the development of (1) empirical equations based on field observations and (2) the solution of equations based on the mechanics of saturated flow in porous media [1],[2].

Later in this chapter, several commonly used infiltration models are discussed. As a preface to that discussion, a brief description of the infiltration process follows. It reviews the principal factors affecting infiltration and points out some of the problems encountered by hydrologic modelers.

We begin our discussion with an ideal case, one in which the soil is homogeneous throughout the profile and all the pores are directly interconnected by capillary passages. Furthermore, it is assumed that the rainfall is uniformly distributed over the area of concern. Under these conditions, the infiltration process may be characterized as one dimensional and the major influencing factors are therefore soil type and moisture content [3].

The soil type characterizes the size and number of the passages through which the water must flow while the moisture content sets the capillary potential and relative conductivity of the soil. Capillary potential is the hydraulic head due to capillary forces. Capillary suction is the same as capillary potential but with opposite sign. Capillary conductivity is the volume rate of flow of water through the soil under a gradient of unity (dependent on soil moisture content). Relative conductivity is the

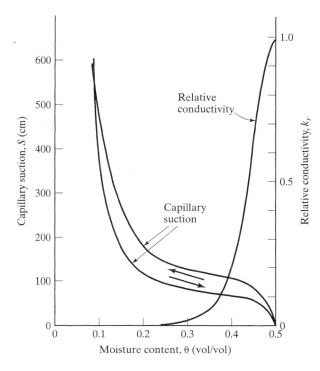

FIGURE 7.1

Typical capillary suction–relative conductivity–moisture content relation.

(After Mein and Larson [9])

capillary conductivity for a specified moisture content divided by the saturated conductivity. Figure 7.1 illustrates the relations among these variables. Note that at low moisture contents, capillary suction is high while relative conductivity is low. At high moisture contents the reverse is true.

With this background, an infiltration event can be examined. Consider that rainfall is occurring on an initially dry soil. As shown in Fig. 7.1, the relative conductivity is low at the outset due to the low soil moisture conditions. Thus, for the water to move downward through the soil, a higher moisture level is needed. As moisture builds up, a wetting front forms with the moisture content behind the front being high (essentially saturated) and that ahead of the front being low. At the wetting front, the capillary suction is high due to the low moisture content ahead of the front.

At the beginning of a rainfall event, the potential gradient that drives soil moisture movement is high because the wetting front is virtually at the soil surface. Initially, the infiltration capacity is higher than the rainfall rate and thus the infiltration rate cannot exceed the rainfall rate. As time advances and more water enters the soil, the wetting zone dimension increases and the potential gradient is reduced. Infiltration capacity decreases until it equals the rainfall rate. This occurs at the time the soil at the land surface becomes saturated. Figures 7.2 and 7.3 illustrate these conditions. Figure 7.2 shows how a moisture profile might develop when a rainstorm of constant intensity occurs. In the diagram the soil moisture at the surface is shown to range from its initial value at the top left to its saturated value at the top right. Thus in moving downward on the left-hand side of the diagram, one can trace the downward progression of the

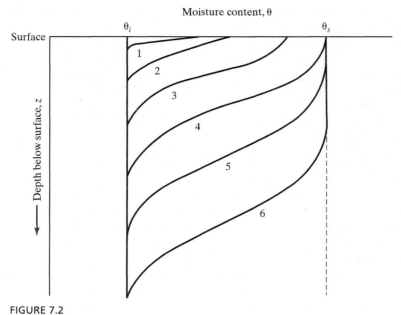

FIGURE 7.2

Typical moisture profile development with a constant rainfall rate.

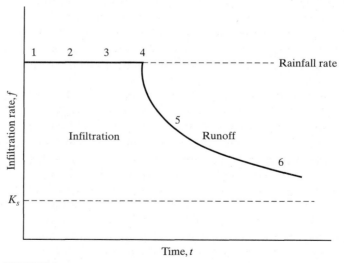

FIGURE 7.3

Infiltration rate versus time for a given rainfall intensity.

(After Mein and Larson [9]).

wetting front for varying levels of soil moisture content at the land surface. Figure 7.3 indicates that until saturation is reached at the surface, the infiltration rate is constant and equal to the rainfall application rate at the surface. At Point 4, a point that corresponds to the time at which saturation occurs at the surface, the infiltration rate begins to proceed at its capacity rate, the maximum rate at which the soil can transmit water across its surface. As time goes on, the infiltration capacity continues to decline until it becomes equal to the saturated conductivity of the soil, the capillary conductivity when the soil is saturated. This ultimate infiltration rate is shown by the dashed line to the right of K_s in Fig. 7.3.

Of particular interest is the determination of Point 4 on the curve of Fig. 7.3. This is the point at which runoff would begin for the conditions specified above. It is also the point at which the actual infiltration rate f becomes equal to the infiltration capacity rate f_p rather than the rainfall intensity rate i. The time of occurrence of this point depends, for a given soil type, on the initial moisture content and the rainfall rate. The shape of the infiltration curve after this point in time is also influenced by these factors.

Another factor that must be dealt with in the infiltration process is that of hysteresis. In Fig. 7.1 it can be seen that the plot of capillary suction versus soil moisture is a loop. The curve is not the same for wetting and drying of the soil. The curves shown in the figure are the boundary wetting and boundary drying curves, curves applicable under conditions of continuous wetting or drying. Between these curves, an infinite number of possible paths exist that depend on the wetting and drying history of the soil. A number of approaches to the hysteresis problem have been reported in the literature [3].

The illustration of the infiltration process presented was based on an ideal soil. Unfortunately, such conditions are not replicated in natural systems. Natural soils are highly variable in composition within regions, and soil cover conditions are also far-ranging. Because of this, no simple infiltration model can accurately portray all the conditions encountered in the field. The search has thus been for models that can be called upon to give acceptable estimates of the rates at which infiltration occurs during rainfall events.

Mein and Larson have described three general cases of infiltration associated with rainfall [3]. The first case is one in which the rainfall rate is less than the saturated conductivity of the soil. Under this condition, shown as (4) in Fig. 7.4, runoff never occurs since all the rainfall infiltrates the soil surface. Nevertheless, this condition must be recognized in continuous simulation processes since the level of soil moisture is affected even though runoff does not occur. The second case is one in which the rainfall rate exceeds the saturated conductivity but is less than the infiltration capacity. Curves (1), (2), and (3) of Fig. 7.4 illustrate this condition. It should be observed that the period from the beginning of rainfall to the time of surface saturation varies with the rainfall intensity. The final case is one in which the rainfall intensity exceeds the infiltration capacity. This condition is illustrated by the infiltration capacity curve of Fig. 7.5 and those portions of infiltration curves (1), (2), and (3) of Fig. 7.4 that are in their declining stages. Only under this condition can runoff occur. All three cases have relevance to hydrologic modeling, particularly when it is continuous over time.

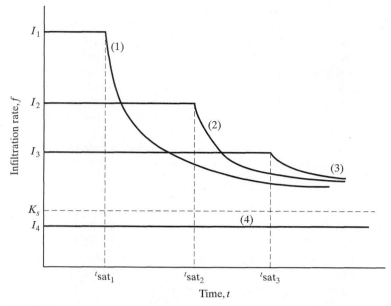

FIGURE 7.4

Infiltration curves for several rainfall intensities.

(After Mein and Larson [9])

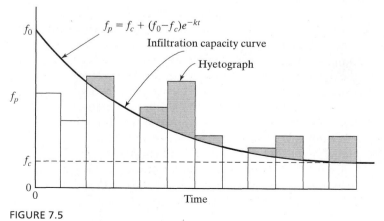

FIGURE 7.5

Horton's infiltration curve and hyetograph.

7.3   HORTON'S INFILTRATION MODEL

The infiltration process was thoroughly studied by Horton in the early 1930s [4]. An outgrowth of his work, shown graphically in Fig. 7.1, was the following relation for determining infiltration capacity:

$$f_p = f_c + (f_0 - f_c)e^{-kt} \tag{7.1}$$

where f_p = the infiltration capacity (depth/time) at some time t
$\quad\quad\quad k$ = a constant representing the rate of decrease in f capacity
$\quad\quad\quad f_c$ = a final or equilibrium capacity
$\quad\quad\quad f_0$ = the initial infiltration capacity

It indicates that if the rainfall supply exceeds the infiltration capacity, infiltration tends to decrease in an exponential manner. Although simple in form, difficulties in determining useful values for f_0 and k restrict the use of this equation. The area under the curve for any time interval represents the depth of water infiltrated during that interval. The infiltration rate is usually given in inches per hour and the time t in minutes, although other time increments are used and the coefficient k is determined accordingly.

By observing the variation of infiltration with time and developing plots of f versus t as shown in Fig. 7.5, we can estimate f_0 and k. Two sets of f and t are selected from the curve and entered in Eq. 7.1. Two equations having two unknowns are thus obtained; they can be solved by successive approximations for f_0 and k.

Typical infiltration rates at the end of 1 hr (f_1) are shown in Table 7.1. A typical relation between f_1 and the infiltration rate throughout a rainfall period is shown graphically in Fig. 7.6a; Fig.7.6b shows an infiltration capacity curve for normal antecedent conditions on turf. The data given in Table 7.1 are for a turf area and must be multiplied by a suitable cover factor for other types of cover complexes. A range of cover factors is listed in Table 7.2.

Total volumes of infiltration and other abstractions from a given recorded rainfall are obtainable from a discharge hydrograph (plot of the streamflow rate versus time) if one is available. Separation of the base flow (dry weather flow) from the discharge hydrograph results in a direct runoff hydrograph (DRH), which accounts for the direct surface runoff, that is, rainfall less abstractions. Direct surface runoff or precipitation excess in inches uniformly distributed over a watershed can readily be calculated by picking values of DRH discharge at equal time increments through the

TABLE 7.1   Typical f_1 Values

Soil group	f_1 (in./hr)	f_1 (mm/h)
High (sandy soils)	0.50–1.00	12.50–25.00
Intermediate (loams, clay, silt)	0.10–0.50	2.50–12.50
Low (clays, clay loam)	0.01–0.10	0.25–2.50

Source: After ASCE Manual of Engineering Practice, *No. 28.*

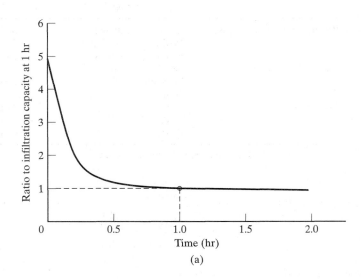

(a)

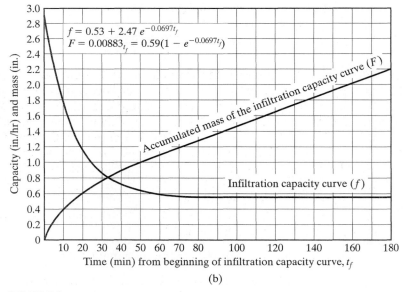

(b)

FIGURE 7.6

(a) Typical infiltration curve. (b) Infiltration capacity and mass curves for normal antecedent conditions of turf areas.

*[After A. L. Tholin and Clint J. Kiefer, "The Hydrology of Urban Runoff," Proc. ASCE J. Sanitary Eng. Div. **84**(SA2), 56 (Mar. 1959).]*

TABLE 7.2   Cover Factors

	Cover	Cover factor
Permanent forest and grass	Good (1 in. humus)	3.0–7.5
	Medium ($\frac{1}{4}$–1 in. humus)	2.0–3.0
	Poor ($<\frac{1}{4}$ in. humus)	1.2–1.4
Close-growing crops	Good	2.5–3.0
	Medium	1.6–2.0
	Poor	1.1–1.3
Row crops	Good	1.3–1.5
	Medium	1.1–1.3
	Poor	1.0–1.1

Source: After ASCE Manual of Engineering Practice, *No. 28.*

hydrograph and applying the formula [5]:

$$P_e = \frac{(0.03719)(\Sigma\ q_i)}{An_d} \tag{7.2}$$

where P_e = precipitation excess (in.)

q_i = DRH ordinates at equal time intervals (cfs)

A = drainage area (mi^2)

n_d = number of time intervals in a 24-hr period

For most cases the difference between the original rainfall and the direct runoff can be considered as infiltrated water. Exceptions may occur in areas of excessive sub-surface drainage or tracts of intensive interception potential. The calculated value of infiltration can then be assumed as distributed according to an equation of the form of Eq. 7.1 or it may be uniformly spread over the storm period. Choice of the method employed depends on the accuracy requirements and size of the watershed.

To circumvent some of the problems associated with the use of Horton's infiltration model, some adjustments can be made [6]. Consider Fig. 7.5. Note that where the infiltration capacity curve is above the hyetograph, the actual rate of infiltration is equal to that of the rainfall intensity, adjusted for interception, evaporation, and other losses. Consequently, the actual infiltration is given by:

$$f(t) = \min\ [f_p(t), i(t)] \tag{7.3}$$

where $f(t)$ is the actual infiltration into the soil and $i(t)$ is the rainfall intensity. Thus the infiltration rate at any time is equal to the lesser of the infiltration capacity $f_p(t)$ or the rainfall intensity.

Commonly, the typical values of f_0 and f_c are greater than the prevailing rainfall intensities during a storm. Thus, when Eq. 7.1 is solved for f_p as a function of time alone, it shows a decrease in infiltration capacity even when rainfall intensities are much less than f_p. Accordingly, a reduction in infiltration capacity is made regardless of the amount of water that enters the soil.

To adjust for this deficiency, the integrated form of Horton's equation may be used:

$$F(t_p) = \int_0^{t_p} f_p\, dt = f_c t_p + \frac{f_0 - f_c}{k}(1 - e^{-kt_p}) \tag{7.4}$$

where F is the cumulative infiltration at time t_p, as shown in Fig. 7.7. In the figure, it is assumed that the actual infiltration has been equal to f_p. As previously noted, this is not usually the case, and the true cumulative infiltration must be determined. This can be done using:

$$F(t) = \int_0^t f(t)\, dt \tag{7.5}$$

where $f(t)$ is determined using Eq. 7.3.

Equations 7.4 and 7.5 may be used jointly to calculate the time t_p, that is, the equivalent time for the actual infiltrated volume to equal the volume under the infiltration capacity curve (Fig. 7.7). The actual accumulated infiltration given by Eq. 7.5 is equated to the area under the Horton curve, Eq. 7.4, and the resulting expression is solved for t_p. This equation:

$$F = f_c t_p + \frac{f_0 - f_c}{k}(1 - e^{-kt_p}) \tag{7.6}$$

cannot be solved explicitly for t_p, but an iterative solution can be obtained. It should be understood that the time t_p is less than or equal to the actual elapsed time t. Thus the available infiltration capacity as shown in Fig. 7.7 is equal to or exceeds that given by

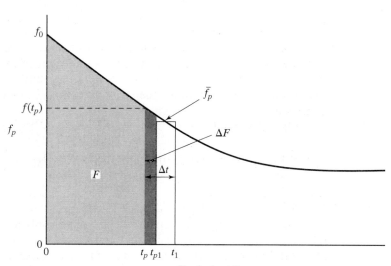

FIGURE 7.7

Cumulative infiltration.

Eq. 7.1. By making the adjustments described, f_p becomes a function of the actual amount of water infiltrated and not just a variable with time as is assumed in the original Horton equation.

In selecting a model for use in infiltration calculations, it is important to know its limitations. In some cases a model can be adjusted to accommodate shortcomings; in other cases, if its assumptions are not realistic for the nature of the use proposed, the model should be discarded in favor of another that better fits the situation.

The first eight chapters of this book deal with the principal components of the hydrologic cycle. In later chapters, the emphasis is on putting these components together in various hydrologic modeling processes. When these models are designed for continuous simulation, the approach is to calculate the appropriate components of the hydrologic equation, Eq. 1.4, continuously over time. A discussion of how infiltration could be incorporated into a simulation model follows. It exemplifies the use of Horton's equation in a storm water management model (SWMM) [6].

First, an initial value of t_p is determined. Then, considering that the value of f_p depends on the actual amount of infiltration that has occurred up to that time, a value of the average infiltration capacity, $\overline{f}_p$, available over the next time step is calculated using:

$$\overline{f}_p = \frac{1}{\Delta t} \int_{t_p}^{t_1 = t_p + \Delta t} f_p \, dt = \frac{F(t_1) - F(t_p)}{\Delta t} \tag{7.7}$$

Equation 4.3 is then used to find the average rate of infiltration, $\overline{f}$:

$$\overline{f} = \begin{cases} \overline{f}_p & \text{if } \overline{i} \geq \overline{f}_p \\ \overline{i} & \text{if } \overline{i} < \overline{f}_p \end{cases} \tag{7.8}$$

where $\overline{i}$ is the average rainfall intensity over the time step.

Following this, infiltration is incremented using the expression:

$$F(t + \Delta t) = F(t) + \Delta F = F(t) + \overline{f} \, \Delta t \tag{7.9}$$

where $\Delta F = \overline{f} \Delta t$ is the added cumulative infiltration (Fig. 7.7).

The next step is to find a new value of t_p. This is done using Eq. 7.6. If $\Delta F = \overline{f}_p \Delta t$, $t_{p1} = t_p + \Delta t$. But if the new t_{p1} is less than $t_p + \Delta t$ (see Fig. 7.7), Eq. 7.6 must be solved by iteration for the new value of t_p. This can be accomplished using the Newton–Raphson procedure [6].

When the value of $t_p \geq 16/k$, the Horton curve is approximately horizontal and $f_p = f_c$. Once this point has been reached, there is no further need for iteration since f_p is constant and equal to f_c and no longer dependent on F.

Example 7.1

Given an initial infiltration capacity f_0 of 2.9 in./hr and a time constant k of 0.28 hr^{-1}, derive an infiltration capacity versus time curve if the ultimate infiltration capacity is 0.50 in./hr. For the first 8 hours, estimate the total volume of water infiltrated in inches over the watershed.

Solution:

1. Using Horton's equation (Eq. 7.1), values of infiltration can be computed for various times. The equation is:

$$f = f_c + (f_0 - f_c)e^{-kt}$$

2. Substituting the appropriate values into the equation yields:

$$f = 0.50 + (2.9 - 0.50)e^{-0.28t}$$

3. For the times shown in Table 7.3, values of f are computed and entered into the table. Using a spreadsheet graphics package, the curve of Fig. 7.8 is derived.

TABLE 7.3   Calculations for Example 7.1

Time (hr)	Infiltration (in./hr)	Time (hr)	Infiltration (in./hr)
0	2.90	5.00	1.09
0.10	2.83	6.00	0.95
0.25	2.74	7.00	0.84
0.50	2.59	8.00	0.76
1.00	2.31	9.00	0.69
2.00	1.87	10.00	0.65
3.00	1.54	15.00	0.54
4.00	1.28	20.00	0.51

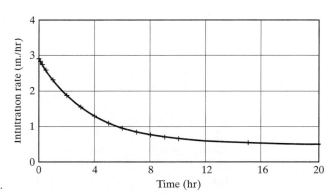

FIGURE 7.8

Infiltration curve for Example 7.1.

4. To find the volume of water infiltrated during the first 8 hours, Eq. 7.1 can be integrated over the range of 0–8:

$$V = \int [0.50 + (2.9 - 0.50)e^{-0.28t}]dt$$

$$V = [0.5t + (2.40 - 0.28)e^{-0.28t}]_0^8$$

$$V = 11.84 \text{ in.}$$

The volume over the watershed is thus 11.84 in.

7.4   GREEN–AMPT MODEL

The Green–Ampt infiltration model, originally proposed in 1911, has had a resurgence of interest [3], [6]–[11]. This approach is based on Darcy's law (see Chapter 10). In its original form, it was intended for use where infiltration resulted from an excess of water at the ground surface at all times. In 1973, Mein and Larson presented a methodology for applying the Green–Ampt model to a steady rainfall input [9]. They also developed a procedure for determining the value of the capillary suction parameter used in the model. In 1978, Chu demonstrated the applicability of the model for use under conditions of unsteady rainfall [10]. As a result of these and other efforts, the Green–Ampt model is now employed as an option in such widely used continuous simulation models as SWMM [6].

The original formulation by Green and Ampt assumed that the soil surface was covered by ponded water of negligible depth and that the water infiltrated a deep homogenous soil with a uniform initial water content (see Fig. 7.9). Water is assumed to enter the soil so as to sharply define a wetting front separating the wetted and unwetted regions as shown in the figure. If the conductivity in the wetted zone is defined as K_s, application of Darcy's law yields the equation:

$$f_p = \frac{K_s(L + S)}{L} \tag{7.10}$$

where L is the distance from the ground surface to the wetting front and S is the capillary suction at the wetting front. Referring to Fig. 7.9, it can be seen that the cumulative infiltration F is equivalent to the product of the depth to the wetting front L and the initial moisture deficit, $\theta_s - \theta_i = $ IMD. Making these substitutions in Eq. 7.10 and

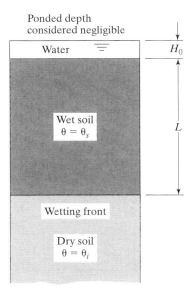

FIGURE 7.9

Definition sketch for Green–Ampt model.

rearranging, we obtain:

$$f_p = K_s\left(1 + \frac{S \times \text{IMD}}{F}\right) \tag{7.11}$$

Considering that $f_p = dF/dt$, we can state:

$$\frac{dF}{dt} = K_s\left(1 + \frac{S \times \text{IMD}}{F}\right) \tag{7.12}$$

Integrating and substituting the conditions that $F = 0$ at $t = 0$, we obtain:

$$F - (S \times \text{IMD}) \times \log_e\left(\frac{F + (\text{IMD} \times S)}{\text{IMD} \times S}\right) = K_s t \tag{7.13}$$

This form of the Green–Ampt equation is more convenient for use in watershed modeling processes than Eq. 7.10 because it relates the cumulative infiltration to the time at which infiltration began. The derivation of this equation assumes a ponded surface so that the actual rate of infiltration is equal to the infiltration capacity at all times. Using Eq. 7.13, we can determine the cumulative infiltration at any time, a feature desirable for continuous systems modeling. All the parameters in the equation are physical properties of the soil–water system and are measurable. The determination of suitable values for the capillary suction S is often difficult, however, particularly for relations such as that shown for a clay-type soil in Fig. 7.10. It can be observed from the figure that for this curve there is a wide variation of capillary suction with soil moisture content [3].

The Mein–Larson formulation using the Green–Ampt model incorporates two stages [3],[6]. The first stage deals with prediction of the volume of water that infiltrates before the surface becomes saturated. The second stage is one in which infiltration capacity is calculated using the Green–Ampt equation. In the widely used storm water management model, the modified Green–Ampt model of infiltration is one of

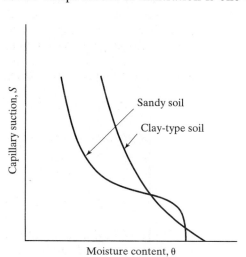

FIGURE 7.10

Capillary suction versus moisture content curves.

the options that can be employed to estimate infiltration [6]. Computations are made using the following equations: for $F < F_s (f = i)$:

$$F_s = \frac{S \times \text{IMD}}{i/K_s - 1} \quad \text{for } i > K_s \tag{7.14}$$

and there is no calculation of F_s for $i \leq K_s$; for $F \geq F_s (f = f_p)$:

$$f_p = K_s \left(1 + \frac{S \times \text{IMD}}{F} \right) \tag{7.11}$$

where f = actual infiltration rate (ft/sec)
f_p = infiltration capacity (ft/sec)
i = rainfall intensity (ft/sec)
F = cumulative infiltration volume in the event (ft)
F_s = cumulative infiltration volume required to cause surface saturation (ft)
S = average capillary suction at the wetting front (ft of water)
IMD = initial moisture deficit for the event (ft/ft)
K_s = saturated hydraulic conductivity of soil (ft/sec)

Equation 7.10 shows that the volume of rainfall needed to saturate the surface is a function of the rainfall intensity. In the modeling process, for each time step for which $i > K_s$, the value of F_s is computed and compared with the volume of rainfall infiltrated to that time. If F equals or exceeds F_s, the surface saturates and calculations for infiltration then proceed using Eq. 7.14. Note that by substituting f for i in Eq. 7.14 and rearranging, the equation takes the same form as Eq. 7.11.

For rainfall intensities less than or equal to K_s, all the rainfall infiltrates and its amount is used only to update the initial moisture deficit, IMD [6]. The cumulative infiltration volume F_s is not altered.

After saturation is achieved at the surface, Eq. 7.11 shows that the infiltration capacity is a function of the infiltrated volume, and thus of the infiltration rates during previous time steps. To avoid making numerical errors over long time steps, the integrated form of the Green—Ampt equation (Eq. 7.13) is used. This equation takes the following form as it is used in SWMM:

$$K_s(t_2 - t_1) = F_2 - C \ln (F_2 + C) - F_1 + C \ln (F_1 + C) \tag{7.15}$$

where C = $\text{IMD} \times S$ (ft of water)
t = times (sec)
$1, 2$ = subscripts indicating the starting and ending of the time steps.

Equation 7.15 must be solved iteratively for F_2, the cumulative infiltration at the end of the time step. A Newton–Raphson routine is used [6].

In the SWMM model, infiltration during time step $t_2 - t_1$ is equal to $(t_2 - t_1)i$ if the surface is not saturated and is equal to $F_2 - F_1$ if saturation has previously occurred and there is a sufficient water supply at the surface. If saturation occurs during an interval, the infiltrated volumes over each stage of the process within the time steps are computed and summed. When the rainfall ends or becomes less than the infiltration capacity, any ponded water is allowed to infiltrate and is added to the cumulative infiltration volume.

7.5   HUGGINS–MONKE MODEL

Several investigators have circumvented the time dependency problem by introducing soil moisture as the dependent variable [2],[10]–[13]. The following equation proposed by Huggins and Monke is an example [2]:

$$f = f_c + A\left(\frac{S - F}{T_p}\right)^P \tag{7.16}$$

where A and P = coefficients

S = the storage potential of a soil overlying the impeding layer (T_p minus antecedent moisture)

F = the total volume of water that infiltrates

T_p = the total porosity of soil lying over the impeding stratum

The coefficients are determined using data from sprinkling infiltrometer studies. The variable F must be calculated for each time increment in the iteration process. At the beginning of a storm, $F = 0$ and f is therefore known. In essence the continuity equation is solved for a block of soil with an inflow rate f (or smaller if the rainfall is less) and an outflow determined according to Eq. 7.17. Expression $(dS/dt)\Delta t$ then gives the change in storage of the soil. When added to the storage at the beginning of the time increment, the total storage is obtained. Equation 7.16 is a modification of one originated by Holtan and Overton [12],[13] and appears to have merit over the form of Eq. 7.1 if the rate of infiltration supply is less than infiltration capacity.

In order to use this relation when the water supply rate only intermittently exceeds the infiltration capacity, the rate at which water drains from the "control zone," which determines the soil moisture content $(S - F)$, must be found. It is evaluated as follows [10]:

1. Where the moisture content of the control zone is less than the field capacity (amount of water held in the soil after excess gravitational water has drained), the drainage rate is considered zero.
2. The drainage rate is assumed equal to the infiltration rate when the soil is saturated and the infiltration rate becomes constant.
3. If the water content is between the field capacity and saturation, the drainage rate is computed as:

$$\text{Drainage rate} = f_c\left(1 - \frac{P_u}{G}\right)^3 \tag{7.17}$$

where P_u = the unsaturated pore volume

G = maximum gravitational water, that is, the total porosity minus the field capacity

Data from sprinkling infiltrometer studies of various watersheds of interest are used to estimate the coefficients in Eq. 7.16 [2].

7.6   HOLTAN MODEL

Another equation for infiltration capacity has been developed by Holtan [14],[15]:

$$f = aS_a^{1.4} + f_c \qquad (7.18)$$

where f = the infiltration capacity (in./hr)

a = the infiltration capacity $[(in./hr)/in.^{1.4}]$ of the available storage (index of surface-connected porosity)

S_a = available storage in the surface layer (A-horizon in agricultural soils, that is, about the first 6 in.) (in. of water equivalent)

f_c = the constant rate of infiltration after long wetting (in./hr)

This equation has been modified somewhat for use in the USDAHL-70 watershed model [16]:

$$f = (GI \times aS_a^{1.4}) + f_c \qquad (7.19)$$

where a is a vegetation parameter and GI is a growth index (see Chapter 6). Information about a is given in Table 7.4.

TABLE 7.4   Tentative Estimates of the Vegetation Parameter a in the Infiltration Equation $f = GI \times aS_a^{1.4} + f_c$

	Basal area rating[a]	
Land use or cover	Poor condition	Good condition
Fallow[b]	0.10	0.30
Row crops	0.10	0.20
Small grains	0.20	0.30
Hay (legumes)	0.20	0.40
Hay (sod)	0.40	0.60
Pasture (bunchgrass)	0.20	0.40
Temporary pasture (sod)	0.40	0.60
Permanent pasture (sod)	0.80	1.00
Woods and forests	0.80	1.00

[a] Adjustments needed for weeds and grazing.
[b] For fallow land only, poor condition means after row crop, and good condition means after sod.

Source: U.S. Department of Agriculture, Agricultural Research Service, 1975.

In Eq. 7.19, it is assumed that the portion of the available storage connected to the surface is a function of the density of plant roots. This is given by the vegetation parameter a, which has been determined at plant maturity as the percentage of the ground surface area occupied by plant stems or root crowns. In this manner, the fraction of porosity in the agricultural A-horizon that is surface connected by mature plant roots to form conduits for air or water is represented.

Example 7.2

Using the USDAHL-70 watershed model equation (Eq. 7.19), calculate the infiltration capacity in inches per hour if the ultimate infiltration capacity is 0.5 in./hr, the crop grown is corn, and the soil is sandy loam. Assume the basal area rating is good, and that it is 16 weeks into the growing season. It has been found that the percentage of the available soil porosity occupied by moisture is 80%.

Solution.

1. Entering Table 7.4 for row crops, the parameter a for the USDAHL-70 equation is estimated to be 0.2. Referring to Fig. 6.6, the growth index is found to be 0.29.
2. Referring to Table 6.5, the value of S_a is determined in the following manner: The percentage of pore volume available to water is $0.80 \times 36.6 = 29.2\%$. In a 6-inch layer of soil, this would represent $6 \times 0.29 = 1.75$ in. of water.
3. Now entering the data in Eq. 7.19, we get:

$$f = (\text{GI} \times aS_a^{1.4}) + f_c$$

$$f = 0.29 \times 0.2(1.75)^{1.4} + 0.5$$

$$f = 0.63 \text{ in./hr}$$

7.7   RECOVERY OF INFILTRATION CAPACITY

The infiltration capacity curve of Fig. 7.5 illustrates that the ability of a soil to infiltrate water decays over time, providing that rainfall is continuous and that it exceeds infiltration capacity. In nature, once rainfall ceases, there is a recovery of infiltration capacity with time. The extent of recovery at any point in time depends on the dryness of the period. This condition of recovery must be recognized and incorporated in continuous simulation modeling processes. For the SWMM model, Huber and coworkers have developed an approach that follows the notation of Fig. 7.11 [6]. The SWMM model regenerates infiltration capacity whenever there are dry time steps, that is, during periods when there is no precipitation or surface water ponding. The equation used in the model is:

$$f_p = f_0 - (f_0 - f_c)e^{-k_d(t-t_w)} \tag{7.20}$$

where k_d = a decay coefficient for the recovery curve (sec^{-1})

t_w = a hypothetical projected time at which $f_p = f_c$ on the recovery curve (sec)

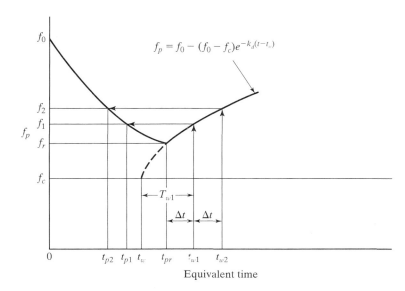

FIGURE 7.11

Equivalent time

Recovery of infiltration capacity.

In the SWMM model, k_d is assumed to be a constant fraction or multiple of k:

$$k_d = Rk \tag{7.21}$$

where R is a constant ratio, considered to be much less than 1.0. This implies a longer drying curve than a wetting curve [6].

Following the sequence shown in Fig. 7.11, new values of t_p are generated. For example, along the recovery curve:

$$f_1 = f_p(t_{w1}) = f_0 - (f_0 - f_c)e^{-k_d T_{w1}} \tag{7.22}$$

where $T_{w1} = t_{w1} - t_w$:

$$T_{w2} = t_{w2} - t_w$$

Then, solving Eq. 7.22 for the initial time difference, T_{wr}:

$$T_{wr} = t_{pr} - t_w = \frac{1}{k_d} \ln \left(\frac{f_0 - f_c}{f_0 - f_r} \right) \tag{7.23}$$

$$T_{w1} = T_{wr} + \Delta t \tag{7.24}$$

where t_{pr} = the value of t_p at the beginning of recovery (sec)
f_r = the corresponding value of f_p (ft/sec)

The value of f_1 (see Fig. 7.11) is found using Eq. 7.22. Then t_{p1} is obtained by the application of Eq. 7.1:

$$t_{p1} = \frac{1}{k} \ln \left(\frac{f_0 - f_c}{f_1 - f_c} \right) \tag{7.25}$$

7.8   TEMPORAL AND SPATIAL VARIABILITY OF INFILTRATION CAPACITY

The infiltration capacity generally varies both in space and time within a given drainage basin [17]–[19]. Spatial variations occur because of differences in soil types and vegetation. The usual procedure used to accommodate this type of variation is to subdivide the total region into components having approximately uniform soil and vegetal cover properties.

The infiltration capacity at a given location in a watershed varies with time as shown in Fig. 7.5. The initial infiltration capacity is a function of antecedent conditions and can be estimated from a knowledge of the area's soil moisture or from an antecedent precipitation index. If precipitation occurs at a rate less than the f capacity rate, the change in f capacity with time will not be that given by the f capacity curve; during periods of no precipitation, the infiltration capacity will recover. Example 7.3 illustrates these concepts.

Example 7.3

Given the rainfall pattern of Fig. 7.12 and the infiltration capacity curve of Fig. 7.13, determine the overland flow supply rate σ. Assume a turf cover and that the OGEE curve of Fig. 5.3 governs. Neglect interception losses.

Solution.   In order to solve the problem, it is necessary to determine P, F, i, and f.

    1. Construct a curve of mass infiltration F versus f capacity. This is done by calculating the areas under the curve in Fig. 7.13a at given times and plotting them versus f capacity as shown in Fig. 7.13b. Calculations to determine cumulative infiltration are shown in Table 7.5. Note that the F values are plotted versus f capacity at the end of the corresponding time interval. For example, the first value, 0.33, is plotted versus $f = 3.4$, which occurs at the end of the 5-min interval.

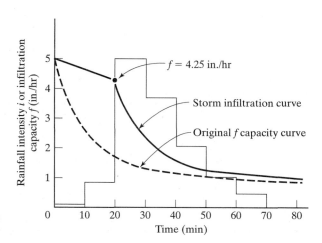

FIGURE 7.12

Storm infiltration capacity curve constructed from original f capacity curve.

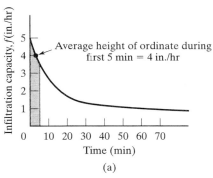
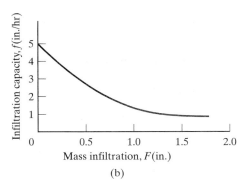

FIGURE 7.13

(a) Infiltration capacity curve and (b) mass infiltration versus f curve for Example 7.3.

2. Determine the storm period infiltration. The storm pattern and original f capacity curve are plotted as shown in Fig. 7.12.

   a. In the first 20 min $f > i$; therefore, all the rainfall is infiltrated:

$$F = (0.1 \times \tfrac{1}{6}) + (0.8 \times \tfrac{1}{6}) = 0.15 \text{ in.}$$

   b. From the F versus f curve (Fig. 7.13b), for $F = 0.15$, $f = 4.25$ in./hr.

   c. Use this as the initial value of f at $t = 20$ min and shift the original f capacity curve to the right to obtain the storm infiltration curve (Fig. 7.12). Note

TABLE 7.5   Determination of Infiltration

Time increment (min)	Average height of ordinate (in./hr)	Cumulative infiltration, F (in.)
0–5	4.00	4 × 5/60 = 0.33
5–15	2.50	2.5 × 10/60 + 0.33 = 0.75
15–30	1.50	1.5 × 15/60 + 0.75 = 1.13
30–60	1.15	1.15 × 30/60 + 1.13 = 1.7

TABLE 7.6   Determination of Cumulative P and F

Time (min)	i (in./hr)	Cumulative precipitation, P (in.)	f (in./hr)	Cumulative infiltration, F (in.)
10	0.10	0.10 × 10/60 = 0.02	0.10	0.10 × 10/60 = 0.02
20	0.80	0.80 × 10/60 + 0.02 = 0.15	0.80	0.8 × 10/60 + 0.02 = 0.15
30	5.00	5 × 10/60 + 0.15 = 0.98	2.90	2.9 × 10/60 + 0.15 = 0.63
40	3.70	3.7 × 10/60 + 0.98 = 1.60	1.80	1.8 × 10/60 + 0.63 = 0.93
50	2.00	2 × 10/60 + 1.60 = 1.93	1.40	1.4 × 10/60 + 0.93 = 1.17
60	1.10	1.1 × 10/60 + 1.93 = 2.12	1.10	1.1 × 10/60 + 1.17 = 1.35
70	0.50	0.5 × 10/60 + 2.12 = 2.20	0.50	0.5 × 10/60 + 1.35 = 1.43

TABLE 7.7   Determination of the Overland Flow Supply Rate

Time (min)	P (in.)	F (in.)	$P - F$ (in.)	i (in./hr)	f (in./hr)	$i - f$ (in./hr)	$\dfrac{\sigma^a}{i - f}$	σ (in./hr)
10	0.02	0.02	0	0.1	0.1	—	0	0
20	0.15	0.15	0	0.8	0.8	—	0	0
30	0.98	0.63	0.35	5.0	2.9	2.1	0.91	1.9
40	1.60	0.93	0.67	3.7	1.8	1.9	1.0	1.9
50	1.93	1.17	0.76	2.0	1.4	0.6	1.0	0.6
60	2.12	1.35	0.77	1.1	1.1	—	1.0	0
70	2.20	1.43	0.77	0.5	0.5	—	1.0	0

[a] From Fig. 5.3.

that this would not have been done if all rainfall intensities had exceeded the original f capacity curve ordinates. Since at the end of 20 min, some f capacity remained unfilled, the curve shift is carried out to accommodate this.

3. Having plots for the storm period infiltration and the rainfall versus time, values of P, F, i, and f can be determined. Calculations for P and F are listed in Table 7.6. Note that the curve of F versus f (Fig. 7.13b) relates to the original f capacity curve and is used to aid in constructing the storm f curve, while the values of F calculated above are related to actual storm conditions. Rainfall intensities (i) are taken from the hyetograph of Fig. 7.12.

Having determined F, P, i, and f, it is now possible to enter Fig. 5.3 using calculated $P - F$ values and determine the ratio of overland flow supply σ to $i - f$. Using this ratio and the calculated values of $i - f$ permits the determination of σ. These operations are tabulated in Table 7.7.

7.9   SCS RUNOFF CURVE NUMBER PROCEDURE

The Soil Conservation Service (SCS) has developed a widely used *curve number procedure* for estimating runoff [20]–[22]. The effects of land use and treatment, and thus infiltration, are embodied in it. The procedure was empirically developed from studies of small agricultural watersheds. While the SCS procedure is not designed to estimate infiltration directly, a look at Fig. 7.14 shows that the method does embody an infiltration estimate.

The SCS procedure consists of selecting a storm and computing the direct runoff by the use of curves founded on field studies of the amount of measured runoff from numerous soil cover combinations. A runoff curve number (CN) is extracted from Table 7.8. Selection of the runoff curve number is dependent on antecedent conditions and the types of cover. Soils are classified A, B, C, or D according to the following criteria:

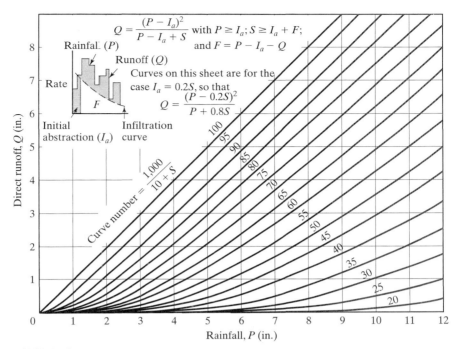

FIGURE 7.14

Solution of direct runoff equation. $S = S' + I_a$, where S is watershed storage in inches, I_a is initial abstraction, and S' is potential maximum retention exclusive of I_a.

A. (Low runoff potential) Soils having high infiltration rates even if thoroughly wetted and consisting chiefly of deep well to excessively drained sands or gravels. They have a high rate of water transmission.

B. Soils having moderate infiltration rates if thoroughly wetted and consisting chiefly of moderately deep to deep, moderately well to well-drained soils with moderately fine to moderately coarse textures. They have a moderate rate of water transmission.

C. Soils having slow infiltration rates if thoroughly wetted and consisting chiefly of soils with a layer that impedes the downward movement of water, or soils with moderately fine to fine texture. They have a slow rate of water transmission.

D. (High runoff potential) Soils having very slow infiltration rates if thoroughly wetted and consisting chiefly of clay soils with a high swelling potential, soils with a permanent high water table, soils with a claypan or clay layer at or near the surface, and shallow soils over nearly impervious material. They have a very slow rate of water transmission.

A composite curve number (CN) for a watershed having more than one land use, treatment, or soil type can be found by weighting each curve number according to its area. If, for example, 80 percent of a watershed has a CN of 75 and the remaining

TABLE 7.8 Runoff Curve Numbers for Hydrologic Soil-Cover Complexes[a]

Land use or cover	Treatment or practice	Hydrologic condition	Hydrologic soil group A	B	C	D
Fallow	Straight row	—	77	86	91	94
Row crops	Straight row	Poor	72	81	88	91
	Straight row	Good	67	78	85	89
	Contoured	Poor	70	79	84	88
	Contoured	Good	65	75	82	86
	Contoured and terraced	Poor	66	74	80	82
	Contoured and terraced	Good	62	71	78	81
Small grain	Straight row	Poor	65	76	84	88
		Good	63	75	83	87
	Contoured	Poor	63	74	82	85
		Good	61	73	81	84
	Contoured and terraced	Poor	61	72	79	82
		Good	59	70	78	81
Close-seeded legumes[b] or rotation meadow	Straight row	Poor	66	77	85	89
	Straight row	Good	58	72	81	85
	Contoured	Poor	64	75	83	85
	Contoured	Good	55	69	78	83
	Contoured and terraced	Poor	63	73	80	83
	Contoured and terraced	Good	51	67	76	80
Pasture or range		Poor	68	79	86	89
		Fair	49	69	79	84
		Good	39	61	74	80
	Contoured	Poor	47	67	81	88
	Contoured	Fair	25	59	75	83
	Contoured	Good	6	35	70	79
Meadow		Good	30	58	71	78
Woods		Poor	45	66	77	83
		Fair	36	60	73	79
		Good	25	55	70	77
Farmsteads		—	59	74	82	86
Roads (dirt)[c]		—	72	82	87	89
(hard surface)[c]		—	74	84	90	92

[a] Antecedent moisture condition II and $I_a = 0.2S$.
[b] Close-drilled or broadcast.
[c] Including right-of-way.

Source: After "Hydrology," Suppl. A to Sec. 4, Engineering Handbook, U.S. Department of Agriculture, Soil Conservation Service, 1968.

20 percent is impervious (CN = 100), then the weighted CN = 0.80 × 75 + 0.20 × 100 = 80.

The curve numbers in Table 7.8 are applicable to average antecedent moisture conditions. Other antecedent moisture conditions (AMCs) are as follows:

AMC I. A condition of watershed soils where the soils are dry but not to the wilting point, and when satisfactory plowing or cultivation takes place. (This condition is not considered applicable to the design flood computation methods presented in this text.)

AMC II. The average case for annual floods, that is, an average of the conditions that have preceded the occurrence of the maximum annual flood on numerous watersheds.

AMC III. If heavy rainfall or light rainfall and low temperatures have occurred during the 5 days previous to the given storm and the soil is nearly saturated.

The corresponding curve numbers for Condition I and Condition III can be obtained from Table 7.9 if the CN for AMC II is known.

The SCS has developed two synthetic 24-hr rainfall distributions from Weather Service rainfall frequency data. The Type I distribution is representative of the mar-

TABLE 7.9   Curve Numbers (CNs) for Wet (AMC III) and Dry (AMC I) Antecedent Moisture Conditions Corresponding to an Average Antecedent Moisture Condition

CN for AMC II	Corresponding CNs	
	AMC I	AMC III
100	100	100
95	87	98
90	78	96
85	70	94
80	63	91
75	57	88
70	51	85
65	45	82
60	40	78
55	35	74
50	31	70
45	26	65
40	22	60
35	18	55
30	15	50
25	12	43
20	9	37
15	6	30
10	4	22
5	2	13

AMC I: Lowest runoff potential. Soils in the watershed are dry enough for satisfactory plowing or cultivation.

AMC II: The average condition.

AMC III: Highest runoff potential. Soils in the watershed are practically saturated from antecedent rains.

Source: After "Hydrology," Suppl. A to Sec. 4, Engineering Handbook, *U.S. Department of Agriculture, Soil Conservation Service, 1968.*

TABLE 7.10   Runoff Curve Numbers for Urban Areas

Cover description		Curve numbers for hydrologic soil group[a]			
Cover type and hydrologic condition	Average percent impervious area[b]	A	B	C	D
Fully developed urban areas (vegetation established)					
Open space (lawns, parks, golf courses, cemeteries, etc.)[c]					
Poor condition (grass cover <50%)		68	79	86	89
Fair condition (grass cover 50–75%)		49	69	79	84
Good condition (grass cover >75%)		39	61	74	80
Impervious areas					
Paved parking lots, roofs, driveways, etc.					
(excluding right-of-way)		98	98	98	98
Streets and roads					
Paved; curbs and storm sewers					
(excluding right-of-way)		98	98	98	98
Paved; open ditches (including right-of-way)		83	89	92	93
Gravel (including right-of-way)		76	85	89	91
Dirt (including right-of-way)		72	82	87	89
Western desert urban areas					
Natural desert landscaping (pervious areas only)[d]		63	77	85	88
Artificial desert landscaping (impervious weed barrier, desert shrub with 1–2-in. sand or gravel mulch and basin borders)		96	96	96	96
Urban districts					
Commercial and business	85	89	92	94	95
Industrial	72	81	88	91	93
Residential districts by average lot size					
$\frac{1}{8}$ acre or less (town houses)	65	77	85	90	92
$\frac{1}{4}$ acre	38	61	75	83	87
$\frac{1}{3}$ acre	30	57	72	81	86
$\frac{1}{2}$ acre	25	54	70	80	85
1 acre	20	51	68	79	84
2 acres	12	46	65	77	82
Developing urban areas					
Newly graded areas (pervious areas only, no vegetation)[e]		77	86	91	94
Idle lands (CNs are determined using cover types similar to those in Table 7.8).					

[a] Average runoff condition, and $I_a = 0.2S$.

[b] The average percent impervious area shown was used to develop the composite CNs. Other assumptions are as follows: impervious areas are directly connected to the drainage system, impervious areas have a CN of 98, and pervious areas are considered equivalent to open space in good hydrologic condition. CNs for other combinations of conditions may be computed using Fig. 15.9 or 15.10.

[c] CNs shown are equivalent to those of pasture. Composite CNs may be computed for other combinations of open space cover type.

[d] Composite CNs for natural desert landscaping should be computed using Fig. 15.9 or 15.10 based on the impervious area percentage (CN = 98) and the pervious area CN. The pervious area CNs are assumed equivalent to desert shrub in poor hydrologic condition.

[e] Composite CNs to use for the design of temporary measures during grading and construction should be computed using Fig. 15.9 or 15.10 based on the degree of development (impervious area percentage) and the CNs for the newly graded pervious areas.

Source: U.S. Soil Conservation Service, "Urban Hydrology for Small Watersheds," Tech. Release 55 (2nd ed.), June 1986.

itime climate, including Hawaii, Alaska, and the coastal side of the Sierra Nevada and Cascade mountains in California, Oregon, and Washington. The Type II distribution represents the remainder of the United States, where high runoff rates are generated from summer thunderstorms. The procedure used in developing the SCS rainfall distributions is given in Ref. 21.

Once a rainfall amount has been determined, the direct runoff resulting from this precipitation can be estimated using an appropriate curve number and Figs. 7.14 and 7.15. These figures are applicable for areas up to 2000 acres. In Fig. 7.14, S is a retention index reflecting the potential storage of the watershed in inches [20].

Estimation of CN Values for Urban Land Uses

Table 7.10 includes CN values for various land use types. For these, the CN is based on a specific percentage of imperviousness. The CN values for commercial land use are, for example, based on an imperviousness of 85%. For urban land uses for which the percentages of imperviousness given in Table 7.10 are not valid, the CN can be estimated

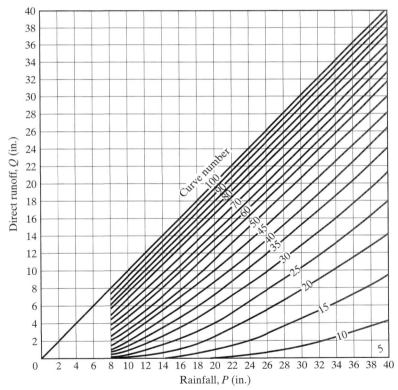

FIGURE 7.15

Solution of direct runoff equation.

(After Ref. 20)

using a weighted curve number. For soil group A, a curve number for the impervious portion would be 98, and for the open areas, it would be 39 (see Table 7.10). Using these values, the following equation can be used to provide a weighted CN estimate for any land use having a given percentage imperviousness and open space:

$$CN_W = CN_P(1 - f) + f(98) \tag{7.26}$$

where f is the fraction of imperviousness, CN_W is the weighted curve number, and CN_P is the CN for the open-space area. If for soil group A, an area is 85% impervious and 15% open space, a CN estimate using Eq. 7.26 would be obtained as follows:

$$CN_W = 39(1 - 0.85) + 0.85(98) = 89$$

Limitations to the CN Method

The SCS method has restrictions that must be complied with if estimates of discharge using the procedure are to be valid. The method should generally be used on watersheds that are homogeneous in CN. If a watershed has several areas with different curve numbers, the weighted curve number can be used. But where the differences in CN are greater than 5, it is better to subdivide the watershed into subareas, analyze them individually, and weight their runoff values rather than the curve numbers. The CN method should only be used when the CN exceeds 50, and the time of concentration is greater than 0.1 hr and less than 10 hr. Furthermore, the computed value of I_a/P should fall within the range of 0.1 to 0.5, where I_a is the initial abstraction (see Fig. 7.14).

Using CN to Estimate Runoff

The SCS method is widely used to estimate runoff (see Fig. 7.14). If a depth of rainfall P is given in inches, the resulting runoff Q may be estimated using the following equation:

$$Q = (P - 0.2S)^2/(P + 0.85) \quad \text{for } P \geq 0.2S \tag{7.27}$$

where S, the potential maximum retention, is determined using the following equation:

$$S = [(1000)/(CN)] - 10 \tag{7.28}$$

Equations 7.27 and 7.28 can be used to estimate Q if the value of P is given and the CN value can be determined. As indicated above, Eq. 7.27 requires that $P \geq 0.2S$.

Example 7.4

A watershed has a soil group C, with row crops on contoured and terraced land in good condition. For a 24-hr, 100-year precipitation of 8 in., estimate the runoff using the SCS CN approach. Do this using Eqs. 7.27 and 7.28, and also make an estimate using Fig. 7.14.

Solution

1. From Table 7.8, CN is found to be 78. Then using Eq. 7.28, we find that:

$$S = [(1000)/(78)] - 10 = 2.82$$

2. Equation 7.27 is then used to estimate Q:

$$Q = [8 - (0.2 \times 2.82)]^2/[8 + (0.8 \times 2.82)] = 5.40 \text{ inches of runoff}$$

Using Fig. 7.14 and interpolating between curve numbers, we get Q = about 5.4 inches of runoff, the same result.

7.10   PHI INDEX

Infiltration indexes generally assume that infiltration occurs at some constant or average rate throughout a storm. Consequently, initial rates are underestimated and final rates are overstated if an entire storm sequence with little antecedent moisture is considered. The best application is to large storms on wet soils or storms where infiltration rates may be assumed to be relatively uniform [1].

The most common index is termed the *phi* (ϕ) *index* for which the total volume of the storm period loss is estimated and distributed uniformly across the storm pattern. Then the volume of precipitation above the index line is equivalent to the runoff (Fig. 7.16). A variation is the W index, which excludes surface storage and retention. Initial abstractions are often deducted from the early storm period to exclude initial depression storage and wetting.

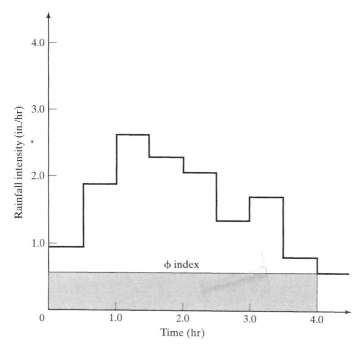

FIGURE 7.16
Representation of a ϕ index.

To determine the ϕ index for a given storm, the amount of observed runoff is determined from the hydrograph, and the difference between this quantity and the total gauged precipitation is then calculated. The volume of loss (including the effects of interception, depression storage, and infiltration) is distributed uniformly across the storm pattern as shown in Fig. 7.16.

Use of the ϕ index for determining the amount of direct runoff from a given storm pattern is essentially the reverse of this procedure. Unfortunately, the ϕ index determined from a single storm is not generally applicable to other storms, and unless it is correlated with basin parameters other than runoff, it is of little value.

SUMMARY

Infiltration is a significant component of hydrologic processes. Soils have varying capacities to infiltrate water. Influencing factors include soil type, degree of saturation, and nature of ground cover. Activities that change the soil surface or alter its properties also have a modifying effect.

When the rainfall intensity is less than the infiltration capacity, all of the water reaching the ground can infiltrate. But if the rainfall intensity exceeds the infiltration capacity, infiltration will occur only at the infiltration capacity rate, and water in excess of that capacity will be stored in depressions, become surface runoff, or evaporate. In general, the initial infiltration capacity of a dry soil is high. As rainfall continues, and as the soil becomes saturated, it diminishes to a relatively constant rate (ultimate capacity).

Infiltration rates have been determined for a variety of soils and ground cover conditions. A number of equations have been developed to serve as models for the infiltration process. They are exemplified by Eqs. 7.1, 7.11, and 7.16.

PROBLEMS

7.1 Given an initial infiltration capacity f_0 of 3.0 in./hr and a time constant k of 0.29 hr^{-1}, derive an infiltration capacity versus time curve if the ultimate infiltration capacity is 0.55 in./hr. For the first 10 hours, estimate the total volume of water infiltrated in inches over the watershed.

7.2 Gross rain intensities during each hour of a 5-hr storm over a 1000-acre basin were 5, 4, 1, 3, and 2 in./hr, respectively. The direct surface runoff from the basin was 375 acre-ft. Determine the basin ϕ index.

7.3 The infiltration rate for excess rain on a small area was observed to be 4.5 in./hr at the beginning of rain, and it decreased exponentially toward an equilibrium of 0.5 in./hr. A total of 30 in. of water infiltrated during a 10-hr interval. Determine the value of k in Horton's equation $f = f_c + (f_0 - f_c)e^{-kt}$.

7.4 Rework Problem 7.2 if the storm occurred over a basin of 2.5 km^2; the rainfall intensities were 12, 10, 3, 8, and 5 cm/hr; and the direct surface runoff was 463,000 m^3.

7.5 Rework Problem 7.3 if the initial infiltration capacity was 10 cm/hr, the ultimate capacity was 1.2 cm/hr, and a total of 33 cm of water infiltrated during the 10-hr interval.

7.6 Precipitation falls on a 500-acre drainage basin according to the following schedule:

30-min period	1	2	3	4
Intensity (in./hr)	4.0	2.0	6.0	5.0

a. Determine the total storm rainfall (in inches).

b. Determine the ϕ index for the basin if the net storm rain is 3.0 in.

7.7 The direct surface runoff volume from a 4.40-mi^2 drainage basin is determined by planimeter from the area under the hydrograph to be 10,080 cfs-hr. The hydrograph was produced by a 1.71-in./hr rainstorm with a duration of 5 hr. Determine (a) the net rain and (b) the ϕ index.

7.8 The following table lists the storm rainfall data and infiltration capacity data for a 24-hr storm beginning at midnight on April 14 of the current year.

a. Plot the rainfall hyetograph and the f capacity curve on rectangular coordinate paper.

b. Determine the total storm precipitation in inches.

c. By counting squares or by planimeter, determine the net storm rain by the f capacity method.

Rainfall Data for a Hypothetical Storm on April 15 of the Current Year (Beginning at midnight on April 14)

Hour	Rainfall intensity (in./hr)	Infiltration capacity at beginning of hour (in./hr)	Hourly deduction for depression storage (in./hr)
1	0.41	0.200	0.20
2	0.49	0.160	0.14
3	0.32	0.125	0.04
4	0.31	0.100	0.02
5	0.22	0.085	0.00
6	0.08	0.070	0.00
7	0.07	0.065	
8	0.09	0.057	
9	0.08	0.052	
10	0.06	0.047	
11	0.11	0.044	
12	0.12	0.040	
13	0.15	0.037	
14	0.23	0.036	
15	0.28	0.035	
16	0.26	0.034	
17	0.21	0.033	
18	0.09	0.033	
19	0.07	0.033	
20	0.06	0.032	
21	0.03	0.032	
22	0.02	0.032	
23	0.01	0.031	
24	0.01	0.031	

7.9 Tabulated below are total rainfall intensities during each hour of a frontal storm over a drainage basin.

   a. Plot the rainfall hyetograph (intensity versus time).

   b. Determine the total storm precipitation amount in inches.

   c. If the net storm rain is 2.00 in., determine the exact ϕ index (in./hr) for the drainage basin. (Note that by definition the area under the hyetograph above the ϕ index line must be 2.00 in.)

   d. Determine the area of the drainage basin (acres) if the net rain is 2.00 in. and the measured volume of direct surface runoff is 2,015 cfs-hr.

   e. Using the ϕ index calculated in part c, determine the volume of direct surface runoff (acre-ft) that would result from the following:

Hour	1	2	3	4
Intensity (in./hr)	0.40	0.05	0.30	0.20

Hour	Rainfall intensity (in./hr)	Hour	Rainfall intensity (in./hr)
1	0.41	13	0.15
2	0.49	14	0.23
3	0.22	15	0.28
4	0.31	16	0.26
5	0.22	17	0.21
6	0.08	18	0.09
7	0.07	19	0.07
8	0.09	20	0.06
9	0.08	21	0.03
10	0.06	22	0.02
11	0.11	23	0.01
12	0.12	24	0.01

7.10 The SCS curve number method of estimating rainfall excess (net rain) is based on the assumption that the curve number for a watershed depends on several factors. Name or describe the factors that are considered when a curve number is determined.

7.11 A 7-mi^2 drainage basin has a composite curve number CN of 50. According to the SCS, exactly how much rain must fall before the direct runoff commences?

7.12 Determine a composite SCS runoff curve number for a 600-acre basin that is totally within soil group C. The land use is 40 percent contoured row crops in poor hydrologic condition and 60 percent native pasture in fair hydrologic condition.

7.13 Which SCS-classified soil would have the highest infiltration rate: A, B, C, or D?

7.14 Rework Problem 7.11 if the drainage basin is 12 km^2 and the CNs are 55 and 80. Give your answers in centimeters.

7.15 Rework Problem 7.12 if the soil group is B and the land use is 50% straight-row crops under poor conditions and 50% meadow under good conditions.

7.16 For the composite curve number found in Problem 7.15, estimate the amount of runoff if the direct rainfall is 20 cm.

REFERENCES

[1] Ven Te Chow (ed.), *Handbook of Applied Hydrology*, New York: McGraw-Hill, 1964.

[2] L. F. Huggins and E. J. Monke, "The Mathematical Simulation of the Hydrology of Small Watersheds," Tech Rept. 1, Purdue University Water Resources Center, Lafayette, IN, Aug. 1966.

[3] R. G. Mein and C. L. Larson, "Modeling the Infiltration Component of the Rainfall-Runoff Process," Bull. 43, Water Resources Research Center, University of Minnesota, Minneapolis, MN, Sept. 1971.

[4] R. E. Horton, *Surface Runoff Phenomena: Part I, Analysis of the Hydrograph*, Horton Hydrol. Lab. Pub. 101, Ann Arbor, MI: Edwards Bros., 1935.

[5] W. D. Mitchell, "Unit Hydrographs in Illinois," Illinois Waterways Division, 1968.

[6] W. C. Huber, J. P. Heaney, S. J. Nix, R. E. Dickinson, and D. J. Polmann, "Storm Water Management User's Manual, Version III," Department of Environmental Engineering Sciences, University of Florida, Gainesville, Nov. 1981.

[7] H. T. Haan, H. P. Johnson, and D. L. Brakensiek (eds.), *Hydrologic Modeling of Small Watersheds*, ASAE Monograph No. 5, St. Joseph, MI: American Society of Agricultural Engineers, 1982.

[8] W. H. Green and G. A. Ampt, "Studies on Soil Physics, 1. The Flow of Air and Water Through Soils," *J. Agric. Sci.* **4**, 11–24(1911).

[9] R. G. Mein and C. L. Larson, "Modeling Infiltration During a Steady Rain," *Water Resources Res.* **9**(2), 384–394(Apr. 1973).

[10] S. T. Chu, "Infiltration During an Unsteady Rain," *Water Resources Res.* **14**(3), 461–466(June 1978).

[11] B. J. Knapp, "Infiltration and Storage of Soil Water," in *Hillslope Hydrology*, M. J. Kirby (ed.), New York: John Wiley, 1978.

[12] H. N. Holtan, "A Concept for Infiltration Estimates in Watershed Engineering," U.S. Department of Agriculture, Agricultural Research Service, 1961, pp. 41–51.

[13] D. C. Overton, "Mathematical Refinement of an Infiltration Equation for Watershed Engineering," ARS 41-99, U.S. Department of Agriculture, Washington, D.C., 1964.

[14] H. N. Holtan, "A Concept for Infiltration Estimates in Watershed Engineering," ARS 41–51. U.S. Department of Agriculture, Agricultural Research Service, Washington, D.C., 1961.

[15] H. N. Holtan, "A Model for Computing Watershed Retention from Soil Parameters," *J. Soil Water Conserv.* **20**(3), 91–94(1965).

[16] H. N. Holtan, G. J. Stiltner, W. H. Henson, and N. C. Lopez, "USDAHL-74 Revised Model of Watershed Hydrology," ARS Tech. Bull. No. 1518, U.S. Department of Agriculture, Washington, D.C., 1975.

[17] N. H. Crawford and R. K. Linsley, Jr., "Digital Simulation in Hydrology: Stanford Watershed Model IV," Tech. Rept. 39, Department of Civil Engineering, Stanford University, July 1966.

[18] Staff, Hydrologic Research Laboratory, "National Weather Service—River Forecast System Forecast Procedures," NWS HYDRO 14, U.S. Department of Commerce, Washington, D.C., Dec. 1972.

[19] A. M. Lumb et al., "GTWS: Georgia Tech Watershed Simulation Model," School of Civil Engineering, ERC-0175, Georgia Institute of Technology, Atlanta, GA, 1975.

[20] "Hydrology," Suppl. A to Sec. 4, *Engineering Handbook*, U.S. Department of Agriculture, Soil Conservation Service, 1968.

[21] Soil Conservation Service, "A Method for Estimating Volume and Rate of Runoff in Small Watersheds," Technical Paper No. 149, USDA-SCS, Washington, D.C., 1973.

[22] Soil Conservation Service, *Urban Hydrology for Small Watersheds*, Tech. Release 55, Washington, D.C., 1975 (Ed. 1), 1986 (Ed. 2).

C H A P T E R   8

Surface Water Hydrology

OBJECTIVES

The purpose of this chapter is to:

- Define terms frequently used in describing runoff and streamflow processes
- Present concepts of the rainfall-runoff process
- Introduce the characteristics of streamflow and hydrographs
- Define elements of drainage basin physiography and geomorphology
- Describe quantitative measures of watershed physiographic characteristics
- Review the empirical and analytical methods available for predicting peak runoff rates from ungauged watersheds
- Describe the physics of snowmelt and present methods of estimating snowmelt
- Provide the reader with references to other standard sources that expand discussions of surface water hydrology beyond the scope of this book.

Surface water hydrology deals with the movement of water along the earth's surface as a result of precipitation and snow melt. The amount of water flowing in surface water courses at any instant of time is small in terms of the earth's total water budget, but it is of considerable importance to those concerned with water resources development, supply, and management. A knowledge of the quantity and quality of streamflow is a requisite for municipal, industrial, agricultural, and other water supply projects; flood control; reservoir design and operation; hydroelectric power generation; water-based recreation; navigation; fish and wildlife management; drainage; the management of natural systems such as wetlands; and water and wastewater treatment.

8.1   STREAMFLOW

Streamflow is the topic of many standard engineering texts [1],[2] as well as several more-recently developed hydrology texts designed for watershed managers [3] and environmental scientists and technicians [4],[5]. This chapter presents most of the fundamental tenets of streamflow and snowmelt hydrology needed by engineers.

Units of Measurement

Two types of units are used in measuring water flowing in streams. They are units of *discharge* and units of *volume*. Discharge, or rate of flow, is the volume of water that passes a particular reference point in a unit of time. The basic units used in connection with stream gauging in the United States are the foot and meter for measurements of dimension and the second for measurements of time. Commonly used units of discharge measurement are cubic feet per second (cfs) and cubic meters per second (m^3/sec). Other units of discharge in use are second-foot per square mile (sec-ft/mi^2), for expressing the average rate of discharge from a drainage basin or defined area, and million gallons per day (mgd), commonly used in potable water supply calculations. Units of volume used are the cubic foot, cubic meter, liter, gallon, and acre-foot (a volume equivalent to 1 ft of water over an acre, 43,560 ft^2, of land). The latter unit is commonly used in irrigation practice in the western United States.

Streamflow Data

The principal sources of streamflow data for the United States are the U.S. Geological Survey (USGS), U.S. Natural Resources Conservation Service (NRCS), U.S. Forest Service, and U.S. Agricultural Research Service (ARS). In addition, the U.S. Army Corps of Engineers (COE), the Tennessee Valley Authority (TVA), and the U.S. Bureau of Reclamation (USBR) make some streamflow measurements and tabulate streamflow data relative to their missions. State agencies, universities, and various research organizations also collect and publish a variety of streamflow data.

The USGS Water Supply Papers (WSPs) are the benchmark for referencing streamflow data in the form of daily discharge. Computerized data are also available from the USGS. *Publications of the Geological Survey*, published every 5 years and supplemented annually, are an excellent source of information. The NRCS historically published data on streamflow from small watersheds and plots in its *Hydrologic Bulletin* series, but much of the data have been republished by ARS. Records from NRCS "pilot watersheds" are published in cooperation with the USGS. U.S. Forest Service streamflow data are published at irregular intervals in various technical bulletins and professional papers.

Statistical Summaries of Streamflow Data

Using many of the methods described in Chapter 3, the U.S. Geological Survey routinely prepares statistical summaries of the data being collected by their hydrographers. Data from every USGS stream gauge and many other gauges operated by cooperating agencies are published both on the Internet and in written reports, usually in the form of annual water supply reports. U.S. Geological Survey software for computing surface water statistics for USGS stream gauges, named SWSTAT, is available at http://water.usgs.gov/software/surface_water.html. In addition to daily discharge records and peak flow data, these and other recurring publications of the USGS provide statistical summaries that are useful for a wide range of water resource evaluations. Among commonly published values are:

■ Magnitudes of monthly and annual flows, including the maximum, minimum, and mean monthly and annual flows, and the standard deviations and coefficients of variation.

■ Magnitudes and frequencies of daily low, high, and peak instantaneous flood flow rates, and annual mean flows.

■ Flow-duration analyses providing the percentage of time daily flow values were equaled or exceeded during the period of record.

■ Maximum, median, and minimum daily mean flows.

Low-flow frequencies of daily discharges, including the annual minimum mean flows, are reported for periods of 1, 3, 7, 14, 30, 60, 90, 120, and 183 consecutive days for recurrence intervals of 2, 5, 10, 20, 50, and 100 years. High-flow frequencies of daily discharges are reported for periods of 1, 3, 7, 15, 30, 60, and 90 consecutive days and for recurrence intervals of 2, 5, 10, 25, 50, and 100 years.

Recurrence intervals for low flows and high flows are generally reported only to twice the period of record, but for records of more than 40 years, the records are extrapolated using a log–Pearson Type III distribution (see Section 3.7) to the 100-year and sometimes 500-year recurrence period. Except for low-flow frequencies, annual values are based on the water year, which ends September 30. This standard was adopted because the growing season is typically at an end and surface-water storage is typically near a minimum, allowing reliable water supply tracking and forecasting.

Flood-frequency information includes the record of magnitudes and dates of instantaneous peak flows and results of applying log–Pearson Type III frequency analyses to the annual or partial-duration series of peak flows. For stations with 10 or more years of record, the Bulletin 17B methods described in Section 3.7 are used. Flood rates for recurrence intervals of 2, 5, 10, 25, 50, and 100 years are tabulated.

Streamflow Measurements by Direct and Indirect Methods

The most accurate measurements of streamflow rates are by *direct* methods, involving measurements of velocity and cross-sectional area, and then determining the discharge as the product [6]. *Indirect* methods include calculation of discharge from high water marks and channel roughness characteristics at a known cross section, or by construction of structures in the stream that have a fixed relationship between water stage and discharge rate, called a *rating curve*. Facilities such as weirs, flumes, and lateral or vertical contractions in the stream such as culverts and bridges can have relatively fixed stage-discharge relationships. Once a stream cross section has been rated, only the stage needs to be measured. River stage is recorded either visually from a staff gauge or continuously using a float or nitrogen bubbler assembly connected to a recorder or satellite transmitter. The rating curve is developed from a number of direct measurements of discharge over a range of flows. These are repeated frequently so that the rating curve can be updated to reflect any shifts in the control or changes in the river cross section.

A = cross-sectional area (ft^2)

R = the hydraulic radius, equal to the area divided by the wetted perimeter

S = the head loss per unit length of channel, approximated by the channel slope

For most streams, Manning's n values range between about 0.01 and 0.75. When reasonable determinations can be made of n, A, R, and S, Eq. 8.1 can be used to estimate the streamflow that occurred during the high-water period.

Discharge from Stage–Discharge Rating Curves

Where flow-measuring devices are used, it is customary to observe the water level (called *stage*) and use a rating curve to translate the stage into discharge [7]. Locations in streams having a relatively strong relationship between stage and discharge, such as at contractions, are called *controls*. Usually, a stage recording is obtained at a gauging site and this record is converted into discharge by one or more of several methods. Rating curves, tables, and formulas are used for this purpose. Figure 8.2 shows a typical stage–discharge rating curve, developed after completing numerous streamflow measurements.

As shown later, a *hydrograph* is a continuous graph of rate of streamflow with respect to time, normally obtained by means of a continuous recorder that shows stage

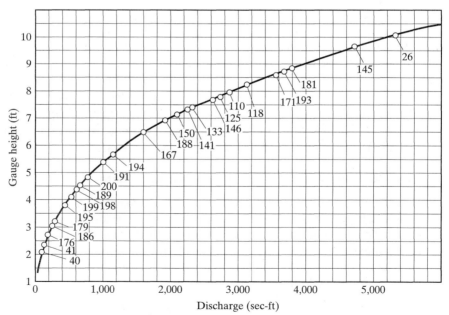

FIGURE 8.2

Station rating curve for Raquette River at Piercefield, New York.

(U.S. Geological Survey)

(depth) versus time (stage hydrograph), which is then transformed into a discharge hydrograph by application of a rating curve. In general, the term *hydrograph* as used herein means a discharge hydrograph.

8.2 RUNOFF

The term *runoff* normally applies to flow over a surface, and the term *streamflow* is used to describe the drainage after it reaches a defined channel. Rain falling on a watershed in quantities exceeding the soil or vegetation uptake becomes *surface runoff*. Water infiltrating the soil may eventually return to a stream and combine with surface runoff in forming the total *drainage* from the basin. The network of overland flow courses and defined drainage channels comprises the watershed. Surface runoff from tracts of land begins its journey as *overland flow*, often called *sheet flow*, before it reaches a defined swale or channel, usually before flowing more than a few hundred feet.

Catchments, Watersheds, and Drainage Basins

Runoff occurs when excess precipitation or snowmelt moves across the land surface— some of which eventually reaches natural or artificial streams and lakes. The land area over which precipitation falls is called the *catchment*, and the land area that contributes surface runoff to any point of interest is called a *watershed* [1],[8]. This can be a few acres in size or thousands of square miles. A large watershed can contain many smaller subwatersheds.

Streams and rivers convey both surface water and groundwater away from high-water areas, preventing surface flooding and rising groundwater problems. The tract of land (both surface and subsurface) drained by a river and its tributaries is called a *drainage basin*. A watershed supplies surface runoff to a river or stream, whereas a drainage basin for a given stream is the tract of land drained of both surface runoff and groundwater discharge. The lines separating the land surface into watersheds are called *divides*. These normally follow ridges and mounds and can be delineated using contour maps, digital elevation models (DEMs), field surveys, or stereograph pairs of aerial photographs to identify gradient directions.

Rainfall-Runoff Process

During a given rainfall, water is continually being abstracted to moisten the upper levels of the soil surface; however, this infiltration is only one of many continuous abstractions. Rainfall is also intercepted by trees, plants, and roof surfaces, and at the same time is evaporated. Once rain falls and fulfills initial requirements of infiltration, natural depressions collect falling rain to form small puddles, creating *depression storage*. In addition, numerous pools of water forming *detention storage* build up on permeable and impermeable surfaces within the watershed. This stored water gathers in small rivulets, which carry the water originating as *overland flow* into small channels, then into larger channels, and finally as *channel flow* to the watershed outlet.

The surface runoff component consists of water that flows overland until a stream channel is reached. During large storms it is the most significant hydrograph

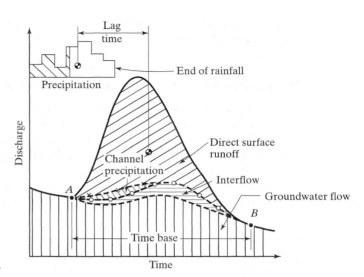

FIGURE 8.3

Components of a hydrograph.

component. Figure 8.3 illustrates the surface runoff and groundwater components of a hydrograph.

Interflow is that part of the subsurface flow that moves at shallow depths and reaches the surface channels in a relatively short period of time and therefore is commonly considered part of the direct surface runoff. Rain falling directly on the stream surface, called *channel precipitation*, can increase the flow rate at any time.

In general, the channel of a watershed possesses a certain amount of *base flow* during most of the year. This flow comes from groundwater or spring contributions and may be considered as the normal day-to-day flow. Direct surface runoff (DSR) from precipitation excess—that is, after abstractions are deducted from the original rainfall—constitutes the direct runoff hydrograph (DRH). Methods of separating the hydrograph into base flow and direct runoff are detailed in Chapter 9. Arrival of direct runoff at the outlet accounts for an initial rise in the DRH. As precipitation excess continues, enough time elapses for progressively distant areas to add to the outlet flow. Consequently, the duration of rainfall dictates the proportionate area of the watershed amplifying the peak, and the intensity of rainfall during this period of time determines the resulting greatest discharge.

Processes involved in forming the DRH can be better understood by visualizing the precipitation excess as partially disposed of immediately by surface runoff while a portion remains held within the watershed boundaries and is released later from storage. Thus the shape and timing of the DRH are integrated effects of the duration and intensity of rainfall and other hydrometeorological factors as well as the effect of the physiographic factors of the watershed upon the storage capacity.

The relation between precipitation and runoff is influenced by various storm and basin characteristics. Because of these complexities and the frequent paucity of adequate runoff data, many approximate formulas have been developed to relate rainfall and runoff. The earliest of these were usually crude empirical statements, whereas the trend now is to develop descriptive equations based on physical processes. These are described in detail in Chapters 9–10 and 12–13.

Quantitative Measures of Drainage Basin Characteristics

The description of a drainage basin in quantitative terms was an important step forward in hydrology and can be traced back in large part to the efforts of Robert E. Horton [9],[10]. Others have expanded his original work [8],[11]. Measures of the surface area contributing to runoff, and the length, gradient, and density of stream segments are principal factors needed for relating drainage basin physiographic and topologic characteristics to runoff patterns and magnitudes. These are briefly described here. Other standard references provide greater detail [1],[2],[8],[13],[14].

Contributing Area   In the majority of analyses, the hydrologist needs to ascertain the total land surface area contributing runoff to some point of interest. Because of variations in topography, the true surface area that receives rainfall is not easily measured. Fortunately, the horizontal projection of any land area is easily obtained and is universally used in hydrologic calculations that require the area over which precipitation falls. The actual watershed surface area is larger than the projected area, but the difference is neglected in practice.

Partial-Area Hydrology   Using a procedure called *partial-area hydrology,* watershed areas are divided into *contributing* (active) and *noncontributing* (passive) subareas [1],[2]. For infrequent but severe storms, larger percentages of the watershed surface may contribute to the peak flow and volume of runoff, both of which are of interest. Significantly smaller portions of some watersheds may contribute for more frequent storms. Other than excluding obvious depressions, noncontributing areas are not easily identified. Consequently, partial-area hydrology is seldom incorporated in design of hydraulic structures, and is of greater interest in water supply and water quality studies.

As will be shown later (Chapter 12), unit hydrograph theory and runoff curve number methods assume that all the subareas of a watershed contribute to runoff in all storms and in proportional amounts at different times in the same storm. Application of these assumptions to watersheds that have significant noncontributing zones could introduce error if the zones are not partially or completely excluded. Most methods in common use reduce the amount of runoff from areas that have numerous depressions, high degrees of vegetative cover, or sandy soils.

To evaluate the assumption that the full area contributes to runoff, Boughton [15] monitored runoff from various locations in a test watershed over a 15-year study period. The entire watershed contributed to runoff on only 3 of 30 events. In about two-thirds of the events, discharge occurred only from the subareas with the lowest capacity to store or infiltrate the precipitation. Other than reducing runoff potential based on land use and soil types, most practitioners do not attempt to acknowledge this phenomenon. Design of hydraulic structures (see Chapter 13) involves estimating runoff for rare, heavy-rainfall events during which the majority of the drainage area above the design location contributes to runoff.

Boughton's study logic is as follows:

1. Watersheds can be idealized as a group of "surface storage capacity" cells, each representing a fraction of the watershed area and each having some capacity to abstract rainfall into storage, infiltration, or evapotranspiration.

2. Runoff from each cell occurs when rain fills the surface storage capacity.

3. Runoff occurs from the cell with the smallest capacity before flowing from the cell with the next largest capacity (this is an assumption by Boughton that has not been fully verified).

4. Using these principles, storm data for the watershed are evaluated first to find those in which runoff occurs only from the area of smallest capacity. This is done using a graphical method outlined in the article that looks at slope changes in the rainfall-runoff graph. Both the capacity of the cell and its area as a percentage of the watershed are estimated.

5. After subtracting the contribution to runoff from the smallest-capacity cell, the capacity and contributing area for the second-smallest-capacity cell are determined by the same procedure.

6. The process is repeated until all the runoff is accounted for, or until 100 percent of the watershed is contributing, whichever occurs first.

Area Relationships   Correlations have been observed between the area, $\overline{A}_u$, of basins and the length of streams, $\overline{L}_u$. These variables are often related by an exponential function. For example, studies of seven streams in the Maryland-Virginia area by Hack have produced the relationship [12]:

$$L = 1.4A^{0.6} \tag{8.2}$$

where L = the stream length measured in miles to the drainage divide
       A = the drainage area (mi^2)

Hack's observations indicate that as the drainage basin increases in size, it becomes longer and narrower; thus precise geometric similarity is not preserved.

Drainage area has long been used as a parameter in precipitation runoff equations or in simple equations indexing streamflow to area or other parameters. Many early empirical equations are of the form [13]:

$$Q = cA^m \tag{8.3}$$

where Q = a measure of flow such as mean annual runoff
       A = the size of the contributing drainage area

Values of c and m are determined by regression analysis (see Chapter 3); Fig. 8.4 illustrates a relation of this form.

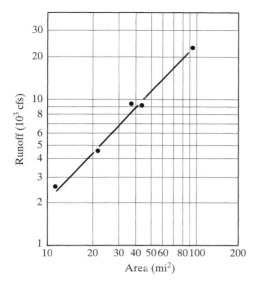

FIGURE 8.4

Runoff–drainage area correlation for five Maryland streams.

Stream Lengths and Stream Order Important measures of lengths of drainage basin streams include overland flow lengths and stream lengths. The concept of stream order is often associated with the dimension of stream length.

If the stream system in a drainage basin is clearly defined on a topographic map, the smallest tributaries are classified as Order 1 [10]. This is illustrated in Fig. 8.5. The

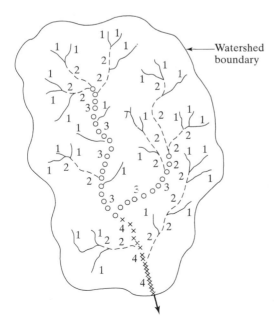

FIGURE 8.5

Sketch indicating definition of stream order.

point at which two first-order streams join is the beginning of a second-order segment. Third-order segments initiate where two second-order streams join, and so on. The main stream channel that carries the flow from the entire tributary area upstream of a point of interest will necessarily be the highest-order stream in that system.

The practical utility of the stream order system is based on the hypothesis that the size of the watershed, its channel dimensions, and streamflow are all proportional to the stream order, provided that a large enough sample is investigated. The order number permits comparisons of drainage systems that are quite different in size because the number is a dimensionless quantity. Such comparisons should be made at locations in the two systems that have a similar geometry, that is, second-order streams, third-order streams, and so forth.

Stream lengths are determined by the measurement of their projections onto a horizontal plane. Topographic maps are useful for obtaining such measurements. If the mean length of a stream segment L_u of order u is defined as $\overline{L}_u$, then it is possible to determine $\overline{L}_u$ using [10]:

$$\overline{L}_u = \frac{\sum_{i=1}^{N_u} L_{ui}}{N_u} \tag{8.4}$$

where N_u is the number of stream segments of stream order u.

Another measure related to stream length is the distance L_{ca} from a point of interest on the main stream to a point on the primary channel that is nearest the center of gravity of the drainage area (center of gravity of the plane area of the drainage basin). Studies of basin lag (time between the centers of mass of effective storm input and the resulting runoff) have made use of this dimension.

Of particular significance in the physiographic development of a drainage basin is the *overland flow length* L_0. This is the distance from the ridge line or drainage divide, measured along the path of surface flow which is not confined in any defined channel, to the intersection of this flow path with an established flow channel. If a drainage basin of the first order is the basic element of a larger drainage system, then a representative overland flow length can be determined for these first-order basins. One approach is to measure a number of possible flow paths from a map of the area and to average these. In some cases (for example, with the rational method, Chapter 12), the use of the longest overland flow length is prescribed, measured from the upstream end of the first-order stream to the most remote point of flow that will terminate at this point.

Gradients The slopes of a drainage basin and its channels have a very strong effect on the surface runoff process of that region. Most stream channel profiles exhibit the characteristic of decreasing slope proceeding in a downstream direction. Figure 8.6 illustrates this particular trait. Also illustrated in the figure are the gross slope, which is the total elevation drop divided by the channel length, and the mean slope, which is determined such that the areas between the average slope line and the stream profile are equal, that is, $A_1 = A_2$ in the figure. The gross slope and the mean slope are not very useful as parameters to describe drainage character due to their generality;

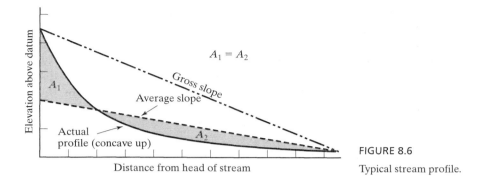

FIGURE 8.6

Typical stream profile.

Fig. 8.6 should make this clear. Some mathematical functions that are used to more fully describe stream profiles are linear, exponential, logarithmic, and power forms. A single numerical value to represent the primary channel slope has been used by Taylor and Schwartz [16]. This factor, known as the equivalent main stream slope S_{st}, is the slope of a uniform channel that is equivalent in length to the longest water course and has the same travel time. This factor has been found to be related to unit hydrograph lag (time from the center of mass of rainfall excess to the center of mass of runoff) and maximum discharge.

In addition to the slope of the stream channel, the overall land slope of the basin is an important topographic factor. A quantitative relation between valley wall slopes and stream channel slopes has been derived by Strahler [13]. A commonly used method of determining the slopes of a basin has been presented by Horton [17].

Drainage Density and Stream Frequency   Other topographic measures include definitions of the basin shape and the density of the drainage network or *drainage density*, defined as the ratio of total channel segment lengths cumulated for all stream orders within a basin to the basin area. The *stream frequency* is defined as the summation of all segments in a drainage basin (total number of segments of all orders) divided by the drainage area.

8.3 FLOODS AND DROUGHTS

Economic considerations for design of hydraulic structures or water supply facilities require that peak streamflow rates or severities of droughts be determined within acceptable levels of risk. Most hydraulic structures in U.S. rivers and streams are designed to withstand effects of the highest discharge rate that would occur during a moderate to large flood. Frequency-based (FB) design is achieved by determining the peak flow rate having a given probability of being equaled or exceeded in any given year (see Chapter 13). For example, the 100-year flood and the 50-year flood have a 1 percent and 2 percent chance, respectively, of occurring in any year.

As noted in Chapter 3, the *recurrence interval* of a hydrologic event is defined as the reciprocal of the exceedance probability. The term *return period* is also used.

Recall that the *frequency* of an event is the annual exceedance probability, though many use it to refer to the recurrence interval. A flood magnitude with a 10-year recurrence interval has a frequency of 10 percent. Some texts define the recurrence interval as the average interval in years, over a long period of time, between events of equal or smaller probability, but this can be misleading because any event can occur in any year.

Bridges, culverts, and flood control structures are considered safe for floods less than the design value, and repairs or replacement of the structures should be anticipated when the design flood is exceeded. Similarly, water supply facilities in the United States are designed to accommodate a given severity of drought corresponding to some acceptable risk. Shortages in water supply occur when a drought is worse than the design drought.

Statistical methods of estimating flood and drought magnitudes were presented in Chapter 3. This section describes several empirical methods of analyzing drought severity and estimating peak flow rates for small rural watersheds. Methods for estimating peak flow rates from urban watersheds are presented in Chapter 12, and analytical methods and computer models for estimating peak flow rates and entire hydrographs from large or complex watersheds are described in Chapters 9, 12, and 13.

Drought Frequency Analysis

Comprehensive descriptions of the analysis of droughts and drought severity exist in standard references [1],[2],[11],[14]. A single example is presented here to illustrate a simple case. Whipple [18] prepared Fig. 8.7 by performing a frequency analysis of low flows for five watersheds in the same region. The graph has considerable scatter but allows an assessment of magnitudes and frequencies of droughts in the general area.

Because droughts are defined by nonexceedance rather than exceedance of given runoff amounts, the runoff amounts were tabulated in ascending rather than descending order. The data revealed that runoff amounts less than 1 in. occurred in 8 of the 136 years of records, thus the Weibull equation predicts that a 1-in. drought has a frequency of $m/(N + 1) = 8/137 = 6$ percent and a recurrence interval of 17 years. The point is highlighted in the figure to illustrate its plotting position.

Using the same illustration, the interpretation is that the risk each year of a 1-in. or less drought is 6 percent. The 1-in. drought has an apparent recurrence interval of 17 years. This type of analysis establishes that the 1-in. drought has a 6 percent chance of nonexceedance each year but cannot predict what will happen this year or even over a period of years.

Peak Flow Methods for Small Watersheds

Several techniques are available to estimate peak flow rates for small rural watersheds, ranging in size from a few to several thousand acres. Some of the more widely used methods are described in this section. Peak flow methods (and hydrograph analyses) for urban watersheds and large, rural watersheds are described in later chapters.

Peak flow methods for small watersheds rely on either statistical analysis of the probability distributions of measured peak flow rates or regression and correlation

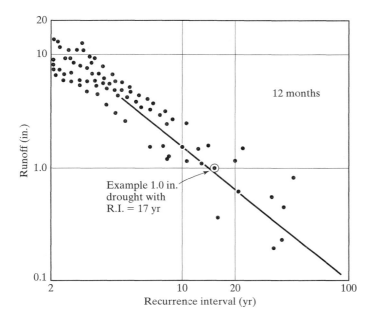

FIGURE 8.7

Low-flow frequency data consolidated for five rivers.

(After Whipple [18])

analysis of the relationships among peak flow and precipitation, basin characteristics, and meteorological conditions (see Chapter 3). Streamflow from small watersheds is seldom measured, especially over sufficiently long periods needed to estimate the distribution statistics for the variable, so regional regression and correlation are often the best available methods.

SCS TP-149 Method In the 1970s, the U.S. Soil Conservation Service (SCS), now called the Agricultural Soil and Conservation Service (ASCS), developed the Technical Paper 149 method to allow estimation of peak flow rates from small (5- to 2000-acre) agricultural watersheds [19]. The technique is still used even though other ASCS methods have been developed (see Chapters 12 and 13). Data from numerous watersheds were subjected to a correlation analysis, resulting in a series of 42 charts relating peak discharges to drainage area and other basin and storm parameters for 24-hour storm depths. Figures 8.8 and 8.9 are typical charts from the paper.

Input to the procedure is the drainage area, average watershed slope, storm distribution type (I or II), watershed composite curve number (see Chapter 7), and depth of rainfall. Shown are Type-I and Type-II curves for moderately sloped watersheds, with CN = 70 for both. Similar charts are available for the combinations given in Table 8.3. Applications of TP-149 to watersheds having curve numbers other than the 5-unit increments of Table 8.3, or for slopes other than 1, 4, or 16 percent, can be accomplished by arithmetic or logarithmic interpolation between adjacent chart values.

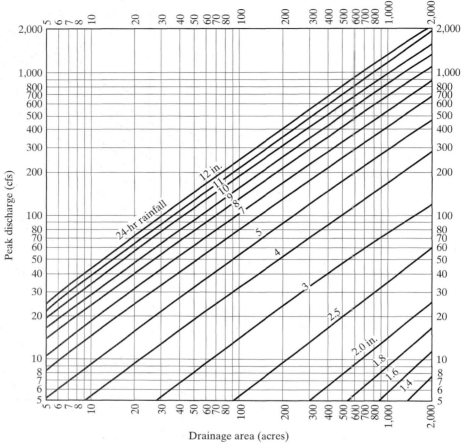

FIGURE 8.8

TP-149 peak rates of discharge for small watersheds, Type-I storms: 24-hr rainfall, moderate slopes, and CN = 70.

(After U.S. Soil Conservation Service, "A Method of Estimating Volume and Rate of Runoff in Small Watersheds," U.S. Department of Agriculture, Jan. 1968)

Example 8.2

Compare the peak flow rates from Type-I and Type-II storms using Figs. 8.8 and 8.9. Assume that only storm type changes and all other conditions are equal.

Solution. A 4-in. rain over 200 acres on a watershed with CN = 70 results in $Q_p = 52$ cfs for a Type-I storm (Fig. 8.8) and $Q_p = 97$ cfs for Type II (Fig. 8.9). Thus the storm distribution type makes a significant difference in results of peak flow estimation using SCS techniques.

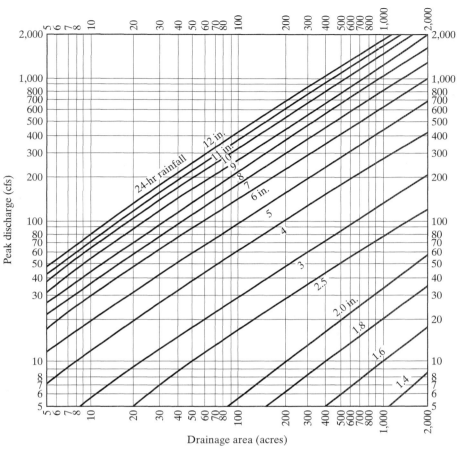

FIGURE 8.9

TP-149 peak rates of discharge for small watersheds, Type-II storms: 24-hr rainfall, moderate slopes, CN = 70.

(After U.S. Soil Conservation Service, "A Method of Estimating Volume and Rate of Runoff in Small Watersheds," U.S. Department of Agriculture, Jan. 1968)

TABLE 8.3   Charts Available in TP-149 for Peak Flow Rates of Small Watersheds

Storm distribution type	Slope type	Slope range (%)	Curve number, CN
I, II	Flat, 1%	0–3	60, 65, 70, 75, 80, 85, 90
I, II	Moderate, 4%	3–8	60, 65, 70, 75, 80, 85, 90
I, II	Steep, 16%	8–30	60, 65, 70, 75, 80, 85, 90

FHWA HDS-2 Peak Flow Design Method The Federal Highway Administration (FHWA) lists in its Hydrologic Design Series 2, HDS-2, a procedure for estimating peak flow rates for homogeneous, small- to medium-sized watersheds having times of concentration between 0.1 and 10 hours [20]. It employs an SCS regression equation that has coefficients determined from data on different rainfall distribution types and ratios of the initial abstraction I_a (see Chapter 5) and total precipitation, P. The peak discharge in metric units is calculated from:

$$q_p = q_u \, AQ \tag{8.5}$$

where q_p is the peak discharge in m^3/sec, A is the drainage area in km^2, Q is the net rain depth in mm, and q_u is the unit peak discharge from:

$$q_u = 0.000431 \{ 10^{[C_0 + C_1 + C_2 (\log t_c)^2]} \} \tag{8.6}$$

in which t_c is the time of concentration in hours, and the regression coefficients are obtained from Table 8.4.

TABLE 8.4   Coefficients for FHWA HDS-2 SCS Peak Discharge Method

Rainfall type	I_a/P	C_0	C_1	C_2
I	0.10	2.30550	−0.51429	−0.11750
	0.20	2.23537	−0.50387	−0.08929
	0.25	2.18219	−0.48488	−0.06589
	0.30	2.10624	−0.45695	−0.02835
	0.35	2.00303	−0.40769	0.01983
	0.40	1.87733	−0.32274	0.05754
	0.45	1.76312	−0.15644	0.00453
	0.50	1.67889	−0.06930	0.0
IA	0.10	2.03250	−0.31583	−0.13748
	0.20	1.91978	−0.28215	−0.07020
	0.25	1.83842	−0.25543	−0.02597
	0.30	1.72657	−0.19826	0.02633
	0.50	1.63417	−0.09100	0.0
II	0.10	2.55323	−0.61512	−0.16403
	0.30	2.46532	−0.62257	−0.11657
	0.35	2.41896	−0.61594	−0.08820
	0.40	2.36409	−0.59857	−0.05621
	0.45	2.29238	−0.57005	−0.02281
	0.50	2.20282	−0.51599	−0.01259
III	0.10	2.47317	−0.51848	−0.17083
	0.30	2.39628	−0.51202	−0.13245
	0.35	2.35477	−0.49735	−0.11985
	0.40	2.30726	−0.46541	−0.11094
	0.45	2.24876	−0.41314	−0.11508
	0.50	2.17772	−0.36803	−0.09525

Source: After Ref. 20

The procedure has the following limitations:

- Use with homogeneous watersheds (CNs from zone to zone should not differ by 5).
- CN should be greater than 50.
- t_c should be between 0.1 and 10 hours.
- I_a/P should be between 0.1 and 0.5.
- t_c should be about the same for any of the main channels, if the watershed has more than one main channel.
- No channel or reservoir routing is allowed.
- No storage facility on the main channel is allowed.
- Watershed area in storage ponds and lakes should be less than 5 percent.

Discharge–Area and Regression Formulas   A multitude of peak-flow formulas relating the discharge rate to drainage area have been proposed and applied. Gray [14] lists 35 such formulas, and Maidment [1] compares many others. Most of these empiric equations are derived using pairs of measurements of drainage area and peak flow rates in a regression equation having the form:

$$Q = CA^m \qquad (8.7)$$

where Q = the peak discharge associated with a given return period
A = the drainage area
C, m = regression constants

Discharge–area formulas in the form of Eq. 8.7 include the Meyers equation [21]:

$$Q = 10,000A^{0.5} \qquad (8.8)$$

where A = the drainage area, which must be 4 mi^2 or more
Q = the ultimate maximum flood flow (cfs)

Cyprus Creek Formula   Extremely flat areas pose particular difficulties to the hydrologist, including estimates of infiltration, runoff volume, and peak runoff rates. Flooding in these areas tends to be shallow and widespread. Flow velocities are low, and water stands on the surface for relatively long periods of time. These areas are often distinguished by networks of straight drainage channels that have been constructed to store and eventually discharge the excess rain.

The SCS developed a procedure [22] to calculate the instantaneous peak flow from small flatland areas based on first calculating the capacity of canals that would be needed to limit flat-area flooding for the design storm to a duration that would prevent excess crop damage, and then to apply a multiplier to this rate to obtain the instantaneous peak for the design of drainage structures. The procedure is illustrated in Fig. 8.10 [22]. The selected duration was 24 hours, considered to be the maximum allowable time for inundation of crops. An equation, called the *Cyprus Creek formula*, was developed to determine the canal design flow rate, called the *24-hr removal rate*. The

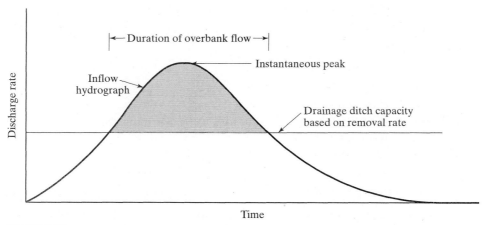

FIGURE 8.10

Illustration of relation between Cyprus Creek "removal" rate and peak instantaneous flow.
(After Soil Conservation Service [22])

equation, based on rainfall depth, contributing drainage area, and the SCS composite curve number, is:

$$Q_{24} = CA^{5/6} \tag{8.9}$$

where Q_{24} = required channel capacity for 24-hr removal (cfs)
   C = drainage coefficient
   A = drainage area (mi^2)

The drainage coefficient, C, for Eq. 8.9 is found from an equation developed by Stephens and Mills [23]:

$$C = 16.39 + (14.75\, Q_{scs}) \tag{8.10}$$

where

   Q_{scs} = the SCS direct runoff (in.) for the 24-hr design event from Fig. 7.14.

Once Eq. 8.9 is solved for the given frequency, the instantaneous peak flow rate is obtained from:

$$Q_p/Q_{24} = 2 - 0.43\,(\log A) \tag{8.11}$$

The procedure is limited to drainage areas from 1 to about 200 square miles. It is suggested that ratios of the peak instantaneous rate to the 24-hr canal removal rate be limited to values greater than or equal to 1.0. For flatland areas that have part of the area in storm sewers, the SCS recommends that the peak flows from Eq. 8.11 be increased by 0.7 times the percentage of the area served by storm sewers. The SCS further recommends restricting use of this procedure to watersheds that have slopes that are less than 0.002. For steeper-slope watersheds, other methods such as TR-20, TR-55, TP-149, or regression equations are recommended.

Example 8.3

Use the Cyprus Creek method to determine the peak 50-yr flow rate from a 1.0 mi^2 drainage area that has a CN = 80, is 50 percent storm sewered, and has a 50-yr, 24-hr rainfall depth of 12.0 inches.

Solution.

1. From Fig. 7.14, the direct runoff for 12 inches of rain is 9.45 in. The drainage coefficient, C, is found from Eq. 8.10:

$$C = 16.39 + (14.75)(9.45) = 155.8$$

2. The 24-hr removal rate is found from Eq. 8.9:

$$Q_{24} = 155.8(1.0)^{5/6} = 155.8 \text{ cfs}$$

From Eq. 8.11 the peak instantaneous rate Q_p is 311.6 cfs if no storm sewers existed. The unsewered area discharge should be increased by 0.7×50 percent for a watershed with 50 percent storm sewers. The final design flow is $1.35 \times 311.6 = 420.7$ cfs.

Paleohydrology

Paleohydrology is the study of floods that occurred prior to the time of direct measurement or historical documentation. The rare, high-hazard floods are studied by examining high-stage indicators or other indirect, preserved evidences of their occurrence. Such studies infer flood magnitudes from the visible effects of ancient floods on landscapes, vegetation (primarily marks on trees), soils, deposited sediments, or even man-made structures. If such evidence of paleofloods exists, the flood stage and peak flow magnitude can be estimated by direct or indirect hydraulic methods or hydraulic modeling techniques [24]. Types of indicators of paleofloods include:

1. Slackwater sand and silt deposits that result from deposition of suspended sediment transported during extreme floods.
2. Stratigraphic anomalies such as organic layers, tributary alluvium, mud cracks, bed forms such as ripples or dunes, or rounded gravel or cobble deposits.
3. Flood-transported litter such as organic materials, buried trees, compounds found only in buried soils, and archaeological debris or refuse.
4. High-water marks from ancient floods such as rock or floating debris scars on tree trunks, floating debris deposits, sediment deposits in elevated caves, and scour lines or other disturbances along canyon or terrace walls.

Paleohydrologic studies can provide important supplemental design flood information, and can most often provide reasonableness checks for the upper limits of the maximum size of floods that have occurred in a river reach [25],[26].

USGS Regional Peak Flow Regression Equations

Early in the 1950s, the U.S. Geological Survey instituted a process of correlating flood flow magnitudes and frequencies with drainage basin characteristics. Using methods described in Section 3.9, sets of regression equations for the 2-, 5-, 10-, 25-, 50-, and 100-year floods have been developed for practically every hydrologically homogeneous region in every state. The work was largely inaugurated to develop methods for estimating peak flow rates for design of highway structures at ungauged basins. Data from gauged sites was evaluated by regional analysis to provide the best fit of regression models to the data.

Continuous water-stage recorders and crest-stage gauge data were consulted to develop frequency curves for all gauged watersheds. Given the frequency curves, a number of tests were made using multiple linear regression to predict the peak flows from various easily obtained independent parameters such as drainage area, basin slope, watershed aspect, elevation, mean temperature during the snowmelt season, and other variables.

Each study was reported by state. The open file or water resource investigation reports are available from the USGS and include discussions of the equations, comments on range of applicability, information on the reliability of the equations, copies of all the gauged basin frequency curves, and sets of equations for estimating floods in ungauged watersheds. Equations for all states have been compiled by the U.S. Geological Survey [27] into a PC software package called NFF (National Flood Frequency), available from the USGS (or FHWA as part of its package, HYDRAIN).

Figure 8.11 shows the six regions for the state of Texas. Regression was conducted independently by region using available gauged station data. As in many of the reports, the Texas manual reveals that different independent variables were selected for each region [28]. The equations developed for Region 2 are:

$$Q_2 = 216\,A^{0.574}S^{0.125} \tag{8.12}$$

$$Q_5 = 322\,A^{0.620}S^{0.184} \tag{8.13}$$

$$Q_{10} = 389\,A^{0.646}S^{0.214} \tag{8.14}$$

$$Q_{25} = 485\,A^{0.668}S^{0.236} \tag{8.15}$$

$$Q_{50} = 555\,A^{0.682}S^{0.250} \tag{8.16}$$

$$Q_{100} = 628\,A^{0.694}S^{0.261} \tag{8.17}$$

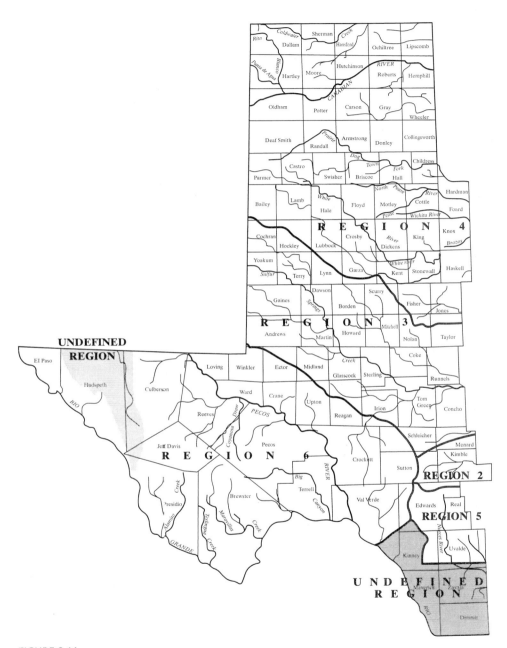

FIGURE 8.11

Hydrologic regions in Texas for 1976 USGS regional regression equations.

(From Ref. 28)

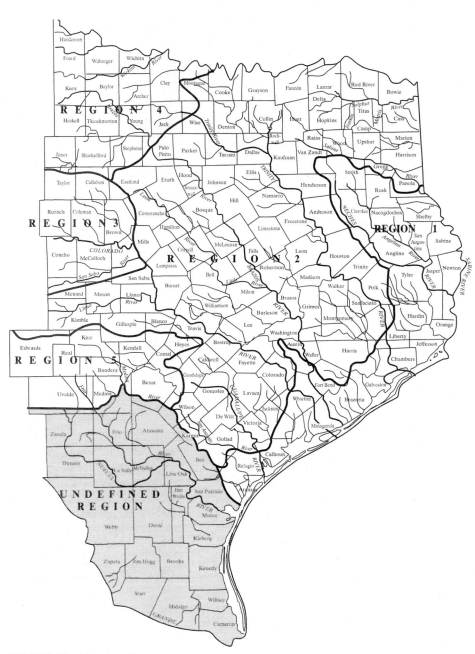

FIGURE 8.11 *(Continued)*

where Q = peak discharge for given frequency, cfs

   A = drainage area, square miles

   S = average slope of the streambed between points 10 and 85 percent of the distance along the main stream channel from the mouth to the basin divide, feet per mile

Equations for other regions in Texas include the mean annual precipitation, P, along with the area and slope terms. The equations apply to rural basins with areas from 0.3 to 5000 square miles. Drainage areas in Region 2 ranged in size from 0.33 to 4255 square miles, and slopes from 1.16 to 108.1 feet per mile. Lack of data prevented the development of regression equations for the southern and western parts of the state.

Example 8.4

Develop estimates of flood peaks for a 200-square-mile rural watershed near Dallas. The mean slope between the 10 and 85 percent points is 3.4 ft per mile.

Solution.   Dallas is in Region 2. Equations 8.12–8.17 give:

$$Q_2 = 216\,A^{0.574}S^{0.125} = 5{,}270 \text{ cfs}$$

$$Q_5 = 322\,A^{0.620}S^{0.184} = 10{,}770 \text{ cfs}$$

$$Q_{10} = 389\,A^{0.646}S^{0.214} = 15{,}490 \text{ cfs}$$

$$Q_{25} = 485\,A^{0.668}S^{0.236} = 22{,}300 \text{ cfs}$$

$$Q_{50} = 555\,A^{0.682}S^{0.250} = 27{,}800 \text{ cfs}$$

$$Q_{100} = 628\,A^{0.694}S^{0.261} = 34{,}170 \text{ cfs}$$

National Flood Frequency (NFF) Program

Since 1973, regression equations like Eqs. 8.12 through 8.17 for estimating flood-peak discharges for rural, unregulated watersheds have been published, at least once, for every state and the Commonwealth of Puerto Rico. In 1993 the USGS, in cooperation with the Federal Highway Administration and the Federal Emergency Management Agency, compiled all of the current statewide and metropolitan area regression equations into a computer program entitled the National Flood Frequency (NFF) Program [27]. This program summarizes regression equations for estimating flood-peak discharges for all 52 states. It also addresses techniques for estimating a typical flood hydrograph for a given recurrence interval or exceedance probability peak discharge for unregulated rural and urban watersheds. The program lists statewide regression equations for rural watersheds and provides much of the reference information and input data needed to run the computer program. Regression equations for estimating urban flood-peak discharges for several metropolitan areas in at least 13 states are also available.

Information on computer specifications and the computer program are available [27]. Instructions for installing the NFF program on a personal computer and a description of the NFF program and the associated database of regression statistics are also given. The program is available as part of the Federal Highway Administration package, HYDRAIN, or by itself. Though the USGS and FHWA do not service the software, information about vendors who provide software sales and service can be obtained by contacting the agencies.

National Flood Insurance Program (NFIP)

In 1968, the U.S. Department of Housing and Urban Development (HUD), later called the Federal Emergency Management Agency (FEMA), initiated the NFIP to identify flood hazard areas and to provide occupants of floodplains with mapping of the flood-prone areas and access to low-cost flood insurance. The NFIP requires local governments to adopt and implement flood management programs that prevent developments in excess of national standards.

Hydrology is a key ingredient in these studies. It is used for identifying peak flow rates; studying effects of dams, open channels, and other water control structures; and determining volumes of floodwaters that need to be safely conveyed by the nation's waterways.

As of 2002 the National Flood Insurance Program had grown to over 20,000 participating communities with over 5 million properties insured, producing over $600 billion in coverage. Though successful, it is estimated that only 25 percent of the 12 million households in the U.S. floodplains are protected by flood insurance. The flood hazard areas have been mapped for these communities and are updated as developments occur. Each of these studies has required either approximate or detailed evaluation of peak flow rates for a range of recurrence intervals. The 100-year discharge, called the *base flood*, has been determined in all cases. The portion of the floodplain occupied by the base flood has been mapped, allowing communities to determine whether a property is in the 100-yr floodplain, and in many cases, what water surface elevation would be experienced at the property during the base flood.

Figure 8.12 illustrates the typical NFIP mapping and floodplain management procedure. Surveyed valley and channel cross sections are used in determining the 100-yr flow depth, allowing the hydrologist to delineate the lateral extent of flooding during the 100-yr flood. Then a *floodway* width is generally determined as that portion of the floodplain that is reserved in order to discharge the 100-year flood without cumulatively increasing the water surface more than 1.0 ft. This procedure is illustrated in Fig. 8.13. The floodway is most often centered over the main stream channel, but can be offset or even split into several zones.

Development within the floodway is allowed only if compensated by relocating the floodway or mitigating the water surface increase due to the development. The *flood fringe* is that portion of the floodplain outside the floodway in which development is allowed, up to a point of full encroachment by buildings, roadbeds, berms, and so forth. As much as 7–10 percent of the total land area of the United States lies within the 100-year floodplain. The largest areas of floodplain are in the southern parts of the country, and the most populated floodplains are along the north Atlantic coast, in the Great Lakes region, and in California.

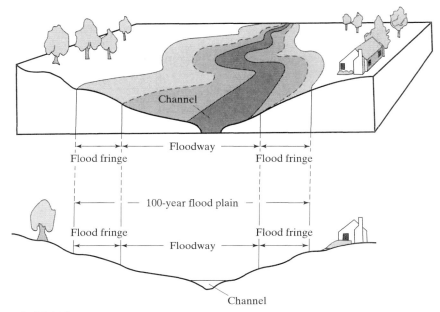

FIGURE 8.12

Definition sketch of floodplain delineations.

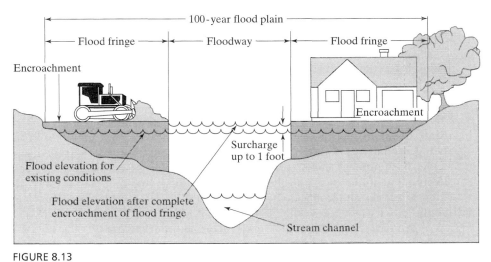

FIGURE 8.13

Procedure for determining the floodway width.

The floodplain mapping effort produced a large amount of data and analyses useful to design hydrologists. The products of the program include:

1. The 10-, 50-, 100-, and 500-year frequency discharge for streams.
2. The 10-, 50-, 100-, and 500-year flood elevations for riverine, coastal, and lacustrine floodplains.
3. The 100- and 500-year mapped floodplain delineations at scales ranging from 1:4,800 to 1:24,000.
4. The 100-year floodway data and mapping.
5. Coastal high-hazard-area mapping (areas subject to significant wave hazards).
6. Floodway flow velocities.
7. Insurance risk zones.

This information is provided in the form of three products:

1. *Flood Insurance Study* (FIS) *Reports* provide general program and community background information; tabulated flood discharge data; tabulated floodway data including velocity, floodway width, and surcharge information; tabulated flood insurance zone data; and profiles of the 10-, 50-, 100-, and 500-year flood elevation versus stream distances for riverine flooding.
2. *Flood Insurance Rate Maps* (FIRMs) provide delineations of the 100- and 500-year floodplains, base flood elevations, coastal high-hazard areas, and insurance risk zones on a planimetric base at a scale between 1:4,800 and 1:24,000.
3. *Flood Boundary Hazard Maps* (FBHMs) provide delineations of the 100- and 500-year floodplains, locations of surveyed floodplain and channel cross sections used in hydraulic analyses, and delineations of the 100-year floodway on a planimetric or topographic base at a scale between 1:4,800 and 1:24,000.

FEMA has a service center that can be accessed on the Internet to determine whether any location in the United States has been mapped and to order copies of available maps. The service center is accessible at http://web1.msc.fema.gov/webapp/commerce/command/ExecMacro/MSC/macros/welcome.d2w/report.

NFIP Map Modernization Program

State-of-the-art technologies for hydrologic, hydraulic, and mapping procedures are being applied by FEMA in modernizing its program for mapping flood hazards nationwide. A significant improvement is the expanded use of the Internet, including the eventual goal of making all the flood hazard maps and mapping-related products and data available online. Products available can be accessed from FEMA's map service center at www.fema.gov/MSC/femahome.htm.

Activities included in the modernization program are:

■ Updating old or inadequate maps.
■ Maintaining existing maps and improving base map standards.

- Verifying the accuracy of maps after floods.
- Transitioning from hard copy to digital maps and from paper reports to CD-ROMs or the Internet.
- Streamlining the map production process.
- Simplifying the process for reviewing letter of map change requests.
- Forming standardized partnerships with state or local agencies for engineering and/or mapping.
- Setting up a customer service telephone line for the mapping program.
- Expanding use of FEMA's Web site for mapping purposes.
- Distributing educational materials.

Flood Warning Systems

In addition to delineating and regulating development in floodplains, many communities increase their emergency preparedness by implementing flood warning systems. Flood warning systems rely on radar and rainfall and streamflow gauges connected by satellite transmitters to feed real-time data to central computers that can be accessed on the Internet. The computers continuously monitor water surface profiles, rain depths, and intensities, and once preset thresholds are reached, the computer combines the real-time data, weather forecasts, and watershed models to predict when and where the water levels will crest [29]. Warnings are automatically sent to various emergency management officials, allowing enactment of emergency operation plans for flooding. Emergency actions can range from barricading roadways that are prone to flooding to evacuating neighborhoods.

Measurements include rain gauges, temperature and humidity sensors, water pressure transducers, and water level sensors that feed data to the computer system. In addition to feeding data to the emergency management computers, the data are also fed in real time to a public Web site. By studying past floods, the agencies can profile under which conditions certain streams overflow and actions can be implemented. An example of one such system installed in the Kansas City, KS, area can be accessed at www.stormwatch.com.

Hydrology for Floodplain Studies

Flood flow frequency estimates for gauged locations in NFIP studies are based on log–Pearson Type III (see Chapter 3) analyses of streamflow records. Annual peak flows and historical data are fitted according to procedures recommended by FEMA.

For ungauged locations, flood flow frequency estimates are developed through regional regression equations, frequency analyses, or rainfall-runoff modeling. Regression equations published by the U.S. Geological Survey relate peak discharges of various frequencies to various drainage basin characteristics such as size, slope, elevation, shape, and land use.

Rainfall-runoff modeling techniques (Chapter 12) use actual or synthetic rainfall hyetographs. Storm-event models, such as the Corps HEC-HMS and SCS TR-20 packages, employ design storms of particular frequencies and then mathematically simulate the physical runoff process. The resulting peak discharge is assumed to have the same frequency as the rainfall.

8.4   SNOWMELT HYDROLOGY

In many regions, snow is the dominant source of water supply. About 90 percent of the yearly water supply in the high elevations of the Colorado Rockies is derived from snowfall [30]. Equally high proportions are also likely in the Sierras of California and numerous regions in the Northwest. A significant but lesser share of the annual water yield in the Northeast and Great Lake states also originates as snow. The annual snowfall distribution in the United States is shown in Fig. 8.14.

It is important that the hydrologist understand the nature and distribution of snowfall and the mechanisms involved in the snowmelt process. Snowmelt usually begins in the spring. The runoff derived is normally out of phase with the periods of greatest water need; therefore, various control schemes such as storage reservoirs have been developed to minimize this problem. Some of the greatest floods result from combined large-scale rainstorms and snowmelt. Streamflow forecasting is highly dependent on adequate knowledge of the extent and characteristics of snow fields within the watershed.

Snowmelt routines have been incorporated in numerous hydrologic models, some of which also include water quality dimensions (Chapter 12). A good accounting of the fundamentals of the snowmelt process and of contemporary snowmelt modeling approaches may be found in Refs. 30–34.

Physical Processes of Snowmelt

Runoff from the snowpack is the last occurrence in a series of events beginning when the snowfall reaches the ground. The time interval from the start to the end of the process might vary from as little as a day or less to several months or more. When milder weather sets in, melting occurs first at the snowpack surface. This initial meltwater moves only slightly below the surface and again freezes through contact with colder underlying snow. During the refreezing process, the heat of fusion released from meltwater raises the snowpack temperature. Heat is also transferred to the snowpack from overlying air and the ground. During persistent warm periods, the temperature of the entire snowpack continually rises and finally reaches 32°F. With continued melting, water begins flowing down through the pack. The initial melt component is retained on snow crystals in capillary films. Once the liquid water-holding capacity of the snow is reached, the snow is said to be *ripe*.

The *water equivalent* is the depth of water that would weigh the same amount as that of the sample. In this way snow can be described in terms of inches of water. *Density* is the percentage of snow volume that would be occupied by its water equivalent. The *quality* of the snow relates to the ice content of the snowpack and is expressed as a decimal fraction. It is the ratio of the weight of the ice content to the total weight. Snow quality is usually about 0.95 except during periods of rapid melt, when it may drop to 0.70–0.80 or less. Newly fallen snow has a density of about 10 percent (the percentage of snow volume its water equivalent would occupy), but as the snow depth

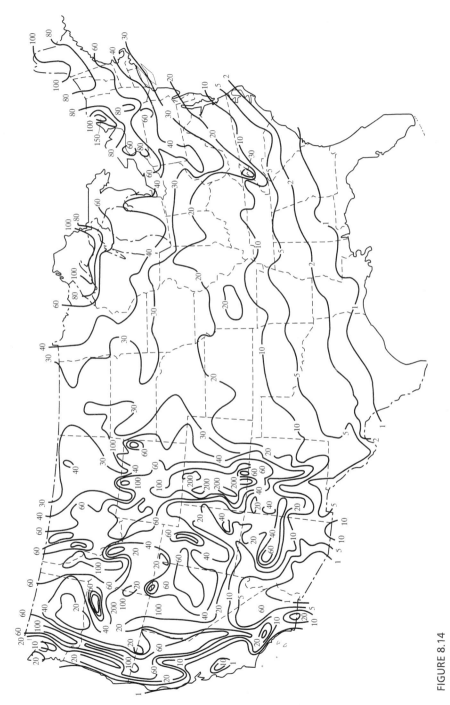

FIGURE 8.14

Mean annual snowfall in the United States in inches, 1899–1938.
(U.S. Weather Bureau)

enlarges, settling and compaction increase the density [35]. Throughout the foregoing process, pack density increases due to the refreezing of meltwater and buildup of capillary films. After the water-holding capacity is reached, the density remains relatively constant with continued melt. Meltwater that exceeds the water-holding capacity will continue to move down through the snowpack until the ground is finally reached. At this point runoff can occur.

The snowmelt process converts ice content into water within the snowpack. Rates differ widely due to variations in causative factors to be discussed later. These divergencies are not as strikingly apparent when considering drainage from the snowpack, however, since the pack itself tends to filter out these nonuniformities so that the drainage exhibits a more consistent rate.

The heat necessary to induce snowmelt is derived from short- and long-wave radiation, condensation of vapor, convection, air and ground conduction, and rainfall. The most important of these sources are convection, vapor condensation, and radiation. Rainfall ranks about fourth in importance while conduction is usually a negligible cause.

Radiation Melt   The net amount of short- and long-wave radiation received by a snowpack can be a very important source of heat energy for snowmelt. Under clear skies, the most significant variables in radiation melt are the *insolation, reflectivity*, or *albedo* of the snow, and air temperature. Humidity effects, while existent, are usually not important. When cloud cover exists, striking changes in the amount of radiation from an open snowfield are in evidence. The general nature of these effects is illustrated in Fig. 8.15 [31]. Combined short- and long-wave radiation exchange as a function of cloud height and cover is represented. Radiation melt is shown to be more significant in the spring than in the winter. It should also be noted that winter radiation melt tends to increase with cloud cover and decreasing cloud height as a result of the more dominant role played by long-wave radiation during that period.

Forest canopies also exhibit important characteristics in regulating radiative heat exchange. These effects differ somewhat from those exhibited by the cloud cover, especially where short-wave radiation is concerned. Clouds and trees both limit insolation, but clouds are very reflective, while a large amount of the intercepted insolation is absorbed by the forest. Consequently, the forest is warmed and part of the incident energy is directly transferred to snow in the form of long-wave radiation; an additional fraction is transferred indirectly by air also heated by the forest.

Solar energy provides an important source of heat for snowmelt. Above the earth's atmosphere, the thermal equivalent of solar radiation normal to the radiation path is 1.97 langleys/min (1 langley is approximately 3.97×10^{-3} Btu/cm^2). The actual amount of radiation reaching the snowpack is modified by many factors such as the degree of cloudiness, topography, and vegetal cover. The importance of vegetal cover in influencing snowmelt, long recognized, has prompted many forest management schemes to regulate snowmelt [36].

Two basic laws are applicable to radiation. Planck's law states that the temperature of a blackbody is related to the spectral distribution of energy that it radiates. Integration of Planck's law for all wavelengths produces Stefan's law:

$$R_a = \sigma T^4 \tag{8.18}$$

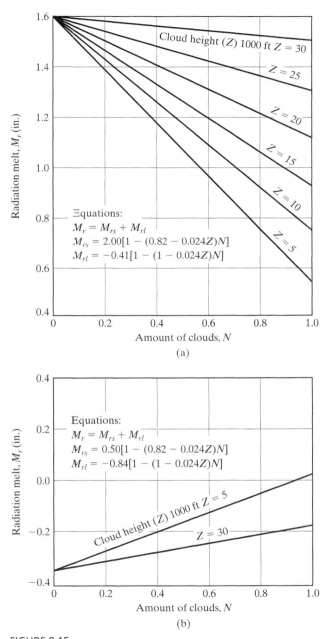

FIGURE 8.15

Daily radiation melt in the open with cloudy skies: (a) during spring, May 20; and (b) during winter, February 15.

(After U.S. Army Corps of Engineers [31])

where R_a = the total radiation
        σ = Stefan's constant [0.813×10^{-10} langley/(min-K^{-4})]
        T = the temperature ($^\circ K$)

Because snow radiates as a blackbody, the amount of radiation is related to its temperature (Planck's law), and total energy radiated is according to Stefan's law. Long-wave radiation by a snowpack is determined in a complex fashion through the interactions of temperature, forest cover, and cloud conditions.

Direct solar short-wave radiation received at the snow surface is not all transferred to sensible heat. Part of the radiation is reflected and thus lost for melt purposes. Short-wave reflection is known as *albedo* and ranges from about 40 percent for melting snow late in the season to approximately 80 percent for newly fallen snow. Values as high as 90 percent have also been reported in several cases [37]. This property of the snowpack to reflect large fractions of the insolation explains why the covers persist and air temperatures remain low during clear, sunny winter periods.

That portion of short-wave radiation not reflected and available for snowmelt may become long-wave radiation or be conducted within the snowpack. Some heat may also be absorbed by the ground with no resultant melt if the ground is frozen.

An expression for hourly short-wave radiation snowmelt is given as [31]:

$$M = \frac{H_m}{203.2Q_t} \tag{8.19}$$

where H_m = the net absorbed radiation (langleys)
        203.2 = a conversion factor for changing langleys to inches of water

When the snow quality is 1, long-wave radiation is exchanged between the snow cover and its surroundings. Snowmelt from net positive long-wave radiation follows Eq. 8.19. If the net long-wave radiation is negative (back radiation), there is an equivalent heat loss from the snowpack.

An approximate method of estimating 12-hr snowmelt D_{12} (periods midnight to noon, noon to midnight) from direct solar radiation has been given by Wilson [38]. The relation is of the form:

$$D_{12} = D_0(1 - 0.75m) \tag{8.20}$$

where D_0 = the snowmelt occurring in a half-day in clear weather
        m = the degree of cloudiness (0 for clear weather, 1.0 for complete overcast)

Suggested values for D_0 are 0.35 in. (March), 0.42 in. (April), 0.48 in. (May), and 0.53 in. (June) within latitudes 40–48° [38].

Condensation   Heat given off by condensing water vapor in a snowpack is often the most important heat source, particularly when temperatures are in the higher ranges (50–60°F). To melt a pound of ice at 32°F, a thermal input of 144 Btu is required. A pound of moisture originating from the condensation process at 32°F produces about 1073 Btu. On this basis, 1 in. of condensate produces approximately 7.5 in. of water from the snow. A total yield of around 8.5 in. of snowmelt including the condensate is thus derived.

A water vapor supply at the snow surface is formed by the turbulent exchange process; consequently, a mass transfer equation similar to those presented for evaporation studies fits the melt process. An equation for hourly snowmelt from condensation takes the form [35]:

$$M = \frac{bV}{Q_t}(e_a - 6.11) \tag{8.21}$$

where b = an empirical constant

   e_a = the vapor pressure of the air (mb)

   6.11 = the saturation vapor pressure (mb) over ice at 32°F (e_a must exceed 6.11)

Also, M, Q_t, and V are as previously defined. The constant b has a value of 0.001 for temperature and wind measurements at 4 and 15 ft, respectively [35].

A similar expression but for 6-hr snowmelt (D) is given as:

$$D = K_1V(e_a - 6.11) \tag{8.22}$$

where the theoretical value of K_1 is said by Light to equal 0.00578 if wind and temperature data are obtained at the 50- and 10-ft levels, respectively, and the snowfield is level and open [39]. Actual figures based on observation are generally lower than this due mainly to forest influences. A value of 0.0032 has been reported by Wilson [38] for three study basins in Wyoming. For condensation melt to occur, the dew-point temperature must exceed 32°F. When it drops below that level, evaporation occurs at the snow surface. An equation for snow evaporation takes the form:

$$E = \frac{kV(e_a - e_s)}{Q_t} \tag{8.23}$$

where E = the hourly evaporation in inches

   e_s = the saturation vapor pressure over the snow

   k = an empirical constant

Also, V, e_a, and Q_t are as defined before [35]. In the expression $k = 0.0001$, temperature and wind measurements are taken as for Eq. 8.22, and the temperature of the air is assumed equal to that of the snow surface for temperatures below 32°F.

Convection   Heat for snowmelt is transferred from the atmosphere to the snowpack by convection. The amount of snowmelt by this process is related to temperature and wind velocity. The following equation can be used to estimate the 6-hr depth of snowmelt in inches by convection [38]:

$$D = KV(T - 32) \tag{8.24}$$

where V = the mean wind velocity (mph)

   T = the air temperature (°F)

On the basis of the theory of air turbulence and heat transfer (turbulent exchange), a theoretical value for the exchange coefficient K of $0.00184 \times 10^{-0.0000156h}$ has been given by Light [39]. In this relation, h, the elevation in feet, is used to reflect the change in barometric pressure due to the difference in altitude. The expression is said to represent conditions for an open, level snowfield where measurements of wind and temperature are made at heights of 50 and 10 ft, respectively, above the snow. Values of the expression $10^{-0.0000156h}$ vary from 1.0 at sea level to 0.70 at 10,000 ft of elevation. The actual values of K are normally less than the theoretical figure due to such factors as forest cover. Empirical 6-hr K values have been reported in the literature [38].

Anderson and Crawford [35] give an expression for the hourly snowmelt due to convection as:

$$M = \frac{cV(T_a - 32)}{Q_t} \tag{8.25}$$

where M = the hourly melt (in.)
   V = the wind velocity (mi/hr)
   T_a = the surface air temperature (°F)
   Q_t = the snow quality
   c = a turbulent exchange coefficient determined empirically

Temperature measurements are at 4 ft, with wind gauged at 15 ft. The corresponding value of c is reported as 0.0002.

Ground Conduction   Major sources of heat energy to the snowpack are radiation, convection, and condensation. Under usual conditions, the reliable determination of hourly or daily melt quantities can be founded on these heat sources plus rainfall if it occurs. An additional source of heat, negligible in daily melt computations but perhaps significant over an entire melt season, is ground conduction.

Ground conduction melt is the result of upward transfer of heat from ground to snowpack due to thermal energy that was stored in the ground during the preceding summer and early fall. This heat source can produce meltwater during winter and early spring periods when snowmelt at the surface does not normally occur. Heat transfer by ground conduction can be expressed by the relation [37]:

$$H_q = K\frac{dT}{dz} \tag{8.26}$$

where K = the thermal conductivity of the soil
   dT/dz = the temperature gradient perpendicular to the soil surface

The snowmelt from ground conduction is generally exceedingly small. Wilson notes that after about 30 days of continuous snow cover, heat transferred from the ground to the snow is insignificant [38]. The amount of snowmelt from ground conduction during a snowmelt season has been estimated at approximately 0.02 in./day [40]. Ground conduction does act to provide moisture to the soil; thus, when other favorable conditions for snowmelt occur, a more rapid development of runoff can be expected.

Rainfall Heat derived from rainfall is generally small, since during those periods when rainfall occurs on a snowpack, the temperature of the rain is probably quite low. Nevertheless, at higher temperatures, rainfall may constitute a significant heat source; it affects the aging process of the snow and frequently is very important in this respect. An equation for hourly snowmelt from rainfall is [35]:

$$M = \frac{P(T_w - 32)}{144Q_t} \tag{8.27}$$

where P = the rainfall (in.)

   T_w = the wet-bulb temperature assumed to be that of the rain (°F)

This equation is based on the relation between heat required to melt ice (144 Btu per pound of ice) and the amount of heat given up by a pound of water when its temperature is decreased by one degree.

Daily snowmelt by rainfall estimates are given by:

$$M_d = 0.007P_d(T_a - 32) \tag{8.28}$$

where M_d = the daily snowmelt (in.)

   P_d = the daily rainfall (in.)

   T_a = the mean daily air temperature (°F) of saturated air taken at the 10-ft level [40]

Figure 8.16 illustrates the process, showing how relative quantities of storage, melt, and runoff adjust over time.

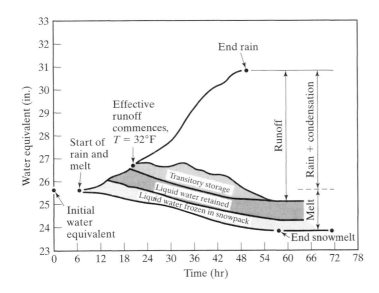

FIGURE 8.16

Water balance in a snowpack during rainfall.

Example 8.5

During a completely cloudy April period of 12 hr, the following averages existed for a ripe snowpack located at 10,000 ft above sea level at a latitude of 44°N: air temperature, 50°F; mean wind velocity, 10 mph; relative humidity, 65%; average rainfall intensity, 0.03 in./hr for 12 hr; wet-bulb psychrometer reading, 48°F. Estimate the snowmelt in inches of water for radiation, condensation, convection, and warm rain for the 12-hr period.

Solution

1. Radiation melt, 12 hr, from Eq. 8.20:

    $$D_{12} = D_0(1 - 0.75m)$$
    $$D_{12} = 0.42 \times [1 - (0.75 \times 1)] = 0.11 \text{ in.}$$

2. Condensation melt, 6 hr, from Eq. 8.22:

    $$D = K_1 V(e_a - 6.11)$$
    $$D = 2 \times 0.00578 \times 10 \times [(12.19 \times 0.65) - 6.11]$$
    $$= 0.21 \text{ in.}$$

3. Convection melt, 6 hr, from Eq. 8.24:

    $$D = KV(T - 32)$$
    $$D = 2 \times 0.7 \times 0.00184 \times 10 \times (50 - 32) = 0.50 \text{ in.}$$

4. Rainfall melt, hourly, from Eq. 8.27:

    $$M = P(T_w - 32)/144Q_t$$
    $$M = [0.03 \times 12 \times (48 - 32)]/(144 \times 0.97) = 0.04 \text{ in.}$$

    Thus, total melt is 0.86 in.

Snowmelt Runoff

Snowmelt runoff estimates are extremely important for many regions of the United States and other countries in (1) forecasting seasonal water yields for a diversity of water supply purposes, (2) regulating rivers and storage works, (3) implementing flood control programs, and (4) selecting design floods for particular watersheds. Maximum floods in many areas are often due to a combination of rainfall and snowmelt runoff. In effect, the determination of snowmelt runoff has the same utility as the calculation of runoff from rainfall. In some areas, snowmelt runoff will, in fact, be the more important of the two.

Various approaches to runoff determination from snowmelt have been developed [2]. They range from relatively simple correlation analyses that completely ignore the physical snowmelt process to relatively sophisticated methods using physical process equations. Most techniques can be considered as based on degree-day correlations, analyses of recession curves, correlation analyses, physical equations, or various indexes. Some of the most widely used methods are described here.

Snowpack Analysis   The manner in which runoff from either rainfall or snowmelt is affected by conditions prevalent within the snowpack is of primary interest to a hydrologist. Various views on storage characteristics of a snowpack have been advanced. These range from the concept that a snowpack can retain large amounts of liquid water to the hypothesis that snowpack storage is negligible. There is no universally applicable relation, and it becomes important to base any runoff considerations on a knowledge of the character of a snowpack at the time of study.

Degree-Day Correlation Method   The atmospheric temperature is an extremely useful parameter in snowmelt determination. It reflects the extent of radiation and the vapor pressure of the air; it is also sensitive to air motion. Frequently, it is the only adequate meteorologic variable regularly on hand, so widespread use has been made of degree-day relations in snowmelt computations.

A *degree day* is defined as a deviation of 1° from a given datum temperature consistently over a 24-hr period. In snowmelt computations, the reference temperature is usually 32°F. If the mean daily temperature is 43°F, for example, this is equivalent to 11 degree days above 32°F. If the temperature does not drop below freezing during the 24-hr period, there will be 24 degree hours for each degree departure above 32°F. In this example there would be 264 degree hours for the day of observation.

Various ways of estimating the mean temperature have enabled investigators to take several approaches. One method is simply to average the maximum and minimum daily temperatures. Bases other than 32°F are also used. Regardless of the particular method employed, a degree hour or degree day is an index to the amount of heat present for snowmelt or other purposes and has proved useful in point-snowmelt and runoff from snowmelt determinations.

The standard practice in developing snowmelt relations on the basis of temperature is to correlate degree days or degree hours with the snowmelt or basin runoff. In some cases, other factors are introduced to define forest cover effects and/or other influences. Another approach often used is to calculate a degree-day factor—the ratio of runoff or snowmelt to accumulated degree days that produced the runoff or melt. Single representative values should be used with caution. Point-degree-day factors for snow-covered basins range from 0.015 to 0.20 in. per degree per day when melting occurs.

Generalized Basin Snowmelt Equations   Extensive studies by the U.S. Army Corps of Engineers at various laboratories in the West have produced several general equations for snowmelt during (1) rain-free periods and (2) periods of rain [41]. When rain is falling, heat transfer by convection and condensation is of prime importance. Solar radiation is slight, and long-wave radiation can readily be determined from theoretical considerations. When rain-free periods prevail, both solar and terrestrial radiation become significant and may require direct evaluation. Convection and condensation are usually less critical during rainless intervals. The equations for these two cases are [31]:

1. Equations for periods with rainfall.

   a. For open (cover below 10 percent) or partly forested (cover from 10 to 60 percent) watersheds:

   $$M = (0.029 + 0.0084kv + 0.007P_r)(T_a - 32) + 0.09 \qquad (8.29)$$

   b. For heavily forested areas (over 80 percent cover):

   $$M = (0.074 + 0.007P_r)(T_a - 32) + 0.05 \qquad (8.30)$$

   where M = the daily snowmelt (in./day)
   P_r = the rainfall intensity (in./day)
   T_a = the temperature of saturated air at the 10-ft level (°F)
   v = the average wind velocity at the 50-ft level (mph)
   k = the basin constant, which includes forest and topographic effects, and represents average exposure of the area to wind. Values of k decrease from about 1.0 for clear plains areas to about 0.2 for dense forests.

2. Equations for rain-free periods.

   a. For heavy forested areas:

   $$M = 0.074(0.53T'_a + 0.47T'_d) \qquad (8.31)$$

   b. For forested areas (cover of 60–80 percent):

   $$M = k(0.0084v)(0.22T'_a + 0.78T'_d) + 0.029T'_a \qquad (8.32)$$

   c. For partly forested areas:

   $$M = k'(1 - F)(0.0040I_i)(1 - a) + k(0.0084v)(0.22T'_a + 0.78T'_d) + F(0.029\,T'_a) \qquad (8.33)$$

   d. For open areas:

   $$M = k'(0.00508I_i)(1 - a) + (1 - N)(0.0212T'_a - 0.84) + N(0.029T'_c) + k(0.0084v)(0.22T'_a + 0.78T'_d) \qquad (8.34)$$

   where M, v, k = as previously described
   T'_a = the difference between the 10-ft air and the snow surface temperatures (°F)
   T'_d = the difference between the 10-ft dew-point and snow-surface temperatures (°F)
   I_i = the observed or estimated insolation (langleys)
   a = the observed or estimated mean snow surface albedo
   k' = the basin short-wave radiation melt factor (varies from 0.9 to 1.1), which is related to mean exposure of open areas compared to an unshielded horizontal surface

F = the mean basin forest-canopy cover (decimal fraction)

T'_c = the difference between the cloud-base and snow-surface temperatures (°F)

N = the estimated cloud cover (decimal fraction)

Example 8.6

    a. Use Eq. 8.29 to estimate the snowmelt at an elevation of 3000 ft in a partly forested area if the rainfall intensity is 0.3 in./day, the wind velocity is 20 mph, and the temperature of the saturated air is 42°F.

    b. Rework your solution for a dense forest cover and a saturated air temperature of 53°F.

Solution

    1. $M = (0.029 + 0.0084kv + 0.007P_r)(T_a - 32) + 0.09$

          $M = [0.029 + (0.0084 \times 0.5 \times 20) + (0.007 \times 0.3)](42 - 32) + 0.09$

          $M = 1.24$ in./day

    2. $M = (0.074 + 0.007P_r)(T_a - 32) + 0.05$

          $M = [0.074 + (0.007 \times 0.3)](53 - 32) + 0.05$

          $M = 1.65$ in./day

SUMMARY

Runoff is probably the most complex, yet most important, hydrologic process needing to be understood by the hydrological scientist or engineer. It has attracted by far the most attention and focus by these groups, and occupies the greatest percentage of most hydrology textbooks and publications.

Runoff and streamflow are the result of storm-period precipitation, snowmelt, and groundwater discharge, and they are a primary source of water for a multitude of in-stream and out-of-stream uses. Tenets of the surface water runoff process presented in this chapter provide the foundation for the study, preservation, and management of this natural resource.

In other regions of the United States and other nations, water derived from snowmelt is the major source of supply. An understanding of snowmelt processes is an important adjunct to the management of watersheds in these cold regions. Water supply engineers using methods described here can understand and quantify snow accumulation, snowmelt, and the translation of snowmelt runoff into streamflow.

PROBLEMS

SECTION 8.1: STREAMFLOW

8.1 Consider that you have obtained a gauge height reading of 4 ft at a gauging site on the Raquette River (Fig. 8.2). What would you estimate the discharge to be in cfs and in m³/sec? If the gauge height had been 9 ft, what would the discharge be? Which of the two estimates do you think would be the most reliable? Why?

8.2 Rework problem 8.1 if the gauge height readings were 5 ft and 7 ft.

8.3 For the major surface water course in your locality, discuss the value of making streamflow forecasts.

8.4 Calculate the discharge at the section given in Fig. 8.1 if the depth measurements at the verticals were 0, 3.8, 5.4, 7.7, 8.1, 7.0, 4.5, and 0 ft, respectively. Give results in cfs and m³/s.

8.5 Calculate the discharge at the section given in Fig. 8.1 if the velocities were 0, 2.3, 2.6, 3.1, 2.9, 2.7, 2.5, and 0 fps, respectively, and the depths of Problem 8.4 apply. Give results in cfs and m³/s.

SECTION 8.2: RUNOFF

8.6 Using any dictionary, plus indexes or glossaries from one or two other hydrology texts, find and compare definitions of the following terms: *runoff, direct runoff, direct surface runoff, surface runoff, surface water, overland flow, streamflow, drainage, watershed, catchment, drainage basin, subbasin, drainage divide.*

8.7 For a drainage basin of your choice, plot the annual precipitation in inches versus the runoff in inches. Does the relation appear to be strong? Under what conditions and for what purposes might you use this?

8.8 Determine the drainage density of the basin shown. Area = 6400 acres. Lengths are in miles.

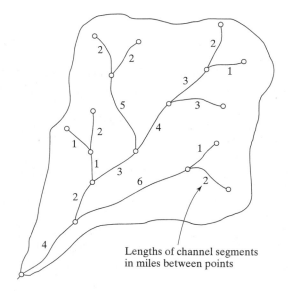

Lengths of channel segments
in miles between points

SECTION 8.3: FLOODS AND DROUGHTS

8.9 Use Fig. 8.7 to determine the frequency of a 0.8-inch drought. What would be the risk of this drought occurring at least once in a 100-yr period? What is the probability of three consecutive years of this drought?

8.10   You are asked to determine the magnitude of the 50-year flood for a small, rural drainage basin (near your town) that has no streamflow records. State the names of at least two techniques that would provide estimates of the desired value.

8.11   A timber railroad bridge in Nebraska at Milepost 271.32 on the railroad system shown in the sketch is to be replaced with a new concrete structure. The 50- and 100-year flood magnitudes are needed to establish the low chord and embankment elevations, respectively. Determine the design flow rates using the SCS TP-149 method. The bridge drains the zone marked, about 45 acres. The moderately sloped basin lies in a Type-II storm region, the curve number is 70, and the 24-hr, 50- and 100-year rainfall depths are 8.6 and 9.4 in., respectively.

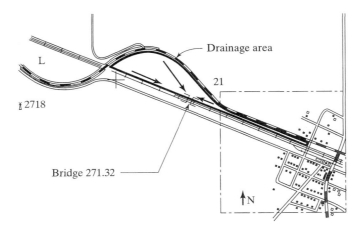

8.12   Repeat Problem 8.11 using the FHWA HDS-2 peak flow SCS design method. The time of concentration is 0.2 hr. Values of I_a can be determined from the relationships in Fig. 7.14. Provide the answers in both metric and English units.

8.13   Use the following data to determine a best-fit relationship in the form of Eq. 8.7 between the drainage area and 50-yr flood for the given region.

Peak Flood Frequency Discharges (cfs) for Stations in the Rappahannock River Basin

Station	Drainage area (mi)	Type of series	2.33 (mean)	Return period in years			
				5	10	25	50
Rappahannock River near	192	Annual	4,150	8,350	9,000	14,000	19,250
Warrenton, VA		Partial	4,600	8,650	9,200	14,000	19,250
Rush River at Washington,	15.2	Annual	530	860	1,290	2,100	3,000
VA		Partial	610	900	1,310	2,100	3,000
Thornton River near Laurel	142	Annual	5,900	11,500	19,900	34,000	48,000
Mills, VA		Partial	7,200	12,500	20,500	34,000	48,000
Hazel River at	286	Annual	7,300	11,800	17,200	25,000	41,000
Rixeyville, VA		Partial	8,300	12,400	18,000	25,500	41,000

(Continued)

Station	Drainage area (mi)	Type of series	2.33 (mean)	Return period in years			
				5	10	25	50
Rappahannock River at Remington, VA	616	Annual	11,000	14,500	18,100	24,500	31,000
		Partial	12,000	15,200	18,900	25,000	31,000
Rappahannock River at Kellys Ford, VA	641	Annual	12,300	19,000	26,800	42,000	57,500
		Partial	14,000	20,000	27,500	42,000	57,500
Mountain Run near Culpeper, VA	14.7	Annual	750	1,750	3,350	6,000	10,000
		Partial	950	1,900	3,550	6,000	10,000
Rapidan River near Ruckersville, VA	111	Annual	3,950	7,100	11,600	21,000	34,000
		Partial	4,700	7,700	12,000	21,000	34,000
Robinson River near Locust Dale, VA	180	Annual	4,600	7,000	9,800	15,400	21,500
		Partial	5,150	7,300	10,100	15,800	21,500
Rapidan River near Culpeper, VA	456	Annual	9,100	16,400	26,900	50,000	78,000
		Partial	10,800	17,600	27,600	50,000	78,000
Rappahannock River near Fredericksburg, VA	1,599	Annual	26,000	39,900	55,000	85,000	117,000
		Partial	29,300	42,000	57,500	85,000	117,000

8.14 From the annual series information given in Problem 8.13, find the relation between the mean annual flow and drainage area. (Note that the functional expression should be of the form $Q_{2.33} = rA^s$.)

8.15 Use the Cyprus Creek method to determine the 25-yr peak discharge for the watershed described in Example 11.3. Assume that the watershed is nearly flat.

8.16 A timber railroad bridge in Hydrologic Region 2 of Texas is to be replaced with a new concrete structure. The 50- and 100-year flood magnitudes are needed to establish the low chord and embankment elevations, respectively. Determine the design flow rates using the USGS regression equations. The drainage area is 0.43 mi^2, and the streambed slope is 62 ft per mi.

8.17 The results of a multiple regression analysis of over 200 flood records in Virginia led to the following regional flood frequency equations:

$$Q_{1.2\text{-yr}} = 9.13A^{.909}S^{.293}$$
$$Q_{2.33\text{-yr}} = 20.8A^{.861}S^{.309}$$
$$Q_{5\text{-yr}} = 38.1A^{.830}S^{.300}$$
$$Q_{10\text{-yr}} = 63.0A^{.802}S^{.283}$$
$$Q_{25\text{-yr}} = 104A^{.779}S^{.266}$$
$$Q_{50\text{-yr}} = 118A^{.795}S^{.279}$$

where the flood discharge for the given frequency is in cfs, A is the drainage area in mi^2, and S is the channel slope in ft/mi (measured between the points that are 10 and 85% of the total river miles upstream of the gauging station to the drainage divide). Devise a method for graphically portraying these regional flood frequency relations. (Note that there are four factors: Q, T, A, and S.)

8.18 Using the regression equations in Problem 8.17, find the predicted floods for the North Fork, Shenandoah River, at Cootes Store. Drainage area = 215 mi² and channel slope = 44.3 ft/mi.

8.19 Compare the predictions from the regression equations in Problem 8.17 with the values estimated by the frequency analysis in Example 3.11. Drainage area = 487 mi² and channel slope = 21.1 ft/mi.

SECTION 8.4: SNOWMELT HYDROLOGY

8.20 Are the snowmelt effects of condensation, convection, radiation, warm rain, and conduction additive? Answer by analyzing the conditions that produce large amounts of snowmelt by each process and examine the conditions to determine if the effects are additive.

8.21 Given a snowpack with a thermal quality of 0.87, determine the snowmelt in inches if the total input is 135 langleys.

8.22 During a partly cloudy April period of 12 hr, the following averages existed for a ripe snowpack located at 10,000 ft above sea level at a latitude of 44°N: air temperature, 50°F; mean wind velocity, 8 mph; relative humidity, 65%; average rainfall intensity, 0.04 in./hr for 12 hr; wet-bulb psychrometer reading, 48°F. Estimate the snowmelt in inches of water for convection, condensation, radiation, and warm rain for the 12-hr period.

8.23 A core sample of a snowpack produces the following information: air temperature, 68°F; relative humidity, 25%; snowpack density, 0.2; snowpack depth, 8 ft; snowpack temperature, 22°F. (a) What is the vapor pressure of the air? (b) Will condensation on the snowpack occur, based on the vapor pressure? (c) What is the cold content of 1 ft² of surface area of the snowpack? (d) Is the snowpack ripe?

8.24 (a) Use Eq. 8.29 to estimate the snowmelt at an elevation of 3,000 ft in a partly forested area if the rainfall intensity is 0.2 in./day, the wind velocity is 15 mph, and the temperature of the saturated air is 44°F. (b) Rework your solution for a dense forest cover and a saturated air temperature of 57°F.

REFERENCES

[1] D. R. Maidment (ed.), *Handbook of Hydrology,* New York: McGraw-Hill, 1993.

[2] W. Viessman and G. L. Lewis, *Introduction to Hydrology, 4th ed.,* New York: HarperCollins College, 1996.

[3] G. A. Burton, *Stormwater Effects Handbook: A Toolbox for Watershed Managers, Scientists, and Engineers,* Boca Raton, FL: CRC Press, 2001.

[4] I. Watson and A. Burnett, *Hydrology—An Environmental Approach,* Boca Raton, FL: Lewis, 1995.

[5] A. D. Ward and W. J. Elliot, *Environmental Hydrology,* Boca Raton, FL: Lewis, 1995.

[6] R. S. Gupta, *Hydrology and Hydraulic Systems,* Prospect Heights, IL: Waveland Press, 1995.

[7] K. C. Patra, *Hydrology and Water Resources Engineering,* Boca Raton, FL: CRC Press, 2000.

[8] S. L. Dingman, *Physical Hydrology,* Englewood Cliffs, NJ: Prentice Hall, 1994.

[9] R. E. Horton, "Erosional Development of Streams and Their Drainage Basins: Hydrophysical Approach to Qualitative Morphology," *Bull. Geol. Soc. Am.* **56**(1945).

[10] R. E. Horton, "Drainage Basin Characteristics," *Trans. Am. Geophys. Union* **13**(1932).

[11] C. T. Haan, B. J. Barfield, and J. C. Hayes, *Design Hydrology and Sedimentology for Small Catchments,* San Diego: Academic Press, 1994.

[12] J. T. Hack, "Studies of Longitudinal Stream Profiles of Small Watersheds," Tech. Rept. 18, Columbia University, Department of Geology, New York, 1959.

[13] A. N. Strahler, "Geology—Part II," in *Handbook of Applied Hydrology,* New York: McGraw-Hill, 1964.

[14] D. M. Gray (ed.), *Handbook on the Principles of Hydrology,* New York: Water Information Center, 1973.

[15] W. C. Boughton, "Systematic Procedure for Evaluating Partial Areas of Watershed Runoff," Proc. American Society of Civil Engineers, *J. Irrigation Drainage Eng.,* Vol. 116, No. 1, February 1990.

[16] A. B. Taylor and H. E. Schwartz, "Unit Hydrograph Lag and Peak Flow Related to Basin Characteristics," *Trans. Am. Geophys. Union* **33**(1952).

[17] R. E. Horton, "Discussion of Paper, Flood Flow Characteristics by C. S. Jarvis," *Trans. ASCE* **89**(1926).

[18] William W. Whipple, Jr., "Regional Drought Frequency Analysis," *Proc. ASCE J. Irrigation Drainage Div.* **92**(IR2), 11–31 (June 1966).

[19] U.S. Soil Conservation Service, "A Method for Estimating Volume and Rate of Runoff in Small Watersheds," Tech. Paper 149, rev., Washington, D.C. 1973.

[20] R. H. McCuen, P. A. Johnson, and R. M. Ragan, *Hydrology, HDS-2,* U.S. Federal Highway Administration, 2nd ed., September 1996.

[21] C. S. Jarvis, "Floods," in *Hydrology* (O. E. Meinzer, ed.), Chap. 11-G, New York: McGraw-Hill, 1942.

[22] Soil Conservation Service, "Guide to Determine Instantaneous Peak Flow for Flatland Areas," Texas SCS Engineering Technical Note No. 210-18-TX8 Hydrology, U.S. Department of Agriculture, February 1985.

[23] Stephens and Mills, "Using the Cypress Creek Formula to Estimate Runoff Rates in the Southern Coastal Plain and Adjacent Flatwoods Land Resource Areas," U.S. Agricultural Research Service Report ARS 41-95, U.S. Department of Agriculture, 1970.

[24] V. R. Baker, "Paleoflood Hydrology and Extraordinary Flood Events," *J. Hydrology,* Vol. 96, 1987.

[25] R. M. Jarrett, "Paleohydrologic Techniques Used to Define the Spatial Occurrence of Floods, in *Geomorphology, 3rd ed.,* Amsterdam: Elsevier Science, 1990.

[26] V. R. Baker, "Magnitude and Frequency of Paleofloods," in *Floods: Hydrological, Sedimentological and Geomorphological Implications,* New York: Wiley, 1989.

[27] M. E. Jennings, W. O. Thomas, and H. C. Riggs, "Nationwide Summary of U.S. Geological Survey's Regional Regression Equations for Estimating Magnitude and Frequency of Floods at Ungaged Sites," USGS WRI 93-1, Reston, VA, 1993.

[28] U.S. Geological Survey, "Technique for Estimating the Magnitude and Frequency of Floods in Texas," WRI Report 77-110, 1977.

[29] J. Peters and D. Easton, "Runoff Simulation Using Radar Rainfall Data," *Water Res. Bull.,* Vol. 32, No. 4, 1996.

[30] B. C. Goodell, "Snowpack Management for Optimum Water Benefits," Conference Preprint 379, ASCE Water Resources Engineering Conference, Denver, CO, May 1966.

[31] Corps of Engineers, "Snow Hydrology," NTIS PB 151 660, North Pacific Division, Corps of Engineers, Portland, OR, June 1956.

[32] Corps of Engineers, "Runoff Evaluation and Streamflow Simulation by Computer," Tech. Rep., North Pacific Division, Corps of Engineers, Portland, OR, 1971.

[33] E. A. Anderson, "National Weather Service River Forecast System—Snow Accumulation and Ablation Model," NOAA Tech. Memo NWS HYDRO-17, U.S. Department of Commerce, Washington, D.C., 1973.

[34] E. A. Anderson, "A Point Energy and Mass Balance Model of a Snow Cover," NOAA Tech. Rep. NWS 19. U.S. Department of Commerce, Washington, D.C., Feb. 1976.

[35] E. A. Anderson and N. H. Crawford, "The Synthesis of Continuous Snowmelt Runoff Hydrographs on a Digital Computer," Department of Civil Engineering, Stanford University, Stanford, CA, Tech. Rep. No. 36, June 1964.

[36] J. Kittredge, *Forest Influences,* New York: McGraw-Hill, 1948.

[37] H. T. Mantis et al., "Review of the Properties of Snow and Ice," U.S. Army Corps of Engineers, Snow, Ice, and Permafrost Research Establishment, SIPRE Rept. 4, July 1951.

[38] W. T. Wilson, "An Outline of the Thermodynamics of Snowmelt," *Trans. Am. Geophys. Union* 22, Part 1 (1941).

[39] P. Light, "Analysis of High Rates of Snow Melting," *Trans. Am. Geophys. Union* 22, Part 1, pp. 195–235(1941).

[40] U.S. Army Corps of Engineers, "Runoff from Snowmelt," Engineering and Design Manuals, EM 1110-2-1406, Jan. 1960.

[41] S. S. Butler, *Engineering Hydrology,* Englewood Cliffs, NJ: Prentice-Hall, 1957.

Hydrographs

OBJECTIVES

The purpose of this chapter is to:

- Introduce the components of hydrographs
- Define the meaning and use of hydrographs
- Describe the time relationships most commonly used in hydrograph analysis
- Define unit hydrographs and show their utility in hydrologic studies and design
- Teach methods of obtaining, analyzing, and synthesizing unit hydrographs
- Describe methods for converting unit hydrographs for one storm duration to other storm durations
- Present techniques for routing hydrographs through river reaches and reservoirs
- Provide the reader with references to other standard sources that expand discussions of hydrograph analysis beyond the scope of this book.

A *streamflow hydrograph* provides the rate of flow at all points in time during and after a storm or snowmelt event [1]. Hydrologists depend on measured or computed (synthesized) hydrographs to provide peak flow rates so that hydraulic structures can be designed to accommodate the flow safely. Because a hydrograph is a plot of rates against time, the area beneath a hydrograph between any two points in time gives the total volume of water passing the point of interest during the time interval. Thus, in addition to peak flows, hydrographs allow analysis of sizes of reservoirs, storage tanks, detention ponds, and other facilities that deal with volumes of runoff.

Hydrograph analysis is the most widely used method of analyzing surface runoff. The rate of discharge from a watershed varies with time, resulting in a hydrograph. Both the peak flow rate and the total runoff volume, defined as the area under the hydrograph, are of interest in watershed studies and design of water supply and flood control facilities. Any hydrograph is the sum of its component parts. For this reason, hydrographs are described in the first part of this chapter by describing their component parts. Next, methods of synthesizing hydrographs from rainfall, called *unit hydrograph methods,* are described in order to demonstrate that methods of synthesizing hydrographs are able to replicate physical processes. The chapter concludes by presenting standard methods, called *routing methods*, of translating the shapes of hydrographs from one point in a stream to a downstream location.

9.1   HYDROGRAPH COMPONENTS

A hydrograph has four component elements: (1) direct surface runoff, (2) interflow, (3) groundwater or base flow, and (4) channel precipitation [2]. The rising portion of a hydrograph is known as the *concentration curve;* the region in the vicinity of the peak is called the *crest segment;* and the falling portion is the *recession* [3]. The shape of a hydrograph depends on precipitation pattern characteristics and basin properties. Figure 9.1 illustrates the definitions presented.

Hydrograph Shape

The first hydrograph in Fig. 9.1 is the result of rain that fell just before or during the rising limb. Following the runoff event (shown as the end of direct runoff), a period of dry-weather flow occurs. The nonzero flow rate reflects discharge from groundwater and release of stored water from the stream channel banks. The shift in slope of the graph indicates that no overland runoff into the channel occurs during this interval. For many watersheds, groundwater discharge normally occurs continuously, making up part of the rising, crest, and falling (also called recession) limbs of the hydrograph. The groundwater discharge forms its own hydrograph, which is termed *base flow*.

The *total runoff hydrograph* (TRH) is thus defined as the sum of base flow plus direct runoff. Direct runoff is the collection of all other contributions to runoff except groundwater. If base flow discharge could be separated from the TRH, a direct runoff hydrograph (DRH) would result. Methods of separating base flow are presented later in this chapter.

The two rainstorms in Fig. 9.1 had short durations. If the rainfall maintains a constant intensity for a long enough period of time, a state of equilibrium discharge is

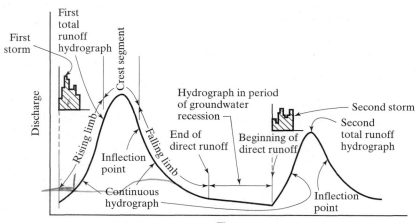

FIGURE 9.1

Hydrograph definition.

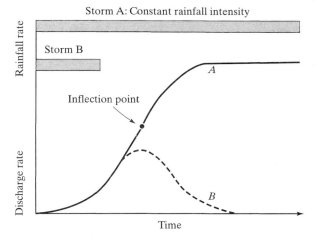

FIGURE 9.2

Equilibrium discharge hydrograph for an infinite storm duration (shown as curve *A*) and DRH for a finite storm duration (curve *B*).

reached, as depicted by curve *A* in Fig. 9.2. The inflection point on curve *A* often indicates the time at which storage in the watershed begins to fill. As rainfall continues, maximum storage capacity is attained and equilibrium [inflow (rainfall) equals outflow (runoff)] is reached. The condition of maximum storage and equilibrium is seldom if ever attained in nature. Extended rainfall may occur, but variations in intensity throughout its duration negate any possibility of a DRH of the theoretical shape for constant rainfall intensity.

A normal single-peak DRH generally possesses the shape shown by curve *B* in Fig. 9.2. The time to peak magnitude of this hydrograph depends on the intensity and duration of the rainfall, and the size, slope, shape, and storage capacity of the watershed. Once peak flow has been reached for a given isolated rainstorm, the DRH begins to descend, its source of supply coming largely from water accumulated within the watershed such as detention and channel storage.

This timed accumulation and release of runoff from depressions, detentions, and channels is illustrated in Fig. 9.3. In this theoretical case, the storm duration ends just at the point of equilibrium. At that time, all storage is full and the rate of infiltration decreases to its ultimate, constant value (Chapter 7).

Because rain stops at time *t*, the runoff rate drops rapidly to the rate matching the release of any temporarily stored water. This residual is known as *detention storage release* because the water was only "detained" and is not permanently held in storage or lost to infiltration or evaporation. The volume of detention storage and its rate of release from storage are reflected in this portion of the hydrograph.

Figures 9.4a–d illustrate how hydrograph shape can be modified by areal variations in rainfall and rainfall intensity and by watershed configuration [4]. Minor fluctuations shown in these hydrographs are linked to variations in storm intensity. In Fig. 9.4a only the delaying effects pertinent to a storm over the upstream section of the area are indicated. Figure 9.4b shows the reverse of this condition. Figures 9.4c and d depict the comparative effects of basin geometry.

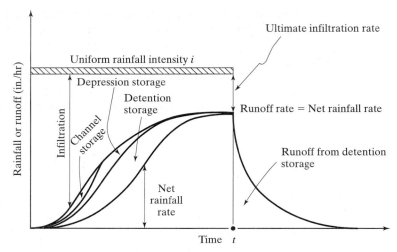

FIGURE 9.3

Distribution of a uniform storm rainfall. All water stored in depressions is ultimately evaporated or infiltrated.

In most hydrograph analyses, interflow and channel precipitation are grouped with surface runoff rather than treated independently. Channel precipitation begins with inception of rainfall and ends with the storm. Its distribution with respect to time is highly correlated with the storm pattern. The relative volume contribution tends to increase somewhat as the storm proceeds, since stream levels rise and the water surface area tends to increase. The fraction of watershed area occupied by streams and lakes is generally small, usually on the order of 5 percent or less, so the percentage of runoff related to channel precipitation is usually minor during important storms.

Base Flow Recession

Several techniques are used to separate a total runoff hydrograph's direct surface runoff and base flow components. Most are based on analysis of groundwater *recession*

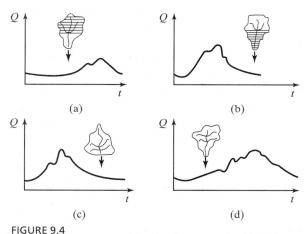

FIGURE 9.4

Effects of storm and basin characteristics on hydrograph shape.

or *depletion* curves. A groundwater recession is characterized by a gradually decreasing rate of base flow. The recession curve shape, shown by the segment connecting the two runoff hydrographs in Fig. 9.1, has been found to approximate an exponential function.

If there is no added inflow to the groundwater reservoir, and if all groundwater discharge from the upstream area is intercepted at the stream-gauging point of interest, then the groundwater discharge recession can be described by either [9],[10]:

$$q_t = q_0\, K^t \tag{9.1}$$

$$\text{or} \quad q_t = q_0 e^{-Kt} \tag{9.2}$$

where q_0 = a specified initial discharge
q_t = the discharge at any time t after flow q_0
K = a recession constant
e = base of natural logarithms

Time units frequently used are days for large watersheds and hours or minutes for small basins. A plot of either yields a straight line on semilogarithmic paper by plotting t on the linear scale.

For most watersheds, groundwater depletion characteristics are approximately stable, since they closely fit watershed geology. Nevertheless, the recession constant varies with seasonal effects such as evaporation and freezing cycles and other factors. Because $q_t\, dt$ is equivalent to $-dS$, where S is the quantity of water obtained from storage, integration of Eq. 9.1 produces:

$$S = \frac{q_t - q_0}{\log_e K} \tag{9.3}$$

This equation determines the quantity of water released from groundwater storage between the times of occurrence of the two discharges of interest, or it can be used to calculate the volume of water still in storage at a time some chosen value of flow occurs. To get the latter, q_t is set equal to zero and q_0 becomes the reference discharge. Figure 9.5 is a plot of Eqs. 9.1 and 9.3.

Groundwater depletion curves can be analyzed by various graphical methods to evaluate the recession constant K. Data from a stream-gauging station are a prerequisite and should reflect rainless periods with no upstream regulation, such as a reservoir, to affect flow at the gauging point.

From the streamflow data, plot a portion of the recession hydrograph to find values of discharge at the beginning and end of selected time intervals. Flows at the beginning of each interval are analogous to q_0, whereas those at the end are analogous to q_1. Next, select several time intervals and plot corresponding q_0's versus q_1's as shown in Fig. 9.6. The time period between consecutive values of q should be identical for each data set. Figures taken from recession curves of times that still reflect surface runoff will usually fall below and to the right of a 45° line drawn on the plot. These values will

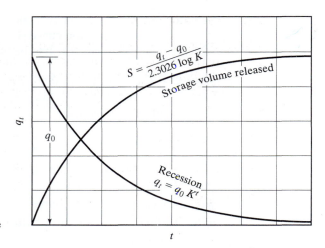

FIGURE 9.5

Base flow recession and cumulative storage release curves.

also be associated with larger numbers for q. Points taken from true groundwater recession periods should approximately describe a straight line. Because $q_1 = q_0 = 0$ when $q_0 = 0$, a straight line can be fitted graphically to the data points. The slope of this line is $q_1/q_0 = K$. Using this value, the depletion curve plots as a straight line on semilogarithmic paper (t is the linear scale variable) or as a curve on arithmetic paper, Fig. 9.5.

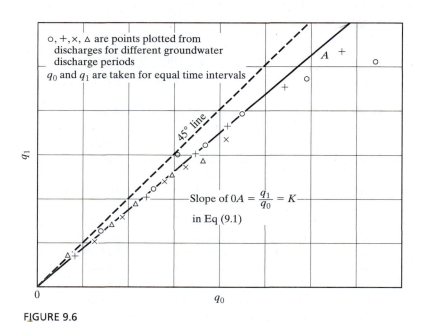

FIGURE 9.6

Graphical method for determining base flow recession constant K.
(U.S. Department of Agriculture, Soil Conservation Service)

Base Flow Separation

Several methods for base flow separation are used when the actual amount of base flow is unknown. During large storms, the maximum rate of discharge is only slightly affected by base flow, and inaccuracies in separation may not be important.

The simplest base flow separation technique is to draw a horizontal line from the point at which surface runoff begins, point A in Fig. 9.7, to an intersection with the hydrograph recession where the base flow rate is the same as at the beginning of direct runoff as indicated by point B. A second method projects the initial recession curve downward from A to C, which lies directly below the peak rate of flow. Then point D on the hydrograph, representing N days after the peak, is connected to point C by a straight line defining the groundwater component. One estimate of N is based on the formula [3]:

$$N = A^{0.2} \tag{9.4}$$

where N = the time in days
        A = the drainage area in square miles

A third procedure is to develop a base flow recession curve using Eq. 9.1 or 9.2 for data from the segment FG, and then back-calculate all base flow to the left of point F, where the computed curve begins to deviate from the actual hydrograph, marking the end of direct runoff. The curve is projected backward arbitrarily to some point E below the inflection point and its shape from C to E is arbitrarily assigned. A fourth widely used method is to draw a line between A and F, and a fifth common method is to project the line AC along the slope to the left of A, and then connect points C and B. All these methods are approximate since the separation of hydrographs is partly a subjective procedure. The U.S. Geological Survey maintains updates of software for performing base flow separation computations. A download of the program HYSEP is available at http://water.usgs.gov/software/surface_water.html.

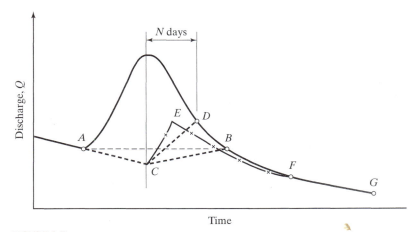

FIGURE 9.7

Illustration of hydrograph separation techniques.

9.2 HYDROGRAPH TIME RELATIONSHIPS

Another important consideration in hydrograph analysis is the relationship among the timing of rain, the time the peak flow occurs, and the time required for direct runoff to end. These and other time relationships are depicted in Fig. 9.8.

Travel time is defined as the time required for direct runoff originating at some point in the channel to reach the outlet. The last drop of direct runoff to pass the outlet conceptually travels over the water surface at the speed of a small surface wave, rather than at a speed equal to the average velocity of flow. The *wave travel time*, defined by this surface wave, is faster than the average flow velocity and varies with channel shape and other factors. For a rectangular channel, the ratio is approximately 5/3 (see Section 9.5 for other wave velocities).

Because of its importance in unit hydrograph theory, the *excess-rainfall release time* is introduced. This is defined as the time required for the last, most remote drop of excess rain that fell on the watershed to pass the outlet, signalling the cessation of direct runoff. It is easily determined as the time interval between the end of rain and the end of direct runoff. Only that part of the outflow which classifies as direct runoff (excess rain) is considered in determining the release time. Watershed outflow normally continues after cessation of direct runoff, in the form of interflow and base flow.

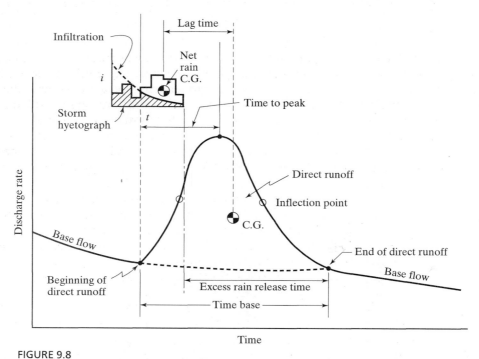

FIGURE 9.8

Hydrograph time relationships.

The *time base* (Fig. 9.8) of a hydrograph is considered to be the time from which the direct runoff begins until the direct-runoff component reaches zero. An equation for time base may take the form:

$$T_b = t_s + t_r \qquad (9.5)$$

where T_b = the time base of the direct runoff hydrograph
   t_s = the duration of runoff-producing rain
   t_r = the excess rainfall release time

Watershed *lag time,* illustrated in Fig. 9.8, is defined as the time from the center of mass of effective rainfall to the center of mass of direct runoff [5]. Other definitions and several equations relating lag time to watershed characteristics are provided here and in other references [3],[6]–[9].

A foundational assumption of unit hydrograph theory [10] is that the watershed excess release time is a constant, regardless of the storm duration, and is related to basin factors rather than meteorological characteristics. The excess release time is also conceptually identical with the time base of an *instantaneous unit hydrograph* (IUH). This is the runoff hydrograph from 1.0 in. of excess rain applied uniformly over the watershed in an instant of time (see Section 9.3). Both wave travel time and excess-rainfall release time are often used synonymously with *time of concentration.*

Time of Concentration

The most common definition of time of concentration originates from consideration of overland flow. If a uniform rain is applied to a tract, the portions nearest the outlet contribute runoff at the outlet almost immediately. As rain continues, the depth of excess on the surface grows and discharge rates increase throughout. Runoff contributions from various points upstream arrive at later times, adding themselves to continuing runoff from nearer points, until flow eventually arrives from all points on the watershed, "concentrating" at the outlet. Thus, concentration time is the time required, with uniform rain, for 100 percent of a tract of land to contribute to the direct runoff at the outlet [11].

As a second popular definition, the concentration time is often equated with either the excess-rainfall release time or the wave travel time because the time for runoff to arrive at the outlet from the most remote point after rain ceases is assumed to be indicative of the time required for 100 percent contribution from all points during any uniform storm having sufficient duration. The latter definition is often preferred because few storm durations exceed the time of concentration, making determination of t_c possible only by examining excess rain recession.

Because time of concentration is conceptually the time required for 100 percent of the watershed to contribute, it is also often defined as the time from the end of excess rainfall to the inflection point on the hydrograph recession limb (e.g., see Fig. 9.8). The reasoning used in this definition is that direct runoff ceases at the point of inflection.

For a small tract of land experiencing uniform rain, the entire area contributes at approximately the same time that the runoff reaches an equilibrium. This gives rise to yet another definition of time of concentration. If rain abruptly ceased, the direct runoff would continue only as long as the excess-rainfall release time t_r. On the basis of the second definition, excess release time and time of concentration can be considered equivalent.

Numerous equations relating time of concentration to watershed parameters have been developed. Table 9.1 summarizes several popular versions. Other variations are presented in Refs. 1, 3, 6, 7, 9, and 12–14.

Sheet Flow Travel Time

Sheet flow occurs as shallow flow over plane surfaces with relatively uniform depth. Such flow occurs over relatively short distances, rarely more than about 300 ft in natural watersheds. The ASCE kinematic wave equation for sheet flow travel time (Table 9.1) has been proven to be very accurate, but requires iterative adjustments in the rainfall intensity, i, until the calculated time of concentration t_c matches the duration of the storm having intensity i for the selected recurrence interval.

As an alternative to the ASCE kinematic wave equation, another iterative solution is suggested by the U.S. Federal Highway Administration [21] using the following equation for sheet flow travel time, also applicable for flow lengths up to 300 ft:

$$t_c = \frac{0.933\,(nL)^{0.6}}{i^{0.4}\,(S)^{0.3}} \qquad (9.6)$$

where t_c is the time of concentration (hr), i is the rainfall intensity (in./hr), L is the overland flow length (ft), n is Manning's roughness coefficient (sec/ft$^{1/3}$), and S is the slope of the surface (ft/ft). Manning's n values are provided in Table 9.2.

To avoid the iterative process, the SCS TR-55 urban runoff model (Chapter 11) recommends the following variation of the ASCE equation [20]:

$$t_c = \frac{0.007\,(nL)^{0.8}}{P_2^{0.5}\,S^{0.4}} \qquad (9.7)$$

where t_c is the time of concentration (hr), P_2 is the 2-year, 24-hour rainfall depth (in.), L is the overland flow length (ft), n is Manning's roughness coefficient (sec/ft$^{1/3}$), and S is the slope of the surface (ft/ft). The SCS also recommends a maximum L of 300 ft.

Basin Lag Time

Though direct runoff begins with the commencement of effective rain (Fig. 9.8), the largest portion of runoff generally lags the rainfall. Basin lag time, t_l, locates the hydrograph's position relative to the causative storm pattern. It is most often defined as the difference in time between the center of mass of effective (net) rainfall and the center of mass of direct runoff produced by the net rain as shown in Fig. 9.8. Two common variations in the definition are (1) the time interval from the maximum rainfall rate to the peak rate of runoff and (2) the time from the center of mass of actual rainfall to the peak rate of runoff.

TABLE 9.1 Summary of Time of Concentration Formulas

Method and date	Formula for t_c (min)	Remarks
Kirpich (1940)	$t_c = 0.0078L^{0.77}S^{-0.385}$ L = length of channel/ ditch from headwater to outlet, ft S = average watershed slope, ft/ft	Developed from SCS data for seven rural basins in Tennessee with well-defined channel and steep slopes (3% to 10%); for overland flow on concrete or asphalt surfaces multiply t_c by 0.4; for concrete channels multiply by 0.2; no adjustments for overland flow on bare soil or flow in roadside ditches.
USBR Design of Small Dams (1973)	$t_c = 60(11.9L^3/H)^{0.385}$ L = length of longest watercourse, mi H = elevation difference between divide and outlet, ft	Essentially the Kirpich formula; developed from small mountainous basins in California (U.S. Bureau of Reclamation, 1973, pp. 67–71) [15].
Izzard (1946) [16]	$t_c = \dfrac{41.025(0.0007i + c)L^{0.33}}{S^{0.333}i^{0.667}}$ i = rainfall intensity, in/hr c = retardance coefficient L = length of flow path, ft S = slope of flow path, ft/ft	Developed in laboratory experiments by Bureau of Public Roads for overland flow on roadway and turf surfaces; values of the retardance coefficient range from 0.0070 for very smooth pavement to 0.012 for concrete pavement to 0.06 for dense turf; solution requires iteration; product i times L should be ≤500.
Federal Aviation Administration (1970) [17]	$t_c = 1.8(1.1 - C)L^{0.50}/S^{0.333}$ C = rational method runoff coefficient L = length of overland flow, ft S = surface slope, %	Developed from air field drainage data assembled by the Corps of Engineers; method is intended for use on airfield drainage problems, but has been used frequently for overland flow in urban basins.
ASCE Kinematic Wave Morgali and Linsley (1965) [18] Aron and Erborge (1973) [19]	$t_c = \dfrac{0.94L^{0.6}n^{0.6}}{(i^{0.4}S^{0.3})}$ L = length of overland flow, ft n = Manning roughness coefficient i = rainfall intensity, in/hr S = average overland slope, ft/ft	Overland flow equation developed from kinematic wave analysis of surface runoff from developed surfaces; method requires iteration since both i (rainfall intensity) and t_c are unknown; superposition of intensity–duration–frequency curve gives direct graphical solution for t_c.
SCS Lag Equation (1972) [11]	$t_c = \dfrac{1.67\ L^{0.8}[(1000/\mathrm{CN}) - 9]^{0.7}}{1900\ S^{0.5}}$ L = hydraulic length of watershed (longest flow path), ft CN = SCS runoff curve number S = average watershed slope, %	Equation developed by SCS from agricultural watershed data; it has been adapted to small urban basins under 2000 acres; found generally good where area is completely paved; for mixed areas it tends to overestimate; adjustment factors are applied to correct for channel improvement and impervious area.
SCS Average Velocity Charts (1975, 1986) [20]	$t_c = \dfrac{1}{60} \Sigma \dfrac{L}{V}$ L = length of flow path, ft V = average velocity in feet per second from Fig. 3-1 of Ref. [20] for various surfaces	Overland flow charts in Ref. 20 provide average velocity as function of watercourse slope and surface cover.

Source: After Ref. 15

TABLE 9.2 Manning's Roughness Coefficients for Overland Sheet Flow

Surface description	n
Smooth asphalt	0.011
Smooth concrete	0.012
Ordinary concrete lining	0.013
Good wood	0.014
Brick with cement mortar	0.014
Vitrified clay	0.015
Cast iron	0.015
Corrugated metal pipe	0.024
Cement rubble surface	0.024
Fallow (no residue)	0.05
Cultivated soils	
Residue cover ≤20%	0.06
Residue cover >20%	0.17
Range (natural)	0.13
Grass	
Short prairie grass	0.15
Dense grasses	0.24
Bermuda grass	0.41
Woods	
Light underbrush	0.40
Dense underbrush	0.80

Source: After Ref. 21

Numerous equations for lag time have been developed from rainfall-runoff data collected on experimental watersheds. The U.S. Soil Conservation Service [11] observed that the lag time is generally shorter than the time of concentration, t_c, and in most cases:

$$t_l = 0.6\, t_c \tag{9.8}$$

Once time of concentration is defined (Table 9.1), the lag time can be approximated from Eq. 9.8. As shown later (Chapters 11 and 12), most computer programs for hydrograph analysis, especially those emulating SCS procedures, default to this relationship to determine lag time when a method for computing t_c is chosen. Similarly, the equation is reversed to compute t_c from t_l if needed by the program.

Because the lag time is governed largely by physiographic features of the watershed, various studies have been conducted to empirically or statistically relate time lag to meaningful parameters such as the shape of the basin, slope of the main channel, distance from the center of gravity of the watershed to the outlet, channel roughness, and channel geometry [22]–[25]. Linsley and Ackerman [25] show that the form taken by many of these relationships is:

$$t_l = C_t \frac{(L\, L_{ca})^a}{S^{0.5}} \tag{9.9}$$

where t_l = lag time (hr)

$\quad\quad$ L = the length of the main stream channel (mi) from the basin outlet to the most remote point of the watershed divide

$\quad\quad$ S = the slope (ft/mi) of the maximum flow distance profile

$\quad\quad$ L_{ca} = the distance along the main stream (mi) from the outlet to a point nearest the center of gravity of the basin

$\quad\quad$ C_t = a coefficient of timing, representing various types of streams and basin shapes and storage capability

$\quad\quad$ a = an exponent, developed from local watershed data

To illustrate, the Denver Urban Drainage and Flood Control District [26] analyzed local urban watersheds (up to 5 square miles) with mild slopes and found that $a = 0.48$. The timing coefficient, C_t, is related to the percent imperviousness, I_p, of the watershed by:

$$C_t = -0.00371\, I_p + 0.163 \quad\quad\quad\quad\quad\quad 1 \le I_p \le 10 \quad\quad\quad (9.10)$$

$$C_t = 0.000023\, I_p^2 - 0.002241\, I_p + 0.146 \quad 10 \le I_p \le 40 \quad\quad\quad (9.11)$$

$$C_t = -0.00371\, I_p^2 + 0.163\, I_p + 0.12 \quad\quad 40 \le I_p \le 100 \quad\quad\quad (9.12)$$

The Riverside County Flood Control and Water Conservancy District in California developed a similar equation for watersheds between 2 and 650 square miles which uses $a = 0.38$ in Eq. 9.9, with $C_t = 1.20$, 0.72, and 0.38, respectively, for mountainous, foothill, and valley terrain. Underground storm sewers are generally constructed in urbanized watersheds resulting in changes in the natural basin lag. Eagleson [27] proposed the use of Eq. 9.9 with $a = 0.39$ and $C_t = 0.32$. His equation is considered applicable for fully sewered urban areas from 0.2 to 7.5 square miles, maximum channel lengths from 1 to 7 miles, slopes from 6 to 20 ft/mi, and impervious cover from 30 to 80 percent. Section 9.4 presents other variations of lag time equations.

9.3   UNIT HYDROGRAPHS

This section defines unit hydrographs and presents methods of deriving them from actual rainfall and runoff records or synthesizing them from watershed physiographic data. Finally, given either an actual or synthetic unit hydrograph, the methods of applying unit hydrographs to determine the direct runoff hydrograph (Section 9.1) for any storm are presented.

Unit Hydrograph Definitions

The concept of a unit hydrograph was first introduced by Sherman [28] in 1932. He defined a unit graph as follows:

> If a given one-day rainfall produces a 1-in. depth of runoff over the given drainage area, the hydrograph showing the rates at which the runoff occurred can be considered a unit graph for that watershed.

Thus, a unit hydrograph is the hydrograph of *direct runoff* (excluding base flow) for any storm that produces exactly 1.0 inch of net rain (the total runoff after abstractions). Such a storm would not be expected to occur, but Sherman's assumption is that the ordinates of a unit hydrograph are $1.0/P$ times the ordinates of the direct runoff hydrograph for an *equal-duration* storm with P inches of net rain.

The term *unit* has to do with the net rain amount of 1.0 inch and does not mean to imply that the duration of rain that produced the hydrograph is one unit, whether an hour, day, or any other measure of time. The storm duration, X, that produced the unit hydrograph must be specified because a watershed has a different unit hydrograph for each possible storm duration. An *X-hour unit hydrograph* is defined as a direct runoff hydrograph having a 1.0-in. volume and resulting from an X-hour storm having a net rain rate of $1/X$ in./hr. A 2-hr unit hydrograph would have a 1.0-in. volume produced by a 2-hr storm, and a 1-day unit hydrograph would be produced by a storm having 1.0 in. of excess rain uniformly produced during a 24-hr period. The value X is often a fraction. Figure 9.9 illustrates a 2-hr, 12-hr, and 24-hr unit hydrograph for a given watershed.

By Sherman's assumption, application of an X-hour unit graph to design rainfall excess amounts other than 1 in. is accomplished simply by multiplying the rainfall excess amount by the unit graph ordinates, since the runoff ordinates for a given duration are assumed to be directly proportional to rainfall excess. A 3-hr storm producing 2.0 in. of net rain would have runoff rates 2 times the values of the 3-hr unit hydrograph. One-half inch in 3 hr would produce flows half the magnitude of the 3-hr unit hydrograph. This principle of proportional flows is expanded in Section 9.4 and applies only to equal-duration storms.

Implicit in deriving the unit hydrograph is the assumption that rainfall is distributed in the same temporal and spatial pattern for all storms. This is generally not true; consequently, variations in ordinates for different storms of equal duration can be expected.

The construction of unit hydrographs for storms with other than integer multiples of the derived duration is facilitated by a method known as the *S-hydrograph* developed by Morgan and Hulinghorst [29]. The procedure, as explained in Section 9.3, employs a unit hydrograph to form an S-hydrograph resulting from a continuous applied rainfall. The need to alter duration of a unit hydrograph led to studies of the shortest possible storm duration—the instantaneous unit rainfall. The concept of *instantaneous unit hydrograph* (IUH) is traced to Clark [30] and can also be used in constructing unit hydrographs for other than the derived duration.

The previous discussion assumes that the analyst has runoff and rainfall data for deriving a unit hydrograph for the subject watershed. The application of unit hydrograph theory to ungauged watersheds received early attention by Snyder [31] and also by Taylor and Schwartz [32], who tried to relate aspects of the unit hydrograph to watershed characteristics. As a result, a full set of synthetic unit-hydrograph methods emerged. A number of these are presented in Section 9.4.

Derivation of Unit Hydrographs From Streamflow Data

Data collection preparatory to deriving a unit hydrograph for a gauged watershed can be extremely time consuming. Fortunately, many watersheds have available records of

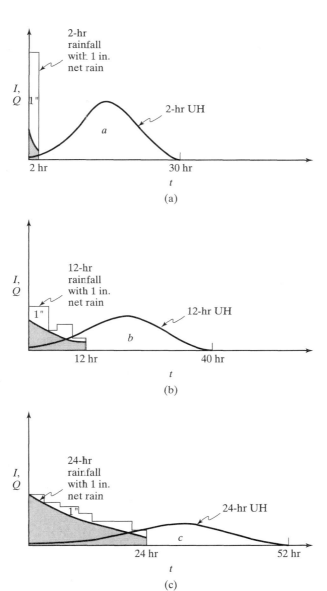

FIGURE 9.9

Illustration of 2-hr, 12-hr, and 24-hr unit hydrographs for the same watershed.

(Note: area a = area b = area c = 1″ × A)

streamflow and rainfall, and these can be supplemented with office records of the Water Resources Division of the U.S. Geological Survey [33]. Rainfall records may be secured from *Climatological Data* [34], published for each state in the United States by the National Oceanic and Atmospheric Administration (NOAA). Hourly rainfall records for recording rainfall stations are published as a *Summary of Hourly Observations* for the location. Summaries are listed for approximately 300 first-order situations in the United States.

To develop a unit hydrograph, it is desirable to acquire as many rainfall records as possible within the study area to ensure that the amount and distribution of rainfall over the watershed are accurately known [35]. Preliminary selection of storms to use in deriving a unit hydrograph for a watershed should be restricted to the following:

1. Storms occurring individually, that is, simple storm structure.
2. Storms having uniform distribution of rainfall throughout the period of rainfall excess.
3. Storms having uniform spatial distribution over the entire watershed.

These restrictions place both upper and lower limits on size of the watershed to be employed. An upper limit of watershed size of approximately 1000 mi^2 is overcautious, although general storms over such areas are not unrealistic and some studies of areas up to 2000 mi^2 have used the unit-hydrograph technique. The lower limit of watershed extent depends on numerous other factors and cannot be precisely defined. A general rule of thumb is to assume about 1000 acres. Fortunately, other hydrologic techniques help resolve unit hydrographs for watersheds outside this range.

The preliminary screening of suitable storms for unit-hydrograph formation should meet more restrictive criteria before further analysis:

1. Duration of rainfall event should be approximately 10–30 percent of the drainage area lag time.
2. Direct runoff for the selected storm should range from 0.5 to 1.75 in.
3. A suitable number of storms with the same duration should be analyzed to obtain an average of the ordinates (approximately five events). Modifications may be made to adjust different unit hydrographs to a single duration by means of S-hydrographs or IUH procedures.
4. Direct runoff ordinates for each hydrograph should be reduced so that each event represents 1 in. of direct runoff.
5. The final unit hydrograph of a specific duration for the watershed is obtained by averaging ordinates of selected events and adjusting the result to obtain 1 in. of direct runoff.

Constructing the unit hydrograph in this way produces the integrated effect of runoff resulting from a representative set of equal-duration storms. Extreme rainfall intensity is not reflected in the determination. If intense storms are needed, a study of records should be made to ascertain their influence upon the discharge hydrograph by comparing peaks obtained utilizing the derived unit hydrograph and actual hydrographs from intense storms.

Essential steps in developing a unit hydrograph for an isolated storm are:

1. Analyze the streamflow hydrograph to permit separation of surface runoff from groundwater flow, accomplished by the methods developed in Section 9.1.
2. Measure the total volume of surface runoff (direct runoff) from the storm producing the original hydrograph. This is the area under the hydrograph, after groundwater base flow has been removed.
3. Divide the ordinates of the direct runoff hydrograph by total direct runoff volume in inches, and plot these results versus time as a unit graph for the basin.
4. Find the effective duration of the runoff-producing rain for this unit graph from the hyetograph (time history of rainfall intensity) of the storm event used.

Procedures other than those listed are required for complex storms or in developing synthetic unit graphs when data are limited. Unit hydrographs can also be transposed from one basin to another under certain circumstances. An example illustrates the derivation of a unit hydrograph.

Example 9.1

Using the total direct runoff hydrograph given in Table 9.3 derive a unit hydrograph for the 1,715-ac drainage area.

Solution

1. Separate the base or groundwater flow to get the total direct runoff hydrograph. A common method is to draw a straight line AC that begins when the hydrograph starts an appreciable rise and ends where the recession curve intersects the base flow curve. The important point here is to be consistent in methodology from storm to storm.

TABLE 9.3 Determination of a 2-hr Unit Hydrograph From an Isolated Storm

(1) Time (hr)	(2) Runoff (cfs)	(3) Base flow (cfs)	(4) Direct runoff, (2)–(3) (cfs)	(5) 2-hr unit hydrograph ordinate, (4) ÷ 1.415 (cfs)
1	110	110	0	0
2	122	110	12	8.5
3	230	110	120	84.8
4	578	110	468	331
4.7	666	110	556	393
5	645	110	535	379
6	434	110	324	229
7	293	110	183	129
8	202	110	92	65.0
9	160	110	50	35.3
10	117	110	7	4.9
10.5	105	105	0	0
11	90	90	0	0
12	80	80	0	0

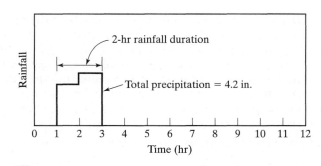

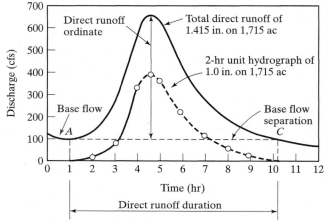

FIGURE 9.10

Illustration of the derivation of a unit hydrograph from an isolated storm.

2. The depth of direct runoff over the watershed is calculated using:

$$\frac{\Sigma \, (DR \times \Delta t)}{area} = \frac{2{,}447 \text{ cfs-hr}}{1{,}715 \text{ ac}} = 1.415 \text{ in.} \qquad (9.13)$$

where DR is the average height of the direct runoff ordinate during a chosen time period Δt (in this case $\Delta t = 1.0$ hr). The values of DR determined from Fig. 9.10 are listed in Table 9.3.

3. Compute the ordinates of the unit hydrograph by using:

$$\frac{Q_s}{V_s} = \frac{Q_u}{1} \qquad (9.14)$$

where Q_s = the magnitude of a hydrograph ordinate of direct runoff having a volume equal to V_s (in.) at some instant of time after start of runoff

Q_u = the ordinate of the unit hydrograph having a volume of 1 in. at some instant of time

In this example the values are obtained by dividing the direct runoff ordinates by 1.415. Table 9.3 outlines the computation of the unit-hydrograph ordinates.

4.   Determine the duration of effective rainfall (rainfall that actually produces surface runoff). As stated previously, the unit hydrograph storm duration should not exceed about 25 percent of the drainage area lag time, but violates this rule for the example. From Fig. 9.10, the rain duration is 2 hr.

5.   Using the values from Table 9.3, plot the unit hydrograph shown in Fig. 9.10.

Unit Hydrograph Applications by Lagging Methods

Once an X-hr unit hydrograph has been derived from streamflow data (or synthesized from basin parameters, Section 9.4) it can be used to estimate the direct runoff hydrograph shape and duration for virtually any rain event. Applications of the X-hr UH to other storms begin with *lagging procedures*, used for storms having durations that are integer multiples of the derived duration. Applications to storms with fractional multiples of X are known as *S-hydrograph* and *IUH* procedures.

Because unit hydrographs are applicable to effective (net) rain, the process of applying UH theory to a storm begins by first abstracting the watershed losses from the precipitation hyetograph, resulting in an effective rain hyetograph. Any of the procedures detailed in Chapter 7 can be applied. The remainder of this discussion assumes that the analyst has already abstracted watershed losses from the storm.

If the duration of another storm is an integer multiple of X, the storm is treated as a series of end-to-end X-hour storms. First, the hydrographs from each X increment of rain are determined from the X-hour unit hydrograph. The ordinates are then added at corresponding times to determine the total hydrograph.

Example 9.2

Discharge rates for the 2-hr unit hydrograph shown in Fig. 9.11 are:

Time (hr)	0	1	2	3	4	5	6
Q (cfs)	0	100	250	200	100	50	0

Develop hourly ordinates of the total hydrograph resulting from a 4-hr design storm having the following excess amounts:

Hour	1	2	3	4
Excess (in.)	0.5	0.5	1.0	1.0

Solution.   The 4-hr duration of the design storm is an integer multiple of the unit-hydrograph duration. Thus, the total hydrograph can be found by adding the contributions of two 2-hr increments of end-to-end rain, as shown in Fig. 9.11c. The first 2-hr storm segment has 1.0 in. of net rain and thus reproduces a unit hydrograph. The second 2-hr storm segment has 2.0 in. of net rain (in 2 hr); thus its ordinates are twice those of a 2-hr unit hydrograph. The total hydrograph, Fig. 9.11e, is found by summing the two contributions at corresponding times. Note in Fig. 9.11d that runoff from the second storm begins when the second rain begins, not at the beginning of the first storm.

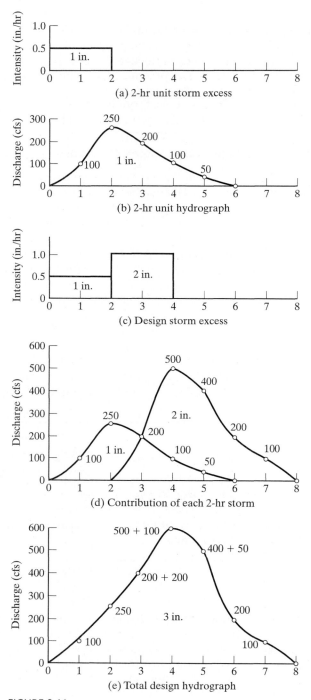

FIGURE 9.11

Example 9.2 derivation of total runoff hydrograph using a 2-hr unit hydrograph.

This method of "lagging" is based on the assumption that linear response of the watershed is not influenced by previous storms—that is, one can superimpose hydrographs offset in time and the flows will be directly additive. The simplest way to develop composite direct runoff hydrographs for multiple-hour storms is in a spreadsheet. Care must be taken, however, in visually confirming, as in Example 9.2, that the start and end points of runoff from each contributing X-hr increment of rain are properly selected. A common error is to lag each additional contributing hydrograph by Δt, the time interval between readings, rather than X, the associated duration with the given unit hydrograph. Also, the multiplier for the UH ordinates must be the net rain occurring in X hours, not the rain occurring in the time increment Δt. Example 9.3 illustrates these points.

Example 9.3

Using the derived 2-hr unit hydrograph in Table 9.3, determine the direct runoff hydrograph for a 4-hr storm having the following excess rain amounts:

Hour	1	2	3	4
Excess rain, in.	0.7	0.7	1.2	1.2

Solution

1. Tabulate the unit hydrograph at intervals of the selected time interval, Δt, as shown in Table 9.4.
2. Determine the correct UH multiplier for each X-hr interval. Because X is 2 hours for this example, the first two hours of the storm produce a total net rain of 1.4 inches. Similarly, the last two hours of the storm produce 2.4 inches of net rain.

TABLE 9.4   Unit Hydrograph Application of Example 9.3

Time (hr)	Effective rainfall (in.)	Unit hydrograph (cfs)	Contrib. of first 2-hr rain, UH × 1.4	Contrib. of second 2-hr rain, UH × 2.4	Total outflow hydrograph (cfs)
0	—	0	0	—	0
1	0.7	8.5	11.9	—	11.9
2	0.7	84.8	119	0	119
3	1.2	331	463	20.4	483
4	1.2	379	531	203	734
5		229	321	794	1,115
6		129	181	910	1,091
7		65	91	550	641
8		35.3	49.4	310	359
9		4.9	6.9	156	163
10		0	0	84.7	84.7
11		—	—	11.8	11.8
12		—	—	0	0

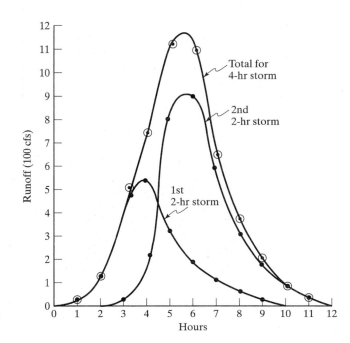

FIGURE 9.12

Synthesized hydrograph for Example 9.3
derived by the unit hydrograph method.

3. Determine the correct start and end times for each of the two hydrographs
    and tabulate the contribution of the 1.4-inch and 2.8-inch rains at the appro-
    priate lag times. Because the second X-hr storm started at $t = 2$ hours, runoff
    for this storm cannot begin until $t = 2$ hours as shown in Table 9.4.

4. Add the contributions at each time to determine the total runoff hydro-
    graphs for the 4-hr storm.

5. Check the tabular solution by plotting each of the two hydrographs and sum
    the ordinates at each t, as shown in Fig. 9.12.

S-Hydrograph Method

The S-hydrograph method overcomes restrictions imposed by the lagging method and
allows construction of any duration unit hydrograph. By observing the lagging system
just described, it is apparent that for a 1-hr unit hydrograph, the 1-in. rainfall excess has
an intensity of 1 in./hr, whereas the 2-hr unit hydrograph is produced by a rainfall
intensity of 0.5 in./hr. Continuous lagging of either one of these unit hydrographs is
comparable to a continuously applied rainfall at either 0.5 in./hr or 1 in./hr intensity,
depending on which unit hydrograph is chosen.

   As an example, using the 1-hr unit hydrograph, continuous lagging represents the
direct runoff from a constant rainfall of 1 in./hr as shown in Fig. 9.13a. The cumulative
addition of the initial unit hydrograph ordinates at time intervals equal to the unit
storm duration results in an S-hydrograph (see Fig. 9.14). Graphically, construction of

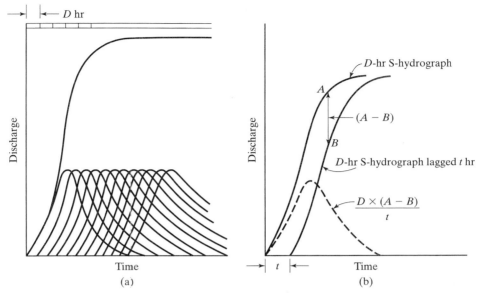

FIGURE 9.13

S-hydrograph method.

an S-hydrograph is readily accomplished with a spreadsheet. The maximum discharge of the S-hydrograph occurs at a time equal to D hours less than the time base of the initial unit hydrograph as shown in Fig. 9.13a.

To construct a pictorial 2-hr unit hydrograph, simply lag the first S-hydrograph by a second S-hydrograph a time interval equal to the desired duration. The difference in S-hydrograph ordinates must then be divided by 2. Any duration t unit hydrograph may be obtained in the same manner once another duration D unit hydrograph is known. Simply form a D-hr S-hydrograph; lag this S-hydrograph t hr, and multiply the difference in S-hydrograph ordinates by D/t. Accuracy of the graphical procedure depends on the scales chosen to plot the hydrographs. Tabular solution of the S-hydrograph method is also employed, but tabulations must be at intervals of the original unit-hydrograph duration.

Example 9.4

Given the following 2-hr unit hydrograph, use S-hydrograph procedures to construct a 3-hr unit hydrograph.

Time (hr)	0	1	2	3	4	5	6
Q (cfs)	0	100	250	200	100	50	0

Solution. The 2-hr unit hydrograph is the runoff from a 2-hr storm of 0.5 in./hr. The S-hydrograph is formed from a net rain rate of 0.5 in./hr lasting indefinitely as shown

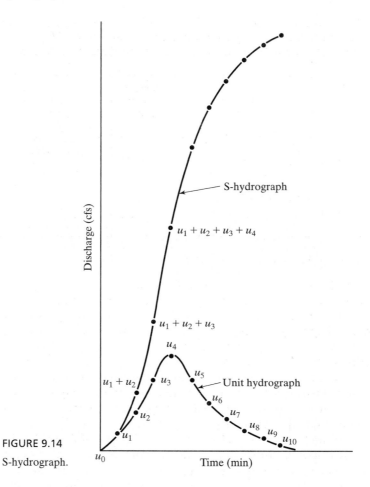

FIGURE 9.14

S-hydrograph.

in Fig. 9.13a. Its ordinates are found by adding the 2-hr unit-hydrograph (UH) runoff rates from each contributing 2-hr block of rain:

Time (hr)	1st 2-hr	2nd 2-hr	3rd 2-hr	4th 2-hr	S-hydrograph
0	0				0
1	100				100
2	250	0			250
3	200	100			300
4	100	250	0		350
5	50	200	100		350
6	0	100	250	0	350
7		50	200	100	350
8		0	100	250	350

To find a 3-hr hydrograph, the S-curve is lagged by 3 hr and subtracted as shown in Fig. 9.13b. This results in a hydrograph from a 3-hr storm of 0.5 in./hr, or 1.5 in. total. Thus the ordinates need to be divided by 1.5 to produce the 3-hr unit hydrograph:

Time (hr)	S-hydrograph	Lagged S-hydrograph	Difference	3-hr unit hydrograph
0	0		0	0
1	100		100	67
2	250		250	167
3	300	0	300	200
4	350	100	250	167
5	350	250	100	67
6	350	300	50	33
7	350	350	0	0

Instantaneous Unit Hydrograph (IUH)

The unit-hydrograph method of estimating a runoff hydrograph can be used for storms of extremely short duration. For example, if the duration of a storm is 1 min and a unit volume of surface runoff occurs, the resulting hydrograph is the 1-min unit hydrograph. The hydrograph of runoff for any 1-min storm of constant intensity can be computed from the 1-min unit hydrograph by multiplying the ordinates of the 1-min unit hydrograph by the appropriate rain depth. A storm lasting for many minutes can be described as a sequence of 1-min storms. The runoff hydrograph from each 1-min storm in this sequence can be obtained as in the preceding example. By superimposing the runoff hydrograph from each of the 1-min storms, the runoff hydrograph for the complete storm can be obtained.

From the unit hydrograph for any duration of uniform rain, the unit hydrograph for any other duration can be obtained. As the duration becomes shorter, the resulting unit hydrograph approaches an instantaneous unit hydrograph. The *instantaneous unit hydrograph* (IUH) is the hydrograph of runoff that would result if 1 in. of water were spread uniformly over an area in an instant and then allowed to run off [36].

To develop an IUH, an I-in./hr S-hydrograph must first be obtained. The resulting S-curve is lagged by the interval Δt to develop a Δt-hour unit hydrograph. The resulting Δt-hour unit graph becomes an IUH when Δt is set to 0 in the limit.

If a continuing I-in./hr excess storm produces the original and lagged S-hydrographs of Fig. 9.13b, the Δt-hour unit hydrograph is the difference between the two curves, divided by the amount of excess rain depth in Δt hours, or:

$$Q_t(\Delta t\text{-hr UH}) = \frac{Q_A - Q_B}{I\,\Delta t} \tag{9.15}$$

The $Q_A - Q_B$ differences are divided by $I\,\Delta t$ to convert from a storm with $I\,\Delta t$ inches in Δt hours to one with 1.0 in. in Δt hours, which is the definition of a Δt-hour unit graph.

As Δt approaches zero, Eq. 9.15 becomes:

$$Q_t(\text{IUH}) = \frac{1}{I}\frac{dQ}{dt} \tag{9.16}$$

which shows that the flow at time t is proportional to the slope of the S-hydrograph at time t. In applications, the slope is approximated by $\Delta Q/\Delta t$, and the IUH ordinates can be estimated from pairs of closely spaced points of the S-hydrograph.

If an IUH is supplied, the above process can be reversed, and any X-hour unit graph can be found by averaging IUH flows at X-hr intervals, or:

$$Q_t(X\text{-hr UH}) \cong \tfrac{1}{2}(\text{IUH}_t + \text{IUH}_{t-X}) \tag{9.17}$$

Use of this approximate equation is allowed for small X values and permits direct calculation of a unit graph from an IUH, bypassing the normal S-hydrograph procedure.

Example 9.5

Given the following 1.0-in./hr S-hydrograph, determine the IUH, and then use it to estimate a 1-hr UH.

Time (hr)	0	0.5	1.0	1.5	2.0	2.5	3.0	3.5	4.0
S-curve (cfs)	0	50	200	450	500	650	700	750	800

Solution.   The IUH is found from Eq. 9.16. The slope at time t is approximated by $(Q_{t+0.5} - Q_{t-0.5})/\Delta t$.

Time	S-curve	IUH $\cong \Delta Q/\Delta t$
0	0	0
0.5	50	200
1	200	400
1.5	450	300
2	500	200
2.5	650	200
3	700	100
3.5	750	100
4	800	50
4.5	800	0
5	800	0

The 1-hr UH is obtained from Eq. 9.17, using readings at 1-hr intervals:

Time	IUH_t	IUH_{t-1}	1-hr UH
0	0	0	0
1	400	0	200
2	200	400	300
3	100	200	150
4	50	100	75
5	0	50	25
6	0	0	0

The reader should verify that the 1-hr UH obtained through use of the IUH is approximately the same as that obtained by lagging the S-hydrograph 1 hr, subtracting, and converting the difference to a 1-hr UH.

9.4 SYNTHETIC UNIT HYDROGRAPHS

Generally, streamflow and rainfall data are not available to allow construction of a unit hydrograph except for relatively few watersheds; therefore, techniques have evolved that allow generation of *synthetic unit hydrographs*. As shown below, the linear characteristics exhibited by unit hydrographs for a watershed are a distinct advantage in constructing more complex storm discharge hydrographs.

Gamma Distribution

Many of the synthetic unit hydrograph procedures result in only three to five points on the hydrograph, through which a smooth curve must be fitted. In addition to the requirement that the curve passes through all the points, the area under the hydrograph must equal the runoff volume from one unit of rainfall excess over the watershed. This latter requirement is often left unchecked and can result in considerable errors in performing calculations through the use of ordinates of a hydrograph that do not represent a "unit" of runoff.

The shapes of hydrographs often closely match a two-parameter gamma function, given by:

$$f(x) = \frac{x^\alpha e^{-x/\beta}}{\beta^{\alpha+1}\Gamma(\alpha + 1)} \tag{9.18}$$

where $0 < x < \infty$. The parameter α is a dimensionless shape factor (must be greater than -1), and β is a positive scale factor having the same units as x and controlling the base length. The product of α and β gives the value x corresponding to the apex, or maximum value of $f(x)$. For $\alpha > 1$, the distribution has a single apex and plots similar to hydrograph shapes, as shown in Fig. 9.15. The distribution mean is $\beta(\alpha + 1)$, and the variance is $\beta^2(\alpha + 1)$.

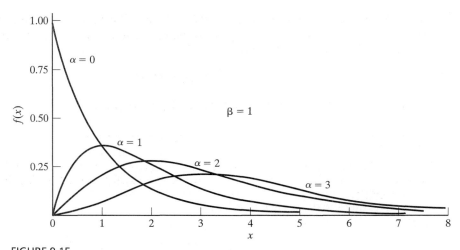

FIGURE 9.15

Gamma function shapes for various values of α when $\beta = 1$.

The most useful feature of the gamma distribution function (explained in greater detail later) is that it guarantees a unit area under the curve. It can conveniently be used to synthesize an entire hydrograph if the calculated peak flow rate Q_p and its associated time t_p are known. This uses a procedure developed by Aron and White [37].

If time t is substituted for x in Eq. 9.18, the time to peak t_p is $\alpha\beta$. At this point, the function $f(t)$ equals the peak flow rate Q_p, or:

$$Q_p = \frac{C_v A \alpha^{\alpha+1}}{t_p e^{\alpha} \Gamma(\alpha + 1)} = \frac{C_v A}{t_p} \phi(\alpha) \tag{9.19}$$

where $C_v A$ is the unit volume of runoff from a basin with area A. The conversion factor $C_v = 1.008$ is selected to make $\phi(\alpha)$ dimensionless.

The function $\phi(\alpha)$ is shown by Aron and White to be related to α by [38]:

$$\alpha = 0.045 + 0.5\phi + 5.6\phi^2 + 0.3\phi^3 \tag{9.20}$$

Collins shows that this can be approximated reasonably well in the range $1 < \alpha < 8$ by [39]:

$$\alpha = 0.5\phi + 5.9\phi^2 \tag{9.21}$$

Combining this with Eq. 9.19 gives:

$$\alpha = 0.5\frac{Q_p t_p}{C_v A} + 5.9\left(\frac{Q_p t_p}{C_v A}\right)^2 \tag{9.22}$$

To fit a unit graph using Eqs. 9.19 and 9.22, the peak flow rate and time must be estimated. Several of the methods described subsequently allow this. Next, $\phi(\alpha)$ is found from Eq. 9.19, and α from Eq. 9.20 or 9.21. The unit hydrograph can now be constructed by calculating Q at any convenient multiple, a, of t_p. Substituting at_p for x in Eq. 9.18 gives the flow at $t = at_p$ as:

$$Q_{at_p} = Q_p a^\alpha e^{(1-a)\alpha} \tag{9.23}$$

which can be solved for all the flow rates of the hydrograph.

Example 9.6

The peak flow rate for the unit hydrograph of a 36,000-acre watershed is 1,720 cfs and occurs 12 hr following the initiation of runoff. Use Eq. 9.18 to synthesize the rest of the hydrograph.

Solution.   From Eq. 9.19:

$$\phi(\alpha) = \frac{1,720(12)}{1.008(36,000)} = 0.57$$

From Eq. 9.20 (and 9.21):

$$\alpha = 2.2$$

The hydrograph is then found from Eq. 9.23:

$$Q_{at_p} = 1,720a^{2.2}e^{2.2(1-a)}$$

Solving, we obtain the following values:

$t = at_p$ (hr)	Q (cfs)
0	0
6	1,125
12	1,720
24	876
60	9
120	0

Sufficient intermediate points should be generated to define the entire shape of the hydrograph.

Snyder's Method

One technique employed by the Corps of Engineers [40] and many others is based on methods developed by Snyder [31] and expanded by Taylor and Schwartz [32]. It allows computation of lag time, time base, unit-hydrograph duration, peak discharge, and hydrograph time widths at 50 and 75 percent of peak flow. By using these seven points, a sketch of the unit hydrograph is obtained (see Fig. 9.16) and checked to see if it contains 1 in. of direct runoff.

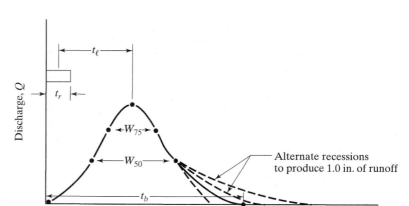

FIGURE 9.16

Snyder's synthetic unit
hydrograph.

Time to Peak   Snyder's method of synthesizing a unit hydrograph assumes that the peak flow rate occurs at the watershed lag, estimated from:

$$t_l = C_t(L_{ca}\,L)^{0.3} \qquad\qquad (9.24)$$

where t_l = the lag time (hr) between the center of mass of the rainfall excess for a specified type of storm and the peak rate of flow

L_{ca} = the distance along the main stream (mi) from the base to a point nearest the center of gravity of the basin

L = length of the main stream channel (mi) from the base outlet to the upstream end of the stream and including the additional distance to the watershed divide

C_t = a coefficient representing variations of types and locations of streams

For the Appalachian Highland area studied, the constant C_t was found to vary from 1.8 to 2.2, with somewhat lower values for basins with steeper slopes. The constant is considered to include the effects of slope and storage. The value of t_l is assumed to be constant for a given drainage area, but allowance is made for the use of different values of lag for different types of storms. The relation is considered applicable to drainage areas ranging in size from 10 to 10,000 mi².

The lag time is shown in Fig. 9.16. The lag time and peak discharge rate are both correlated with various physiographic watershed characteristics. For the lag time, the variables L and L_{ca} for Eq. 9.24 are estimated from map measurements, and C_t is developed for the locale, using Snyder's estimates or other sources. Table 9.5 summarizes a variety of C_t values for various regions.

It is assumed that lag time is a constant for a particular watershed—that is, uninfluenced by variations in rainfall intensities or similar factors. The use of L_{ca} accounts for the watershed shape, and C_t takes care of wide variations in topography, from plains to mountainous regions.

Steeper slopes tend to generate lower values of C_t, with extremes of 0.4 noted in Southern California and 8.0 along the Gulf of Mexico and Rocky Mountains. When

TABLE 9.5   Typical Snyder's Coefficients for U.S. Localities

Location	Range of C_t	Average C_t	Range of C_p	Average C_p
Appalachian Highlands	1.8–2.2	2.0	0.4–0.8	0.6
Western Iowa	0.2–0.6	0.4	0.7–1.0	0.8
Southern California	—	0.4	—	0.9
Ohio	0.6–0.8	0.7	0.6–0.7	0.6
Eastern Gulf of Mexico	—	8.0	—	0.6
Central Texas	0.4–2.3	1.1	0.3–1.2	0.8
North and mid-Atlantic states	—	$0.6/\sqrt{S}^a$	—	—
Sewered urban areas	0.2–0.5	0.3	0.1–0.6	0.3
Mountainous watersheds	—	1.2	—	—
Foothills areas	—	0.7	—	—
Valley areas	—	0.4	—	—
Eastern Nebraska	0.4–1.0	0.8	0.5–1.0	0.8
Corps of Engineers training course	0.4–8.0	0.3–0.9	—	—
Great Plains	0.8–2.0	1.3	—	—
Rocky Mountains	1.5–8.8	5.4	—	—
SW desert	0.7–1.9	1.4	—	—
NW coast and Cascades	2.0–4.4	3.1	—	—
21 urban basins	0.3–0.9	0.6	—	—
Storm-sewered areas	0.2–0.3	0.2	—	—

^{a} Channel slope S.

snowpack accumulations influence peak discharge, values of C_t will be one-sixth to one-third of Snyder's values.

Time Base   The time base of a synthetic unit hydrograph (see Fig. 9.16) by Snyder's method is:

$$t_b = 3 + \frac{t_l}{8} \tag{9.25}$$

where t_b = the base time of the synthetic unit hydrograph (days)
t_l = the lag time (hr)

Equation 9.25 gives reasonable estimates for large watersheds but will produce excessively large values for smaller areas. A general rule of thumb for small areas is to use three to five times the time to peak as a base value when sketching a unit hydrograph. In any event, the time base should be adjusted as shown in Fig. 9.16 until the area under the unit hydrograph is 1.0 in.

Duration   The duration of rainfall excess for Snyder's synthetic unit-hydrograph development is a function of lag time:

$$t_r = \frac{t_l}{5.5} \tag{9.26}$$

where t_r = duration of the unit rainfall excess (hr)

   t_l = the lag time from the centroid of unit rainfall excess to the peak of the unit hydrograph

This synthetic technique always results in an initial unit-hydrograph duration equal to $t_l/5.5$. Because changes in lag time occur with changes in duration of the unit hydrograph, the following equation was developed to allow lag time and peak discharge adjustments for other unit-hydrograph durations:

$$t_{lR} = t_l + 0.25(t_R - t_r) \tag{9.27}$$

where t_{lR} = the adjusted lag time (hr)

   t_l = the original lag time (hr)

   t_R = the desired unit-hydrograph duration (hr)

   t_r = the original unit-hydrograph duration = $t_l/5.5$ (hr)

Peak Discharge   If one assumes that a given duration rainfall produces 1 in. of direct runoff, the outflow volume is some relatively constant percentage of inflow volume. A simplified approximation of outflow volume is $t_l \times Q_P$, and the equation for peak discharge can be written:

$$Q_P = \frac{640\, C_P A}{t_{lR}} \tag{9.28}$$

where Q_P = the peak discharge (cfs)

   C_P = the coefficient accounting for flood wave and storage conditions; it is a function of lag time, duration of runoff-producing rain, effective area contributing to peak flow, and drainage area

   A = the watershed size (mi^2)

   t_{lR} = the lag time (hr)

Thus peak discharge can be calculated given lag time and coefficient of peak discharge C_P. Values for C_P range from 0.4 to 0.8 and generally indicate retention or storage capacity of the watershed. Larger values of C_P are generally associated with smaller values of C_t. Typical values are shown in Table 9.5.

Hydrograph Construction   To construct the UH use Eqs. 9.24, 9.25, 9.26, and 9.28 to plot three points for the unit hydrograph and sketch a synthetic unit hydrograph, remembering that total direct runoff amounts to 1 in. An analysis by the Corps of Engineers (see Fig. 9.17) gives additional assistance in plotting time widths for points on the hydrograph located at 50 and 75 percent of peak discharge [40]. As a general rule of thumb, the time width at W_{50} and W_{75} ordinates should be proportioned each side of the peak in a ratio of 1:2 with the short time side on the left of the synthetic unit-hydrograph peak. As noted earlier, for smaller watersheds, Eq. 9.25 gives unrealistic values for the base time. If this occurs, a value can be estimated by multiplying total

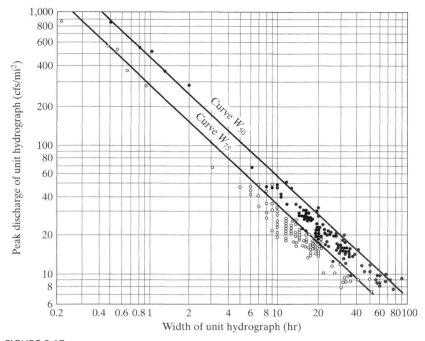

FIGURE 9.17

Unit hydrograph width at 50 and 75 percent of peak flow. ● = observed value of W_{50}. ○ = observed value of W_{75}.

time to the peak by a value of from 3 to 5. This ratio can be modified based on the amount and time rate of depletion of storage water within the watershed boundaries.

The width envelope curves in Fig. 9.17 are defined by:

$$W_{50} = 830/(Q_p/A)^{1.1} \tag{9.29}$$

$$W_{75} = 470/(Q_p/A)^{1.1} \tag{9.30}$$

The seven points formed through the use of these equations can be plotted and a smooth curve drawn. To assure a unit hydrograph, the curve shape and ordinates should be adjusted until the area beneath the curve is equivalent to one unit of direct runoff depth over the watershed area. This can be done by hand-fitting and using a planimeter, or by curve-fitting.

The application of Snyder's synthetic unit-hydrograph method to areas other than the original study area should be preceded by a reevaluation of coefficients C_t and C_P in Eqs. 9.24 and 9.28. This analysis can be accomplished by the use of unit hydrographs in the region under study which have the proper lag time–rainfall duration ratio; that is, $t_r = t_l/5.5$. If another rainfall duration is selected, variations of C_t and C_P can be expected.

Dimensionless SCS Unit Hydrograph

A method developed by the Soil Conservation Service [11],[41] for constructing synthetic unit hydrographs is based on a dimensionless hydrograph (Fig. 9.18). This dimensionless graph is the result of an analysis of a large number of natural unit hydrographs from a wide range in size and geographic locations. The method requires only the determination of the time to peak and the peak discharge as follows:

$$t_p = \frac{D}{2} + t_l \tag{9.31}$$

where t_p = the time from the beginning of rainfall to peak discharge (hr)
    D = the duration of rainfall (hr)
    t_l = the lag time from the centroid of rainfall to peak discharge (hr)

The ratios corresponding to Fig. 9.18 are listed in Table 9.6. The peak flow for the hydrograph is developed by approximating the unit hydrograph as a triangular shape with base time of $\frac{8}{3}t_p$ and unit area. The reader should verify that this produces:

$$Q_P = \frac{484A}{t_p} \tag{9.32}$$

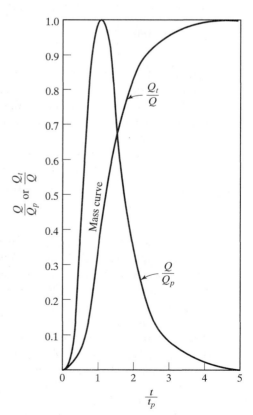

FIGURE 9.18

SCS dimensionless unit hydrograph and mass curve.
(After Mockus [41])

TABLE 9.6   Coordinates of the SCS
Dimensionless Unit
Hydrograph of Figure 9.18

t/t_p	Q/Q_P	t/t_p	Q/Q_P
0	0	1.4	0.75
0.1	0.015	1.5	0.66
0.2	0.075	1.6	0.56
0.3	0.16	1.8	0.42
0.4	0.28	2.0	0.32
0.5	0.43	2.2	0.24
0.6	0.60	2.4	0.18
0.7	0.77	2.6	0.13
0.8	0.89	2.8	0.098
0.9	0.97	3.0	0.075
1.0	1.00	3.5	0.036
1.1	0.98	4.0	0.018
1.2	0.92	4.5	0.009
1.3	0.84	5.0	0.004

where Q_P = peak discharge (cfs)
   A = drainage area (mi^2)
   t_p = time to peak (hr)

The time base of $\frac{8}{3}t_p$ is based on empirical values for average rural experimental water-sheds and should be reduced (causing increased peak flow) for steep conditions or increased (causing decreased peak flow) for flat conditions. The resulting coefficient in Eq. 9.32 ranges from nearly 600 for steep mountainous conditions to 300 for flat swampy conditions.

   A relation of t_l to size of watershed can be used to estimate lag time. Typical relations from two geographic regions are:

$$t_l = 1.44A^{0.6} \quad \text{Texas} \tag{9.33a}$$

$$t_l = 0.54A^{0.6} \quad \text{Ohio} \tag{9.33b}$$

The average lag time is $0.6t_c$, where t_c is the time of concentration, defined by the SCS as either the time for runoff to travel from the furthermost point in the watershed (called the upland method) or the time from the end of excess rain to the inflection of the unit hydrograph. For the first case:

$$t_c = 1.7t_p - D \tag{9.34}$$

The dimensionless unit hydrograph, Fig. 9.18, has a point of inflection at approximately $1.7t_p$. If the lag time of $0.6t_c$ is assumed, Eqs. 9.31 and 9.34 give:

$$D = 0.2t_p \tag{9.35}$$

or

$$D = 0.133t_c \tag{9.36}$$

A small variation in D is permissible, but it should not exceed $0.25t_p$ or $0.17t_c$. Once the $0.133t_c$-hour unit hydrograph is developed, unit hydrographs for other durations can be developed using S-hydrograph or IUH procedures.

By finding a value of t_l, a synthetic unit hydrograph of chosen duration D is obtained from Fig. 9.18.

Another equation used by the SCS is:

$$t_l = \frac{l^{0.8}(S + 1)^{0.7}}{1,900Y^{0.5}} \tag{9.37}$$

where t_l = lag time (hr)
   l = length to divide (ft)
   Y = average watershed slope (%)
   S = potential maximum retention (in.) = $(1,000/CN) - 10$, where CN is a curve number described in Chapter 4

The lag time from Eq. 9.37 is adjusted for imperviousness or improved watercourses, or both, if the watershed is in an urban area. The multiple to be applied to the lag time is:

$$M = 1 - P[-6.8 \times 10^{-3} + (3.4 \times 10^{-4})CN - (4.3 \times 10^{-7})CN^2$$
$$- (2.2 \times 10^{-8})CN^3] \tag{9.38}$$

where CN is the curve number for urbanized conditions, and P can be either the percentage impervious or the percentage of the main watercourse that is hydraulically improved from natural conditions. If part of the area is impervious and portions of the channel are improved, two values of M are determined, and both are multiplied by the lag time.

Example 9.7

For a drainage area of 70 mi² having a lag time of $8\frac{1}{2}$ hr, derive a unit hydrograph of duration 2 hr. Use the SCS dimensionless unit hydrograph.

Solution

1.   Using Eq. 9.31 we obtain:

$$t_p = \frac{2}{2} + 8\frac{1}{2} = 9\frac{1}{2} \text{ hr}$$

2.   From Eq. 9.32:

$$Q_P = \frac{484 \times 70}{9.5}$$
$$Q_P = 3,564 \text{ cfs occurring at } t = 9\frac{1}{2} \text{ hr}$$

3.   Using Fig. 9.18, we find the following:

a. The peak flow occurs at $t/t_p = 1$ or at $t = 9\frac{1}{2}$ hr.
b. The time base of the hydrograph = $5t_p$ or 47.5 hr.

c. The hydrograph ordinates are:

1. At $t/t_p = 0.5$, $Q/Q_P = 0.43$; thus at $t = 4.75$ hr, $Q = 1{,}531$ cfs.
2. At $t/t_p = 2$, $Q/Q_P = 0.32$; thus at $t = 19$ hr, $Q = 1{,}139$ cfs.
3. At $t/t_p = 3$, $Q/Q_P = 0.07$; thus at $t = 28.5$ hr, $Q = 249$ cfs.

4. Check $D/t_p = 0.21$; OK.

Espey 10-Minute Synthetic Unit Hydrograph

A regional analysis of 19 urban watersheds was conducted by Espey and Altman [42] and resulted in a set of regression equations that provide seven points of a 10-min hydrograph. The entire hydrograph is developed by fitting a smooth curve through the points using eye-fitting or curve-fitting procedures. In either case, a unit area is necessary.

The equations for time to peak (minutes), peak discharge (cfs), time base (minutes), and width at 50 and 75 percent of the peak flow rate are:

$$T_p = 3.1L^{0.23}S^{-0.25}I^{-0.18}\phi^{1.57} \tag{9.39}$$

$$Q_P = 31.62 \times 10^3 A^{0.96}T_p^{-1.07} \tag{9.40}$$

$$T_B = 125.89 \times 10^3 AQ_p^{-0.95} \tag{9.41}$$

$$W_{50} = 16.22 \times 10^3 A^{0.93}Q_p^{-0.92} \tag{9.42}$$

$$W_{75} = 3.24 \times 10^3 A^{0.79}Q_p^{-0.78} \tag{9.43}$$

where L = total distance (ft) along the main channel from the point being considered to the upstream watershed boundary

S = main channel slope (ft/ft) defined by $H/0.8L$, where H is the difference in elevation between the point on the channel bottom at a distance of $0.2L$ downstream from the upstream watershed boundary and a point on the channel bottom at the downstream point being considered

I = impervious area within the watershed (%)

ϕ = dimensionless watershed conveyance factor

A = watershed drainage area (mi^2)

T_p = time of rise of the unit hydrograph (min)

Q_P = peak flow of the unit hydrograph (cfs)

T_B = time base of the unit hydrograph (min)

W_{50} = width of the unit hydrograph at 50% of Q_P (min)

W_{75} = width of the unit hydrograph at 75% of Q_P (min)

The coefficients of determination (explained in Chapter 3) for the five equations ranged from 80 to 94 percent. The watershed conveyance factor is found from Fig. 9.19. The W_{50} and W_{75} widths are normally drawn with one-third of the calculated width placed to the left of the peak and two-thirds to the right.

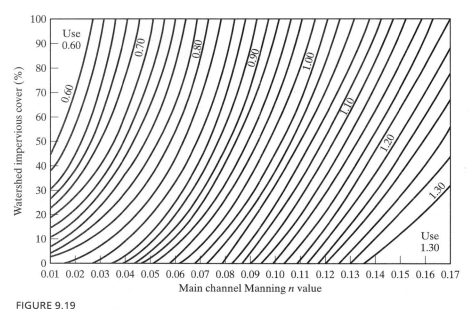

FIGURE 9.19

Watershed conveyance factor ϕ as a function of percent watershed impervious cover I and weighted main channel Manning n value, for the Espey method.

Clark's Time–Area IUH Method

A synthetic unit hydrograph that utilizes an instantaneous unit hydrograph (IUH) was developed in 1945 by Clark [30]. It has been widely used, is often called the time–area method, and has appeared in several computer programs for hydrograph analysis (see Chapters 11 and 12) [43].

The technique recognizes that the discharge at any point in time is a function of the translation and storage characteristics of the watershed. The translation is obtained by estimating the overland and channel travel time of runoff, which is then combined with an estimate of the delay caused by the storage effects of a watershed.

The translation of excess rainfall from its point of falling to the watershed mouth is accomplished using the time–area curve for the watershed. This is a histogram of incremental runoff versus time, constructed as shown in Fig. 9.20. The dashed lines in Fig. 9.20a subdivide the basin into several areas. Each line identifies the locus of points having equal travel times to the outlet. The isochrones are drawn equal "times" apart, and sufficient zones are selected to fully define the time–area relation.

The time–area graph of Fig. 9.20b is a form of unit hydrograph. The area beneath the curve integrates to 1.0 unit of rain depth over the total area A, and it has a translation hydrograph shape if sufficient subareas are delineated.

If one unit of net rain is placed on the watershed at $t = 0$, the runoff from A_1 would pass the outlet during the first Δt period at an average rate of A_1 units of runoff per unit of time. The volume discharged would be A_1 units of area times one unit of rain. After all areas contribute, one unit of rainfall over the entire area would have passed the outlet.

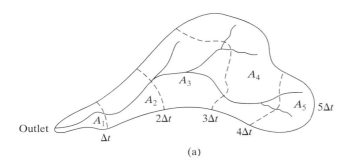

(a)

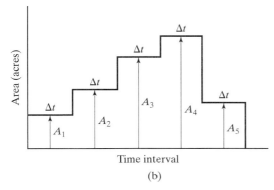

Time interval

(b)

FIGURE 9.20

Development of time–area histogram for use with Clark's method: (a) isochrones spaced Δt apart (shown as dashed lines) and (b) time–area histogram.

The impact of watershed storage on the translation hydrograph is incorporated by routing the time–area histogram through a hypothetical linear reservoir located at the watershed outlet, having a retardance coefficient K equivalent to that of the watershed. For the simplest form of reservoir, the storage S_t at time t is linearly related to the outflow Q_t at time t, or:

$$S_t = KQ_t \tag{9.44}$$

where K is a constant of proportionality called the storage coefficient. It has units of time and is often approximated by the lag time of the watershed. Others estimate it as $0.6t_c$ to $2.0t_c$.

From continuity, the inflow, storage, and outflow for the reservoir are related by:

$$I_t - Q_t = \frac{dS_t}{dt} = K\frac{dQ}{dt} \tag{9.45}$$

If the differential is discretized to $\Delta Q/\Delta t$, and if Q_2 and Q_1 are the flows at t and $t-1$, then Eq. 9.45 becomes:

$$\overline{I}_{\Delta t} - \overline{Q}_{\Delta t} = K\frac{Q_2 - Q_1}{\Delta t} \tag{9.46}$$

Because $\overline{Q} = (Q_1 + Q_2)/2$, the flow at the end of any Δt is:

$$Q_2 = C_0\overline{I} + C_1Q_1 \tag{9.47}$$

where

$$C_0 = \frac{2 \, \Delta t}{2K + \Delta t} \qquad (9.48)$$

and

$$C_1 = \frac{2K - \Delta t}{2K + \Delta t} \qquad (9.49)$$

The IUH is found from Eq. 9.47 by solving for Q_2 at the end of each successive time interval.

Example 9.8

Given the following 15-min time–area curve, find the IUH for the 1,000-acre watershed. Then determine the 15-min synthetic unit hydrograph. The storage coefficient K is 30 min.

Time interval (min)	Area between isochrones (acres)
0–15	100
15–30	300
30–45	500
45–60	100

Solution.   From Eqs. 9.48 and 9.49, $C_0 = 0.4$ and $C_1 = 0.6$. Routing is most easily accomplished using a table or spreadsheet as follows:

Time (hr)	$\bar{I}$ (acre-in./Δt)	$\bar{I}$ (cfs)	$C_0 \bar{I}$ 1 $C_1 Q_1$	IUH (cfs)
0	100	400	160 + 0	0
0.25	300	1,200	480 + 96	80
0.50	500	2,000	800 + 346	368
0.75	100	400	160 + 688	861
1.00	0	0	0 + 509	997
1.25	0	0	0 + 305	679
1.50	0	0	0 + 407	$\vdots$
$\vdots$	$\vdots$	$\vdots$	$\vdots$	

The IUH has a characteristically long recession due to the magnitude of K for this example. Note that after 1.25 hr, the flow becomes 0.6 times the previous flow and continues to decay at this rate indefinitely. As discussed in Section 9.3, the time base of the IUH should equal the excess-runoff release time, which is one definition of time of concentration. Clark's method often produces prolonged runoff because of this shortcoming.

The 15-min unit hydrograph is found using Eq. 9.17, or:

$$Q_t(\text{15-min UH}) = \tfrac{1}{2}(\text{IUH}_t + \text{IUH}_{t-15})$$

This results in:

Time (hr)	IUH (cfs)	15-min UH (cfs)
0	0	0
0.25	160	80
0.50	576	368
0.75	1,146	861
1.00	848	997
1.25	509	679
1.50	⋮	⋮
⋮		

If the watershed lag time is not available, the K value can also be estimated by recognizing that $Q_t = K\,dQ/dt$ when the inflow is zero in Eq. 9.45. This occurs at approximately the inflection point on the recession of Fig. 9.10, when inflow to the channel ceases. If hydrograph data are available, the estimate of the K value is the ratio of the flow rate to the slope of the hydrograph at this particular point on the hydrograph.

Several other synthetic unit hydrograph methods have been developed [9],[44]–[46]. The methods presented here are the most widely used by practitioners, and most are incorporated in standard single-event storm simulation models (Chapters 11 and 12).

9.5   HYDROGRAPH ROUTING

Hydrograph routing is used to predict the temporal and spatial variations of a hydrograph as it traverses a river reach or reservoir. Flood forecasting, reservoir design, watershed simulation modeling, and comprehensive flood control planning studies generally use some form of routing technique. Routing methods are classified as hydrologic and hydraulic, depending, respectively, on whether the methods are based on empirical or physical process equations of motion.

Hydrologic routing employs the equation of continuity with either a linear or curvilinear relation between storage and discharge within a river or reservoir. Hydraulic routing, on the other hand, uses both the equation of continuity and the equation of motion, customarily the momentum equation. This particular form utilizes the partial differential equations for unsteady flow in open channels. It more adequately describes the dynamics of flow than does the hydrologic routing technique. Only hydrologic river and reservoir methods are presented here. Discussions of hydraulic routing techniques are available in other standard literature [6],[7],[12]–[14].

Applications of hydrologic routing techniques to problems of flood prediction, evaluations of flood control measures, and assessments of the effects of urbanization are numerous. Most flood warning systems instituted by NOAA and the Corps of Engineers incorporate this technique to predict flood stages in advance of a severe storm. It is the method most frequently used to size spillways for small, intermediate, and large dams. Hydrologic river and reservoir routing and hydraulic river routing techniques are presented in separate sections of this chapter.

Hydrologic River Routing

The first reference to routing a flood hydrograph from one river station to another was by Graeff in 1883 [47]. The technique was based on the use of wave velocity and a rating curve of stage versus discharge. Hydrologic river routing techniques are all founded upon the equation of continuity:

$$I - O = \frac{dS}{dt} \tag{9.50}$$

where I = the inflow rate to the reach
            O = the outflow rate from the reach
       dS/dt = the rate of change of storage within the reach

Three of the most popular hydrologic river routing techniques are described in subsequent paragraphs.

Muskingum Method   Storage in a stable river reach can be expected to depend primarily on the discharge into and out of a reach and on hydraulic characteristics of the channel section. The storage within the reach at a given time can be expressed as [48]:

$$S = \frac{b}{a}[XI^{m/n} + (1 - X)O^{m/n}] \tag{9.51}$$

Constants a and n reflect the stage discharge characteristics of control sections at each end of the reach, and b and m mirror the stage-volume characteristics of the section. The factor X defines the relative weights given to inflow and outflow for the reach.

The Muskingum method assumes that $m/n = 1$ and lets $b/a = K$, resulting in:

$$S = K[XI + (1 - X)O] \tag{9.52}$$

where K = the storage time constant for the reach
         X = a weighting factor that varies between 0 and 0.5.

Application of this equation has shown that K is usually reasonably close to the wave travel time through the reach and X averages about 0.2.

Behavior of the flood wave due to changes in the value of the weighting factor X is readily apparent from examination of Fig. 9.21. The resulting downstream flood wave is commonly described by the amount of translation—that is, the time lag—and by the amount of attenuation or reduction in peak discharge. As can be noted from Fig. 9.21, the value $X = 0.5$ results in a pure translation of the flood wave.

Application of Eqs. 9.50 and 9.52 to a river reach is a straightforward procedure if K and X are known. The routing procedure begins by dividing time into a number of equal increments, Δt, and expressing Eq. 9.50 in finite difference form, using subscripts 1 and 2 to denote the beginning and ending times for Δt. This gives:

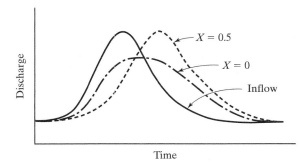

FIGURE 9.21

Effect of weighting factor.

$$\frac{I_1 + I_2}{2} - \frac{O_1 + O_2}{2} = \frac{S_2 - S_1}{\Delta t} \tag{9.53}$$

The routing time interval Δt is normally assigned any convenient value between the limits of $K/3$ and K.

The storage change in the river reach during the routing interval from Eq. 9.52 is:

$$S_2 - S_1 = K[X(I_2 - I_1) + (1 - X)(O_2 - O_1)] \tag{9.54}$$

and substituting this into Eq. 9.53 results in the Muskingum routing equation:

$$O_2 = C_0 I_2 + C_1 I_1 + C_2 O_1 \tag{9.55}$$

in which

$$C_0 = \frac{-KX + 0.5\Delta t}{K - KX + 0.5\Delta t} \tag{9.56}$$

$$C_1 = \frac{KX + 0.5\Delta t}{K - KX + 0.5\Delta t} \tag{9.57}$$

$$C_2 = \frac{K - KX - 0.5\Delta t}{K - KX + 0.5\Delta t} \tag{9.58}$$

Note that K and Δt must have the same time units and also that the three coefficients sum to 1.0.

Theoretical stability of the numerical method is accomplished if Δt falls between the limits $2KX$ and $2K(1 - X)$. The theoretical value of K is the time required for an elemental (kinematic) wave to traverse the reach. It is approximately the time interval between inflow and outflow peaks, if data are available. If not, the wave velocity can be estimated for various channel shapes as a function of average velocity V for any representative flow rate Q. Velocity for steady uniform flow can be estimated by either the Manning or Chézy equation. The approximate wave velocities for different channel shapes are given in Table 9.7.

Since I_1 and I_2 are known for every time increment, routing is accomplished by solving Eq. 9.55 for successive time increments using each O_2 as O_1 for the next time increment. Example 9.9 illustrates this row-by-row computation.

TABLE 9.7 Kinematic Wave Velocities for Various Channel Shapes

Channel shape	Manning equation	Chézy equation
Wide rectangular	$\frac{5}{3}V$	$\frac{3}{2}V$
Triangular	$\frac{4}{3}V$	$\frac{5}{4}V$
Wide parabolic	$\frac{11}{9}V$	$\frac{7}{6}V$

Example 9.9

Perform the flood routing for a reach of river given $X = 0.2$ and $K = 2$ days. The inflow hydrograph with $\Delta t = 1$ day is shown in Table 9.8, column 1. Assume equal inflow and outflow rates on March 16.

TABLE 9.8 Solution to Example 9.9

Date	(1) Inflow	(2) $C_0 I_2$	(3) $C_1 I_1$	(4) $C_2 Q_1$	(5) Computed outflow
3-16	4,260	—	—	—	4,260
17	7,646	364	1,823	2,232	4,419
18	11,167	532	3,272	2,315	6,119
19	16,730	798	4,779	3,206	8,783
20	21,590	1,029	7,160	4,602	12,791
21	20,950	999	9,240	6,702	16,941
22	26,570	1,267	8,966	8,877	19,110
23	46,000	2,194	11,371	10,013	23,578
24	59,960	2,860	19,688	12,355	34,903
25	57,740	2,754	25,662	18,289	46,705
26	47,890	2,284	24,712	24,473	51,469
27	34,460	1,643	20,496	26,970	49,109
28	21,660	1,033	14,748	25,733	41,514
29	34,680	1,654	9,270	21,753	32,677
30	45,180	2,155	14,843	17,122	34,120
31	49,140	2,343	19,337	17,879	39,559
4-1	41,290	1,969	21,031	20,729	43,729
2	33,830	1,613	17,672	22,914	42,199
3	20,510	978	14,479	22,112	37,569
4	14,720	702	8,778	19,686	29,166
5	11,436	545	6,300	15,283	22,128
6	9,294	443	4,894	11,595	16,932
7	7,831	373	3,977	8,872	13,222
8	6,228	297	3,351	6,928	10,576
9	6,083	290	2,665	5,542	8,497

Solution. If $\Delta t = 1$ day, $X = 0.2$, and $K = 2$ days, then Eqs. 9.56 to 9.58 give $C_0 = 0.0477$, $C_1 = 0.428$, and $C_2 = 0.524$. Row-by-row computation is given in Table 9.8.

Determination of Muskingum K and X Values of K and X for Muskingum routing are commonly estimated using K equal to the travel time in the reach and an average value of $X = 0.2$. If inflow and outflow hydrograph records are available for one or more floods, the routing process is easily reversed to provide better values of K and X for the reach. To illustrate the latter method, instantaneous values of S versus $XI + (1 - X)O$ are first graphed for several selected values of X as shown in Example 9.10. Because S and $XI + (1 - X)O$ are assumed to be linearly related via Eq. 9.52, the accepted value of X is that which gives the best linear plot (the narrowest loop). After plotting, the value for K is determined as the reciprocal of the slope through the narrowest loop, since from Eq. 9.52:

$$K = \frac{S}{XI + (1 - X)O} \tag{9.59}$$

Instantaneous values of S for the graphs in Example 9.10 were determined by solving for S_2 in Eq. 9.53 for successive time increments. A value of $S_1 = 0$ was used for the initial increment, but the value is arbitrary since only the slope and not the intercept of Eq. 9.52 is desired. The S_2 values are plotted against average weighted discharges, $X\bar{I} + (1 - X)\bar{O}$ in Table 9.9. A preferable method would be to plot S_2 values

---

TABLE 9.9 Solution to Example 9.10

Date	$\bar{I} = \dfrac{I_1 + I_2}{2}$ (cfs)	$\bar{O} = \dfrac{O_1 + O_2}{2}$ (cfs)	S_2^a (10^3 cfs-days)	Weighted discharge (cfs) $X\bar{I} + (1 - X)\bar{O}$		
				$X = 0.1$	$X = 0.2$	$X = 0.3$
3-16	5,870	4,180	1.7	4,350[b]	4,520	4,690
17	9,310	6,970	4.0	7,200	7,440	7,670
18	12,900	7,560	9.4	8,090	8,630	9,160
19	20,500	14,200	15.7	14,800	15,500	16,100
20	21,000	18,300	18.4	18,600	18,800	19,100
21	23,400	18,500	23.3	19,000	19,500	20,000
22	32,500	21,300	34.5	22,400	23,500	24,700
23	55,400	29,300	60.6	31,900	34,500	37,100
24	62,700	39,700	83.6	42,000	44,300	46,600
25	52,600	48,700	97.5	49,100	49,500	50,000
3-26	43,200	53,300	87.4	52,300	51,300	50,300
27	25,200	48,700	73.9	46,400	44,000	41,700
28	22,800	37,100	59.6	35,700	34,200	32,800
29	41,200	35,800	65.0	36,300	36,900	37,400
30	50,400	35,800	79.6	37,300	38,700	40,200
31	45,300	35,800	89.1	36,800	37,700	38,600
4-1	38,800	42,700	85.2	42,300	41,900	41,500
2	27,000	44,100	68.0	42,400	40,800	39,000
3	16,200	35,400	48.9	33,500	31,600	29,600
4	12,400	25,200	36.1	23,900	22,600	21,400
5	10,200	16,400	29.9	15,800	15,200	14,500
6	8,080	11,500	26.5	11,200	10,800	10,500
7	6,010	9,380	23.1	9,040	8,710	8,370
8	5,050	7,860	20.3	7,300	7,300	7,020

[a] Note: $S_2 \cong S_1 + \bar{I}\,\Delta t - \bar{O}\,\Delta t$ [see Eq. 9.53].
[b] Example: $4,350 = 0.1(5,870) + (1 - 0.1)(4,180)$.

against corresponding values of instantaneous (rather than average) values of $XI_2 + (1 - X)O_2$, using recorded values of inflow and outflow (not provided).

Example 9.10

Given inflow and outflow hydrographs on the Muckwamp River, determine K and X for the river reach. (See Table 9.9.)

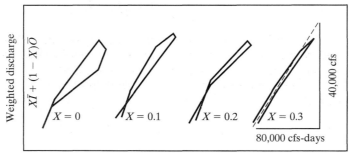

Storage, S

Solution.   Selecting the narrowest loop gives $X = 0.3$; $K = 80,000$ cfs-days$/40,000$ cfs $= 2.0$ days. These values could now be used to route other floods through the reach as in Example 9.9.

Inherent in this procedure is the postulate that the water surface in the reach is a uniform unbroken surface profile between upstream and downstream ends of the section. Additionally, it is presupposed that K and X are constant throughout the range of flows. If significant departures from these restrictions are present, it may be necessary to work with shorter reaches of the river or to employ a more sophisticated approach.

Muskingum–Cunge Method   Several attempts to overcome the limitations of the Muskingum method have not been totally successful because of computational complexity or difficulties in physically interpreting the routing parameters [49],[50]. The Muskingum parameters are best derived from streamflow measurements and are not easily related to channel characteristics.

Cunge [51] blended the accuracy of the diffusion wave method with the simplicity of the Muskingum method, resulting in one of the most recommended techniques for general use. It is classified as a hydrologic method, yet it gives results comparable with hydraulic methods.

Cunge showed that the finite-difference form of the Muskingum equation becomes the diffusion wave equation if the parameters for both methods are appropriately related. From Eqs. 9.50 and 9.52, the Muskingum equation is:

$$K\frac{d}{dt}[XI + (1 - X)O] = \overline{I} - \overline{O} \tag{9.60}$$

Substituting Q_i for I and Q_{i+1} for O, and rewriting in finite-difference form, we obtain:

$$\frac{K}{\Delta t}[XQ_i^{t+1} + (1 - X)Q_{i+1}^{t+1} - XQ_i^t - (1 - X)Q_{i+1}^t]$$
$$= \tfrac{1}{2}(Q_i^{t+1} - Q_{i+1}^{t+1} + Q_i^t - Q_{i+1}^t) \qquad (9.61)$$

If K is set equal to $\Delta x/c$, Eq. 9.61 is also the finite-difference form of:

$$\frac{\partial Q}{\partial t} + c\frac{\partial Q}{\partial x} = 0 \qquad (9.62)$$

which is called the *kinematic wave equation* and can be derived by combining the continuity and momentum (or friction) equations. The variable Δx denotes an increment of distance along the stream axis and c is the wave speed.

The equation to be used for routing is obtained from Eq. 9.61 by solving for the unknown flow rate:

$$Q_{i+1}^{t+1} = C_0 Q_i^{t+1} + C_1 Q_i^t + C_2 Q_{i+1}^t \qquad (9.63)$$

where

$$C_0 = \frac{\Delta t/K - 2X}{2(1 - x) + \Delta t/K} \qquad (9.64)$$

$$C_1 = \frac{\Delta t/K + 2X}{2(1 - x) + \Delta t/K} \qquad (9.65)$$

$$C_2 = \frac{2(1 - x) - c\,\Delta t/\Delta x}{2(1 - x) + \Delta t/K} \qquad (9.66)$$

Because $K = \Delta x/c$, it represents the time for a wave to travel the routing reach length Δx, moving at velocity c. Cunge shows that the velocity c is the celerity of a kinematic wave previously described (Table 9.7).

When $X = 0.5$ and $c\,\Delta t/\Delta x = 1.0$, the routing equation produces translation without attenuation. When $\Delta x = 0$ (zero reach length), no translation or attenuation occurs.

If previous flood data are available, the routing parameter c can be extracted by reversing the routing calculations. Estimates of the parameters can also be obtained from flow and channel measurements.

The value of X for use in Cunge's formulation is:

$$X = \frac{1}{2}\left(1 - \frac{q_0}{S_0 c\,\Delta x}\right) \qquad (9.67)$$

where S_0 = channel bottom slope (dimensionless)

   q_0 = discharge per unit width (cfs/ft), normally determined for the peak rate

The value of celerity c can be estimated as a function of the average velocity V by:

$$c = mV \qquad (9.68)$$

where V is the average velocity Q/A, and m is about $\frac{5}{3}$ for wide natural channels. The coefficient m comes from the uniform flow equation:

$$Q = bA^m \tag{9.69}$$

which reduces, by taking partial derivatives, to:

$$\frac{\partial Q}{\partial A} = m\frac{Q}{A} = mV \tag{9.70}$$

Substituting this into the continuity equation:

$$\frac{\partial Q}{\partial x} + \frac{\partial A}{\partial t} = 0 \tag{9.71}$$

gives Eq. 9.62 if $c = mV$. If discharge data are available, m can be estimated from Eq. 9.69. Values for common shape channels are given in Table 9.7.

The routing can now be done using either constant m and c parameters (i.e., using a single average velocity) or variable parameters (using each new velocity V). Equation 9.68 is solved for c, the value X is derived from Eq. 9.67, and Eqs. 9.64 to 9.66 are solved using $K = \Delta x/c$.

When using this method, the values of Δx and Δt should be selected to assure that the flood wave details are properly routed. Nominally, the time to peak of inflow is broken into 5 or 10 time increments Δt. To give both temporal and spatial resolution, the total reach length L can be divided into several increments of Δx length, and outflow from each is treated as inflow to the next.

Example 9.11

Use the Muskingum–Cunge method to route the hydrograph from Example 9.9. Use $S_0 = 0.0001$, $\Delta x = 545$ mi, flow cross-sectional area at $Q = 59{,}960$ is $5{,}996$ ft^2, width at $Q = 59{,}960$ is 60 ft, and $\Delta t = 1.0$ day (as in Example 9.9).

Solution. From the inflow, the peak rate of 59,960 cfs gives:

$$q_0 = \frac{Q_p}{T_p} = \frac{59{,}960}{60} = 1{,}000 \text{ cfs/ft}$$

$$V_p = \frac{Q_p}{A_p} = \frac{59{,}960}{5{,}996} = 10 \text{ fps}$$

$$c = \tfrac{5}{3}V_p = 16.7 \text{ fps}$$

From Eq. 9.67:

$$X = \frac{1}{2}\left[1 - \frac{1,000}{0.0001(16.7)545(5,280)}\right] = 0.4$$

$$K = \frac{\Delta x}{c} = \frac{545(5,280)}{16.7} = 172,800 \text{ sec}$$

and

$$C_0 = -0.1765$$
$$C_1 = 0.7647$$
$$C_2 = 0.2941$$

The routing for a portion of the hydrograph is given in Table 9.10.

TABLE 9.10   Solution to Example 9.11

Date, t	$C_0 Q_{inflow}^{t+1}$	$C_1 Q_{inflow}^t$	$C_2 Q_{outflow}^t$	$Q_{outflow}^{t+1}$
3-16	−1350	3260	0	1910 (3-17)
17	−1970	5850	560	4440
18	−2950	8540	1310	6900
19	−3810	12,790	2030	11,010
20	−3700	16,510	3240	16,050
21	−4690	16,020	4720	16,050
22	−8120	20,320	4720	16,920
23	−10,580	35,180	4980	29,580
24	−10,190	45,850	8700	44,360
25	−8450	44,150	13,050	48,750
26	−6080	36,620	14,340	44,880
27	−3820	26,350	13,200	35,730
3-28	−6120	16,560	10,510	20,950 (3-29)

Note that the peak outflow of 48,750 cfs on March 26 occurs on the same date as in Example 9.9 but has experienced slightly greater attenuation from the Muskingum–Cunge example.

The value C_1 is always positive, and negative values of C_2 are not particularly troublesome. Although C_0 is negative in this example, this condition should be avoided in practice. As seen from Eq. 9.64, negative values of C_0 are avoided when:

$$\frac{\Delta t}{K} > 2X \tag{9.72}$$

SCS Att-Kin TR-20 Method   In 1983, the SCS replaced the convex method with the modified *att-kin* (*att*enuation-*kin*ematic) method as the agency's preferred channel routing method [52]. The 1964 SCS TR-20 (Chapter 12) single-event simulation model used the convex method but was subsequently modified to route by the att-kin method.

The procedure is a blend of the storage indication and kinematic wave methods. Figure 9.22 shows the two-step process of simulating attenuation first by means of storage routing and then translating the wave in time by the kinematic wave method to account for the fact that routed flow rates not only decrease in magnitude but also require time to traverse the length of the routing reach. The storage routing portion provides attenuation with instantaneous translation, and the kinematic wave routing provides translation and distortion but does not attenuate the peak. Both are needed to produce the desired effect. The full dynamic equations [6] simultaneously account for both effects but are difficult to solve.

Figure 9.22 helps to visualize the process. The same volume V_1 of water flowing into the reach during time t_1 would flow out of a hypothetical storage reservoir during interval t_2. This same volume would translate and distort downstream by kinematic action, flowing out of the reach during interval t_3.

Through theoretical development and selection of routing coefficients in the att-kin method, its equations satisfy the physical propagation and timing of the peak flow rate first. Conservation of mass is also assured (areas under the three hydrographs of Fig. 9.22 are equal).

The actual process routes the inflow hydrograph through storage, then translates the peak flow rate, without attenuation, to its final time location in the outflow hydrograph. The location in time of the peak outflow is assumed equal to that corresponding to the maximum storage in the reach during passage of the flood.

Because celerity changes with storage, the other flows of the storage-routed hydrograph are translated, each by a different celerity, to their respective final times and values.

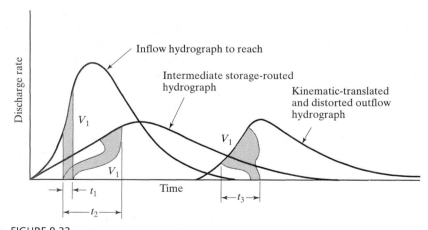

FIGURE 9.22
Routing principles used in the SCS att-kin method.

The storage indication routing is accomplished by substitution of the relation:

$$Q = KS^m \tag{9.73}$$

into the continuity equation, Eq. 9.71, where S is the storage and K and m are coefficients. Kinematic routing solves the unsteady flow equation with:

$$Q = bA^m \tag{9.74}$$

where b and m are input coefficients, and A is cross-sectional area. If L is the length of routing reach, and if the cross-sectional area throughout L is relatively constant, the storage is given by:

$$S = LA \tag{9.75}$$

These equations are combined in an iterative fashion to assure that the peak flow resulting from the kinematic routing equals the peak resulting from storage routing, and simultaneously ensuring that the time of the peak outflow occurs at the time of maximum storage in the reach, or:

$$Q_p = KS_p^m \tag{9.76}$$

Input to the method requires selection of a reach length and estimates of b and m. As discussed for Table 9.7 and Eq. 9.68, m can be shown to be a factor relating average velocity (under bank-full conditions) with wave celerity, or:

$$m = \frac{c}{V} \tag{9.77}$$

The larger m becomes, the shorter the travel time. A value of $m < 1.0$ would incorrectly make the celerity slower than the average flow velocity. Studies by SCS resulted in a recommendation of $\frac{5}{3}$ for general use. Significant errors resulted for m values greater than 2.0. Equation 9.74 is appropriate for cross sections having a single channel with regular shape. Complex cross sections are more diffcult to evaluate, but m values can be developed from a rating table for the stream [52].

As the coefficient b decreases, attenuation of the peak flow increases due to reduced velocity and increased storage in the reach. The value b can be estimated by plotting Q and A on log-log paper and fitting the linear form of Eq. 9.74. The slope would be m and the intercept at $A = 1$ would be b. The SCS has also developed nomographs for estimating b and m [52].

As a general guideline, the reach length L should be increased to a value that results in a kinematic wave travel time c greater than the selected time increment, or:

$$L_R \geq c\,\Delta t \tag{9.78}$$

where L_R is the recommended length, $c = mV$, and Δt is the time increment. The minimum recommended L_R is that giving a wave travel time equal to about half the time increment. This lower value may result in analytical difficulty when lengthy inflow hydrographs or steep streams are encountered. It also results in the peak outflow time being rounded up to the full time increment Δt. If several reaches were routed, this incremental time error would accumulate. Thus a reach length between $c\,\Delta t$ and $c\,\Delta t/2$ is acceptable, but a length greater than $c\,\Delta t$ is recommended. Figure 9.23 provides the range of minimum acceptable and minimum recommended routing reach lengths.

Numerous applications of hydraulic routing techniques appear in the literature; each is generally structured for a specific situation. The need to perform hydraulic routing is frequently undertaken in conjunction with a simulation study as will be further discussed in Chapter 12. Material presented here and in Chapter 12 is by no

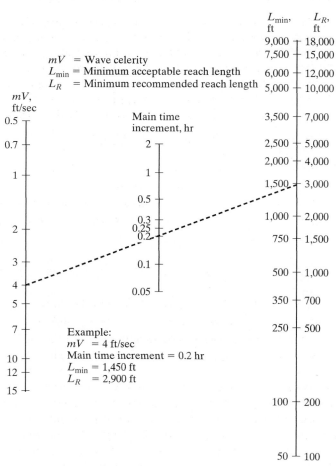

FIGURE 9.23

SCS nomograph for determining reach length for att-kin method of routing.
(After U.S. Soil Conservation Service, "Computer Program for Project Formulation," Technical Release 20, Revised, Appendix G, 1983)

means exhaustive but rather is presented so that an interested student can understand the structuring processes of hydrologic modeling.

Other Methods Other hydrologic river-routing procedures have been developed, including the working R&D method, straddle-stagger method, Tatum method, and multiple storage method. The U.S. Geological Survey maintains updates of software for performing streamflow routing in upland channels or channel networks. A download of the program DAFLOW is available at http://water.usgs.gov/software/surface_water.html. The program FEQ, capable of performing full dynamic wave routing [6] is available at the same Web site.

Hydrologic Reservoir Routing by Modified Puls Method

The *modified Puls* method of routing a hydrograph through a reservoir is also called the *storage indication* method [53]. A flood wave passing through a storage reservoir is both delayed and attenuated as it enters and spreads over the pool surface. Water stored in the reservoir is gradually released as pipe flow through turbines or outlet works, called *principal* spillways, or in extreme floods, over an *emergency* spillway.

Flow over an ungated emergency spillway weir section can be described from energy, momentum, and continuity considerations by the form:

$$O = CYH^x \qquad (9.79)$$

where O = the outflow rate (cfs)
Y = the length of the spillway crest (ft)
H = the deepest reservoir depth above the spillway crest (ft)
C = the discharge coefficient for the weir or section, theoretically 3.0
x = an exponent, theoretically $\frac{3}{2}$

Flow through a free outlet discharge pipe is similarly described by Eq. 9.79

where Y = the cross-sectional area of the discharge pipe (ft^2)
H = the head above the free outlet elevation (ft)
C = the pipe discharge coefficient, theoretically $\sqrt{2g}$
x = an exponent, theoretically $\frac{1}{2}$

Flow equations for other outlet conditions are available in hydraulics textbooks. Storage values for various pool elevations in a reservoir are readily determined from computations of volumes confined between various pool areas measured from topographic maps. Since storage and outflow both depend only on pool elevation, the resulting storage-elevation curve and the outflow-elevation relation (Eq. 9.79) can easily be combined to form a storage-outflow graph. Storage in a reservoir depends only on the outflow, contrasted to the dependence on the inflow and outflow in river routing (Eq. 9.52).

For convenience, S is often defined as the *surcharge storage* or the storage above the emergency spillway crest. Normally the overflow rate is zero when S is zero. If the graphed storage-outflow relation is found to be linear, and if the slope of the line is defined as K, then:

$$S = KO \qquad (9.80)$$

and the reservoir is called a *linear reservoir*. Routing through a linear reservoir is a special case of Muskingum river routing shown in Fig. 9.21 using $x = 0$ in Eq. 9.52. Note also that the outflow rate in Fig. 9.21 is increasing only while the inflow exceeds the outflow. This observation is consistent with the assumptions that the inflow immediately goes into storage over the entire pool surface and that the outflow depends only on this storage.

Routing through a linear reservoir is easily accomplished by first dividing time into a number of equal increments and then substituting $S_2 = KO_2$ in Eq. 9.53 and solving for O_2, which is the only remaining unknown for each time increment.

To route a flood through a *nonlinear* reservoir, the storage-outflow relation and the continuity equation, Eq. 9.53, are combined to determine the outflow and storage at the end of each time increment Δt. Equation 9.53 can be rewritten as:

$$I_n + I_{n+1} + \left(\frac{2S_n}{\Delta t} - O_n\right) = \frac{2S_{n+1}}{\Delta t} + O_{n+1} \qquad (9.81)$$

in which the only unknown for any time increment is the term on the right side. Pairs of trial values of S_{n+1} and O_{n+1} could be generated that satisfy Eq. 9.81 and checked in the storage-outflow curve for confirmation. Rather than resort to this trial procedure, a value of Δt is selected and points on the storage outflow curve are replotted as the "storage indication" curve shown in Fig. 9.24. This graph allows a *direct* determination of the outflow O_{n+1} once a value of the ordinate $(2S_{n+1}/\Delta t) + O_{n+1}$ has been calculated from Eq. 9.81. The second unknown, S_{n+1}, can be read from the S-O curve (which

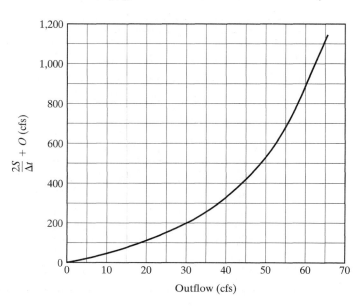

FIGURE 9.24

Curve of $(2S/\Delta t) + O$ versus O.

could also be plotted on the graph in Fig. 9.24) or found from Eq. 9.81. This row-by-row numerical integration of Eq. 9.81 with Fig. 9.24 is illustrated using $\Delta t = 1$ hr in Example 9.12.

Example 9.12

Given the triangular-shaped inflow hydrograph and the $(2S/\Delta t) + O$ curve of Fig. 9.24 find the outflow hydrograph for the reservoir assuming it to be completely full at the beginning of the storm. (See Table 9.11.)

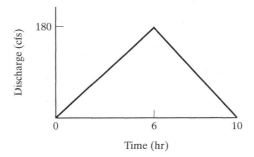

Solution. In selecting a routing period Δt, generally at least five points on the rising limb of the inflow hydrograph are employed in the calculations. An increased number of points on the rising limb, that is, a small Δt, improves the accuracy, since as $\Delta t \rightarrow 0$ the numerical integration approaches the true limit of the function being integrated, in this case dS/dt.

TABLE 9.11 Routing Table for Example 9.12

(1)	(2)	(3)	(4)	+	(5)	=	(6)	(7)	(8)
Time	n	I_n	$I_n + I_{n+1}$		$\dfrac{2S_n}{\Delta t} - O_n$		$\dfrac{2S_{n+1}}{\Delta t} + O_{n+1}$	O_{n+1}	S_{n+1}
(hr)		(cfs)	(cfs)		(cfs)		(cfs)	(cfs)	(cfs-hr)
0	1	0	30		0		30	5	12.5
1	2	30	90		20		110	18	46
2	3	60	150		74		224	32	96
3	4	90	210		160		370	43	164
4	5	120	270		284		554	52	250
5	6	150	330		450		780	58	361
6	7	180	315		664		979	63	458
7	8	135	225		853		1078	65	506
8	9	90	135		948		1085	65	510
9	10	45	45		953		998	64	467
10	11	0	0		870		870	62	404
11	12	0	0		746		746	58	344
12	13	0	0		630		630	54	288

Column 3 in Table 9.11 comes from the given inflow hydrograph, column 4 is simply $I_n + I_{n+1}$, and columns 5 and 7 are initially zero, since in this problem the reservoir is assumed full at the commencement of inflow. Therefore, there is no available storage.

The starting value for $n = 1$ in column 6 is computed as the sum of columns 4 and 5 from Eq. 9.81:

$$(I_1 + I_2) + \left(\frac{2S_1}{\Delta t} - O_1\right) = \frac{2S_2}{\Delta t} + O_2$$

$$30 + 0 = \frac{2S_2}{\Delta t} + O_2$$

Entering the ordinate of Fig. 9.24 with the value 30 from column 6 gives a value for O_2 of 5 cfs, which is recorded in column 7. The corresponding end-of-time-interval storage, S_2, is calculated from columns 6 and 7 and recorded in column 8. Moving to the second row, a value of the term in column 5 can now be found for $n = 2$ using S_2 and O_2 from columns 7 and 8.

The stepwise procedure used to get outflow figures for all values of n can be summarized as

1. Entries in columns 1 and 3 are known from the given inflow hydrograph.
2. Entries in column 4 are the values of $I_n + I_{n+1}$ from column 3.
3. The initial value of the term in column 5 is zero, though it could also be based on any arbitrary starting storage value, and columns 4 and 5 are added to produce the value in column 6.
4. The $(2S/\Delta t) + O$ versus O plot is entered with known values of $(2S/\Delta t) + O$ to find values of O for column 7.
5. Columns 6 and 7 are solved for S_{n+1}, which is recorded in column 8.
6. Advance to the next row and calculate the next value for column 5 using the values in the preceding row for O and S from columns 7 and 8.
7. Add the value in column 5 to the advanced sum in column 4 and enter the result in column 6 for the new period under consideration.
8. The new outflow for column 7 is again found from the relation of $(2S/\Delta t) + O$ as in Fig. 9.24.
9. The corresponding new storage in column 8 is found by solving from columns 6 and 7.
10. Steps 6 through 9 are repeated until the entire outflow hydrograph is generated.

SUMMARY

Understanding the structure of hydrographs is important to many flood control and water supply studies and designs. Hydrograph analysis is one of the most-used procedures in determining waterway openings under bridges, designing flood protection levees, and establishing the lateral extent of flooding. Similarly, the volume of runoff into a reservoir or wetland, or the volume passing a water supply diversion, is determined from the area under the hydrograph. Accurate estimates of these peak flow rates and runoff volumes are important to the design of bridges, dams, reservoirs, irrigation water diversions, and numerous other structures.

Unit hydrograph methods allow the hydrologist to estimate runoff volumes and rates for virtually any storm. By far the greatest number of practical problems is solved using unit hydrograph procedures. The most widely applied hydrograph and hydrologic routing procedures were presented in this chapter.

All significant federal agency public domain computer routines for storm event modeling are described in Chapters 11 and 12. Most of the current computer models use unit hydrograph procedures. These simulation models are simply computer programs that perform the unit hydrograph synthesis and arithmetic lagging methods described in this chapter. Computer software for hydrograph synthesis and routing is available from numerous public and private vendors. Some models, such as HEC-HMS (Chapter 12) include the majority of the procedures described, in addition to incorporating Geographic Information System (GIS) routines.

Any software user should understand the origin, applicability, and parameter estimation procedures for each unit hydrograph method selected. The most successful uses of the computer models will result from a thorough familiarity with the processes described in this chapter.

PROBLEMS

SECTION 9.1: HYDROGRAPH COMPONENTS

9.1 Tabulated below are total hourly discharge rates at a cross section of a stream. The drainage area above the section is 1.0 acre. (a) Plot the hydrograph on rectangular coordinate paper and label the rising limb (concentration curve), the crest segment, and the recession limb. (b) Determine the hour of cessation of the direct runoff using a semilog plot of Q versus time. (c) Use the base flow portion of your semilog plot to determine the groundwater recession constant K. (d) Carefully construct and label base flow separation curves on the graph of part (a), using two different methods.

Time (hr)	Q (cfs)	Time (hr)	Q (cfs)
0	102	8	210
1	100	9	150
2	98	10	105
3	220	11	75
4	512	12	60
5	630	13	54
6	460	14	48.5
7	330	15	43.5

9.2 On a sketch of a typical total runoff hydrograph, show or dimension the (a) storm hyetograph, (b) beginning of direct runoff, (c) cessation time of direct runoff, (d) base flow separation assuming that *additional* contributions to base flow are negligible during the period of rise, and (e) crest segment of the hydrograph.

9.3 A 10.0-mi² drainage basin has a time to equilibrium of 100 min and a φ index of 0.25 in./hr. If rain falls uniformly over the basin at a rate of 2.75 in./hr for a duration of 200 min, sketch the approximate hydrograph and estimate the maximum discharge rate (cfs) at the basin outlet.

9.4 A hydrograph for a 4,250-acre basin is shown in the accompanying sketch. The given hydrograph actually appeared as a direct runoff hydrograph from the basin, caused by net rain falling at an intensity of 0.20 in./hr for a duration of 5 hr, beginning at $t = 0$.

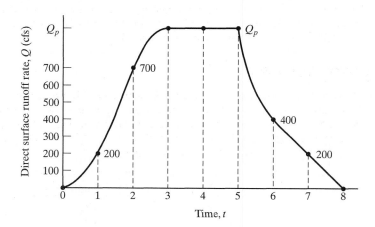

(a) Determine the excess release time of the basin.

(b) What percentage of the drainage basin was contributing to direct runoff 4 hr after rain began ($t = 4$)?

(c) Use your response to part (b) to determine Q_p, as shown in the sketch. Do not scale Q_p from the drawing.

(d) Note that rain continued to fall between $t = 3$ and $t = 5$. Why did the hydrograph form a plateau between $t = 3$ and $t = 5$, rather than continue to rise during those 2 hours?

9.5 Obtain streamflow data for a water course of interest. Plot the hydrograph for a major runoff event and separate the base flow.

SECTION 9.2: HYDROGRAPH TIME RELATIONSHIPS

9.6 For the event of Problem 9.5, tabulate the precipitation causing the surface runoff and determine the duration of runoff-producing rain. Estimate the time of concentration and use Eq. 9.5 to estimate the time base of the hydrograph. Compare this with the time base computed from the hydrograph.

9.7 Discharge rates for a flood hydrograph passing the point of concentration for a 600-acre drainage basin are given in the table below. The flood was produced by a uniform rainfall rate of 2.75 in./hr, which started at 9 a.m., abruptly ended at 11 a.m. and resulted in 5.00 in. of direct surface runoff. The base flow (derived from influent seepage) prior to, during, and after the storm was 100 cfs.

Time	8 a.m.	9	10	11	12	1 p.m.	2	3	4	5	6	7
Measured discharge	100	100	300	500	700	800	600	400	300	200	100	100

(a) At what times did direct runoff begin and cease?

(b) Determine the ϕ index (in./hr) for the basin.

(c) Estimate the time of concentration (excess release time) for the basin.

(d) At what time would direct surface runoff cease if the rainfall of 2.75 in./hr had begun at 9 a.m. and had lasted for 8 hours rather than 2?

9.8 For a watershed assigned by your instructor, obtain measures of the watershed area, length, and slope, and compare estimates of the time of concentration using the Kirpich, USBR, FAA, and SCS lag equations in Table 9.1.

SECTION 9.3: UNIT HYDROGRAPHS

9.9 Recorded flow rates for a net rain of 1.92 inches in 12 hours are shown in the table. If the base flow is 375 cfs throughout the storm, determine the 12-hr unit hydrograph, and convert it to a 6-hr unit hydrograph. Then apply the 6-hr unit hydrograph to determine the total hydrograph (including 400 cfs base flow) for a 24-hr storm having four 6-hr blocks of net rain at rates of 0.7, 3.8, 10.8, and 1.8 in. per hour.

Time (hr)	Observed flow (cfs)
0	375
6	825
12	2200
18	3650
24	3900
30	3200
36	2375
42	1725
48	1250
54	900
60	650
66	490
72	410
78	375

9.10 Using U.S. Geological Survey records, or other data, select a streamflow hydrograph for a large, preferably single-peaked runoff event. Separate the base flow and determine a unit hydrograph for the area.

9.11 Measured total hourly discharge rates (in cfs) from a 2.48-mi^2 drainage basin are tabulated below. The hydrograph was produced by a rainstorm having a uniform intensity of 2.60 in./hr starting at 9 a.m. and abruptly ending at 11 a.m. The base flow from 8 a.m. to 3 p.m. was a constant 100 cfs.

Time	8 a.m.	9	10	11	12	1	2	3 p.m.
Discharge (cfs)	100	100	300	450	300	150	100	100

(a) At what time did direct runoff begin?

(b) Determine the gross and net rain depths (in inches).

(c) Derive a 2-hr unit hydrograph for the basin by tabulating time in hours and discharge in cubic feet per second.

(d) Derive a 4-hr unit hydrograph for the basin.

(e) Derive a 1-hr unit hydrograph for the basin.

9.12 Use the following 4-hr unit hydrograph for a basin to determine the peak discharge rate (cfs) resulting from a net rain of 3.0 in./hr for a 4-hr duration followed immediately by 2.0 in./hr for a 4-hr duration.

Time (hr)	0	2	4	6	8	10
Q (cfs)	0	200	300	100	50	0

9.13 The ordinate for a 5-hr unit hydrograph is 300 cfs at a time 4 hr after the beginning of net rainfall. A storm with a uniform intensity of 3 in./hr and a duration of 5 hr occurs over the basin. What is the runoff rate after 4 hr if the ϕ index is 0.5 in./hr?

9.14 A drainage basin has a time of concentration of 8 hr and produces a peak Q of 4,032 cfs for a 10-hr storm with a net intensity of 2 in./hr. Determine the peak flow rate and the time base (duration) of the direct surface runoff for a net rain of 4 in./hr lasting (a) 12 hr, (b) 8 hr, and (c) 4 hr. State any assumptions used.

9.15 Measured total hourly discharge rates (in cfs) from a 3.10-mi^2 drainage basin are listed in the accompanying table. The hydrograph was produced by a rainstorm having a uniform intensity of 2.60 in./hr starting at 9 a.m. and abruptly ending at 11 a.m. The base flow from 8 a.m. to 3 p.m. was a constant 100 cfs. The volume of direct runoff, determined as the area under the direct surface runoff hydrograph, is 1,000 cfs-hr.

Time	8 a.m.	9	10	11	12	1 p.m.	2	3
Measured discharge	100	100	300	600	400	200	100	100

(a) At what time did the direct runoff begin?

(b) Determine the net rain (in inches) corresponding to the volume of the direct surface runoff of 1,000 cfs-hr.

(c) Determine the ϕ index for the basin.

(d) Derive a 2-hr unit hydrograph for the basin by tabulating time in hours and discharge in cfs.

(e) What is the excess release time of the basin?

(f) For the same basin, use the derived 2-hr unit hydrograph to determine the direct runoff rate (cfs) at 4 p.m. on a day when excess (net) rainfall began at 1 p.m. and continued at a net intensity of 2 in./hr for 4 hr, ceasing abruptly at 5 p.m.

9.16 The 2-hr unit hydrograph for a basin is given by the following table:

Time (hr)	0	1	2	3	4	5	6	7	8	9
Q (cfs)	0	60	200	300	200	120	60	30	10	0

(a) Determine the hourly discharge values (in cfs) from the basin for a net rain of 5 in./hr and a rainfall duration of 2 hr.

(b) Determine the direct runoff (in inches) for the storm of part (a). What is the direct runoff for a net rain of 0.5 in./hr and a duration of 2 hr?

(c) Rain falls on the basin at a rate of 4.5 in./hr for a 2-hr period and abruptly increases to a rate of 6.5 in./hr for a second 2-hr period. Convert these actual intensities to net rain intensities using a ϕ index of 0.5 in./hr. Construct a table that properly lags and amplifies the 2-hr unit hydrograph, and determine the hourly ordinates (in cfs) of direct runoff for the storm. The derived direct runoff hydrograph should begin and end with zero discharge values.

9.17 Given the following storm pattern and assuming a triangular unit hydrograph for one time unit, determine the composite hydrograph.

Storm pattern				
Time unit	1	2	3	4
Rainfall	1	1	4	2

Unit hydrograph base length = 6 time units, time of rise = 2 time units, and maximum ordinate = $\frac{1}{3}$ rainfall unit height.

9.18 Given the following 2-hr unit hydrograph for a drainage basin, determine hourly ordinates of the 4-hr unit hydrograph:

Time (hr)	0	1	2	3	4	5	6
Q (cfs)	0	50	300	400	200	50	0

9.19 Given below is a 3-hr unit hydrograph for a watershed. The ϕ index is 1.5 in./hr. Create the DRH for an 18-hr storm having six successive 3-hr rainfall rates of 2.5, 3.5, 1.5, 4.0, 6.5, and 2.5 in./hr.

Time (hr)	0	1	2	3	4	5	6	7	8	9	10	11	12	13	14
Q (UH)	0	10	40	60	80	100	90	70	60	50	40	30	20	10	0

9.20 Use the following 2-hr unit hydrograph to determine the peak direct-runoff discharge rate (in cfs) resulting from a net rain of 2.0 in./hr for 5 hr.

Time (hr)	0	1	2	3	4	5	6	7
Q (cfs)	0	50	200	300	200	150	100	0

9.21 A 2-hr unit hydrograph for a basin is shown in the sketch.

  (a) Determine the peak discharge (in cfs) for a net rain of 5.00 in./hr and a duration of 2 hr.

  (b) What is the total direct surface runoff (in inches) for the storm described in part (a)?

  (c) A different storm with a net rain of 0.50 in./hr lasts for 4 hr. What is the discharge at 8 p.m. if the rainfall started at 4 p.m.?

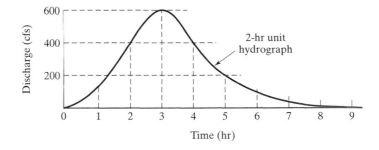

9.22 For the unit hydrograph of Problem 9.17, construct an S-hydrograph.

9.23 Given below is an IUH for a watershed. Use the IUH to find hourly DRH rates for a net rain of 4 in. in a 2-hr period.

Time (hr)	0	1	2	3	4	5	6	7	8	9	10
Q (IUH)	0	10	40	50	60	80	100	80	20	10	0

9.24 Describe two methods that could be used to construct a 2-hr unit hydrograph using a 1-hr unit hydrograph for a basin.

SECTION 9.4: SYNTHETIC UNIT HYDROGRAPHS

9.25 Given a watershed of 100 mi^2, assume that $C_t = 1.8$, the length of the main stream channel is 18 mi, and the length to a point nearest the centroid is 10 mi. Use Snyder's method to find (a) the time lag, (b) the duration of the synthetic unit hydrograph, and (c) the peak discharge of the unit hydrograph.

9.26 Apply Snyder's method to the determination of a synthetic unit hydrograph for a drainage area of your choice.

9.27 By calculus, show that the maximum value of $f(x)$ in Eq. 9.18 occurs when $x = \alpha\beta$, for $\alpha \geq 1$. Also solve for the centroidal distance by taking the first moment about the y axis.

9.28 Use Fig. 9.18 and Table 9.6 to determine a 2-hr unit hydrograph if the drainage area is 60 mi^2 and Eq. 9.33a is applicable.

9.29 Solve Problem 9.28 using Eq. 9.33b.

9.30 Compare the time from the peak to the end of runoff for the SCS triangular unit hydrograph with the time of concentration, t_c. Discuss.

9.31 Prove that the area under the rising limb of the SCS basic dimensionless hydrograph equals that of the triangular unit hydrograph, that is, 37.5% of the total.

9.32 Using the SCS dimensionless unit hydrograph, determine the peak discharge for a net storm of 10 in. in 2 hr on a 400-acre basin with a time to peak of 4 hr and a lag time of 3 hr. Compare with Eq. 9.28.

9.33 Using the peak flow for the SCS dimensionless unit hydrograph, determine the peak discharge for a net storm of 10 in. in 2 hr on a 400-acre basin with a time to peak of 4 hr and a lag time of 3 hr.

9.34 Starting with a triangular-shaped unit hydrograph with a base length of $2.67t_p$ and a height of q_p, derive Eq. 9.32. State the units of each term used in the derivation.

9.35 The SCS synthetic unit hydrograph is derived by computing the peak discharge rate (in cubic feet per second) from $q_p = 484A/t_p$. In the derivation, it was actually assumed that q_p in./hr $= 0.75V/t_p$, where V is the volume of direct runoff (inches), t_p is the time to peak flow (hours), and A is the basin area (square miles). Derive the first equation from the second.

9.36 Which of the techniques for synthesizing a unit hydrograph requires the least computational effort in developing the entire unit hydrograph? Which probably requires the most?

SECTION 9.5: HYDROGRAPH ROUTING

9.37 The Muskingum river routing equation, $O_2 = C_0I_2 + C_1I_1 + C_2O_1$, was derived by substituting the storage equation $S_i = K[XI_i + (1 - X)O_i]$, where $S_1 = K[XI_1 +$

$(1 - X)O_1]$ and $S_2 = K[XI_2 + (1 - X)O_2]$ into the continuity equation $\bar{I} = \bar{O} + (\Delta S/\Delta t)$ and combining like terms. In these equations, I_1, O_1, and S_1 are the inflow, outflow, and storage, respectively, at the beginning of the time period, and I_2, O_2, and S_2 are, respectively, the corresponding values at the end of the time period. The terms $\bar{I}$ and $\bar{O}$ are the average inflow and outflow during the time period, and ΔS is the change in storage. Perform the described derivation and verify the equations for C_0, C_1, and C_2.

9.38 A flood hydrograph is to be routed by the Muskingum method through a 10-mi reach with $K = 2$ hr. Into how many subreaches must the 10-mi river reach be divided in order to use $\Delta t = 0.5$ hr and still satisfy the stability criteria $K/3 \le \Delta t \le K$.

9.39 If the Muskingum K value is 12 hr for a reach of a river, and if the X value is 0.2, what would be a reasonable value of Δt for routing purposes?

9.40 A river reach has a storage relation given by $S_i = aI_i + bO_i$. Derive a routing equation for O_2 analogous to the Muskingum equation (9.55). Give equations for the coefficients of I_1, O_1, and I_2.

9.41 List the steps (starting with a measured inflow and outflow hydrograph for a river reach) necessary to determine the Muskingum K and X values. If the inflow and outflow are recorded in cubic feet per second, state the units that would result for K and X if your list of steps is followed.

9.42 Given the following inflow hydrograph:

Hour	Inflow (cfs)	Outflow (cfs)
6 a.m.	100	100
Noon	300	
6 p.m.	680	
Midnight	500	
6 a.m.	400	
Noon	310	
6 p.m.	230	
Midnight	100	

Assume that the outflow hydrograph at a section 3-mi downstream is desired.

(a) Compute the outflow hydrograph by the Muskingum method using values of $K = 11$ hr and $X = 0.13$.

(b) Plot the inflow and outflow hydrographs on a single graph.

(c) Repeat steps (a) and (b) using $X = 0$.

9.43 Repeat Problem 9.42(a) by dividing the 3-mi reach into two subreaches with equal K values of 5.5 hr. Compare the results.

9.44 Given the following values of measured discharges at both ends of a 30-mi river reach:

(a) Determine the Muskingum K and X values for this reach.

(b) Holding K constant (at your determined value), use the given inflow hydrograph to determine and plot three outflow hydrographs for values of X equal to the computed value, 0.5, and 0. Plot the actual outflow and numerically compare the root mean square of residuals when each of the three calculated hydrographs is compared with the measured outflow.

Time	Inflow (cfs)	Outflow (cfs)
6 a.m.	10	10
Noon	30	12.9
6 p.m.	68	26.5
Midnight	50	43.1
6 a.m.	40	44.9
Noon	31	41.3
6 p.m.	23	35.3
Midnight	10	27.7
6 a.m.	10	19.4
Noon	10	15.1
6 p.m.	10	12.7
Midnight	10	11.5
6 a.m.	10	10.8

9.45 Discuss the problems associated with the use of a reservoir-routing technique such as the storage indication method in routing a flood through a river reach.

9.46 Select a stream in your geographic region that has runoff records. Use the Muskingum method of routing to find K and X.

9.47 Precipitation began at noon on June 14 and caused a flood hydrograph in a stream. As the hydrograph passed, the following measured streamflow data at cross sections A and B were obtained:

Time, June 14–17	Inflow, section A (cfs)	Outflow, section B (cfs)
6 a.m.	10	10
Noon	10	10
6 p.m.	30	13
Midnight	70	26
6 a.m.	50	43
Noon	40	45
6 p.m.	30	41
Midnight	20	35
6 a.m.	10	28
Noon	10	19
6 p.m.	10	15
Midnight	10	13
6 a.m.	10	11
Noon	10	10

(a) Determine the Muskingum K and X values for the river reach.

(b) Determine the hydrograph at section B if a different storm produced the following hydrograph at section A:

Time	Inflow (cfs)	Time	Inflow (cfs)
6 a.m.	100	Noon	400
Noon	100	6 p.m.	300
6 p.m.	200	Midnight	200
Midnight	500	6 a.m.	100
6 a.m.	600	Noon	100

9.48 Precipitation began at noon on June 14 and caused a flood hydrograph in a stream. As the storm passed, the following streamflow data at cross sections A and B were obtained:

Time, June 14–17	Inflow section A (cfs)	Outflow section B (cfs)
6 a.m.	10	10
Noon	10	10
6 p.m.	30	13
Midnight	70	26
6 a.m.	50	43
Noon	40	45
6 p.m.	30	41
Midnight	20	35
6 a.m.	10	28
Noon	10	19
6 p.m.	10	15
Midnight	10	13
6 a.m.	10	11
Noon	10	10

(a) Determine the Muskingum K and X values for the river reach.

(b) Determine the hydrograph at section B if a different storm produced the following hydrograph at section A (continue computations until outflow falls below 101 cfs):

Time	Inflow (cfs)	Time	Inflow (cfs)
6 a.m.	100	Noon	400
Noon	100	6 p.m.	300
6 p.m.	200	Midnight	200
Midnight	500	6 a.m.	100
6 a.m.	600	Noon	100

9.49 Use the following values of measured discharges at both ends of a 30-mi river reach:

Time	Inflow (cfs)	Outflow (cfs)
6 a.m.	10	10
Noon	30	12.9
6 p.m.	68	26.5
Midnight	50	43.1
6 a.m.	40	44.9
Noon	31	41.3
6 p.m.	23	35.3
Midnight	10	27.7
6 a.m.	10	19.4
Noon	10	15.1
6 p.m.	10	12.7
Midnight	10	11.5
6 a.m.	10	10.8

(a) Determine the Muskingum K and X values for this reach.

(b) Holding K constant (at your determined value), use the given inflow hydrograph to determine and plot three outflow hydrographs for values of x equal to the computed value, 0.5, and 0.

9.50 Use the following inflow hydrograph:

Time	Inflow (cfs)	Outflow (cfs)
6 a.m.	10	10
Noon	30	
6 p.m.	68	
Midnight	50	
6 a.m.	40	
Noon	31	
6 p.m.	23	
Midnight	10	

Assume that the outflow hydrograph at a section 3-mi downstream is desired.

(a) Compute the outflow hydrograph by the Muskingum method using values of $K = 11$ hr and $X = 0.13$.

(b) Plot the inflow and outflow hydrographs on a single graph.

(c) Repeat steps (a) and (b) using $X = 0$.

9.51 If the Muskingum K value is 12 hr for a reach of a river, and if the X value is 0.2, what would be a reasonable value of Δt for routing purposes?

9.52 The outflow rate (cfs) and storage (cfs-hr) for an emergency spillway of a certain reservoir are linearly related by $O = S/3$, where the number 3 has units of hours. Use this and the continuity equation to determine the peak outflow rate from the reservoir for the following inflow event:

Time (hr)	I (cfs)	S (cfs-hr)	O (cfs)
0	0	0	0
2	400		
4	600		
6	200		
8	0		

9.53 A simple reservoir has a linear storage–indication curve defined by the equation:

$$O = \frac{O}{2} + \frac{S}{\Delta t}$$

where Δt is equal to 1.0 hr. If S at 8 a.m. is 0 cfs-hr, use the continuity equation to route the following hydrograph through the reservoir:

Time	8 a.m.	9 a.m.	10 a.m.	11 a.m.	Noon	1 p.m.
I (cfs)	0	200	400	200	0	0

9.54 For a vertical-walled reservoir with a surface area A show how the two routing equations (Eq. 9.79) and (Eq. 9.81) could be written to contain only O_2, S_2, and known values (computed from O_1, S_1, and so on). Eliminate H from all the equations. How could these two equations be solved for the two unknowns?

9.55 The following data are given: vertical-walled reservoir, surface area = 1,000 acres; emergency spillway width = 97.1 ft (ideal spillway); H = water surface elevation (in ft) above the spillway crest; and initial inflow and outflow are both 100 cfs.

(a) In acre-ft and cfs-days, determine the values for reservoir storage S corresponding to the following values of H: 0, 0.5, 1, 1.5, 2, 3, and 4 ft.

(b) Determine the values of the emergency spillway Q corresponding to the depths named in part (a).

(c) Carefully plot and label the discharge–storage curve (cfs versus cfs-days) and the storage–indication curve (cfs versus cfs, Fig. 9.24) on rectangular coordinate graph paper.

(d) Determine the outflow rates over the spillway at the ends of successive days corresponding to the following inflow rates (instantaneous rates at the ends of successive days): 100, 400, 1200, 1500, 1100, 700, 400, 300, 200, 100, 100, and 100. Use a routing table similar to the one used in Example 9.10 and continue the routing procedure until the outflow drops below 10 cfs.

(e) Plot the inflow and outflow hydrographs on a single graph. Where should these curves cross?

9.56 Route the given inflow hydrograph through the reservoir by assuming the initial water level is at the emergency spillway level (1,160 ft) and that the principal spillway is plugged with debris. The reservoir has a 500-ft-wide ideal emergency spillway (C = 3.0) located at the 1,160-ft elevation. Storage–area–elevation data are:

Elevation (ft)	Area of pool ($ft^2 \times 10^6$)	Storage ($ft^3 \times 10^6$)
1110	0	0
1120	0.85	4.25
1140	3.75	50.25
1158	9.8	172.15
1160	10.8	192.75
1162	11.8	215.35
1164	12.8	239.95
1166	13.8	266.55
1168	14.85	295.20
1180	25.0	528.55

The inflow hydrograph data are:

Time (hr)	I (cfs)
0.0	0
0.5	3,630
1.0	10,920
1.5	10,720
2.0	5,030
2.5	1,600
3.0	460
3.5	100
4.0	10
4.5	0

9.57 Given the following data, route a storm hydrograph through a full reservoir and plot on the same graph the inflow and resulting outflow hydrograph for the Green Acre watershed. The bottom of the rectangular spillway is placed at an elevation of 980.0 ft. The following data are given: area $= 0.64$ mi^2; length $= 1.10$ mi; $L_{ca} = 0.53$ mi; $C_t = 2.00$; $C_p = 0.62$; outflow $= CYH^{3/2}$; $C = 3.5$; $L = 10$ ft; and the following storage–elevation curve table.

(a) Find the 15-min unit hydrograph by Snyder's method.

(b) Find the 30-min unit hydrograph.

(c) Find the hydrograph that results from 1.8 in. of rain for the first 30 min and 0.63 in. for the next 15 min.

(d) Develop a $(2S/\Delta t) + O$ versus O curve using a routing period of 15 min and the outflow and storage curves provided.

(e) Route the storm hydrograph through the reservoir assuming it is full to the bottom of the spillway elevation 980 ft.

(f) Indicate the maximum height of the water in the reservoir.

(g) At what elevation should the top of the dam be placed to obtain 5 ft of free-board?

Elevation (ft)	Incremental storage (10^{-4} ft^3)	Total storage (10^{-4} ft^3)
960-		0
	40	
970		40
	210	
980		250
	590	
990		840
	1240	
1000		2080

9.58 Repeat Problem 9.57 with the reservoir initially empty.

9.59 By taking derivatives of the reservoir routing equations, prove that the peak outflow rate intersects the falling limb of the inflow hydrograph.

REFERENCES

[1] American Society of Civil Engineers, *Hydrology Handbook,* Manuals of Engineering Practice, No. 28. New York: ASCE, 1957.

[2] Donn G. DeCoursey, "A Runoff Hydrograph Equation," U.S. Department of Agriculture, Agricultural Research Service, Feb. 1966, pp. 41–116.

[3] R. K. Linsley, M. A. Kohler, and J. L. H. Paulhus, *Applied Hydrology,* New York: McGraw-Hill, 1949.

[4] R. J. M. DeWiest, *Geohydrology,* New York: Wiley, 1965.

[5] A. B. Taylor and H. E. Schwartz, "Unit Hydrograph Lag and Peak Flow Related to Basin Characteristics," *Trans. Am. Geophys. Union* **33**(1952).

[6] W. Viessman and G. L. Lewis, *Introduction to Hydrology, 4th ed.,* New York: HarperCollins College, 1996.

[7] D. M. Gray (ed.), *Handbook on the Principles of Hydrology,* Port Washington, NY: Water Information Center, 1970.

[8] Dodson and Associates Inc., *The Dodson Professional HEC-1 System, PROHEC-1 Documentation,* 1992.

[9] C. T. Haan, B. J. Barfield, and J. C. Hayes, *Design Hydrology and Sedimentology for Small Catchments,* San Diego: Academic Press, 1994.

[10] L. K. Sherman, "Streamflow from Rainfall by the Unit-Graph Method," *Eng. News-Rec.* **108**(1932).

[11] U.S. Soil Conservation Service, *National Engineering Handbook,* Sec. 4, Hydrology, U.S. Department of Agriculture, Washington, D.C., 1972.

[12] P. S. Eagleson, *Dynamic Hydrology,* New York: McGraw-Hill, 1970.

[13] V. M. Ponce, *Engineering Hydrology, Principles and Practices,* Englewood Cliffs, NJ: Prentice Hall, 1989.

[14] D. R. Maidment (ed.), *Handbook of Hydrology,* New York: McGraw-Hill, 1993.

[15] D. F. Kibler, "Desk-top methods for Urban Stormwater Calculation," Chap. 4 in *Urban Stormwater Hydrology, Water Resources Monograph No. 7,* American Geophysical Union, Washington, D.C., 1982.

[16] C. F. Izzard, "Hydraulics of Runoff from Developed Surfaces," *Proceedings, 26th Annual Meeting of the Highway Research Board,* Vol. 26, December 1946.

[17] Federal Aviation Administration, "Circular on Airport Drainage," Report A/C 050-5320-5B, Washington, D.C., 1970.

[18] J. R. Morgali and R. K. Linsley, "Computer Analysis of Overland Flow," *J. Hydraulics Division,* ASCE, Vol. 91, No. HY3, May 1965.

[19] G. Aron and C. E. Erborge, "A Practical Feasibility Study of Flood Peak Abatement in Urban Areas," U.S. Army Corps of Engineers, Sacramento, CA, March 1973.

[20] Soil Conservation Service, "Urban Hydrology for Small Watersheds," Technical Release No. 55, Washington, D.C., 1986.

[21] R. H. McCuen, P. A. Johnson, and R. M. Ragan, *Hydrology, HDS-2,* U.S. Federal Highway Administration, 2nd ed., September 1996.

[22] F. F. Snyder, "Synthetic Unit Graphs," *Trans. Am. Geophys. Union* **19**, 447–454(1938).

[23] Peter S. Eagleson, "Characteristics of Unit Hydrographs for Sewered Areas," paper presented before the ASCE, Los Angeles, CA, 1959, unpublished.

[24] W. W. Horner and F. L. Flynt, "Relation Between Rainfall and Runoff from Small Urban Areas," *Trans. ASCE* **62**(101), 140–205(Oct. 1956).

[25] R. K. Linsley, Jr. and W. C. Ackerman, "Method of Predicting the Runoff from Rainfall," *Trans. ASCE* **107**(1942).

[26] Wright-McLaughlin Engineers, *Urban Storm Drainage Criteria Manual,* Denver Regional Council of Governments, 2001.

[27] P. S. Eagleson, "Unit Hydrograph Characteristics of Sewered Areas," *Proceedings of the ASCE,* No. HY2, 1962.

[28] L. K. Sherman, "Stream-Flow from Rainfall by the Unit-Graph Method," *Eng. News-Rec.* **108**, 501–505(Apr. 1932).

[29] Rand Morgan and D. W. Hulinghorst, "Unit Hydrographs for Gauged and Ungauged Watersheds," U.S. Engineers Office, Binghamton, NY, July 1939.

[30] C. O. Clark, "Storage and the Unit Hydrograph," *ASCE Trans.* **110**, 1419–1446(1945).

[31] F. F. Snyder, "Synthetic Unit Graphs," *Trans. Am. Geophys. Union* **19**, 447–454(1938).

[32] A. B. Taylor and H. E. Schwartz, "Unit Hydrograph Lag and Peak Flow Related to Basin Characteristics," *Trans. Am. Geophys. Union* **33**, 235–246(1952).

[33] *Water Supply Papers,* U.S. Geological Survey, Water Resources Division, Washington, D.C.: U.S. Government Printing Office, 1966–1970.

[34] *Hourly Precipitation Data*, National Oceanic and Atmospheric Administration, Washington, D.C.: U.S. Government Printing Office, 1971.

[35] V. P. Singh, *Hydrologic Systems,* Englewood Cliffs, NJ: Prentice Hall, 1989.

[36] John C. Schaake, Jr., "Synthesis of the Inlet Hydrograph," Tech. Rept. No. 3, Department of Sanitary Engineering and Water Resources, The Johns Hopkins University, Baltimore, MD, 1965.

[37] G. Aron and E. White, "Fitting a Gamma Distribution over a Synthetic Unit Hydrograph," *Water Resources Bull.* **18**(1) (Feb. 1982).

[38] G. Aron and E. White, "Reply to Discussion," *Water Resources Bull.* **19**(2) (Apr. 1983).

[39] M. Collins, "Discussion—Fitting a Gamma Distribution over a Synthetic Unit Hydrograph," *Water Resources Bull.* **19**(2) (Apr. 1983).

[40] "Flood-Hydrograph Analysis and Computations," U.S. Army Corps of Engineers, *Engineering and Design Manuals,* Em1110-2-1405, Washington, D.C.: U.S. Government Printing Office, Aug. 1959.

[41] V. Mockus, "Use of Storm and Watershed Characteristics in Synthetic Hydrograph Analysis and Application," U.S. Department of Agriculture, Soil Conservation Service, 1957.

[42] W. H. Espey and D. G. Altman, "Nomographs for Ten-minute Unit Hydrographs for Small Urban Watersheds," Environmental Protection Agency, Rept. EPA-600/9-78-035, Washington, D.C., 1978.

[43] D. Kull, and A. Feldman, "Evolution of Clark's Unit Graph Method to Spatially Distributed Runoff," *J. Hydrol. Eng.,* Vol. 3, No. 1, 1998.

[44] J. E. Nash, "The Form of the Instantaneous Unit Hydrograph," IASH Publ. No. 45, Vol. 3, 1957.

[45] D. M. Gray, "Synthetic Unit Hydrographs for Small Drainage Areas," *Proc. ASCE J. Hyd. Div.* **87**(HY4) (July 1961).

[46] University of Colorado at Denver, "First Short Course on Urban Storm Water Modeling Using Colorado Urban Hydrograph Procedures," Department of Civil Engineering, June 1985.

[47] Graeff, "Traité d'hydraulique," Paris, 1883, pp. 438–443.

[48] V. T. Chow, *Open Channel Hydraulics,* New York: McGraw-Hill, 1959.

[49] S. Hayami, *On the Propagation of Flood Waves,* Bulletin 1, Kyoto, Japan: Disaster Prevention Institute, 1951.

[50] T. N. Keefer and R. S. McQuivey, "Multiple Linearization Flow Routing Model," *Proc. ASCE J. Hyd. Div.* **100**(HY7) (July 1974).

[51] J. A. Cunge, "On the Subject of a Flood Propagation Method," *J. Hyd. Res. IAHR* 7(2), 205–230(1967).

[52] U.S. Department of Agriculture, Soil Conservation Service, "Simplified Dam-Breach Routing Procedure," Tech. Release 66, Mar. 1979.

[53] U.S. Department of the Interior, "Water Studies," *Bureau of Reclamation Manual,* Vol. IV, Sec. 6.10. Washington, D.C.: U.S. Government Printing Office, 1947.

CHAPTER 10

Groundwater Hydrology

OBJECTIVES

The purpose of this chapter is to:

- Introduce the subject of groundwater hydrology
- Indicate the relationship between surface water and groundwater systems
- Describe the relevant hydrodynamic equations
- Relate the mechanics of groundwater flow to modeling regional groundwater systems
- Present methods for calculating confined and unconfined steady radial flow toward a well
- Describe procedures for dealing with unsteady groundwater flow conditions
- Introduce the principles of finite-difference approaches to modeling regional groundwater systems.

10.1   INTRODUCTION

The amount of water stored below ground in the United States exceeds by a significant amount all aboveground storage in streams, rivers, reservoirs, and lakes including the Great Lakes [1]. This enormous reservoir sustains streamflow during precipitation-free periods and constitutes the major source of fresh water for many localities. Figure 10.1 indicates the distribution and nature of primary groundwater areas of the United States.

The quantification of the volume and rate of flow of groundwater in various regions is a difficult task because volumes and flow rates are determined to a considerable extent by the geology of the region. The character and arrangement of rocks and soils are important factors, and these are often highly variable within a groundwater reservoir. An additional difficulty is the inability to directly measure many critical geologic and hydraulic reservoir characteristics.

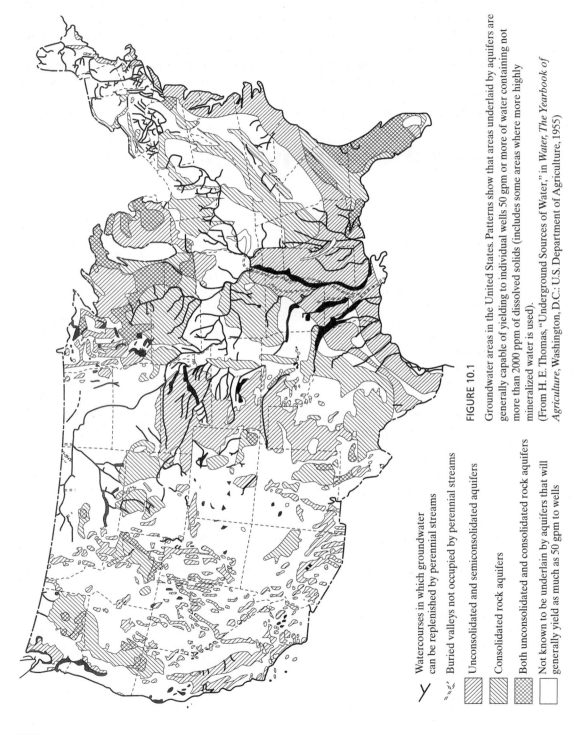

FIGURE 10.1

Groundwater areas in the United States. Patterns show that areas underlaid by aquifers are generally capable of yielding to individual wells 50 gpm or more of water containing not more than 2000 ppm of dissolved solids (includes some areas where more highly mineralized water is used).

(From H. E. Thomas, "Underground Sources of Water," in *Water, The Yearbook of Agriculture*, Washington, D.C.: U.S. Department of Agriculture, 1955)

Watercourses in which groundwater can be replenished by perennial streams

Buried valleys not occupied by perennial streams

Unconsolidated and semiconsolidated aquifers

Consolidated rock aquifers

Both unconsolidated and consolidated rock aquifers

Not known to be underlain by aquifers that will generally yield as much as 50 gpm to wells

The difficulties associated with determining the quantitative aspects of groundwater resources are paled, however, by those associated with their quality. And in many localities, it is the quality dimension that is most critical. More evidence is being uncovered indicating that many aquifers have been contaminated, at least locally, by the improper disposal of chemical and other wastes, by leachates from solid waste disposal sites, and from infiltrating storm water discharges. As a result, protection of groundwater quality has become a national policy, and in many locations it has become more important than overdrafts of groundwater supplies. Today, the hydrologist must be concerned with both the quality and quantity aspects of groundwater. Furthermore, an increasing specialization is emerging in groundwater quality modeling. Such modeling is generally beyond the scope of the text but information on this topic may be found in Refs. 2–6.

Groundwater Flow—General Properties

Understanding the movement of groundwater requires a knowledge of the time and space dependencies of the flow, the nature of the porous medium and fluid, and the boundaries of the flow system.

Groundwater flows are usually three-dimensional. Unfortunately, the solution of such problems by analytic methods is complex unless the system is symmetric [7],[8]. In other cases, space dependency in one of the coordinate directions may be so slight that assumption of two-dimensional flow is satisfactory. Many problems of practical importance fall into this class. Sometimes one-dimensional flow can be assumed, thus further simplifying the solution.

Fluid properties such as velocity, pressure, temperature, density, and viscosity often vary in time and space. When time dependency occurs, the issue is termed an *unsteady flow problem* and solutions are usually difficult. On the other hand, situations where space dependency alone exists are *steady flow problems*. Only homogeneous (single-phase) fluids are considered here. For a discussion of multiple phase flow, Refs. 5 and 8 are recommended.

Boundaries to groundwater flow systems may be fixed geologic structures or free water surfaces that are dependent for their position on the state of the flow. A hydrologist must be able to define these boundaries mathematically if the groundwater flow problems are to be solved.

Porous media through which groundwaters flow may be classified as isotropic, anisotropic, heterogeneous, homogeneous, or several combinations of these. An *isotropic* medium has uniform properties in all directions from a given point. *Anisotropic* media have one or more properties that depend on a given direction. For example, permeability of the medium might be greater along a horizontal plane than along a vertical one. *Heterogeneous* media have nonuniform properties of anisotropy or isotropy, while *homogeneous* media are uniform in their characteristics.

Subsurface Distribution of Water

Groundwater distribution may generally be categorized into zones of aeration and saturation. The saturated zone is one in which all voids are filled with water under hydrostatic pressure. In the zone of aeration, the interstices are filled partly with air, partly with water. The saturated zone is commonly called the *groundwater zone*. The zone of aeration may ideally be subdivided into several subzones. Todd classifies these as follows [9]:

1. *Soil water zone.* A soil water zone begins at the ground surface and extends downward through the major root band. Its total depth is variable and dependent on soil type and vegetation. The zone is unsaturated except during periods of heavy infiltration. Three categories of water classification may be encountered in this region: hygroscopic water, which is adsorbed from the air; capillary water, held by surface tension; and gravitational water, which is excess soil water draining through the soil.

2. *Intermediate zone.* This belt extends from the bottom of the soil-water zone to the top of the capillary fringe and may change from nonexistence to several hundred feet in thickness. The zone is essentially a connecting link between a near-ground surface region and the near-water-table region through which infiltrating fluids must pass.

3. *Capillary zone.* A capillary zone extends from the water table (Fig. 10.2) to a height determined by the capillary rise that can be generated in the soil. The capillary band thickness is a function of soil texture and may fluctuate not only from region to region but also within a local area.

4. *Saturated zone.* In the saturated zone, groundwater fills the pore spaces completely and porosity is therefore a direct measure of storage volume. Part of this water (specific retention) cannot be removed by pumping or drainage because of molecular and surface tension forces. Specific retention is the ratio of volume of water retained against gravity drainage to gross volume of the soil.

Water that can be drained from a soil by gravity is known as the *specific yield*. It is expressed as the ratio of the volume of water that can be drained by gravity to the gross volume of the soil. Values of specific yield depend on the soil particle size, shape and distribution of pores, and degree of compaction of the soil. Average values for alluvial aquifers range from 10 to 20 percent. Meinzer and others have developed procedures for determining the specific yield [12].

Geologic Considerations

The determination of groundwater volumes and flow rates requires a knowledge of the geology of a groundwater basin. In bedrock areas, hydrologic characteristics of the rocks, that is, their location, size, orientation, and ability to store or transmit water, must be known. In unconsolidated rock areas, basins often contain hundreds to thousands of feet of semiconsolidated to unconsolidated fill deposits that originated from the erosion of headwater areas. Such fills often contain extensive quantities of stored water. The characteristics of these basin fills must be evaluated.

A knowledge of the distribution and nature of geohydrologic units such as *aquifers, aquifuges*, and *aquicludes* is essential to proper planning for development or management of groundwater supplies. In addition, bedrock basin boundaries must be located and an evaluation made of their leakage characteristics.

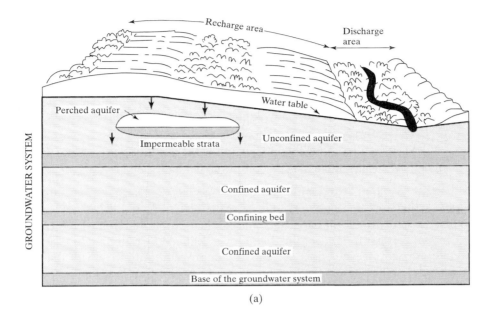

(a)

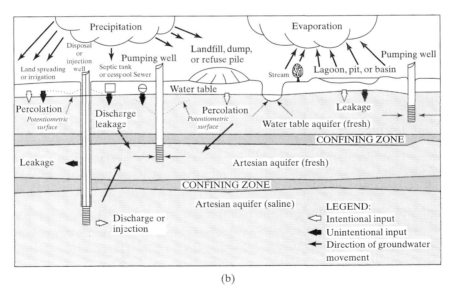

(b)

FIGURE 10.2

Definition sketches of groundwater systems and mechanisms for recharge and withdrawal:
(a) aquifer notation [10] and (b) components of the hydrologic cycle affecting groundwater [11].

An aquifer is a water-bearing stratum or formation that is capable of transmitting water in quantities sufficient to permit development. Aquifers may be considered as falling into two categories, confined and unconfined, depending on whether a water table or free surface exists under atmospheric pressure. Storage volume within an aquifer is changed whenever water is recharged to, or discharged from, an aquifer. In the case of an unconfined aquifer this may easily be determined as:

$$\Delta S = S_y \, \Delta V \tag{10.1}$$

where ΔS = change in storage volume

S_y = average specific yield of the aquifer

ΔV = volume of the aquifer lying between the original water table and the water table at some later specific time

For saturated, confined aquifers, pressure changes produce only slight modifications in the storage volume. In this case, the weight of the overburden is supported partly by hydrostatic pressure and somewhat by solid material in the aquifer. When hydrostatic pressure in a confined aquifer is reduced by pumping or other means, the load on the aquifer increases, causing its compression, with the result that some water is forced out. Decreasing the hydrostatic pressure also causes a small expansion, which in turn produces an additional release of water. For confined aquifers, water yield is expressed in terms of a *storage coefficient* S_c, defined as the volume of water an aquifer takes in or releases per unit surface area of aquifer per unit change in head normal to the surface. Figure 10.2 illustrates the classifications of aquifers.

In addition to water-bearing strata exhibiting satisfactory rates of yield, there are also non-water-bearing and impermeable strata that may contain large quantities of water but whose transmission rates are not high enough to permit effective development. An aquifuge is a formation impermeable and devoid of water; an aquiclude is an impervious stratum.

Topography

To understand how a groundwater system operates, it is essential to know something about the region's surface. A topographic map should be compiled showing all surface water bodies, including streams, lakes, and artificial channels and/or ponds, as well as land surface contours. Furthermore, an inventory of pumping wells, observation wells, and exploration wells should be made for purposes such as identifying types of soils and rocks, pinpointing discharge locations and rates, and determining water table elevations.

Subsurface Geology

The geologic structure of a groundwater basin governs the occurrence and movement of the groundwater within it. Specifically, the number and types of water-bearing formations, their vertical dimensions, interconnections, hydraulic properties, and outcrop patterns must be understood before the system can be analyzed [6]. Once the subsurface conditions have been identified, contour maps of the upper and lower boundaries of aquifers, water table contour maps, and maps of aquifer characteristics can be prepared. Well-driller logs, experimental test wells, and other geophysical exploration methods can be used to obtain the needed geologic data [5]–[9],[13],[14].

Fluctuations in Groundwater Level

Any circumstance that alters the pressure imposed on underground water will also cause a variation in the groundwater level. Seasonal factors, changes in stream and river stages, evapotranspiration, atmospheric pressure changes, winds, tides, external loads, various forms of withdrawal and recharge, and earthquakes all may produce fluctuations in the water table level or piezometric surface, depending on whether the aquifer is free or confined [9].

Groundwater–Surface Water Relations

Surface water and groundwater are interdependent. Changes in one component can have far-reaching effects on the other. Coordinated development and management of the combined resource are critical. Linkage between surface waters and groundwaters should be investigated in all regional studies so that adverse effects can be noted if they exist and opportunities for joint management understood.

Underground reservoirs are often extensive and can serve to store water for a multitude of uses. If withdrawals from these reservoirs consistently exceed recharge, *mining* occurs, and ultimate depletion of the resource results. By properly coordinating the use of surface water and groundwater supplies, optimum regional water resource development seems most likely to be assured [15],[16].

Hydrostatics

Water located in pore spaces of a saturated medium is under pressure (called *pore pressure*), which can be determined by inserting a piezometer in the medium at a point of interest. If location A (Fig. 10.3) is considered, it can be seen that pore pressure is given by:

$$p = h_a \gamma \tag{10.2}$$

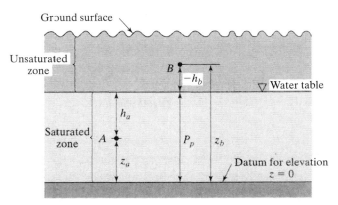

FIGURE 10.3

Definition sketch showing hydrostatic pressures in a porous medium.

where p = pore pressure (gauge pressure)

h_a = head measured from the point to the water table

γ = specific weight of water

Pore pressure is considered positive or negative, depending on whether the pressure head is measured above (positive) or below (negative) the point under consideration. If an arbitrary datum is established, the total head or piezometric head above the datum is:

$$P_p = z + h \qquad (10.3)$$

where P_p is known as the piezometric potential. In Fig. 10.3 this is equal to $h_a + z_a$ for point A in the saturated zone and $z_b - h_b$ for point B in the unsaturated zone. The term h_a is the pore pressure of A while $-h_b$ denotes tension or vacuum (negative pore pressure) at B.

10.2   GROUNDWATER FLOW

Analogies can be drawn between flow in pipes under pressure and in fully saturated confined aquifers. The flow of groundwater with a free surface is also similar to that in an open channel. A major difference is the geometry of a groundwater system flow channel as compared with common hydraulic pipe flow or channel systems. The problem can easily be recognized by envisioning a discharging cross section composed of a number of small openings, each with its own geometry, orientation, and size so that the flow velocity issuing from each pore varies in both magnitude and direction. Difficulties in analyzing such systems are apparent. Computations are usually based on macroscopic averages of fluid and medium properties over a given cross-sectional area.

Unknown quantities to be determined in groundwater flow problems are density, pressure, and velocity if constant temperature conditions are assumed to exist [5]–[8],[17]–[28]. In general, water is considered incompressible, so the number of working variables is reduced. An exception to this is discussed later relative to the storage coefficient for a confined aquifer. Primary emphasis here will be placed on the flow of water in a saturated porous medium.

Darcy's Law

Darcy's law for fluid flow through a horizontal permeable bed is stated as [17]:

$$Q = -KA\frac{dh}{dx} \qquad (10.4)$$

where A = total cross-sectional area including the space occupied by the porous material

K = hydraulic conductivity of the material

Q = flow across the control area A

In Eq. 10.4:

$$h = z + \frac{p}{\gamma} + C \tag{10.5}$$

where h = piezometric head
$\quad\quad z$ = elevation above a datum
$\quad\quad p$ = hydrostatic pressure
$\quad\quad C$ = an arbitrary constant

If the specific discharge q = Q/A is substituted in Eq. 10.4:

$$q = -K\frac{d}{dx}\left(z + \frac{p}{\gamma}\right) \tag{10.6}$$

Note that q also equals the porosity n multiplied by the pore velocity V_p. Darcy's law is widely used in groundwater flow problems. Several applications are illustrated in later sections.

Darcy's law is limited in applicability to cases where the Reynolds number is on the order of 1. For Reynolds numbers less than 1, Darcy's law may be considered valid. Deviations from Darcy's law have been shown to occur at Reynolds numbers as low as 2, depending on such factors as grain size and shape. The Reynolds number N_R is defined herein as:

$$N_R = \frac{\rho q d}{\mu} \tag{10.7}$$

where q = specific discharge
$\quad\quad d$ = mean grain diameter
$\quad\quad \rho$ = fluid density
$\quad\quad \mu$ = dynamic viscosity

For many conditions of practical importance (zones lying adjacent to collecting devices are an exception), Darcy's law has been found to apply.

Of special interest is the fact that the Darcy equation is analogous to Ohm's law:

$$i = \left(\frac{1}{R}\right)E \tag{10.8}$$

where i = current
$\quad\quad R$ = resistance
$\quad\quad E$ = voltage

Current and velocity are analogous, as are K and $1/R$, and E and dh/dx. The similarity of the two equations is the basis for electric analog models of groundwater flow systems [8],[18].

Example 10.1

The water temperature in an aquifer is 70°F and the rate of water movement is 0.75 ft/day. The mean particle diameter in the porous medium is found to be 0.07 in. Calculate the Reynolds number and indicate whether Darcy's law is applicable.

Solution

1. Equation 10.7 gives the Reynolds number as:

$$N_R = \frac{\rho q d}{\mu}$$

This may also be written as:

$$N_R = \frac{qd}{\nu}$$

2. From Table A.2 in Appendix A, ν is found to be 1.06×10^{-5} ft/sec. Converting the velocity q into units of ft/sec gives $q = 0.75/86,400 = 8.7 \times 10^{-6}$. The mean grain diameter in feet $= 0.07/12 = 0.0058$. Substituting these values in the equation above, we obtain:

$$N_R = \frac{(8.7 \times 10^{-6}) \times 0.0058}{1.06 \times 10^{-5}} = 0.0048$$

Since $N_R < 1.0$, Darcy's law applies.

Permeability

The hydraulic conductivity K is an important parameter that is often separated into two components, one related to the medium, the other to the fluid. The product:

$$k = Cd^2 \tag{10.9}$$

called the *specific* or *intrinsic permeability*, is a function of the medium only. In Eq. 10.9, d represents the mean grain diameter of the particles and C is a constant shape factor associated with packing, size distribution, and other factors [8],[9]. By using this definition, hydraulic conductivity, also known as the *coefficient of permeability*, can be written:

$$K = \frac{k\gamma}{\mu} \tag{10.10}$$

Dimensions of intrinsic permeability are L^2. Since values of k given as ft^2 or cm^2 are extremely small, a unit of measure known as the *darcy* has been widely adopted:

$$1 \text{ darcy} = 0.987 \times 10^{-8} \text{ cm}^2 \quad \text{or} \quad 1.062 \times 10^{-11} \text{ ft}^2$$

Several ways of expressing hydraulic conductivity are reported in the literature. The U.S. Geological Survey has defined the standard hydraulic conductivity, K_s, as the number of gallons per day of water passing through 1 ft^2 of medium under a unit-hydraulic gradient at a temperature of 60°F. Another measure, called the field

TABLE 10.1 Some Values of the Standard Hydraulic Conductivity and Intrinsic Permeability for Several Classes of Materials

Material	Approximate range K_s (gpd/ft^2)	Approximate range k (darcys)
Clean gravel	10^6–10^4	10^5–10^3
Clean sands; mixtures of clean gravels and sands	10^4–10	10^3–1
Very fine sands; silts; mixtures of sands, silts, clays; stratified clays	10–10^{-3}	1–10^{-4}
Unweathered clays	10^{-3}–10^{-4}	10^{-4}–10^{-5}

hydraulic conductivity, K_f, is defined as:

$$K_f = K_s \left(\frac{\mu_{60}}{\mu_f} \right)$$
(10.11)

where μ_{60} = dynamic viscosity of water at 60°F

μ_f = dynamic viscosity at the prevailing field temperature

Since the temperature effect on the hydraulic conductivity is small over the range of temperatures normally encountered for groundwater in practice, the correction specified by Eq. 10.11 is seldom used [5].

Table 10.1 gives the values of intrinsic permeability and hydraulic conductivity for several classes of materials. Table 10.2 gives a set of conversion factors for the units of k and K commonly used. Because considerable variation in these parameters within the materials classifications shown in Table 10.1 can occur, it is important that geologic surveys be conducted to support groundwater modeling efforts.

For many groundwater analyses, it is convenient to use the coefficient of transmissivity:

$$T = K_f b$$
(10.12)

TABLE 10.2 Conversion Factors for Permeability and Hydraulic Conductivity Units[a]

	Permeability, k			Hydraulic conductivity, K		
	cm^2	ft^2	darcy	m/s	ft/s	U.S. gal/day/ft^2
cm^2	1	1.08×10^{-3}	1.01×10^8	9.80×10^2	3.22×10^3	1.85×10^9
ft^2	9.29×10^2	1	9.42×10^{10}	9.11×10^5	2.99×10^6	1.71×10^{12}
darcy	9.87×10^{-9}	1.06×10^{-11}	1	9.66×10^{-6}	3.17×10^{-5}	1.82×10^1
m/s	1.02×10^{-3}	1.10×10^{-6}	1.04×10^5	1	3.28	2.12×10^6
ft/s	3.11×10^{-4}	3.35×10^{-7}	3.15×10^4	3.05×10^{-1}	1	6.46×10^5
U.S. gal/day/ft^2	5.42×10^{-10}	5.83×10^{-13}	5.49×10^{-2}	4.72×10^{-7}	1.55×10^{-6}	1

Note: To obtain k in ft^2, for example, multiply k in cm^2 by 1.08×10^{-3}

[a]R Allan Freeze and John A. Cherry, *Groundwater*, 1979, p. 29. Reprinted by permission of Prentice-Hall, Englewood Cliffs, New Jersey.

where K_f is the field hydraulic conductivity and b is the saturated thickness of the aquifer. The coefficient of transmissivity is widely used in the water well industry. If K is expressed in gal/day/ft^2, then T has units of gal/day/ft. The range of values of T can be found by multiplying the pertinent K values from Table 10.1 by the range of expected aquifer thicknesses. It has been determined that aquifers worth considering for water supply development have values of T greater than about 100,000 gal/day/ft (0.015 ft^2/s) [5].

Example 10.2

Laboratory tests on an aquifer material indicate a standard hydraulic conductivity $K_s = 1.08 \times 10^3$ gpd/ft^2. If the field temperature is 70°F, find the field hydraulic conductivity K_f.

Solution.   Using Eq. 10.11:

$$K_f = K_s\left(\frac{\mu_{60}}{\mu_f}\right)$$

and the values of the kinematic viscosity given in Table A.2 in Appendix A for 60°F and 70°F, 1.21×10^{-5} and 1.06×10^{-5}, respectively, we get:

$$K_f = \frac{(1.08 \times 10^3) \times (1.21 \times 10^{-5})}{1.06 \times 10^{-5}} = 1232.8 \text{ gpd/ft}^2$$

Note that the absolute viscosity and the kinematic viscosity are related as shown in the following equation:

$$\nu = \frac{\mu}{\rho}$$

and given that the density of water over the range of temperatures in this case is virtually constant, values for the kinematic viscosity may be used in place of those for the absolute velocity in Eq. 10.11.

Velocity Potential

Potential theory is directly applicable to groundwater flow computations. The *velocity potential* ϕ is a scalar function of time and space. The potential is defined by:

$$\phi(x, y, z) = -K\left(z + \frac{p}{\gamma}\right) + C \tag{10.13}$$

where C is an arbitrary constant. By definition, its derivative with respect to any given direction is the velocity of flow in that direction. Thus we can write:

$$u = \frac{\partial\phi}{\partial x} \quad v = \frac{\partial\phi}{\partial y} \quad w = \frac{\partial\phi}{\partial z} \tag{10.14}$$

where u, v, and w are the velocities in the x, y, and z directions, respectively, and K is assumed constant. In vector notation this becomes:

$$V = \text{grad } \phi = \nabla \phi \tag{10.15}$$

with V the combined velocity vector and:

$$\text{grad } \phi = \frac{\partial \phi}{\partial x} \mathbf{i} + \frac{\partial \phi}{\partial y} \mathbf{j} + \frac{\partial \phi}{\partial z} \mathbf{k} = \nabla \phi \tag{10.16}$$

Hydrodynamic Equations

The determination of values for the variables u, v, w, and h is the target of most groundwater flow problems. The first three variables are the specific discharge components in the x, y, and z directions, respectively, while h is the total head at a specified point in the flow domain. To effect a solution, four equations involving these variables are needed. These are the equations of motion in each direction plus the continuity equation.

The equations of motion are based on Newton's second law:

$$F = ma \tag{10.17}$$

where F = the force
m = the mass
a = the acceleration

Considering forces acting on a fluid element, accelerations in the three coordinate directions may be determined according to Eq. 10.17. If frictionless flow is assumed (reasonable for many cases of flow in porous media), the body forces plus the surface force (pressure) must be equivalent to the total force in each direction. In the manner of Harr [19], the following equations (Euler's equations) in the three coordinate directions are obtained:

$$\frac{\partial u}{\partial t} + u \frac{\partial u}{\partial x} + v \frac{\partial u}{\partial y} + w \frac{\partial u}{\partial z} = X - \frac{1}{\rho} \frac{\partial p}{\partial x} \tag{10.18}$$

$$\frac{\partial v}{\partial t} + u \frac{\partial v}{\partial x} + v \frac{\partial v}{\partial y} + w \frac{\partial v}{\partial z} = Y - \frac{1}{\rho} \frac{\partial p}{\partial y} \tag{10.19}$$

$$\frac{\partial w}{\partial t} + u \frac{\partial w}{\partial x} + v \frac{\partial w}{\partial y} + w \frac{\partial w}{\partial z} = Z - \frac{1}{\rho} \frac{\partial p}{\partial z} - g \tag{10.20}$$

where X, Y, Z, and g are body forces per unit mass. For steady flow [u, v, w, and $h \neq f(t)$], the first terms on the left-hand side of each equation vanish. With laminar groundwater flow in the range of validity of Darcy's law, velocities are small (often on the order of 5 ft/yr to 5 ft/day) [2]. Thus for steady laminar flow, Eqs. 10.18–10.19 reduce to:

$$X = \frac{1}{\rho} \frac{\partial p}{\partial x} \quad Y = \frac{1}{\rho} \frac{\partial p}{\partial y} \quad Z = \frac{1}{\rho} \frac{\partial p}{\partial z} + g \tag{10.21}$$

In most groundwater flow problems the velocity head is negligible; thus p may be given as $\rho g(h - z)$. Then Eq. 10.21 becomes:

$$X = g\frac{\partial h}{\partial x} \quad Y = g\frac{\partial h}{\partial y} \quad Z = g\frac{\partial h}{\partial z} \tag{10.22}$$

Remembering that Darcy's law defines $\partial h/\partial x = u/K$, and so on, it follows that:

$$X = -\frac{gu}{K} \quad Y = -\frac{gv}{K} \quad Z = -\frac{gw}{K} \tag{10.23}$$

For steady laminar flow, the body forces are linear functions of velocity and Eqs. 10.18–10.20 may be written:

$$g\frac{\partial h}{\partial x} = -g\frac{u}{K} \tag{10.24}$$

$$g\frac{\partial h}{\partial y} = -g\frac{v}{K} \tag{10.25}$$

$$g\frac{\partial h}{\partial z} = -g\frac{w}{K} \tag{10.26}$$

where

$$u = -K\frac{\partial h}{\partial x} \quad v = -K\frac{\partial h}{\partial y} \quad w = -K\frac{\partial h}{\partial z} \tag{10.27}$$

This demonstrates that the equations of motion fit Darcy's law for steady laminar flow. The continuity equation may be stated as:

$$\frac{\partial \rho}{\partial t} + \frac{\partial(\rho u)}{\partial x} + \frac{\partial(\rho v)}{\partial y} + \frac{\partial(\rho w)}{\partial z} = 0 \tag{10.28}$$

This equation is valid for a compressible fluid with time-dependent properties. In steady compressible flow the first term becomes zero, and for steady incompressible flow the equation becomes:

$$\frac{\partial u}{\partial x} + \frac{\partial v}{\partial y} + \frac{\partial w}{\partial z} = 0 \tag{10.29}$$

Now since $u = \partial\phi/\partial x$, and so on, Eq. 10.29 becomes:

$$\nabla^2\phi = \frac{\partial^2\phi}{\partial x^2} + \frac{\partial^2\phi}{\partial y^2} + \frac{\partial^2\phi}{\partial z^2} = 0 \tag{10.30}$$

which is known as the *Laplace equation*. With steady-state laminar flow, groundwater motion is completely described by the continuity equation subject to appropriate boundary conditions.

If the hydraulic conductivity K is constant, Eq. 18.30 can be written:

$$\nabla^2 h = 0 \tag{10.31}$$

which is the expression of steady incompressible flow in a homogeneous isotropic porous medium.

For unsteady flow, the compressibility of both aquifer and water are pertinent. Consider a small element of porous medium that has a volume $\Delta x \, \Delta y \, \Delta z$. Then the term in a continuity equation representing a change in storage is defined by:

$$\frac{\partial(\rho n \, \Delta x \, \Delta y \, \Delta z)}{\partial t} \tag{10.32}$$

Presupposing that compressive forces are predominant in the vertical (z) direction, we can neglect lateral changes. Thus in terms of the element described, only Δz is considered variable. A storage expression written as the sum of three terms involving partial derivatives of the variables Δz, ρ, and porosity n is:

$$\frac{\partial(\rho n \, \Delta x \, \Delta y \, \Delta z)}{\partial t} = \left(n\rho \frac{\partial(\Delta z)}{\partial t} + \rho \, \Delta z \frac{\partial n}{\partial t} + n \, \Delta z \frac{\partial \rho}{\partial t} \right) \Delta x \, \Delta y \tag{10.33}$$

The three elements on the right can be expressed in terms of pore pressure p, the aquifer compressibility α, and the fluid compressibility β [8],[9].

Fluid compressibility is defined as the reciprocal of its bulk modulus of elasticity. It is given by:

$$\beta = -\frac{\partial V / V}{\partial p} \tag{10.34}$$

where V = the volume

p = the pore pressure

If the piezometric surface of a confined aquifer is lowered a distance of one unit, the amount of water released from a column of aquifer of unit horizontal cross-sectional area is defined as the storage coefficient S (also known as the storativity). Storativities are dimensionless, and for confined aquifers they range from 0.005 to 0.00005 [5]. This is analogous to the specific yield S_y of an unconfined aquifer. In Eq. 10.34 S is equivalent to ∂V. Furthermore, if the aquifer column is of height b, $V = b$. The change in pressure ∂p is equivalent to the negative product of the change in head (one unit) and specific weight of water. Making these substitutions in Eq. 10.34 we find that:

$$\beta = \frac{S}{\gamma b} \tag{10.35}$$

Now if the aquifer material is considered elastic, that is, if Δz and n can be modified, the volume change can be expressed in terms of alteration in the density of the material due to the difference in packing. Thus:

$$\frac{\partial V}{V} = -\frac{\partial \rho}{\rho} \tag{10.36}$$

Introducing Eqs. 10.35 and 10.36 into Eq. 10.34 gives:

$$\partial\rho = \frac{\rho S}{b\gamma}\partial p \tag{10.37}$$

Next, substituting this expression for $\partial\rho$ in Eq. 10.28 we obtain:

$$\frac{\partial(\rho u)}{\partial x} + \frac{\partial(\rho v)}{\partial y} + \frac{\partial(\rho w)}{\partial z} = -\frac{\rho S}{b\gamma}\frac{\partial p}{\partial t} \tag{10.38}$$

The left-hand side of this equation can be expanded to:

$$\rho\left(\frac{\partial u}{\partial x} + \frac{\partial v}{\partial y} + \frac{\partial w}{\partial z}\right) + \left(u\frac{\partial\rho}{\partial x} + v\frac{\partial\rho}{\partial y} + w\frac{\partial\rho}{\partial z}\right) \tag{10.39}$$

The second term is normally very small compared with the first and can be neglected. The validity of this assumption improves as the flow angle decreases. By using Eq. 10.39 and the foregoing assumption, Eq. 10.38 becomes:

$$\frac{\partial u}{\partial x} + \frac{\partial v}{\partial y} + \frac{\partial w}{\partial z} = -\frac{S}{b\gamma}\frac{\partial p}{\partial t} \tag{10.40}$$

or if isotropic conditions prevail:

$$K\nabla^2 h = \frac{S}{b\gamma}\frac{\partial p}{\partial t} \tag{10.41}$$

since from Eq. 10.27 $u = -K\,\partial h/\partial x$, and so on. Inserting γh for p and the transmissivity T for Kb produces:

$$\nabla^2 h = \frac{S}{T}\frac{\partial h}{\partial t} \tag{10.42}$$

which is the general equation for unsteady flow in a confined aquifer of constant thickness b.

The storage coefficient S and the transmissivity are commonly called the *formation constants* of a confined aquifer. For an unconfined aquifer Eq. 10.42 reverts to:

$$\nabla^2 h = \frac{S}{Kb}\frac{\partial h}{\partial t} \tag{10.43}$$

since b is a function of the change in head. The unsteady flow equation for an unconfined aquifer is nonlinear in form. The solution of such an equation is discussed by Jacob [21]. Where variations in saturated thickness of unconfined aquifers are minor, Eq. 10.42 may be used as an approximation [9].

For unconfined aquifers, the right-hand side of Eq. 10.43 is often negligible so that the equation:

$$\nabla^2 h = 0 \tag{10.31}$$

is frequently valid for both steady and unsteady flow.

Flow Lines and Equipotential Lines

Many problems of practical interest in groundwater hydrology can be considered two-dimensional flow problems. The equation of continuity for steady incompressible flow in an isotropic medium then becomes:

$$\frac{\partial u}{\partial x} + \frac{\partial v}{\partial y} = 0 \tag{10.44}$$

$$\nabla^2 h = \frac{\partial^2 h}{\partial x^2} + \frac{\partial^2 h}{\partial y^2} = 0 \tag{10.45}$$

and

$$\nabla^2 \phi = \frac{\partial^2 \phi}{\partial x^2} + \frac{\partial^2 \phi}{\partial y^2} = 0 \tag{10.46}$$

The Laplace equation is satisfied by two conjugate harmonic functions ϕ and ψ [9],[19]. Curves $\phi(x, y) =$ constant are orthogonal to the curves $\psi(x, y) =$ constant. The function $\phi(x, y)$ is the velocity potential, and the function $\psi(x, y)$ is known as the *stream function* and is defined by:

$$u = \frac{\partial \psi}{\partial y} \quad v = -\frac{\partial \psi}{\partial x} \tag{10.47}$$

Substituting Eq. 10.47 into Eq. 10.48 yields:

$$\frac{\partial^2 \psi}{\partial x\, \partial y} - \frac{\partial^2 \psi}{\partial y\, \partial x} = 0 \tag{10.48}$$

It has already been shown that:

$$u = \frac{\partial \phi}{\partial x} \quad v = \frac{\partial \phi}{\partial y}$$

so we can write:

$$\frac{\partial \phi}{\partial x} = \frac{\partial \psi}{\partial y} \quad \frac{\partial \phi}{\partial y} = -\frac{\partial \psi}{\partial x} \tag{10.49}$$

These are known as the *Cauchy-Riemann equations.* The stream function satisfies both the equation of continuity and the equations of Cauchy-Riemann. It can also be shown that the Laplace equation is satisfied and therefore [8],[19]:

$$\nabla^2 \psi = \frac{\partial^2 \psi}{\partial x^2} + \frac{\partial^2 \psi}{\partial y^2} = 0 \tag{10.50}$$

Refer now to Fig. 10.4. If V is a velocity vector tangent to a particle flow path 3–4, then it can be decomposed into two components u and v [20]. By geometry of the figure:

$$\frac{v}{u} = \frac{dy}{dx} = \tan \alpha \tag{10.51}$$

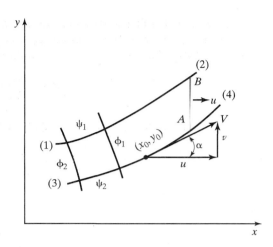

FIGURE 10.4

Definition sketch for a stream function.

and thus:

$$v\,dx - u\,dy = 0 \tag{10.52}$$

If Eqs. 10.47 are substituted into Eq. 10.51, then:

$$\frac{\partial \psi}{\partial x}\,dx + \frac{\partial \psi}{\partial y}\,dy = 0 \tag{10.53}$$

The total differential $d\psi$ is equal to zero, and ψ must be a constant. A series of curves $\psi(x, y)$ equal to a succession of constants can be drawn and will be tangent at all points to the velocity vectors. These curves trace the flow path of a fluid particle and are known as *streamlines* or *flow lines*. An important property of the stream function is demonstrated with the aid of Fig. 10.4. Consider the flow crossing a vertical section AB between streamlines defined as ψ_1 and ψ_2. If the discharge across the section is designated as Q, it is apparent that:

$$Q = \int_{\Psi_2}^{\Psi_1} u\,dy \tag{10.54}$$

or

$$Q = \int_{\Psi_2}^{\Psi_1} d\psi \tag{10.55}$$

and

$$Q = \psi_1 - \psi_2 \tag{10.56}$$

Equation 10.56 illustrates the important property that flow between two streamlines is constant. Streamline spacing reveals the relative magnitudes of flow velocities between them. Higher values are associated with narrower spacings, and vice versa.

The curves in Fig. 10.4, designated as ϕ_1 and ϕ_2 and called *equipotential lines*, are determined by velocity potentials $\phi(x, y)$ = constant. These curves intersect the flow lines at right angles, illustrated in the following way. The total differential $d\phi$ is given by:

$$d\phi = \frac{\partial \phi}{\partial x}\, dx + \frac{\partial \phi}{\partial y}\, dy \qquad (10.57)$$

Substituting for terms $\partial \phi / \partial x$ and $\partial \phi / \partial y$ their equivalents u and v gives us:

$$u\, dx + v\, dy = 0 \qquad (10.58)$$

and

$$\frac{dy}{dx} = -\frac{u}{v} \qquad (10.59)$$

Thus equipotential lines are normal to flow lines. The system of flow lines and equipotential lines forms a flow net.

One significant point of difference between ϕ and ψ functions is that equipotential lines exist only when the flow is irrotational. For two-dimensional flow the condition of irrotationality is said to exist when the z component of vorticity ζ_z is zero, or:

$$\zeta_z = \left(\frac{\partial v}{\partial x} - \frac{\partial u}{\partial y} \right) = 0 \qquad (10.60)$$

Proof of this is given by Eskinazi. Substituting for u and v in Eq. 10.60 in terms of ϕ, we obtain [20]:

$$\frac{\partial^2 \phi}{\partial x\, \partial y} - \frac{\partial^2 \phi}{\partial y\, \partial x} = 0 \qquad (10.61)$$

This indicates that when the velocity potential exists, the criterion for irrotationality is satisfied.

Once either streamlines or equipotential lines in a flow domain are determined, the other is automatically known because of the relations in Eq. 10.49. Thus:

$$\psi = \int \left(\frac{\partial \phi}{\partial x}\, dy - \frac{\partial \phi}{\partial y}\, dx \right) \qquad (10.62a)$$

and

$$\phi = \int \left(\frac{\partial \psi}{\partial y}\, dx - \frac{\partial \psi}{\partial x}\, dy \right) \qquad (10.62b)$$

It is enough then to determine only one of the functions, since the other can be obtained using relations Eqs. 10.62a and 10.62b. The complex potential is given by:

$$w = \phi + i\psi \qquad (10.63)$$

where i, the square root of -1 (hydraulic gradient), is widely used in analytic flow net analyses [8],[18]. Of special importance is the fact that:

$$\nabla^2 w = \nabla^2 \phi + i\nabla^2 \psi = 0 \qquad (10.64)$$

satisfies the conditions of continuity and irrotationality simultaneously.

Equations presented in this section have been limited to the case of two-dimensional flow. Extension to three dimensions would be obtained in a similar fashion.

Boundary Conditions

To solve groundwater flow problems it is necessary to specify appropriate boundary conditions. Some of the more commonly encountered ones are described in this section; more comprehensive discussions are found elsewhere [22],[23].

Boundary conditions discussed can be categorized as follows: impervious boundaries, surfaces of seepage, constant-head boundaries, and lines of seepage (free surfaces).

Impervious boundaries may be artificial objects such as concrete dams, rock strata, or soil strata that are highly impervious. In Fig. 10.5 the impervious boundary AB represents such a limit. Since flow cannot cross an impervious boundary, velocity components normal to it vanish, and the impervious boundary is a streamline. In other words, at the boundary, $\psi = $ constant.

Next look at the upstream face of the earth dam BC. At any point of elevation y along BC the pressure can be assumed hydrostatic, or:

$$p = \gamma(h - y) \qquad (10.65)$$

The definition of a velocity potential states that:

$$\phi = -K\left(\frac{p}{\gamma} + y\right) + C \qquad (10.66)$$

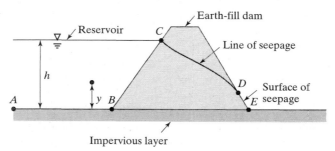

FIGURE 10.5

Some common boundary conditions.

Substituting for pressure in Eq. 10.66 yields:

$$\phi = -K\left[\frac{\gamma(h - y)}{\gamma} + y\right] + C \tag{10.67}$$

and

$$\phi = -Kh + C \tag{10.68}$$

Thus for a constant reservoir level h and an isotropic medium:

$$\phi = \text{constant}$$

and surface BC, often termed a *reservoir boundary*, is an equipotential line.

The free surface or line of seepage CD in Fig. 10.5 is seen to be a boundary between the saturated and unsaturated zones. Since flow does not occur across this boundary, it is obviously also a streamline. Pressure along this free surface must be constant, and therefore along CD:

$$\phi + Ky = \text{constant} \tag{10.69}$$

This is a linear relation in ϕ, and therefore equal vertical falls along CD must be associated with successive equipotential drops. One important groundwater flow problem is to determine the location of the line of seepage.

The surface of seepage DE of Fig. 10.5 represents the location at which water seeps through the downstream face of the dam and trickles toward point E. The pressure along DE is atmospheric. The surface of seepage is neither a flow line nor an equipotential line.

Flow Nets

Flow nets, or graphical representations of families of streamlines and equipotential lines, are widely used in groundwater studies to determine quantities, rates, and directions of flow. The use of flow nets is limited to steady incompressible flow at constant viscosity and density for homogeneous media or for regions that can be compartmentalized into homogeneous segments. Darcy's law must be applicable to the flow conditions.

The manner in which a flow net can be used in problem-solving is best explained with the aid of Fig. 10.6. This diagram shows a portion of a flow net constructed so that each unit bounded by a pair of streamlines and equipotential lines is approximately square. The reason for this will be clear later.

A flow net can be determined exactly if functions ϕ and ψ are known beforehand. This is often not the case, and as a result, graphically constructed flow nets are widely used. The preparation of a flow net requires application of the concept of square elements and adherence to boundary conditions. Graphical flow nets are usually difficult for a beginner to create, but with reasonable practice an acceptable net

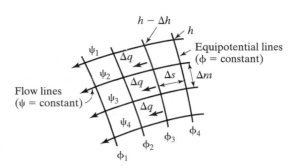

FIGURE 10.6

Segment of an orthogonal flow net.

can be drawn. Various mechanical methods for graphical flow net construction are presented in the literature and are not discussed here [19],[23].

After a flow net has been constructed, it can be analyzed using the geometry of the net and by applying Darcy's law.

Remembering that $h = p/\gamma + z$, we find that Fig. 10.6 shows that the hydraulic gradient G_h between two equipotential lines is given by:

$$G_h = \frac{\Delta h}{\Delta s} \tag{10.70}$$

Then by applying Darcy's law, in the manner of Todd [9], the flow increment between adjacent streamlines is:

$$\Delta q = K \, \Delta m \left(\frac{\Delta h}{\Delta s} \right) \tag{10.71}$$

where Δm represents the cross-sectional area for a net of unit width normal to the plane of the diagram. If the flow net is constructed in an orthogonal manner and composed of approximately square elements:

$$\Delta m \approx \Delta s \quad \text{and} \quad \Delta q = K \Delta h \tag{10.72}$$

Now if there are n equipotential drops between the equipotential lines, it is evident that:

$$\Delta h = \frac{h}{n}$$

where h is the total head loss over the n spaces. If the flow is divided into m sections by the flow lines, then the discharge per unit width of the medium is:

$$Q = \sum_{i=1}^{m} \Delta q = \frac{Kmh}{n} \tag{10.73}$$

When the medium's hydraulic conductivity is known, the discharge can be computed using Eq. 10.73 and a knowledge of flow net geometry.

Where the flow net has a free surface or line of seepage, the entrance and exit conditions given in Fig. 10.5 are useful. A more comprehensive discussion of these conditions is given in Ref. 24.

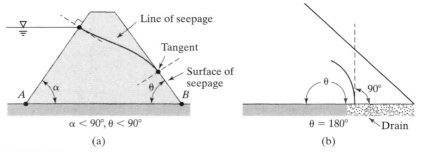

$\alpha < 90°, \theta < 90°$

(a)

$\theta = 180°$ Drain

(b)

FIGURE 10.7

Some entrance and exit conditions for the line of seepage.
(After Casagrande [24])

Some trouble arises in flow net construction at locations where the velocity becomes infinite or vanishes. Such points are known as *singular points* and according to DeWiest may be placed in three separate categories [8]. In the first classification flow lines and equipotential lines do not intersect at right angles. Such a situation often occurs when a boundary coincides with a flow line; point A in Fig. 10.7 is an example.

The second classification has a discontinuity along the boundary that abruptly changes the slope of the streamline. In Fig. 10.8 points A, B, and C represent such discontinuities. At points A and C the velocity is infinite, while at point B it is zero. If the angle of discontinuity measured in a counterclockwise direction inside the flow field is less than 180°, the velocity is zero; if larger than 180°, it is infinite. The angle at A is 270°, for example.

The third category includes the case where a source or sink exists in the flow net. Under these circumstances the velocity is infinite, since squares of the flow net approach zero size as the source or sink is approached. Wells and recharge wells represent sinks and sources in a practical sense and are discussed later.

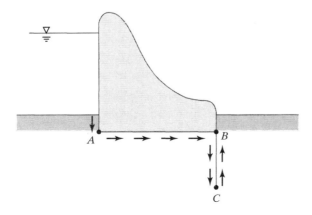

FIGURE 10.8

Flow line slope discontinuities.

Variable Hydraulic Conductivity

It is common for flow within a porous medium of one hydraulic conductivity to enter another region with a different hydraulic conductivity. When such a boundary is crossed, flow lines are refracted. The change in direction that occurs can be determined as a function of the two permeabilities involved in the manner of Todd and DeWiest [8],[9]. Figure 10.9 illustrates this.

Consider two soils of permeabilities K_1 and K_2 that are separated by the boundary LR shown in Fig. 10.9. The directions of the flow lines before and after crossing the boundary are defined by angles θ_1 and θ_2.

For continuity to be preserved, the velocity components in media K_1 and K_2, which are normal to the boundary, must be equal, since the cross-sectional area at the boundary is AB for a unit depth. Using Darcy's law and noting the equipotential drops h_a and h_b, we find:

$$K_1 \frac{\Delta h_a}{AC} \cos \theta_1 = K_2 = \frac{\Delta h_b}{BD} \cos \theta_2 \qquad (10.74)$$

From the geometry of the figure it is apparent that:

$$AC = AB \sin \theta_1$$
$$BD = AB \sin \theta_2$$

The head loss between A and B is shown on the figure to be equal to both Δh_a and Δh_b, and since there can be only a single value:

$$\Delta h_a = \Delta h_b$$

Introducing these expressions in Eq. 10.74 produces:

$$\frac{K_1}{\tan \theta_1} = \frac{K_2}{\tan \theta_2} \qquad (10.75)$$

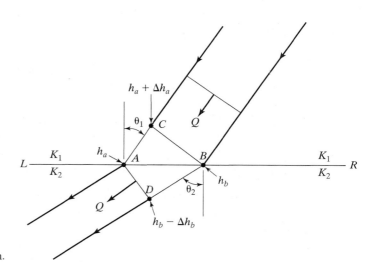

FIGURE 10.9

Flow line refraction.

For refracted flow in a saturated porous medium, the ratio of the tangents of angles formed by the intersection of flow lines with normals to the boundary is given by the ratio of hydraulic conductivities. As a result of refraction, the flow net on the K_2 side of the boundary will no longer be squares if the equipotential line spacing DB is maintained. To adjust the net on the K_2 side, the relation:

$$\frac{\Delta h_b}{\Delta h_a} = \frac{K_1}{K_2} \qquad (10.76)$$

can be used where $\Delta h_b \neq \Delta h_a$.

Equipotential lines are also refracted in crossing permeability boundaries. The relation for this is:

$$\frac{K_1}{K_2} = \frac{\tan \alpha_2}{\tan \alpha_1} \qquad (10.77)$$

where α is the angle between the equipotential line and a normal to the boundary of permeability [8].

Anisotropy

In many cases hydraulic conductivity is dependent on the direction of flow within a given layer of soil. This condition is said to be anisotropic. Sedimentary deposits often fit this aspect, with flow occurring more readily along the plane of deposition than across it. Where the permeability within a plane is uniform but very small across it as compared to that along the plane, a flow net can still be used after proper adjustments are made. A discussion of this is given elsewhere [5],[8],[19],[25]. Nonhomogeneous aquifers require special consideration but may sometimes be analyzed by using representative or average parameters. A detailed study is outside the scope of this book [5],[8],[18],[19].

Dupuit's Theory

Groundwater flow problems in which one boundary is a free surface can be analyzed on the basis of Dupuit's theory of unconfined flow. This theory is founded on two assumptions made by Dupuit in 1863 [26]. First, if the line of seepage is only slightly inclined, streamlines may be considered horizontal and, correspondingly, equipotential lines will be essentially vertical. Second, slopes of the line of seepage and the hydraulic gradient are equal. When field conditions are known to be satisfactorily represented by these assumptions, the results obtained according to Dupuit's theory compare very favorably with those arrived at by more rigorous techniques.

Figure 10.10 is useful in translating the foregoing assumptions into a mathematical statement. Consider an element given in the figure which has a base area $dx\, dy$ and a vertical height h. Writing the continuity equation in the x direction and considering steady flow to be the case:

$$\text{inflow}_{x_0} = \text{velocity}_{x_0} \times \text{area}_{x_0} \qquad (10.78)$$

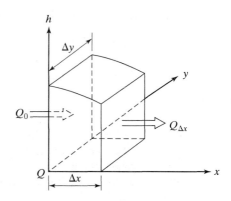

FIGURE 10.10

Definition sketch for development of Dupuit's equation.

The velocity at $x = 0$ is given by Darcy's law as:

$$u_{x_0} = -K\frac{\partial h}{\partial x}$$

(10.79)

Thus the discharge across the element at $x = 0$ is:

$$Q_0 = -K\frac{\partial h}{\partial x}\,h\,dy$$

(10.80)

The outflow at $x = dx$ is obtained by a Taylor's series expansion as:

$$Q_{dx} = -K\frac{\partial h}{\partial x}\,h\,dy + dx\frac{\partial}{\partial x}\left(-K\frac{\partial h}{\partial x}\,h\,dy\right) + \cdots$$

(10.81)

Subtracting the outflow from the inflow if K is considered constant, we obtain:

$$I_x - O_x = K\,dx\,dy\,\frac{\partial}{\partial x}\left(h\frac{\partial h}{\partial x}\right)$$

(10.82)

or

$$I_x - O_x = \frac{K\,dx\,dy}{2}\frac{\partial}{\partial x}\left(\frac{\partial h^2}{\partial x}\right)$$

(10.83)

where dx and dy are considered fixed lengths. A similar consideration in the y direction yields:

$$I_y - O_y = \frac{K\,dx\,dy}{2}\frac{\partial}{\partial y}\left(\frac{\partial h^2}{\partial y}\right)$$

(10.84)

Assuming that there is no movement in the vertical direction, these are the only components of the inflow and outflow. Furthermore, still dealing with steady flow, the change in storage must be zero. As a result:

$$\frac{K\,dx\,dy}{2}\frac{\partial}{\partial x}\left(\frac{\partial h^2}{\partial x}\right) + \frac{K\,dx\,dy}{2}\frac{\partial}{\partial y}\left(\frac{\partial h^2}{\partial y}\right) = 0$$

(10.85)

and since $(K\,dx\,dy)/2$ is constant, this reduces to:

$$\frac{\partial^2 h^2}{\partial x^2} + \frac{\partial^2 h^2}{\partial y^2} = 0 \tag{10.86}$$

or

$$\nabla^2 h^2 = 0 \tag{10.87}$$

Consequently, according to Dupuit's assumptions, Laplace's equation for the function h^2 must be satisfied [27].

In the particular case where recharge is occurring as a result of infiltrated water reaching the water table, a simple adjustment may be made to Eq. 10.86. If the recharge intensity (dimensionally LT^{-1}) is specified as R, then the total recharge to the element of Fig. 10.10 is $R\,dx\,dy$ and the continuity equation for steady flow becomes:

$$K\frac{dx\,dy}{2}\left(\frac{\partial^2 h^2}{\partial x^2} + \frac{\partial^2 h^2}{\partial y^2}\right) + R\,dx\,dy = 0 \tag{10.88}$$

or more simply:

$$\nabla^2 h^2 + \frac{2}{K}R = 0 \tag{10.89}$$

Now, applying Dupuit's theory to the flow problem illustrated in Fig. 10.11, and assuming one-dimensional flow in the x direction only, we obtain the discharge per

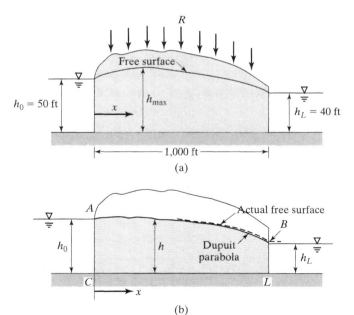

FIGURE 10.11

Steady flow in a porous medium between two water bodies: (a) free surface with infiltration and (b) free surface without infiltration.

unit width of the aquifer given by Darcy's law:

$$Q = -Kh \frac{dh}{dx} \tag{10.90}$$

In this instance h is the height of the line of seepage at any position x along the impervious boundary. For the one-dimensional example considered here, Eq. 10.86 becomes:

$$\frac{d^2h^2}{dx^2} = 0 \tag{10.91}$$

Upon integration:

$$h^2 = ax + b \tag{10.92}$$

where a and b are constants.

Then for boundary conditions at $x = 0, h = h_0$:

$$b = h_0^2 \tag{10.93}$$

Differentiation of Eq. 10.92 yields:

$$2h \frac{dh}{dx} = a \tag{10.94}$$

Also from Darcy's equation, $h \, dh/dx = -Q/K$. Making this substitution, we obtain:

$$a = \frac{-2Q}{K} \tag{10.95}$$

and inserting the values of the constants in Eq. 10.92, we obtain:

$$h^2 = -2 \frac{Q}{K} x + h_0^2 \tag{10.96}$$

This is the equation of a free surface. It is a parabola (often called *Dupuit's parabola*). If the existence of a surface of seepage at B is ignored, and noting that at $x = L, h = h_L$, we find that Eq. 10.96 becomes:

$$h_L^2 = -\frac{2QL}{K} + h_0^2 \tag{10.97}$$

or

$$Q = \frac{K}{2L} (h_0^2 - h_L^2) \tag{10.98}$$

which is known as the *Dupuit equation*.

Example 10.3

Refer to Fig. 10.11a. Given the dimensions shown and a recharge intensity R of 0.01 ft/day, find the discharge at $x = 1,000$ ft using Dupuit's equation. Assume that $K = 8$.

Solution.   Note that:

$$\frac{dQ}{dx} = R$$

or

$$Q = Rx + C$$

At $x = 0$:

$$Q = Q_0$$

therefore

$$Q = Rx + Q_0$$

Also

$$Q = -Kh\frac{dh}{dx}$$

$$-Kh\frac{dh}{dx} = Rx + Q_0$$

Integrating yields:

$$\frac{-Kh^2}{2}\bigg|_{h_0}^{h_L} = \frac{Rx^2}{2}\bigg|_0^L + Q_0x\bigg|_0^L$$

and inserting the limits, we obtain:

$$\frac{-K(h_L^2 - h_0^2)}{2} = \frac{RL^2}{2} + Q_0 L$$

$$Q_0 = \frac{K(h_0^2 - h_L^2)}{2L} - \frac{RL}{2}$$

Then since $Q = Rx + Q_0$:

$$Q = R\left(x - \frac{L}{2}\right) + \frac{K(h_0^2 - h_L^2)}{2L}$$

$$R = 0.01 \times 7.5 = 0.075 \text{ gpd/ft}^2$$

$$Q = 0.075(1{,}000 - 500) + \frac{8(50^2 - 40^2)}{2{,}000}$$

$$= (0.075 \times 500) + \frac{8 \times 900}{2{,}000}$$

$$= 37.5 + 3.6$$

$$= 41.1 \text{ gpd/ft}^2$$

10.3   FLOW TO WELLS

A well system can be considered as composed of three elements—the well structure, pump, and discharge piping [28],[29]. The well itself contains an open section through which water enters and a casing to transport the flow to the ground surface. The open section is usually a perforated casing or slotted metal screen permitting water to enter and at the same time preventing collapse of the hole. Occasionally, gravel is placed at the bottom of the well casing around the screen.

When a well is pumped, water is removed from the aquifer immediately adjacent to the screen. Flow then becomes established at locations some distance from the well in order to replenish this withdrawal. Because of flow resistance offered by the soil, a head loss results and the piezometric surface adjacent to the well is depressed, producing a cone of depression (Fig. 10.12), which spreads until equilibrium is reached and steady-state conditions are established.

The hydraulic characteristics of an aquifer (which are described by the storage coefficient and aquifer permeability) can be determined by laboratory or field tests. The three most commonly used field methods are the application of tracers, the use of

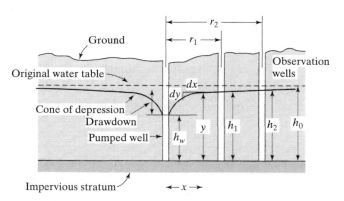

FIGURE 10.12

Well in an unconfined aquifer.

field permeameters, and aquifer performance tests [9]. A discussion of aquifer performance tests is given here along with the development of flow equations for wells [29]–[31].

Aquifer performance tests may be either equilibrium or nonequilibrium tests. In an equilibrium test the cone of depression must be stabilized for a flow equation to be derived. For a nonequilibrium test the derivation includes a condition that steady-state conditions have not been reached. Adolph Thiem published the first performance tests based on equilibrium conditions in 1906 [32].

Steady Unconfined Radial Flow Toward a Well

The basic equilibrium equation for an unconfined aquifer can be derived using the notation of Fig. 10.12. Here flow is assumed to be radial; the original water table is considered to be horizontal; the well is presumed to fully penetrate the aquifer of infinite areal extent; and steady-state conditions must prevail. Then flow toward the well at any distance x away must equal the product of the cylindrical element of area at that section and the flow velocity. With Darcy's law this becomes:

$$Q = 2\pi x y K_f \frac{dy}{dx} \tag{10.99}$$

where $2\pi x y$ = the area through any cylindrical shell (ft^2), with the well as its axis
  K_f = the hydraulic conductivity (ft/sec)
  dy/dx = the water table gradient at any distance x
  Q = the well discharge (ft^3/sec)

Integrating over the limits specified, we find that:

$$\int_{r1}^{r2} Q \frac{dx}{x} = 2\pi K_f \int_{h1}^{h2} y \, dy \tag{10.100}$$

$$Q \ln \frac{r_2}{r_1} = \frac{2\pi K_f (h_2^2 - h_1^2)}{2} \tag{10.101}$$

and

$$Q = \frac{\pi K_f (h_2^2 - h_1^2)}{\ln(r_2/r_1)} \tag{10.102}$$

Converting K_f to the field units of gpd/ft^2, Q to gpm, and ln to log, we can rewrite Eq. 10.102 as:

$$K_f = \frac{1055Q \log(r_2/r_1)}{h_2^2 - h_1^2} \tag{10.103}$$

If the drawdown in the well does not exceed one-half of the original aquifer thickness h_0, reasonable estimates of Q or K_f can be obtained by using Eq. 10.102 or 10.103, even if the height h_1 is measured at the well periphery where $r_1 = r_w$, the radius of the well boring.

Example 10.4

A 20-in. well fully penetrates an unconfined aquifer of 100-ft depth. Two observation wells located 90 and 240 ft from the pumped well are known to have drawdowns of 23 and 21.5 ft, respectively. If the flow is steady and $K_f = 1{,}400$ gpd/ft^2, find the discharge from the well.

Solution.   Equation 10.102 is applicable, and for the given units this is:

$$Q = \frac{K(h_2^2 - h_1^2)}{1{,}055 \log (r_2/r_1)}$$

$$\log(r_2/r_1) = \log(240/90) = 0.42651$$

$$h_2 = 100 - 21.5 = 78.5 \text{ ft}$$

$$h_1 = 100 - 23 = 77 \text{ ft}$$

$$Q = \frac{1{,}400(78.5^2 - 77^2)}{1{,}055 \times 0.42651}$$

$$= 725.7 \text{ gpm}$$

Steady Confined Radial Flow Toward a Well

The basic equilibrium equation for a confined aquifer can be obtained in a similar manner, using the notation of Fig. 10.13. The same assumptions apply. Mathematically,

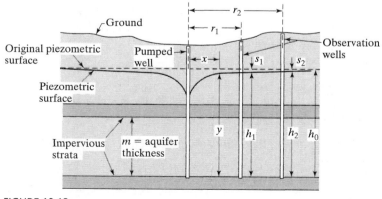

FIGURE 10.13

Radial flow to a well in a confined aquifer.

the flow in ft³/sec is found from:

$$Q = 2\pi x m K_f \frac{dy}{dx} \qquad (10.104)$$

Integrating, we obtain:

$$Q = 2\pi K_f m \frac{h_2 - h_1}{\ln(r_2/r_1)} \qquad (10.105)$$

The coefficient of permeability may be determined by rearranging Eq. 10.105 to the form:

$$K_f = \frac{528Q \log(r_2/r_1)}{m(h_2 - h_1)} \qquad (10.106)$$

where

$Q = $ (gpm)

$K_f = $ the permeability (gpd/ft²)

$r, h = $ (ft)

Example 10.5

Find the permeability of an artesian aquifer being pumped by a fully penetrating well. The aquifer is 100 ft thick and composed of medium sand. The steady-state pumping rate is 1,000 gpm. The drawdown at an observation well 50 ft away is 10 ft; in a second observation well 500 ft away, it is 1 ft.

Solution

$$K_f = \frac{528Q \log(r_2/r_1)}{m(h_2 - h_1)}$$

$$= \frac{528 \times 1,000 \times 1}{100 \times (10 - 1)}$$

$$= 586.7 \text{ gpd/ft}^2$$

Well in a Uniform Flow Field

For a steady-state well in a uniform flow field where the original piezometric surface is not horizontal, a somewhat different situation from that previously assumed prevails. Consider the artesian aquifer shown in Fig. 10.14. The heretofore assumed circular area of influence becomes distorted in this case. A solution is possible by applying potential theory, by using graphical means, or, if the slope of the piezometric surface is very slight, Eq. 10.105 may be employed without serious error.

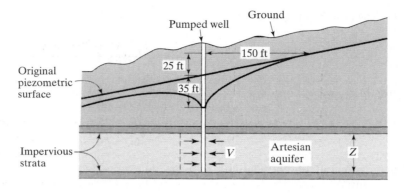

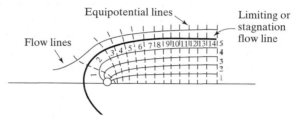

FIGURE 10.14

Well in a uniform flow field and flow net definition.

Figure 10.14 provides a graphical solution to a uniform flow field problem. First, an ortholgonal flow net consisting of flow lines and equipotential lines must be constructed. This should be done so that the completed flow net will be composed of a number of elements that approach little squares in shape. Once the net is complete, it can be analyzed by considering the net geometry and using Darcy's law in the manner of Todd [9].

Example 10.6

Find the discharge to the well of Fig. 10.14 by using an applicable flow net. Consider the aquifer to be 35 ft thick, $K_f = 3.65 \times 10^{-4}$ fps, and other dimensions as shown.

Solution.   Using Eq. 10.73, we find that:

$$q = \frac{Kmh}{n}$$

where $h = 35 + 25 = 60$ ft
$m = 2 \times 5 = 10$
$n = 14$
$$q = \frac{(3.65 \times 10^{-4}) \times 60 \times 10}{14}$$
$= 0.0156$ cfs per unit thickness of the aquifer

The total discharge Q is thus:

$$Q = 0.0156 \times 35 = 0.55 \text{ cfs or } 245 \text{ gpm}$$

Well Fields

When more than one unit in a well field is pumped, there is a composite effect on the free water surface. This consequence is illustrated by Fig. 10.15 in which the cones of depression are seen to overlap. The drawdown at a given location is equal to the sum of the individual drawdowns.

If, within a particular well field, pumping rates of the pumped wells are known, the composite drawdown at a point can be determined. In like manner, if the drawdown at one point is known, the well flows can be calculated.

If the drawdown at a given point is designated as m, and subscripts $1, 2, \ldots, n$ are used to relate this drawdown to a particular well (e.g., m_1 refers to the drawdown for W_1), for the total drawdown m_T at some location [9]:

$$m_T = \sum_{i=1}^{n} = m_i \tag{10.107}$$

The number of wells, their rate of pumping, and well-field geometry and characteristics determine the total drawdown at a specified location.

Again considering Eq. 10.107, we obtain:

$$h_0^2 - h^2 = \frac{Q}{\pi K} \ln \frac{r_0}{r} \tag{10.108}$$

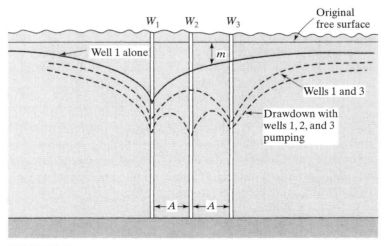

FIGURE 10.15

Combined effect of pumping several wells at equal rates.

It can be seen that the drawdown for a well pumped at rate Q can be computed if h_0, r_0, and r are known. It follows then from Eq. 10.107 that for n pumped wells in an unconfined aquifer:

$$h_0^2 - h^2 = \sum_{i=1}^{n} \frac{Q_i}{\pi K} \ln \frac{r_{0i}}{r_i} \qquad (10.109)$$

where h_0 = the original height of the water table
   h = the combined-effect height of the water table after pumping n wells
   Q_i = the flow rate of the ith well
   r_{0i} = the distance of the ith well to a location at which the drawdown is considered negligible
   r_i = the distance from well i to the point at which the drawdown is being investigated

Todd indicates that values of r_0 used in practice often range from 500 to 1,000 ft [9]. The impact of this assumption is softened because Q in Eq. 10.108 is not very sensitive to r_0. Equation 10.109 should be used only where drawdowns are relatively small.
   For flow in a confined aquifer the expression for combined drawdown becomes:

$$h_0 - h = \sum_{i=1}^{n} \frac{Q_i}{2\pi Km} \ln \frac{r_{0i}}{r_i} \qquad (10.110)$$

Equations for well flow covering a variety of particular well-field patterns are reported in the literature [5],[9]. Those given here are applicable for steady flow in a homogeneous isotropic medium.

The Method of Images

Some groundwater flow problems subjected to boundary conditions negating the direct use of radial flow equations can be transformed into infinite systems fitting these equations by applying the method of images [29],[33],[34].
   When a stream is located near a pumped well and the stream and aquifer are interconnected, the drawdown curve of a pumped well may be affected as shown in Fig. 10.16. Another boundary condition often affecting the drawdown of a well is an impervious formation that limits the extent of the aquifer. The cone of depression of a pumped well is not affected until the boundary is intersected. After that, the shape of the drawdown curve will be changed by the boundary. Boundary effects can frequently be evaluated by means of "image wells." The boundary condition is replaced by either a recharging or a discharging well that is pumped or recharged at a rate equivalent to that of the pumped well. That is, in an infinite aquifer, drawdowns of the real and image wells would be identical. The image well is located at a distance from the boundary equal to that of the real well but on the opposite side (Fig. 10.16). Streams are replaced by recharge wells while impermeable boundaries are supplanted by pumped image wells. Computations for the case of a well and impervious boundary directly follow the procedures outlined under the section on well fields. For the well and stream system,

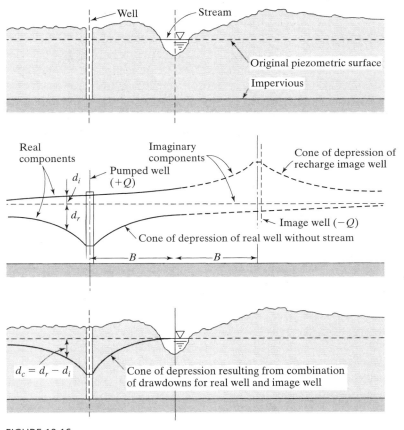

FIGURE 10.16

Drawdown in a pumping well whose aquifer is connected to a stream.

the recharge image well is considered to have a negative discharge. The heads are then added according to this sign convention.

The procedure for combining drawdown curves of real and image wells to obtain an actual drawdown curve is illustrated graphically for the example shown in Fig. 10.16. More detailed information on other cases can be found elsewhere [8],[34].

Unsteady Flow

When a new well is first pumped, a large portion of the discharge comes directly from the storage volume released as the cone of depression develops. Under these circumstances the equilibrium equations overestimate permeability and therefore the yield of the well. When steady-state conditions are not encountered—as is usually the situation in practice—a nonequilibrium equation must be used. Two approaches can be taken, the rather rigorous method of C. V. Theis or a simplified procedure such as that proposed by Jacob [35],[36].

In 1935 Theis published a nonequilibrium approach that takes into consideration time and the storage characteristics of an aquifer. The method provides a solution to Eq. 10.42 for given initial and boundary conditions [35]. Application of the method is appropriate for confined aquifers of constant thickness. For use under conditions of unconfined flow, vertical components of flow must be negligible, and changes in aquifer storage through water expansion and aquifer compression must also be negligible relative to the gravity drainage of pores as the water table drops as a result of pumping [37].

Theis states that the drawdown(s) in an observation well located at a distance r from the pumped well is given by:

$$S = \frac{Q}{4\pi T} \int_u^\infty \frac{e^{-u}}{u} \, du \qquad (10.111)$$

where Q = constant pumping rate (L^3T^{-1} units), T = aquifer transmissivity (L^2T^{-1} units), and u is a dimensionless variable defined by:

$$u = r^2 \frac{S_c}{4tT} \qquad (10.112)$$

where r is the radial distance from the pumping well to an observation well, S_c is the aquifer storativity (dimensionless), and t is time. The integral in Eq. 10.111 is commonly called the *well function of u* and is written as $W(u)$. It can be evaluated from the infinite series:

$$W(u) = -0.577216 - \ln u + u - \frac{u^2}{2 \times 2!} + \frac{u^3}{3 \times 3!} \cdots \qquad (10.113)$$

Using this notation, Eq. 10.110 may be written as:

$$s = \frac{QW(u)}{4\pi T} \qquad (10.114)$$

The basic assumptions of the Theis equation are generally the same as those in Eq. 10.105 except for the nonsteady-state condition. Some values of the well function of u are given in Table 10.3.

In American practice, Eqs. 10.111 and 10.112 commonly appear in the following form:

$$s = \frac{114.6Q}{T} \int_u^\infty \frac{e^{-u}}{u} \, du \qquad (10.115)$$

$$u = \frac{1.87r^2S_c}{Tt} \qquad (10.116)$$

where T is in units of gpd/ft, Q has units of gpm, and t is the time in days since the start of pumping.

TABLE 10.3   Values of $W(u)$ for Various Values of u

u	$W(u)$	u	$W(u)$	u	$W(u)$	u	$W(u)$
1×10^{-10}	22.45	7×10^{-8}	15.90	4×10^{-5}	9.55	1×10^{-2}	4.04
2	21.76	8	15.76	5	9.33	2	3.35
3	21.35	9	15.65	6	9.14	3	2.96
4	21.06	1×10^{-7}	15.54	7	8.99	4	2.68
5	20.84	2	14.85	8	8.86	5	2.47
6	20.66	3	14.44	9	8.74	6	2.30
7	20.50	4	14.15	1×10^{-4}	8.63	7	2.15
8	20.37	5	13.93	2	7.94	8	2.03
9	20.25	6	13.75	3	7.53	9	1.92
1×10^{-9}	20.15	7	13.60	4	7.25	1×10^{-1}	1.823
2	19.45	8	13.46	5	7.02	2	1.223
3	19.05	9	13.34	6	6.84	3	0.906
4	18.76	1×10^{-6}	13.24	7	6.69	4	0.702
5	18.54	2	12.55	8	6.55	5	0.560
6	18.35	3	12.14	9	6.44	6	0.454
7	18.20	4	11.85	1×10^{-3}	6.33	7	0.374
8	18.07	5	11.63	2	5.64	8	0.311
9	17.95	6	11.45	3	5.23	9	0.260
1×10^{-8}	17.84	7	11.29	4	4.95	1×10^{0}	0.219
2	17.15	8	11.16	5	4.73	2	0.049
3	16.74	9	11.04	6	4.54	3	0.013
4	16.46	1×10^{-5}	10.94	7	4.39	4	0.0038
5	16.23	2	10.24	8	4.26	5	0.0011
6	16.05	3	9.84	9	4.14	6	0.0004

Source: After L. K. Wenzel, "Methods for Determining Permeability of Water-Bearing Materials with Special Reference to Discharging Well Methods," U.S. Geological Survey, Water Supply Paper 887, Washington, D.C., 1942

Equations 10.111 and 10.112 can be solved by comparing a log–log plot of u versus $W(u)$, known as a *type curve*, with a log–log plot of the observed data r^2/t versus s. In plotting type curves, $W(u)$ and s are ordinates, and u and r^2/t are abscissas. The two curves are superimposed and moved about until segments coincide. In this operation the axes must remain parallel. A coincident point is then selected on the matched curves and both plots marked. The type curve then yields values of u and $W(u)$ for the desired point. Corresponding values of s and r^2/t are determined from a plot of the observed data. Inserting these values in Eqs. 10.111 and 10.112 and rearranging, values for transmissibility T and storage coefficient S_c can be found.

Often this procedure can be shortened and simplified. When r is small and t large, Jacob found that values of u are generally small [36]. Thus terms in the series of Eq. 10.113 beyond the second one become negligible and the expression for T becomes:

$$T = \frac{264Q(\log t_2 - \log t_1)}{h_0 - h} \qquad (10.117)$$

which can be further reduced to:

$$T = \frac{264Q}{\Delta h} \qquad (10.118)$$

where Δh = drawdown per log cycle of time $[(h_0 - h)/(\log t_2 - \log t_1)]$
           Q = well discharge (gpm)
     h_0, h = as defined in Fig. 10.13
           T = transmissibility (gpd/ft)

Field data on drawdown $(h_0 - h)$ versus t are drafted on semilogarthmic paper. The drawdown is plotted on an arithmetic scale (see Fig. 10.17). This plot forms a straight line whose slope permits computing formation constants using Eq. 10.118 and:

$$S_c = \frac{0.3Tt_0}{r^2} \tag{10.119}$$

with t_0 being the time corresponding to zero drawdown. Equation 10.119 is obtained through manipulation of Eq. 10.111.

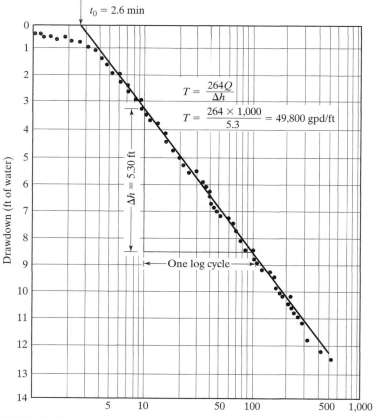

FIGURE 10.17

Pumping test data, Jacob method.

Example 10.7

Using the following data, find the formation constants for an aquifer using a graphical solution to the Theis equation. Discharge equals 540 gpm.

Distance from pumped well, r (ft)	r^2/t	Average drawdown, s (ft)
50	1,250	3.04
100	5,000	2.16
150	11,250	1.63
200	20,000	1.28
300	45,000	0.80
400	80,000	0.51
500	125,000	0.33
600	180,000	0.22
700	245,000	0.15
800	320,000	0.10

Solution.   Plot s versus r^2/t and $W(u)$ versus u as shown in Fig. 10.18. Determine the match point as noted and compute S_c and T using Eqs. 10.115 and 10.116:

$$T = \frac{114.6Q}{s} W(u)$$

$$= \frac{114.6 \times 540}{1.28} \times 1.9 = 91,860 \text{ gpd/ft}$$

$$S_c = \frac{uT}{1.87r^2/t}$$

$$= \frac{0.09 \times 91,860}{1.87 \times 20,000} = 0.22$$

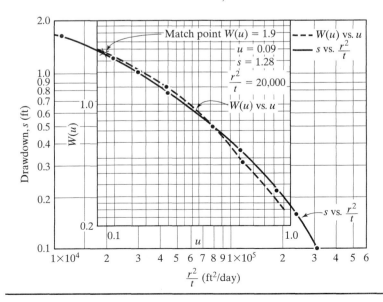

FIGURE 10.18

Graphical solution to the Theis equation.

Example 10.8

Using the data given in Fig. 10.17, find the coefficient of transmissibility T and storage coefficient S_c for an aquifer, given $Q = 1,000$ gpm and $r = 300$ ft.

Solution.   Find the value of Δh from the graph, 5.3 ft. Then by Eq. 10.118:

$$T = \frac{264Q}{\Delta h} = \frac{264 \times 1,000}{5.3}$$

$$= 49,800 \text{ gpd/ft}$$

Using Eq. 10.119, we find that:

$$S_c = \frac{0.3Tt_0}{r^2}$$

Note from Fig. 10.17 that $t_0 = 2.6$ min. Converting to days, we find that this becomes:

$$t_0 = 1.81 \times 10^{-3} \text{ days}$$

and

$$S_c = \frac{0.3 \times 49,800 \times (1.81 \times 10^{-3})}{(300)^2}$$

$$= 0.0003$$

Example 10.9

Find the drawdown at an observation point 300 ft away from a pumping well. It has been found that $T = 2.8 \times 10^4$ gpd/ft, the pumping time is 15 days, the storativity = 2.7×10^{-4}, and $Q = 275$ gpm.

Solution

1.   From Eq. 10.116, u can be computed:

$$u = \frac{1.87r^2 S_c}{Tt}$$

$$u = [1.87 \times (300)^2 \times (2.7 \times 10^{-4})]/[(2.8 \times 10^4) \times 15] = 1.08 \times 10^{-4}$$

2.   Referring to Table 10.3 and interpolating, we estimate $W(u)$ to be 8.62. Then using Eq. 10.115, the drawdown is found to be:

$$s = \frac{114.6Q}{T} \int_{u}^{\infty} \frac{e^{-u}}{u} \, du$$

$$s = (114.6 \times 275 \times 8.62)/(2.8 \times 10^4) = 9.70 \text{ ft}$$

Example 10.10

A well is being pumped at a constant rate of 0.004 m³/s. Given that $T = 0.0025$ m²/s, $r = 100$ meters, and the storage coefficient $= 0.00087$, find the drawdown in the observation well for a time period of (a) 15 minutes, and (b) 20 hours.

Solution

1. Using Eq. 10.112, u can be computed as follows:

$$u = \frac{r^2 S}{4tT}$$

$$u = (100 \times 100 \times 0.00087)/(4 \times 15 \times 60 \times 0.0025)$$

$$u = 0.97$$

   Then from Table 10.3, $W(u)$ is found to be 0.23.
   Applying Eq. 10.114, the drawdown can be determined:

$$s = \frac{QW(u)}{4\pi t}$$

$$s = (0.004 \times 0.23)/(4 \times \pi \times 0.0025)$$

$$s = 0.029 \text{ m}$$

2. Following the procedure used in part 1:

$$u = (100 \times 100 \times 0.00087)/(4 \times 72{,}000 \times 0.0025)$$

$$u = 0.0121$$

   Then from Table 10.3, $W(u)$ is found to be 8.49.
   Applying Eq. 10.114, the drawdown can be determined:

$$s = (0.0004 \times 8.49)/(4 \times \pi \times 0.0025)$$
$$s = 1.08 \text{ m}$$

Leaky Aquifers

The foregoing analyses have dealt with free aquifers or those confined between impervious strata. In reality, many cases exist wherein the confining strata are not completely impervious and water is actually transferred from them to the productive aquifer. The flow regime is altered and computations must include leakage. Since about 1930, leaky aquifers have been the subject of research by investigators such as De Glee, Jacob, Hantush, DeWiest, Walton, Neuman and Witherspoon, and others [18],[25],[38]–[48]. A thorough treatment of their work is beyond the scope of this book; interested readers should consult the indicated references.

Partially Penetrating Wells

In many situations, there is only partial penetration of the well. The question then arises as to the applicability of procedures developed previously for full penetration.

Numerous studies of this problem have been conducted [5],[49],[50]. In 1957 Hantush reported that steady flow to a well just penetrating an infinite leaky aquifer becomes very nearly radial at a distance from the well of about 1.5 times the aquifer thickness [50]. As depth of penetration increases, the approach to radial flow becomes increasingly apparent. Therefore, computations of drawdowns for partially penetrating wells are made using equations for total penetration with relative safety, provided that the distance from the pumped well is greater than 1.5 times the aquifer thickness. At points closer to the well, it is frequently possible to use a flow net or other relations developed for this region.

Flow to an Infiltration Gallery

An infiltration gallery may be defined as a partially pervious conduit constructed across the path of the local groundwater flow such that all or part of this flow will be intercepted. These galleries are often built in a valley area parallel to a stream so that they can convey the collected flow to some designated location under gravity-flow conditions. Figure 10.19 shows a typical cross section through a gallery with one pervious face.

Computation of discharge to an infiltration gallery with one pervious wall (Fig. 10.19) is accomplished in the manner outlined by Dupuit [26]. Several assumptions must be made to effect the solution. They are that the sine and tangent of the angle of inclination of the water table are interchangeable; that the velocity vectors are everywhere horizontal and uniformly distributed; that the soil is incompressible and isotropic; and that the gallery is of sufficient length that end effects are negligible.

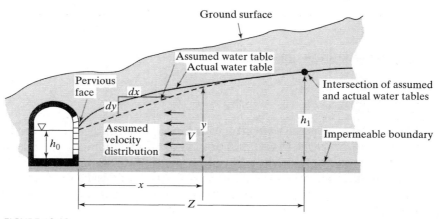

FIGURE 10.19

Cross section through an infiltration gallery.

While permitting a solution of the problem, these assumptions do limit the utility of the results.

Based on these assumptions, Eq. 10.98 can be used to calculate the discharge per unit width. Using the nomenclature of Fig. 10.19, Eq. 10.98 becomes:

$$q = \frac{K}{2Z} (h_1^2 - h_0^2)$$

This equation indicates that the computed water table is parabolic. This is often called Dupuit's parabola. Figure 10.19 shows that the computed water table differs from the actual water table in an increasing manner as the gallery face is approached. It is therefore apparent that the computed parabola does not accurately describe the real water table. The differences, however, are small except near the point of outflow, providing the initial assumptions are satisfied. The calculated discharge approximates the true discharge more closely as the ratio of Z/h_1 increases.

Example 10.11

A stratum of clean sand and gravel 30 ft deep has a coefficient of permeability $K = 3.5 \times 10^{-3}$ ft/sec, and is supplied with water from a channel that penetrates to the bottom of the stratum. If the water surface in an infiltration gallery is 4 ft above the bottom of the stratum and its distance to the channel is 100 ft, what is the flow into a foot of an infiltration gallery?

Solution. Refer to Fig. 10.19 and use Eq. 10.98:

$$q = \frac{K(h_0^2 - h_L^2)}{2L}$$

$$q = 0.5(3.5 \times 10^{-3})[(30 \times 30) - (4 \times 4)]/100$$

$$q = 0.015 \text{ cfs, the flow into one foot of the infiltration gallery}$$

10.4 SALTWATER INTRUSION

The contamination of fresh groundwater by the intrusion of salt water often presents a serious quality problem. Islands and coastal regions are particularly vulnerable. Aquifers located inland sometimes contain highly saline waters as well. Fresh water is lighter than salt water (specific gravity of the latter is about 1.025) and forms a freshwater layer above the underlying salt water. This equilibrium is disturbed when an aquifer is pumped, since salt water replaces the fresh water removed. Under equilibrium conditions, a drawdown of 1 ft in a freshwater table corresponds to a rise of about 40 ft by salt water. Wells subjected to saltwater intrusion obviously have limited pumping rates.

Recharge wells have been drilled in coastal areas to maintain a head sufficient to preclude seawater intrusion, a practice employed effectively in southern California.

10.5   GROUNDWATER BASIN DEVELOPMENT

To use groundwater resources efficiently while simultaneously permitting the maximum development of the resource, equilibrium must be established between withdrawals and replenishments. Economic, legal, political, social, and water quality aspects require full consideration.

Lasting supplies of groundwater will be assured only when long-term withdrawals are balanced by recharge during the corresponding period. The potential of a groundwater basin can be assessed by employing the water budget equation:

$$\sum I - \sum O = \Delta S$$

where the inflow $\sum I$ includes all forms of recharge, the total outflow $\sum O$ includes every kind of discharge, and ΔS represents the change in storage during the accounting period. The most significant forms of recharge and discharge are those listed in Table 10.4.

A groundwater hydrologist must be able to estimate the quantity of water that can be economically and safely produced from a groundwater basin in a specified time period. He or she should also be competent to evaluate the consequences of imposing various rates of withdrawal on an underground supply.

Development of groundwater basins should be based on careful study, since groundwater resources are finite and exhaustible. If the various types of recharge balance the withdrawals from a basin over a period of time, no difficulty will be encountered. Excessive drafts, however, can deplete underground water supplies to a point where economic development is not feasible. The mining of water will ultimately deplete the entire supply.

10.6   REGIONAL GROUNDWATER MODELS

Groundwater systems models may be analog or digital (mathematical) [5],[6],[9], [51]–[73]. The focus here is on digital models, the type commonly employed at present. These models are characterized by a set of equations representing the physical processes occurring in an aquifer. The models may be deterministic or probabilistic in nature, but only deterministic models are discussed here. They describe the cause–effect relations stemming from known features of the physical system to be modeled.

A conceptual groundwater model for a study area is formulated based on a knowledge of the characteristics of the region and an understanding of the mechanics of groundwater flow. Next, the conceptual model is translated into a mathematical

TABLE 10.4   Some Forms of Recharge and Discharge

Recharge	Discharge
Seepage from streams, ponds, lakes	Seepage to lakes, streams, springs
Subsurface inflows	Subsurface outflows
Infiltrated precipitation	Evapotranspiration
Water recharged artificially	Pumping or other artificial means of collection

model, usually represented by a partial differential equation or set of equations accompanied by appropriate boundary and initial conditions. Conditions of continuity and conservation of momentum, generally as described by Darcy's law, are incorporated in the model. Other model features include artesian or water table condition designation and dimensionality (one-, two-, or three-dimensional). If water quality and/or heat transfer considerations are to be incorporated, additional equations describing conservation of mass, for the chemical constituents involved, and conservation of energy are required. Typically used relations are Fick's law for chemical diffusion and Fourier's law for heat transfer.

Once the mathematical model has been formulated, it can be applied to the study area. This requires converting the governing equations into forms that facilitate solution. This is ordinarily achieved through the use of numerical methods such as finite differences or finite elements to represent the applicable partial differential equations. In using a finite-difference approach, the region is divided into grid elements, and the continuous variables are represented as discrete variables at the nodal points. In this manner, the governing differential equation is replaced by a finite number of algebraic expressions that can be solved in an iterative way. Models of this type find wide application in the estimation of site-specific aquifer behavior. They have proven to be effective under irregular boundary conditions, where there are heterogeneities, and where highly variable pumping or recharge rates are expected [51]. Several types of groundwater models and their applications are summarized in Table 10.5.

The first step in modeling a targeted groundwater region is to define the boundaries. These may be physical, such as an impervious layer, or arbitrary, such as politically or otherwise defined subregion. Next, the region is divided into discrete elements by superimposing a rectangular or polygonal grid (see Fig. 10.20).

Once the grid is set, the controlling aquifer parameters (S_c and T) and the initial conditions are set for each grid element. If solute transport is included in the model, additional parameters such as hydrodynamic dispersion properties must also be specified. After all of these specifications have been met, the model can be operated and its output compared with recorded history (called history-matching). Comparisons of recorded values of head and other features with counterpart model predictions permit parameter adjustments to be made until observed and computed data are considered by the modeler to be in close agreement.

TABLE 10.5   Groundwater Models and Applications

Model types	Model applications
Groundwater flow	Water supply, aquifer studies, well development, groundwater-surface water interchanges
Solute transport	Saltwater intrusion, waste management, landfills, groundwater pollution studies
Heat transfer	Geothermal development, thermal pollution, heat pumps
Deformation	Land subsidence

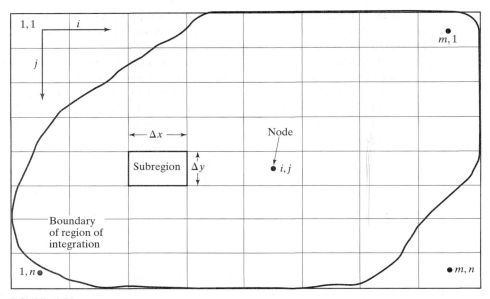

FIGURE 10.20

Subdivision of a region of integration into computational elements for a finite-difference problem formulation.

Upon completion of calibration, the model can be applied to analyze a variety of management and/or development options. The model's prediction of the outcomes of these alternative strategies can be a valuable aid in decision-making processes. The types of problems that can be addressed include the ability of an aquifer to support various levels of use; the impact on an aquifer of varying natural and artificial recharge rates; the effects on underground storage of well location, spacing, and pumping rate; the rate of movement of subsurface contaminants; and the extent of saltwater intrusion.

While numerical groundwater models have their benefits, caution must be exercised to ensure that they are used and interpreted appropriately. Prickett notes that inappropriate use and misinterpretation of results can cause problems [55]. To avoid these pitfalls, both the modeler and user must understand the underlying assumptions upon which the model was founded, its limitations, and its sources of errors. Used wisely, models can be powerful decision-making aids. Used inappropriately, they can lead to erroneous and sometimes damaging proposals.

Finite-Difference Methods

Digital simulation requires an adequate mathematical description of the physical processes to be modeled. For groundwater flow this description consists of a partial differential equation and accompanying boundary and initial conditions. The governing equation is integrated to produce a solution that gives the water levels or heads associated with the aquifer being studied at selected points in space and time. The model can simulate years of physical activity in a span of seconds, so that the consequences of proposed actions can be evaluated before decisions involving construction or social change are implemented. The expectation is that the model runs will lead to wiser and more cost-effective decisions.

The finite-difference method is based on the subdivision of an aquifer into a grid and the analysis of flows associated with zones of the aquifer. The equation that must be solved is derived from continuity considerations and Darcy's law for groundwater motion. This yields the following partial differential equation (a version of Eq. 10.42), describing flow through an areally extensive aquifer. Note that the equation presented here describes the two-dimensional case:

$$\frac{\partial(T\,\partial h/\partial x)}{\partial x} + \frac{\partial(T\,\partial h/\partial y)}{\partial y} = S\frac{\partial h}{\partial t} + W \qquad (10.120)$$

where h = total hydraulic head (L)
   x = x direction in a cartesian coordinate system (L)
   y = y direction in a cartesian coordinate system (L)
   S = specific yield of the aquifer (dimensionless)
   T = transmissivity of the aquifer (L^2/T)
   W = source and sink term (L/T)

In the above equation, vertical flow velocities are considered to be negligible everywhere in the aquifer. The following assumptions are implicit in the derivation: the flow is two-dimensional; fluid density is constant in time and space; hydraulic conductivity is uniform within the aquifer; flow obeys Darcy's law; and the specific yield of the aquifer is constant in space and time. Equation 10.120 is nonlinear for unconfined aquifers because transmissivity is a function of head and thus the dependent variable.

In order to integrate Eq. 10.120, initial values of head, transmissivity, saturated thickness of the aquifer, and the amounts of water produced by sources and sinks must be identified for every point in the region of the integration. The specific yield and location of geometric boundaries must also be defined. Unfortunately, analytic solutions to Eq. 10.120 are impossible to obtain except for the most trivial cases. It is thus necessary to resort to numerical integration techniques to obtain the desired answers [37],[57]–[61].

Application of finite-difference techniques to groundwater flow problems requires that the region of concern be divided into many small subregions or elements (Fig. 10.20). For each of these elements, characteristic values of all the variables in Eq. 10.120 are specified. These values are assigned to the centers of the elements, which are called *nodes*. The heads in adjacent nodes are related through a finite-difference equation, which is derived from Eq. 10.120. These difference equations can be derived by an appropriate Taylor's series expansion or by mass balance considerations [37]. The resulting algebraic equations can then be solved simultaneously to yield the heads at each node for each time step considered.

It should be understood that the simulation methods presented in this chapter are pointed toward the analysis of regional rather than localized groundwater problems such as the prediction of the drawdown at a particular well. Here we are concerned with water level or head changes that might occur over a large area due to prescribed water-use practices.

Boundary Conditions

In order to integrate Eq. 10.120, the governing boundary conditions must be specified. Two types of boundary condition are discussed here.

Where the region of integration is limited by a political or arbitrarily chosen boundary, it is often the policy to employ a constant-gradient boundary condition [57]. In this case, an assumption is made that the gradient of the water table will not change along the boundary even though the water level may rise or fall. Where streams with interconnections to the groundwater system are encountered, stream boundary conditions are employed. Constant-gradient boundaries are expressed mathematically as:

$$\frac{\partial h}{\partial s} = g(x, y) \tag{10.121}$$

where $g(x, y)$ = a constant specified at the location x, y throughout the period of simulation (dimensionless)
   h = hydraulic head (L)
   s = direction perpendicular to the boundary (L)

Stream boundaries are expressed as:

$$h = f(x, y, t) \tag{10.122}$$

where $f(x, y, t)$ = an unknown function of time at the location x, y (dimensionless)
   h = hydraulic head (L)

The volumetric rate of flow across the constant-head boundaries described by Eq. 10.121 can be modeled at each time step using the Darcy equation [57]:

$$Q = T\frac{\partial h}{\partial s}\Delta l \tag{10.123}$$

where h = head (L)
   Δl = dummy variable denoting the length of the side of the subregion perpendicular to s (L)
   s = dummy variable denoting the direction of flow perpendicular to the boundary (L)
   Q = volumetric discharge (L^3/T)
   T = transmissivity at the boundary (L^2/T)

Use of this equation at a boundary is illustrated by the notation of Fig. 10.21. Consider the flow from left to right in the x direction across the left-hand side of the elemental region depicted. The node $i - 1, j$ lies outside the region of integration and

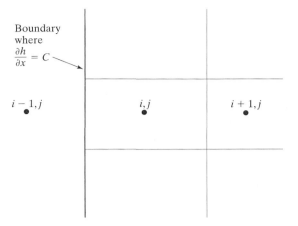

FIGURE 10.21

Subregions adjacent to a constant-gradient boundary.

thus it may be assumed that no information about it is available. An assumption may be made to circumvent this problem. It is that the transmissivity across the boundary is uniform and equal to $T_{i,j}$.

In finite-difference form, the head change term in Eq. 10.27 can be stated as:

$$\frac{\partial h}{\partial x} = \frac{h_{i,j} - h_{i-1,j}}{\Delta x} \tag{10.124}$$

But the head $h_{i-1,j}$ does not exist, and another approximation is required:

$$h_{i,j} - h_{i-1,j} \simeq h_{i+1,j} - h_{i,j} \tag{10.125}$$

These two expressions are then substituted in Eq. 10.123 to yield:

$$Q_{i-1/2,j} \simeq T_{i,j} \frac{h_{i+1,j} - h_{i,j}}{\Delta x} \Delta y \tag{10.126}$$

At the beginning of each time step, a new volumetric flux is calculated along each constant-gradient boundary. This is accomplished by using the heads and transmissivities computed in the previous time interval.

Surface streams are sometimes treated as constant-head boundaries in groundwater problems. The assumption is adequate where the water level in the surface body is expected to remain unchanged during the time period of the modeling process. In many instances, however, surface flows, and hence heads, are significantly affected by withdrawals or recharges to the interconnected groundwater system. They may then be a limited source of water supply for the groundwater system. To accommodate the surface water–groundwater linkage, a leakage term may be applied [57]. This expression may take the form:

$$\text{leakage}_{i,j,k} = -\frac{K_{i,j}}{b_{i,j}} (h_{i,j,k} - h_{i,j,0}) \tag{10.127}$$

where $b_{i,j}$ = thickness of the streambed (L)

   $h_{i,j,k}$ = head in the aquifer at node i, j at time k; $k = 0$ indicates initial conditions (L)

   $k_{i,j}$ = hydraulic conductivity at node i, j (L/T).

When Eq. 10.127 is used, the stream is considered to cover the entire area represented by the related node. After each time step the leakage from the stream to the aquifer is calculated and streamflows are depleted accordingly. If the streamflow at a particular node becomes zero, the model can be made to note that the stream is dry and break the hydraulic connection at that point [57].

Time Steps and Element Dimensions

The success of any finite-difference scheme depends on the incremental values assigned the element dimensions and the time steps. In general, the smaller the dimensions of elements and time increments, the closer the finite-difference approximation to the differential equation. However, as these partitions are made smaller, a price in computational costs and data needs must be paid. Furthermore, oversubdivision may even bring about computational intractability. Thus the object is to select the degree of definition that results in an adequate representation of the system while keeping data and computational costs at a minimum. There are procedures for making such selections, but, except for a brief discussion in the following section, they are not presented here [57]–[61].

One-Dimensional Flow Model

To illustrate the finite-difference approach to groundwater problem-solving, a one-dimensional conceptualization is discussed. Although most practical-scale models are two- or three-dimensional in character, their development is only an extension of the one-dimensional case. For details of some of the more complex models the reader should consult the appropriate references [5],[37],[56]–[62]. The book by McWhorter and Sunada is easy to read and includes excellent example problems [37]. The treatment of one-dimensional flow taken here follows the approach of that reference.

Consider a one-dimensional flow in a confined aquifer system such as that illustrated by Fig. 10.22. It is assumed that the flow is unsteady and that the flow lines are parallel and not time dependent. On this basis, a unit width of the aquifer can be studied and observations made about it can easily be translated to the total system. As shown in the figure, the unit width of the aquifer is Δx. The flow region is overlaid by a grid, and for each grid element, values of hydraulic conductivity K_i, element length y_i, aquifer thickness b_i, storage coefficient S_i, and the initial values of head h_i must be specified. The mass balance for grid element i requires that the inflow ($Q_{i-1 \rightarrow i}$) from element $i - 1$ to element i minus the outflow ($Q_{i \rightarrow i+1}$) from element i to element $i + 1$ must be balanced by the rate of change in storage which occurs in element i, $\Delta V_i / \Delta t$.

To simplify the problem, let us further consider that the aquifer is of uniform thickness and that it is homogeneous and isotropic. Thus the values of K, S, b, and Δy

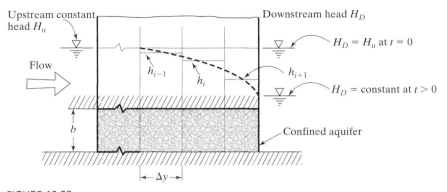

FIGURE 10.22

Grid notation for a one-dimensional groundwater flow case.
(After McWhorter and Sunada [37])

are constant, and we shall consider that from studies of the aquifer properties, they are also known. Therefore it may be stated that:

$$K_1 = K_2 = \cdots = K_m = K$$
$$S_1 = S_2 = \cdots = S_m = S$$
$$b_1 = b_2 = \cdots = b_m = b$$
$$\Delta y_1 = \Delta y_2 = \cdots = \Delta y_m = \Delta y \tag{10.128}$$

where the subscript m represents the total number of grid elements. Using this notation and Fig. 10.22, we can see that the flow from element $i - 1$ to i is:

$$Q_{i-1 \to i} = -KA \frac{h_i^n - h_{i-1}^n}{\Delta y} \tag{10.129}$$

where i = the element number
$\quad\quad\quad n$ = the selected time

Equation 10.129 is recognized as Darcy's equation. It is assumed in this representation that the head generating the flow at time n is the difference between the average heads at the two adjacent elements divided by the distance between their centers (nodes). This approximation approaches exactness as Δy diminishes to zero.

The area A appearing in Eq. 10.129 is the cross-sectional area of flow and is obtained as the product of Δx and b. Since we are dealing with a unit width of aquifer, $\Delta x = 1$ and since b is a constant by definition here, Eq. 10.129 may be written:

$$Q_{i-1 \to i} = -T \frac{h_i^n - h_{i-1}^n}{\Delta y} \tag{10.130}$$

where $T = Kb$. A similar expression for the flow from element i to $i + 1$ may be obtained:

$$Q_{i \to i+1} = -T \frac{h_{i+1}^n - h_i^n}{\Delta y} \qquad (10.131)$$

Equations 10.130 and 10.131 represent the inflow and outflow from element i. Considering that continuity conditions must be met, this change in flow across the element must be balanced by the change in storage that occurs during the time step. This is given as:

$$\frac{\Delta V_i}{\Delta t} = S \, \Delta y \left(\frac{h_i^{t+\Delta t} - h_i^t}{\Delta t} \right) \qquad (10.132)$$

Now inserting these three expressions in the continuity equation (inflow − outflow = change in storage), we get:

$$\left(-T \frac{h_i^n - h_{i-1}^n}{\Delta y} \right) - \left(-T \frac{h_{i+1}^n - h_i^n}{\Delta y} \right) = S \, \Delta y \left(\frac{h_i^{t+\Delta t} - h_i^t}{\Delta t} \right) \qquad (10.133)$$

By rearrangement, the equation becomes:

$$h_{i+1}^n - 2h_i^n + h_{i-1}^n = \frac{S}{T} \frac{(\Delta y)^2}{\Delta t} (h_i^{t+\Delta t} - h_i^t) \qquad (10.134)$$

which is known as the explicit or forward-difference form of the finite-difference equation if n is designated as the current value of time. If, on the other hand, n is defined as $t + \Delta t$, then the equation is the implicit or backward-difference equation. Each of these forms has its own solution techniques [37]. The explicit solution to Eq. 10.134 will be discussed here.

By letting $n = t$ in Eq. 10.134 and rearranging, one obtains:

$$h_i^{t+\Delta t} = \frac{T \Delta t}{S(\Delta y)^2} (h_{i+1}^t + h_{i-1}^t) + h_i^t \left[1 - \frac{2T \Delta t}{S(\Delta y)^2} \right] \qquad (10.135)$$

In this case the space derivatives are centered at the beginning of the time step and the single unknown is $h_i^{t+\Delta t}$. Equation 10.135 can be solved explicitly at each element for the head at the next period of time. The solution depends only on a knowledge of the heads in adjacent elements at the beginning of a time step.

It must be recognized that a solution obtained using Eq. 10.135 is only an approximation to the exact solution. The correspondence with the exact solution is related to the choice of Δy and Δt. If the selected values are too large, the difference between the approximate and exact solution can grow as t increases, bringing about an unstable condition. In the one-dimensional homogeneous case discussed here, stability is assured if:

$$\frac{T \Delta t}{S(\Delta y)^2} < \frac{1}{2} \qquad (10.136)$$

The equation shows that the choice of time and space increments is not independent. Satisfaction of Eq. 10.136 does not guarantee an accurate approximation, however; it only provides for a stable solution [37].

Example 10.12

Refer to the one-dimensional flow problem of Fig. 10.23. Let us assume that the element length is 4 m and that the thickness of the confined aquifer is 2 m. It is further assumed that the head at the left and right sides of the region is 8 m at $t = 0$ and that the head on the right side takes on the value 2 m for all t greater than zero. $K = 0.5$ m/day and $S = 0.02$. As shown in the figure, there are five elements. Using the notation of Eq. 10.135, the initial condition is $h_4^0 = 8.0$ m. Use the explicit method to determine future heads.

Solution

1. First a determination must be made of the time step to use. This may be accomplished using Eq. 10.136:

$$\Delta t < \frac{1}{2}\frac{S(\Delta y)^2}{T} = \frac{1}{2}\frac{(0.02)(4)^2}{1.0} = 0.16 \text{ days}$$

The value of T used in the above expression was obtained using the relation $T = Kb$:

$$T = 0.5 \times 2.0 = 1.0$$

To be on the safe side, we shall choose a time step of 0.1 days, although any value less than 0.16 would have assured stability.

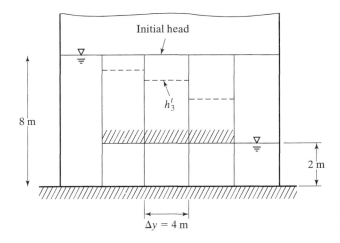

8 m

2 m

$\Delta y = 4$ m

Initial head

h_3^t

FIGURE 10.23

Sketch for Example 10.12.

2. For the first time step, $t = 0.1$, we can calculate $h_4^{t+\Delta t}$ and corresponding heads for the other elements using Eq. 10.135. Thus:

$$h_4^{t+\Delta t} = \frac{T \, \Delta t}{S(\Delta y)^2} (h_5^0 + h_3^0) + h_4^0 \left[1 - 2\frac{T \, \Delta t}{S(\Delta y)^2} \right]$$

and substituting numerical values, we get:

$$h_4^{0.1} = \frac{1.0(0.1)}{(0.02)(4)^2} (2.0 + 8.0) + 8.0 \left[1 - \frac{2(1.0)(0.1)}{(0.02)(4)^2} \right]$$

$$= 3.1 + 3.0 = 6.13 \, \text{m}$$

Since h_4^0 and $h_2^0 = 8.0$ m and since $h_1 = 8.0$ by definition, it can easily be shown using Eq. 10.135 that the values of $h_3^{0.1}$ and $h_2^{0.1}$ are not changed from their original level of 8.0 during the first time step.

3. Now consider the second time step, $t + \Delta t = 0.2$ days. For element 4:

$$h_4^{0.2} = \frac{T \, \Delta t}{S(\Delta y)^2} (h_5^{0.1} + h_3^{0.1}) + h_4^{0.1} \left[1 - \frac{2T \, \Delta t}{S(\Delta y)^2} \right]$$

$$= 0.31(2.0 + 8.0) + 6.13(0.37) = 5.4 \, \text{m}$$

For element 3:

$$h_3^{0.2} = \frac{T \, \Delta t}{S(\Delta y)^2} (h_4^{0.1} + h_2^{0.1}) + h_3^{0.1} \left[1 - \frac{2T \, \Delta t}{S(\Delta y)^2} \right]$$

$$= 0.31(6.13 + 8.0) + 8.0(0.37) = 7.4 \, \text{m}$$

Element 2 does not have a head change until the third time step.

4. The process demonstrated is repeated until the heads have been calculated for the total time period of interest. For this example, they will ultimately reach equilibrium conditions.

This example illustrates the mechanics of the finite-difference procedure. Problems of practical scale would require the use of a computer, but the approach would be the same.

Finite-Element Methods

The most widely used numerical techniques for solving groundwater flow problems are the finite-difference and finite-element methods. The finite-element method is similar to the finite-difference method in that both approaches lead to a set of N equations in N unknowns that can be solved by relaxation [5]. Nodes in the finite-element method are usually the corner points of an irregular triangular or quadrilateral mesh for two-dimensional applications, while for three-dimensional applications, bricks or tetrahedrons are commonly used. The size and shape of the elements selected are arbitrary. They are chosen to fit the application at hand. They differ from the regular rectangular grid elements used in finite-difference modeling. Elements that are closest to points of flow concentration, such as wells, are usually smaller than those farther removed from such influences. Aquifer parameters such as hydraulic conductivity may be kept constant for a given element but may vary from one to another. To minimize the variational function, its partial derivative with respect to head is evaluated for each node and equated to zero. The procedure results in a set of algebraic equations that can be solved by iteration, matrix solution, or a combination of these methods [6]. Finite-element modelers must understand partial differential equations and the calculus of variations [5].

The finite-element approach offers some advantages over the finite-difference technique. Often, a smaller nodal grid is required, and this offers economies in computer effort. The finite-element approach can also accommodate one condition that the finite-difference approach is unable to handle [5]. When using the finite-difference method, the principal directions of anisotropy in an anisotropic formation are parallel to the coordinate directions. In cases where two anisotropic formations having different principal directions occur in a flow field, the finite-difference approach cannot produce a solution, whereas the finite-element approach can. The finite-element technique can be used to simulate transient aquifer performance. A detailed discussion of the finite-element technique is beyond the scope of this book, but there are many good references for the interested reader [5],[6],[63]–[66].

Model Applications

To illustrate how simulation models can be used to provide insights into water management schemes, a model analysis of the Upper Big Blue basin aquifer in Nebraska is presented. The study was conducted by the Conservation and Survey Division of the University of Nebraska under the direction of Huntoon [57].

The use of groundwater for irrigation in the Upper Big Blue basin was observed to be rapidly increasing and by 1972 about 3.3 wells/mi^2 were in operation. At that time farmers were becoming concerned about the progressive decline of water levels and were seeking guidance about the efficiency of implementing some form of basin-wide water management program. The University of Nebraska designed a model to evaluate the situation and to explore various proposals for recharging the aquifer and for estimating the long-term consequences of several scenarios of water use in the basin.

The study area is shown in Fig. 10.24. Generally the water table is free in the region of interest. Figure 10.25 shows the configuration of the water table as observed in 1953. For modeling purposes, the water-level contours shown were considered to be representative of predevelopment conditions. This assumption was based on the fact that groundwater withdrawals before this time were not extensive. It was also surmised that the contours represented a water table in which an equilibrium existed between natural recharge and discharge in the region. Transmissivities were estimated from drill-hole sample logs recorded in the area. These values are needed for modeling and are also important indices of the potential yield of wells that might be constructed.

As might be suspected, the information of most concern to the local landowners and water planners was the rate of decline of the water table. In particular, it was

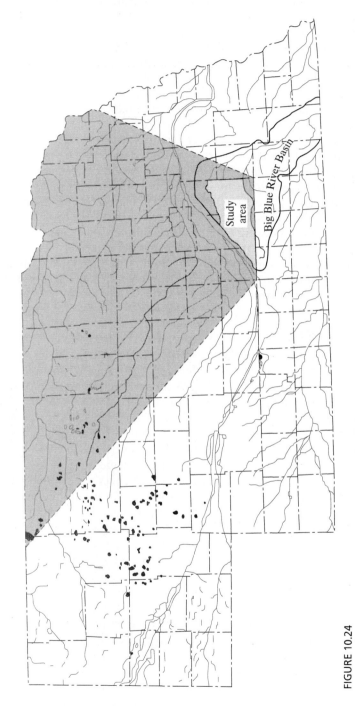

FIGURE 10.24

Big Blue River basin, Nebraska.

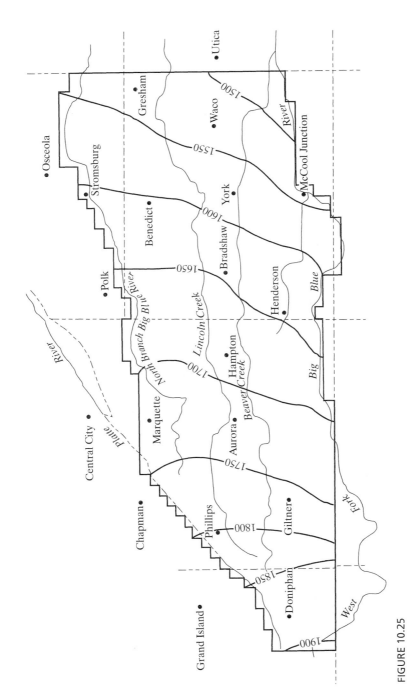

FIGURE 10.25

Elevation of the water table in 1953, in feet above mean sea level.
(After Huntoon [57])

desired to know how rapidly the groundwater resource would be depleted, where and when water level declines would pose an economic constraint on water use, and what impacts future developments and/or management would have on the rate of decline.

The model developed to explore these features was a two-dimensional representation of flow through an areally extensive aquifer [57]. Equation 10.120, along with the appropriate boundary conditions, constituted the model. The region shown in Fig. 10.24 was divided into a finite-difference grid and, after substitution of the nodal values of T and S, the model was operated to predict water-level changes to the year 2020 for various policies of recharge and for several levels of development. Calibration of the model was accomplished using historic data. The model was operated over the period 1953–1972 using the known distribution of wells and the average net pumpage per well to establish a match between observed and estimated water-level changes. Once this was accomplished, the simulation of future trends proceeded. Figures 10.26 and 10.27 show the correspondence achieved in the matching process.

On the basis of the model studies, it was determined that water levels in the study area would continue to decline even if development was limited to the 1972 level. It was further predicted that some parts of the area would experience severe groundwater shortages by the year 2000. It was found, however, that by employing artificial recharge methods, permanent groundwater supplies could be assured. To assess the effects of artificial recharge, two water delivery systems were modeled. Both of these delivered water from Platte River Valley sources to recharge wells located in the project area. Using these two water delivery systems, three recharge schemes were simulated. The gross effect of introducing the recharge wells was the cancellation of the effects of the proportionate number of pumping wells. Figure 10.28 shows the computed water-level changes at one location under a graduated development plan (projected on the basis of the 1972 rate of development) with no recharge and then with graduated development for each of the three recharge schemes. The continual downward trend in water level with no recharge (curve 1) clearly shows the nature of the problem in the Upper Big Blue basin. The other curves depicting the three artificial recharge options show that stability can be achieved if such an approach is taken.

While the costs of implementing artificial recharge might be excessive, it is apparent that any long-term solution to the declining water table problem, short of reducing use, would require a supplemental source of water.

Operation of the model provided useful insights into the nature of the water table problem and suggested that irrigators should be making some important water management decisions about their future mode of operation.

The modeling of groundwater systems is complex [10]–[25]. In structuring models such as that just discussed, simplifying assumptions must usually be made. These have to do with aquifer parameters such as transmissivity, specific yield (for unconfined aquifers), and storage coefficient (for confined systems). Furthermore, the boundary conditions are normally approximations of what occurs in the physical system, and assumptions about the uniformity of materials in various subsurface strata are sometimes crude. This does not mean that groundwater models cannot be expected

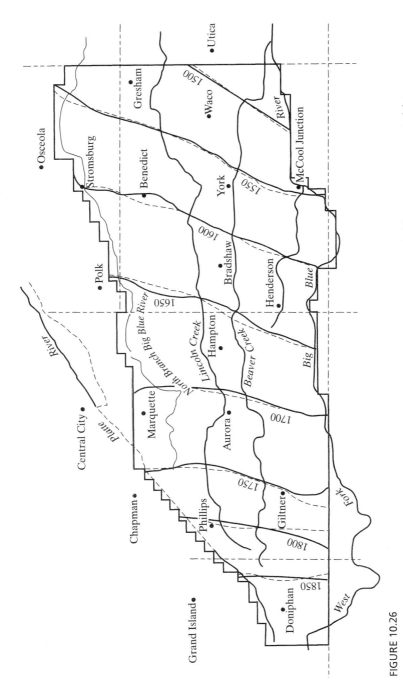

FIGURE 10.26

Measured and computed water table elevations in 1972. Solid line represents measured data; dashed line represents computed data.

(After Huntoon [57])

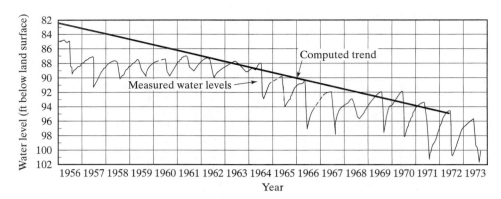

FIGURE 10.27

Measured and computed water-level trends.
(After Huntoon [57])

to yield useful results. It does imply that the users of the models must be cautious about how they interpret the output. For example, an areally extensive aquifer model such as that developed by Huntoon for analyzing the Blue River problem can be expected to give reliable information about water-level trends for various configurations of development. It should not, on the other hand, be considered an accurate predictive tool for monitoring the water-level change at some specific point in the region of concern. This type of information could be derived only from a more detailed

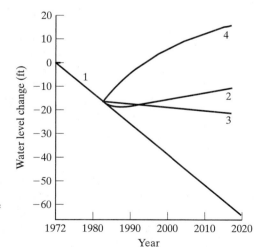

FIGURE 10.28

Computed water-level changes under a plan of graduated development for conditions of (1) no recharge, (2) recharge under scheme 1, (3) recharge under scheme 2, and (4) recharge under scheme 3.
(After Huntoon [57])

modeling of the locality surrounding the point. The information provided by the Blue River model was targeted to show local landowners what the future might hold for several development levels and for several management options. The actual water levels predicted by the model were not of central concern; what was of interest was the determination that unless future development was restricted and supplemental water provided, or unless current uses could be significantly reduced, the outlook in the next 50 years was not good for irrigated farming.

The model thus provided the basis for making some quantitative observations about the future. It also provided insights into the relief that might be expected from artificial recharge. Beyond that, it could be used to model other possible management options. A model such as this, carefully used and properly interpreted, can thus add a powerful dimension to decision-making processes.

10.7   JOINT SURFACE–WATER–GROUNDWATER SYSTEMS

In many cases, surface water and groundwater systems are physically inseparable—groundwater provides the base flow to streams and rivers and is in turn recharged by infiltration from water flowing over the earth's surface. Too often, however, these systems are regulated and managed as if they were unconnected. Many laws and regulations implemented in the past do not recognize the physical relationship that exists, and now they exacerbate this problem. Ideally, wherever possible, surface water systems and groundwater systems should be managed and operated jointly so as to take advantage of the special features of each system. Use of such joint operations is commonly known as *conjunctive use.*

In practice, this concept has been applied to the joint operation of surface reservoirs and nearby groundwater aquifers. It takes advantage of the fact that surface waters are often available seasonally, but at an uncertain time and in an uncertain amount. Surface waters can be collected and stored for future use in surface reservoirs, but a significant quantity of water may be lost by evaporation. Groundwater storage, on the other hand, is not subject to such losses and provides a natural storage reservoir. By operating the two systems jointly, greater and more economical yields can often be realized. This involves impoundment of surface waters when such supplies are at a peak and allocating a portion of them to recharge aquifer storage capacity (also known as aquifer storage and recovery, ASR). During dry periods, groundwater storage can then be withdrawn for use and later replenished during wet periods when surface water surpluses occur. If the system is operated properly, groundwater storage will fluctuate but not be diminished over long periods of time.

Conjunctive use requires careful planning and full understanding of the hydrogeological properties of the region. There must be sufficient storage volume available in the aquifer to permit storing quantities of water sufficient to meet water demands during periods of drought. In addition, surface reservoirs and facilities for recharging aquifers must be provided.

SUMMARY

The importance of groundwater to the health and well-being of humans is well documented. Groundwater is a major source of freshwater for public consumption, industrial uses, and the irrigation of crops. For example, more than half of the fresh water used in Florida for all purposes comes from groundwater sources, and about 90 percent of that state's population depends on groundwater for its potable water supply. The need to husband this resource is clear. Quantity and quality dimensions are both important.

Groundwater protection and management practices must be based on an understanding of groundwater sources; the manner in which groundwater is distributed below the earth's surface; geologic, topographic, and soil characteristics of the region; and the interconnections between groundwater and surface water sources. The rate of movement of water through the ground is of a different magnitude than that through natural or artificial channels or conduits. Typical flow rates range from 5 ft/day to a few feet per year. These low rates of flow exacerbate the impact of contaminant spills on groundwater sources and complicate cleanup since natural flushing from the site may take many years to occur.

Understanding the movement of groundwater requires a knowledge of the time and space dependency of the flow, the nature of the porous medium and fluid, and the boundaries of the flow system. In particular, groundwater development and management depend on understanding the storage properties of the associated soils and rocks and the ability of these subsurface materials to transmit water. Fundamental to the mechanics of groundwater flow is Darcy's law. Using Darcy's equation along with a knowledge of the hydraulic conductivity K, estimates of flow can be made. The hydrodynamic equations presented in this chapter serve as models for a variety of groundwater flow calculations.

The collection of groundwater is accomplished primarily through the construction of wells. Some situations are amenable to solution through the utilization of relatively simple mathematical expressions. Others depend upon sophisticated application of the hydrodynamic equations under various conditions of nonuniformity of aquifer materials and a variety of boundary conditions. The reader is cautioned not to be misled by the simplicity of some of the solutions presented and should observe that many of these relate to special conditions and are not applicable to all groundwater-flow situations.

Natural hydrologic states may be significantly affected by human activities. Aquifer depletions having regional and national economic implications are not uncommon. Depletion of the Ogallala aquifer in the central United States by long-term and extensive water withdrawals for irrigation is a good example. On the other hand, water levels have been made to rise, sometimes inadvertently, by human intervention. Leaky irrigation canals in central Nebraska were at one time responsible for groundwater level rises in some farming locations of a magnitude sufficient to jeopardize use of the land. Once major problems of depletion or overreplenishment occur, they are not easily dealt with. In general, a safe-yield policy for groundwater management has merit and should be considered [5],[74]. Safe yield is the amount of water that can be withdrawn annually without the ultimate depletion of the aquifer.

Regional groundwater flow problems are usually modeled by an equation combining Darcy's law and the equation of continuity. The resulting partial differential equation, or set of equations, describes the hydraulic relations within the aquifer. To effect a solution to the governing equation(s), the aquifer's hydraulic features, geometry, and initial and boundary conditions must be determined. Unfortunately, many groundwater problems exist for which exact analytic solutions cannot be obtained. In such cases, it is necessary to rely on numerical methods for modeling. Under such circumstances, an approximate solution is obtained by replacing the basic differential equations by another set of equations that can be solved iteratively on a computer. Both finite-difference and finite-element methods are applicable.

The finite-difference approach described in this chapter replaces the governing partial differential equations with a set of algebraic equations. These can be solved on the computer to produce a set of water table elevations at a finite number of locations in the aquifer.

Once a groundwater model has been calibrated, it can be used to predict the outcomes (impacts) of alternative development and/or management strategies proposed for an aquifer. Such analyses are valuable adjuncts to decision-making processes. Models can, for example, simulate the effects of opening new well fields, analyze changed operating practices for existing well fields, explore schemes for artificial recharge, and predict the impacts of proposed irrigation development plans. Groundwater models can be applied to unconfined aquifers, semiconfined aquifers, confined aquifers, or any combination thereof. They can accommodate large variations in aquifer parameters, such as hydraulic conductivity and storage coefficient, and they can be used to analyze unsteady as well as steady flow problems.

PROBLEMS

10.1 The water temperature in an aquifer is 60°F and the velocity is 1.0 ft/day. The average particle diameter of the soil is 0.06 in. Find the Reynolds number and indicate whether Darcy's law applies.

10.2 The water temperature in an aquifer is 60°F and the rate of water movement is 1.2 ft/day. The average particle diameter of the porous medium is 0.08 in. Find the Reynolds number and indicate whether Darcy's law applies.

10.3 Laboratory tests of an aquifer material give a standard hydraulic conductivity of 3.78×10^2 gpd/ft². If the prevailing field temperature is 50°F, find the field hydraulic conductivity.

10.4 A laboratory test of a soil gives a standard hydraulic conductivity of 3.8×10^2 gpd/ft². If the prevailing field temperature is 60°F, find the field hydraulic conductivity.

10.5 Given the well and flow net data in the following figure, find the discharge using a flow net solution. The well is fully penetrating; $K = 2.87 \times 10^{-4}$ ft/sec, $a = 180$ ft, $b = 43$ ft, and $c = 50$ ft.

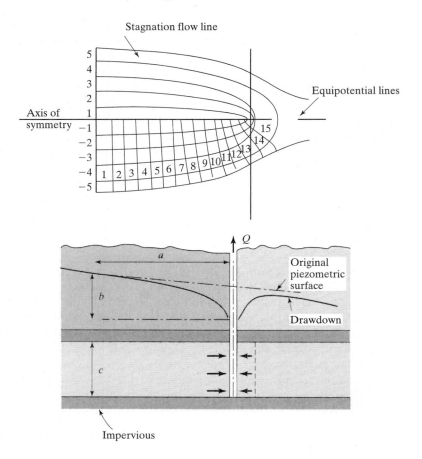

10.6 Rework problem 10.5 if $K = 8.4 \times 10^{-5}$ m/sec, $a = 100$ m, $b = 22$ m, and $c = 35$ m.

10.7 A stratum of clean sand and gravel 20 ft deep has a hydraulic conductivity of $K = 3.25 \times 10^{-3}$ ft/sec and is supplied with water from a channel that penetrates to the bottom of the stratum. If the water surface in an infiltration gallery is 3 ft above the bottom of the stratum and its distance to the channel is 50 ft, what is the flow into a foot of gallery? Use Eq. 10.98.

10.8 A 12-in. well fully penetrates a confined aquifer 100 ft thick. The coefficient of permeability is 600 gpd/ft^2. Two test wells located 45 and 120 ft away show a difference in drawdown between them of 8 ft. Find the rate of flow delivered by the well.

10.9 Determine the permeability of an artesian aquifer being pumped by a fully penetrating well. The aquifer is composed of medium sand and is 100 ft thick. The steady-state pumping rate is 1,200 gpm. The drawdown in an observation well 75 ft away is 14 ft, and the drawdown in a second observation well 500 ft away is 1.2 ft. Find K in gallons per day per square foot.

10.10 Consider a confined aquifer with a coefficient of transmissibility $T = 680$ ft^3/day/ft. At $t = 5$ min, the drawdown $s = 5.6$ ft; at 50 min, $s = 23.1$ ft; and at 100 min, $s = 28.2$ ft. The observation well is 75 ft away from the pumping well. Find the discharge of the well.

10.11 Use the following data: $Q = 59{,}000\ \text{ft}^3/\text{day}$, $T = 630\ \text{ft}^3/\text{day}$, $t = 30\ \text{days}$, $r = 1\ \text{ft}$, and $S_c = 6.4 \times 10^{-4}$. Consider this to be a nonequilibrium problem. Find the drawdown s. Note that for

$$u = 8.0 \times 10^{-9} \quad W(u) = 18.06$$
$$u = 8.2 \times 10^{-9} \quad W(u) = 18.04$$
$$u = 8.6 \times 10^{-9} \quad W(u) = 17.99$$

10.12 Find the hydraulic conductivity of an artesian aquifer being pumped by a fully penetrating well. The aquifer is 130 ft thick and is composed of medium sand. The steady-state pumping rate is 1,300 gpm. The drawdown in an observation well 65 ft away is 12 ft, and in a second well 500 ft away, it is 1.2 ft. Find K_f in gpd/ft^2.

10.13 A well is being pumped at a constant rate of $0.004\ \text{m}^3/\text{s}$. Given that $T = 0.0028\ \text{m}^2/\text{s}$, $r = 100$ meters, and the storage coefficient $= 0.001$, find the drawdown in the observation well for a time period of (a) 1 hr, and (b) 24 hours.

10.14 A well is being pumped at a constant rate of $0.003\ \text{m}^3/\text{s}$. Given that $T = 0.0028\ \text{m}^2/\text{s}$, the storage coefficient $= 0.001$, and the time since pumping began is 12 hours, find the drawdown in an observation well for a radial distance of (a) 150 m, and (b) 500 m.

10.15 An 18-in. well fully penetrates an unconfined aquifer of 100-ft depth. Two observation wells located 100 and 235 feet from the pumped well are known to have drawdowns of 22.2 and 21 ft, respectively. If the flow is steady and $K_f = 1{,}320\ \text{gpd/ft}^2$, what would be the discharge?

10.16 Determine the hydraulic conductivity of an artesian aquifer being pumped by a fully penetrating well. The aquifer is 90 ft thick and composed of medium sand. The steady-state pumping rate is 850 gpm. The drawdown at an observation well 50 ft away is 10 ft; in a second observation well 500 ft away, it is 1 ft.

10.17 Find the drawdown at an observation well 200 ft away from a pumping well. It has been found that $T = 3.0 \times 10^4\ \text{gpd/ft}$, the pumping time is 12 days, the storativity is 3×10^{-4}, and $Q = 300\ \text{gpm}$.

10.18 A well is being pumped at a constant rate of $0.0038\ \text{m}^3/\text{s}$. Given that $T = 0.0028\ \text{m}^2/\text{s}$, $r = 90$ meters, and the storage coefficient $= 0.00098$, find the drawdown in the observation well for a time period of (a) 1,000 seconds, and (b) 20 hours.

10.19 A 12-in. well fully penetrates a confined aquifer 100 ft thick. The hydraulic conductivity is 600 gpd/ft^2. Two test wells located 40 and 120 ft away show a difference in drawdown between them of 9 ft. Find the rate of flow delivered by the well.

10.20 Find the drawdown at an observation point 250 ft away from a pumping well given that $T = 3.1 \times 10^4\ \text{gpd/ft}$, the pumping time is 10 days, $S_c = 3 \times 10^{-4}$, and $Q = 280\ \text{gpm}$.

10.21 An 18-in. well fully penetrates an unconfined aquifer 100 ft deep. Two observation wells located 90 and 235 ft from the pumped well are known to have drawdowns of 22.5 ft and 20.6 ft, respectively. If the flow is steady and $K_f = 1{,}300\ \text{gpd/ft}^2$, what would be the discharge?

10.22 Assume that an aquifer being pumped at a rate of 300 gpm is confined and pumping test data are given as follows. Find the coefficient of transmissibility T and the storage coefficient S. Assume $r = 55\ \text{ft}$.

Time since pumping started (min)	1.3	2.5	4.2	8.0	11.0	100.0
Drawdown, s (ft)	4.6	8.1	9.3	12.0	15.1	29.0

10.23 We are given the following data:

$$Q = 60{,}000\ \text{ft}^3/\text{day} \qquad t = 30\ \text{days} \qquad r = 1\ \text{ft}$$
$$T = 650(\text{day})/(\text{ft}) \qquad S_c = 6.4 \times 10^{-4}$$

Assume this to be a nonequilibrium problem. Find the drawdown s. Note for

$$u = 8.0 \times 10^{-9} \quad W(u) = 18.06$$
$$u = 8.2 \times 10^{-9} \quad W(u) = 18.04$$
$$u = 8.6 \times 10^{-9} \quad W(u) = 17.99$$

10.24 A confined aquifer 80 ft deep is being pumped under equilibrium conditions at a rate of 700 gpm. The well fully penetrates the aquifer. Water levels in observation wells 150 and 230 ft from the pumped well are 95 and 97 ft, respectively. Find the field coefficient of permeability.

10.25 A well is pumped at the rate of 500 gpm under nonequilibrium conditions. For the data listed, find the formation constants S and T. Use the Theis method.

r^2/t	Average drawdown, h (ft)
1,250	3.24
5,000	2.18
11,250	1.93
20,000	1.28
45,000	0.80
80,000	0.56
125,000	0.38
180,000	0.22
245,000	0.15
320,000	0.10

10.26 We are given a well pumping at a rate of 590 gpm. An observation well is located at $r = 180$ ft. Find S and T using the Jacob method for the following test data.

Drawdown (ft)	Time (min)	Drawdown (ft)	Time (min)
0.43	26	2.00	661
0.94	78	2.06	732
1.08	99	2.12	843
1.20	131	2.15	926
1.34	173	2.20	1,034
1.46	218	2.23	1,134
1.56	266	2.28	1,272
1.63	303	2.30	1,351
1.68	331	2.32	1,419
1.71	364	2.36	1,520
1.85	481	2.38	1,611
1.93	573		

10.27 A 24-in. diameter well penetrates the full depth of an unconfined aquifer. The original water table and a bedrock aquifuge were located 50 and 150 ft, respectively, below the land surface. After pumping at a rate of 1,700 gpm continuously for 1,920 days, equilibrium drawdown conditions were established, and the original water levels in observation wells located 1,000 and 100 ft from the center of the pumped well were lowered 10 and 20 ft, respectively.

a. Determine the field permeability (gpd/ft^2) of the aquifer.

b. For the same well, zero drawdown occurred outside a circle with a 10,000-ft radius measured from the center of the pumped well. Inside the circle, the average drawdown in the water table was observed to be 10 ft. Determine the coefficient of storage of the aquifer.

10.28 A well fully penetrates the 100-ft depth of a saturated unconfined aquifer. The drawdown at the well casing is 40 ft when equilibrium conditions are established using a constant discharge of 50 gpm. What is the drawdown when equilibrium is established using a constant discharge of 66 gpm?

10.29 After a long rainless period, the flow in Wahoo Creek decreases by 8 cfs from Memphis downstream 8 mi to Ashland. The stream penetrates an unconfined aquifer, where the water table contours near the creek parallel the west bank and slope to the stream by 0.00020, while on the east side the contours slope away from the stream toward the Lincoln wellfield at 0.00095. Compute the transmissivity of the aquifer knowing $Q = TIL$, where I is the slope and L is the length.

10.30 The time–drawdown data for an observation well located 300 ft from a pumped artesian well (500 gpm) are given in the following table. Find the coefficient of storage (ft^3 of water/ft^3 of aquifer) and the transmissivity (gpd/ft) of the aquifer by the Theis method. Use 3 × 3–cycle log paper.

Time (hr)	Drawdown (ft)	Time (hr)	Drawdown (ft)
1.8	0.27	9.8	1.09
2.1	0.30	12.2	1.25
2.4	0.37	14.7	1.40
3.0	0.42	16.3	1.50
3.7	0.50	18.4	1.60
4.9	0.61	21.0	1.70
7.5	0.84	24.4	1.80

10.31 Over a 100-mi^2 surface area, the average level of the water table for an unconfined aquifer has dropped 10 ft because of the removal of 128,000 acre-ft of water from the aquifer. Determine the storage coefficient for the aquifer. The specific yield is 0.2 and the porosity is 0.22.

10.32 Over a 100-mi^2 surface area, the average level of the piezometric surface for a confined aquifer in the Denver area has declined 400 ft as a result of long-term pumping. Determine the amount of the water (acre-ft) pumped from the aquifer. The porosity is 0.3 and the coefficient of storage is 0.0002.

REFERENCES

[1] J. G. Ferris, "Ground Water," *Mech. Eng.* (Jan. 1960).

[2] The Conservation Foundation, "Groundwater Protection," Final Report of the National Groundwater Policy Forum, Washington, D.C., 1987.

[3] E. F. Wood, Raymond A. Ferrara, William G. Gray, and George F. Pinder, *Groundwater Contamination from Hazardous Wastes*, Englewood Cliffs, NJ: Prentice-Hall, 1984.

[4]   V. I. Pye, R. Patrick, and J. Quarles, *Groundwater Contamination in the United States,* Philadelphia: University of Pennsylvania Press, 1983.

[5]   R. A. Freeze and J. A. Cherry, *Groundwater,* Englewood Cliffs, NJ: Prentice-Hall, 1979.

[6]   D. R. Maidment (ed.), *Handbook of Hydrology,* New York: McGraw-Hill, 1993.

[7]   J. Boonstra and N. A. de Ridder, "Numerical Modeling of Groundwater Basins," International Institute for Land Reclamation and Improvement, The Netherlands, 1981.

[8]   R. J. M. DeWiest, *Geohydrology,* New York: Wiley, 1965.

[9]   D. K. Todd, *Groundwater Hydrology,* New York: Wiley, 1960.

[10] R. C. Heath, "Groundwater Regions of the United States," Geological Survey Water Supply Paper No. 2242, Washington, D.C.: U.S. Government Printing Office, 1984.

[11] U.S. Environmental Protection Agency, "The Report to Congress: Waste Disposal Practices and Their Effects on Groundwater," Executive summary, U.S. EPA, PB 265–364, 1977.

[12] O. E. Meinzer, "The Occurrence of Groundwater in the United States," U.S. Geological Survey, Water-Supply Paper No. 489, 1923.

[13] H. H. Cooper, Jr. and C. E. Jacob, "A Generalized Graphical Method for Evaluating Formation Constants and Summarizing Well-Field History," *Trans. Am. Geophys. Union* **27**, 526–534(1946).

[14] O. E. Meinzer, "Outline of Methods for Estimating Groundwater Supplies," U.S. Geological Survey, Water-Supply Paper 638-C, Washington, D.C., 1932.

[15] Nathan Buras, "Conjunctive Operation of Dams and Aquifers," *Proc. ASCE J. Hyd. Div.* **89**(HY6) (Nov. 1963).

[16] F. B. Clendenen, "A Comprehensive Plan for the Conjunctive Utilization of a Surface Reservoir with Underground Storage for Basin-Wide Water Supply Development: Solano Project California," Doctor of Eng. thesis, University of California, Berkeley, 1959.

[17] Henri Darcy, *Les fontaines publiques de la ville de Dijon,* Paris: V. Dalmont, 1856.

[18] William C. Walton, *Groundwater Resource Evaluation,* New York: McGraw-Hill, 1970.

[19] M. E. Harr, *Groundwater and Seepage,* New York: McGraw-Hill, 1962.

[20] Salamon Eskinazi, *Principles of Fluid Mechanics,* Boston: Allyn and Bacon, 1962.

[21] C. E. Jacob, "Flow of Groundwater," in *Engineering Hydraulics* (Hunter Rouse, ed.), New York: Wiley, 1950.

[22] M. Muskat, *The Flow of Homogeneous Fluids Through Porous Media,* Ann Arbor, MI: J. W. Edwards, 1946.

[23] D. W. Taylor, *Fundamentals of Soil Mechanics,* New York: Wiley, 1948.

[24] A. Casagrande, "Seepage Through Dams," in *Contributions to Soil Mechanics, 1925–1940.* Boston: Boston Society of Civil Engineers, 1940.

[25] M. S. Hantush, "Modification of the Theory of Leaky Aquifers," *J. Geophys. Res.* **65**, 3713–3725(1960).

[26] Jules Dupuit, *Etudes théoriques et pratiques sur le mouvement des eau dans les canaux de couverts et à travers les terrains perméables,* 2nd ed., Paris: Dunod, 1863.

[27] P. Ya. Polubarinova-Kochina, *Theory of Groundwater Movement,* Princeton, NJ: Princeton University Press, 1962.

[28] J. W. Clark, W. Viessman, Jr., and M. J. Hammer, *Water Supply and Pollution Control,* 2nd ed., New York: Thomas Y. Crowell, 1965.

[29] John F. Hoffman, "Field Tests Determine Potential Quantity, Quality of Ground Water Supply," *Heating, Piping, and Air Conditioning* (Aug. 1961).

[30] John F. Hoffman, "How Underground Reservoirs Provide Cool Water for Industrial Uses," *Heating, Piping, and Air Conditioning* (Oct. 1960).

[31] John F. Hoffman, "Well Location and Design," *Heating, Piping, and Air Conditioning* (Aug. 1963).

[32] A. Thiem, *Hydrologische Methodern,* Leipzig: Gebhardt, 1906, p. 56.

[33] C. E. Jacob, "Flow of Groundwater," in *Engineering Hydraulics* (Hunter Rouse, ed.), New York: Wiley, 1950.

[34] C. O. Wisler and E. F. Brater, *Hydrology,* New York: Wiley, 1959.

[35] C. V. Theis, "The Relation Between the Lowering of the Piezometric Surface and the Rate and Duration of Discharge of a Well Using Ground Water Storage," *Trans. Am. Geophys. Union* **16**, 519–524(1935).

[36] H. H. Cooper, Jr. and C. E. Jacob, "A Generalized Graphical Method for Evaluating Formation Constants and Summarizing Well-Field History," *Trans. Am. Geophys. Union* **27**, 526–534(1946).

[37] D. B. McWhorter and D. K. Sunada, *Ground Water Hydrology and Hydraulics,* Fort Collins, CO: Water Resources Publications, 1977.

[38] C. E. Jacob, "Radial Flow in a Leaky Artesian Aquifer," *Trans. Am Geophys. Union* **27**, 198–205(1946).

[39] M. S. Hantush, "Plain Potential Flow of Groundwater with Linear Leakage," Ph.D. dissertation, University of Utah, 1949.

[40] M. S. Hantush and C. E. Jacob, "Nonsteady Radial Flow in an Infinite Leaky Aquifer and Nonsteady Green's Functions for an Infinite Strip of Leaky Aquifer," *Trans. Am. Geophys. Union* **36**, 95–112(1955).

[41] M. S. Hantush and C. E. Jacob, "Flow to an Eccentric Well in a Leaky Circular Aquifer," *J. Geophys. Res.* **65**, 3425–3431(1960).

[42] M. S. Hantush, "Analysis of Data from Pumping Tests in Leaky Aquifers," *Trans. Am. Geophys. Union* **37**, 702–714(1956).

[43] R. J. M. DeWiest, "On the Theory of Leaky Aquifers," *J. Geophys. Res.* **66**, 4257–4262(1961).

[44] R. J. M. DeWiest, "Flow to an Eccentric Well in a Leaky Circular Aquifer with Varied Lateral Replenishment," *Geofis, Pura Aplic.* **54**, 87–102(1963).

[45] W. C. Walton, "Leaky Artesian Aquifer Conditions in Illinois," Report of Investigation No. 39, Illinois State Water Survey, 1960.

[46] S. P. Neuman and P. A. Witherspoon, 1969a, "Theory of flow in a confined two-aquifer system," *Water Resources Res.,* 5, pp. 803–816.

[47] S. P. Neuman and P. A. Witherspoon, 1969b, "Applicability of current theories of flow in leaky aquifers," *Water Resources Res.,* 5, pp. 817–829.

[48] S. P. Neuman and P. A. Witherspoon, 1972, "Field determination of the hydraulic properties of leaky multiple-aquifer systems," *Water Resources Res.,* 8, pp. 1284–1298.

[49] D. Kirkham, "Exact Theory of Flow into a Partially Penetrating Well," *J. Geophys. Res.* **64**, 1317–1327(1959).

[50] M. S. Hantush, "Nonsteady Flow to a Well Partially Penetrating an Infinite Leaky Aquifer," *Proc. Isaqi Sci. Soc.* **1**, 10–19(1957).

[51] J. W. Mercer and C. R. Faust, *Ground-Water Modeling,* Worthington, OH: National Water Well Association, 1981.

[52] C. A. Appel and J. D. Bredehoeft, "Status of Groundwater Modeling in the U.S. Geological Survey," *U.S. Geol. Survey Circular* 737(1976).

[53] Y. Bachmat, B. Andres, D. Holta, and S. Sebastian, "Utilization of Numerical Groundwater Models for Water Resource Management," U.S. Environmental Protection Agency Report EPA-600/8-78-012.

[54] J. E. Moore, "Contribution of Ground-Water Modeling to Planning," *J. Hydrol.* 43(Oct. 1979).

[55] T. A. Prickett, "Ground-Water Computer Models—State of the Art," *Ground Water* **17**(2), 121–128(1979).

[56] G. D. Bennett, *Introduction to Ground Water Hydraulics,* book 3, *Applications of Hydraulics,* Washington, D.C.: U.S. Geological Survey, U.S. Government Printing Office, 1976.

[57] P. W. Huntoon, "Predicted Water-Level Declines for Alternative Groundwater Developments in the Upper Big Blue River Basin, Nebraska," Resource Rep. No. 6, Conservation and Survey Div., University of Nebraska, Lincoln, 1974.

[58] D. W. Peacemen and H. H. Rachford, Jr., "The Numerical Solution of Parabolic and Elliptic Differential Equations," *Soc. Indust. Appl. Math. J.* **3**, 28–41(1955).

[59] G. F. Pinder and J. D. Bredehoeft, "Application of the Digital Computer for Aquifer Evaluation," *Water Resources Res.* **4**(4), 1069–1093(1968).

[60] I. Remson, G. M. Hornberger, and F. J. Molz, *Numerical Methods in Subsurface Hydrology,* New York: Wiley, 1971.

[61] T. A. Prickett and C. G. Lonnquist, *Selected Digital Computer Techniques for Groundwater Resource Evaluation,* Illinois State Water Survey Bull. No. 55, 1971.

[62] R. R. Marlette and G. L. Lewis, "Digital Simulation of Conjunctive-Use of Groundwater in Dawson County, Nebraska," Civil Engineering Rep., University of Nebraska, Lincoln, 1973.

[63] C. S. Desai, and J. F. Abel, 1972, *Introduction to the Finite Element Method,* New York: Van Nostrand Reinhold.

[64] G. F. Pinder, and W. G. Gray, 1977, *Finite Element Simulation in Surface and Subsurface Hydrology,* New York: Academic Press, 295 pp.

[65] I. Remson, G. M. Hornberger, and F. J. Molz, 1971, *Numerical Methods in Subsurface Hydrology,* New York: Wiley-Interscience.

[66] O. C. Zienkiewicz, 1967, *The Finite-Element Method in Structural and Continuum Mechanics,* New York: McGraw-Hill.

[67] R. A. Young and J. D. Bredehoeft, "Digital Computer Simulation for Solving Management Problems of Conjunctive Groundwater and Surface Water Systems," *Water Resources Res.* **8**(3) (June 1972).

[68] H. W. Crooke, "Ground Water Replenishment in Orange County, California," *J. AWWA* (July 1961).

[69] P. van der Heijde, Y. Bachmat, J. Bredehoeft, B. Andrews, D. Holtz, and S. Sebastian, *Groundwater Management: The Use of Numerical Models,* Washington, D.C.: American Geophysical Union, 1985.

[70] J. Boonstra and N. A. de Ridder, *Numerical Modeling of Groundwater Basins,* Wageningen, the Netherlands: International Institute for Land Reclamation and Improvement/LRI, 1981.

CHAPTER 11

Urban Hydrology

OBJECTIVES

The purpose of this chapter is to:

- Describe the effects of urbanization on runoff from undeveloped watersheds
- Present a representative sample of equations and charts used for estimating peak rates of runoff and entire hydrographs for urban and small watersheds
- Explain and illustrate the popular rational method
- Illustrate how hydrograph procedures are used in urban hydrology
- Assist the reader in selecting the appropriate methods or models for urban watershed analysis and design
- Describe in detail the commonly used computer packages for simulation of urban rainfall-runoff processes
- Provide a "shopper's guide" to commercial and public domain software for urban stormwater analysis and design.

11.1 APPROACHES TO URBAN HYDROLOGY

Methods used in estimating quantities of storm water runoff from urban drainage areas and other small watersheds are classified as *empirical methods* and *physical-process* methods. Empirical methods result in relationships derived from observations of rainfall-produced runoff. Physical-process methods focus on replicating the process using laws of physics and equations of motion.

Empirical Lumped-Parameter Approaches to Urban Hydrology

Empirical formulas for urban runoff were historically the principal mechanism for estimating flow rates from urbanized areas. Most are lumped-parameter approaches (Chapter 12), characterized by (1) consideration of the entire drainage area as a single unit, (2) estimation of flow at only the most downstream point, and (3) the assumption that rainfall is uniformly distributed in time and space over the watershed. The foremost example of this approach is the rational method, described later.

A second example of the lumped-parameter approach is the unit-hydrograph method, described in Chapter 9. Once any X-hr unit hydrograph is developed for a watershed, the method allows the user to construct a hydrograph for a storm of any duration and magnitude. Originally, the unit-hydrograph concept was applied mainly to large river basins, but it is now used for urban watershed applications as well. The concept of the instantaneous unit hydrograph (IUH) (also described in Chapter 9) has numerous applications to small drainage areas. A number of models using an IUH or an approximation of it have been reported in urban and small watershed applications [1]–[5].

Physical-Process Approach to Urban Hydrology

This approach is characterized by an attempt to mathematically describe all pertinent physical phenomena from the input (rainfall) to the output (runoff) [1],[6]–[10]. This usually involves the following steps: (1) determine a design storm; (2) deduct losses from the design storm to arrive at an excess rainfall rate; (3) determine the flow to a gutter or some defined channel by overland flow equations; (4) route these gutter or small-channel flows to the main channel; (5) route the flow through the principal conveyance system (pipe, canal, or stream); and (6) determine the outflow hydrograph. The result obtained is affected by the accuracy of calculating losses and hydraulic phenomena and the validity of the simplifying assumptions. If errors are small and noncumulative, the prediction of the runoff is valid.

In the past, most physical-process procedures dealt solely with individual storm events. With the advent of modern computers, the trend has been more toward the continuous simulation of hydrologic processes [1],[11].

11.2   EFFECTS OF URBANIZATION ON RUNOFF

Chapter 8 described basin characteristics affecting runoff. The effects of slope, area size, soil and rock structure, and other factors were illustrated. From these discussions it is easy to understand that modifications of the land surface have varying effects on the runoff characteristics of a given drainage area.

If an undeveloped area is converted to cropland or pasture, the soil is disturbed and the overlying absorptive cover is changed. The result is increased runoff volume and a change in the timing of flows. When lands are urbanized and storm drains installed, the flooding characteristics of these areas are modified. The drains serve to remove the water at an accelerated rate, thus increasing the peak flows and runoff volumes. Inasmuch as there is usually a significant linkage between low, swampy areas and the underlying underground system, this relation is changed as well. The rapid removal of water from the drained area decreases the time—and consequently the opportunity for infiltration—and the net effect is usually a lowering of the underlying water table. Changes in the vegetal cover affect the infiltration capacities of soils, and land-use changes that modify the nature of vegetation can have significant impact on the timing and volume of flows.

Urbanization of the land usually results in the highly accelerated removal of storm water with corresponding increases in the volume and peak rate of runoff. Both effects are described below. In many cases, infiltration might be all but eliminated and a very high percentage of the storm rainfall becomes runoff. On the other hand, by increasing an area's storage capacity and delaying the outflow, it is possible to increase the timing and delay the peak rate of runoff. For example, a shopping center parking lot can be graded and its drains sized to permit several inches of ponding during intense storms. This delays the downstream arrival of flows from the area and significantly reduces the hydrograph peaks. By understanding the effects of land-use change on the hydrology of an area, it is possible to put this knowledge to beneficial use. Several aspects of this are discussed in Refs. 12–19.

A summary of the major hydrologic effects of changes in watershed due to man's activities is presented in Table 11.1 [18]. The principal effects of urbanization have been classified by Leopold as [16]: (1) changes in peak flow characteristics, (2) changes in total runoff, (3) changes in water quality, and (4) changes in hydrologic amenities (the appearance or impression a watercourse and its environment leave with the observer).

Change in Runoff Characteristics

Land-use changes can increase or decrease the volume of runoff and the maximal rate and timing of flow from a given area. The most influential factors affecting flow

TABLE 11.1   Summary of the Major Hydrologic Effects of Land-Use Change

Land-use change	Component affected	Principal hydrologic process involved	Geographic scale and likely magnitude of effect
Afforestation (deforestation has converse effect except where disturbance caused by forest clearance may be of overriding importance)	Annual flow	Increased interception in wet periods. Increased transpiration in dry periods through increased water availability to deep root systems	Basin scale; magnitude proportional to forest cover, world average is 34-mm/yr reduction for 10% increase in forest cover
	Seasonal flow	Increased interception and increased dry period transpiration will increase soil moisture deficits and reduce dry season flow	Basin scale; can be of sufficient magnitude to stop dry season flows
		Drainage activities associated with planting may increase dry season flows through initial dewatering and also through long-term effects of the drainage system	Basin scale; drainage activities will increase dry season flows

(Continued)

TABLE 11.1   (*Continued*)

Land-use change	Component affected	Principal hydrologic process involved	Geographic scale and likely magnitude of effect
		Cloud water (mist or fog) deposition will augment dry season flows	High-altitude basins only; increased cloud water deposition may have a significant effect on dry season flows
	Floods	Interception reduces floods by removing a proportion of the storm rainfall and by allowing buildup of soil moisture storage	Basin scale; effect is generally small but greatest for small storm events
		Management activities: cultivation, drainage, and road construction all increase floods	Basin scale; increased floods for all sizes of storm events
	Water quality	Leaching of nutrients is less from forests through reduced surface runoff and reduced fertilizer applications	Basin scale; variable but leaching can be an order of magnitude less than from agricultural land
		Deposition of most atmospheric pollutants is higher to forests because of reduced aerodynamic resistance	Basin scale; leads to acidification of catchments and runoff
	Erosion	High infiltration rates in natural, mixed forests reduce surface runoff and erosion	Basin scale; reduces erosion
		Slope stability is enhanced by reduced soil pore water pressure and binding of forest roots	Basin scale; reduces erosion
		Wind throw of trees and weight of tree crop reduce slope stability	Basin scale; increases erosion
		Soil erosion, through splash detachment, is increased from forests without an understory of shrubs or grass	Basin scale; increases erosion
		Management activities: cultivation, drainage, road construction, and felling all increase erosion	Basin scale; management activities are often more important than the direct effect of the forest

TABLE 11.1   (*Continued*)

Land-use change	Component affected	Principal hydrologic process involved	Geographic scale and likely magnitude of effect
	Climate	Increased evaporation and reduced sensible heat fluxes from forests affect climate	Micro, meso, and global scale; forests generally cool and humidify the atmosphere; a 2°C increase in regional temperature is predicted for Amazonia if deforestation continues
Agricultural intensification	Water quantity	Alternation of transpiration rates affects runoff	Basin scale; effect is marginal
		Timing of storm runoff altered through land drainage	Basin scale; significant effect
	Water quality: fertilizers	Application of inorganic fertilizers	Basin scale; increased nutrient concentrations in surface waters and groundwaters
	Pesticides	Application of nonselective and persistent pesticides poses health risks to humans and animal life	Basin, regional, and global scale; effects can be long-lasting
	Farm wastes	Inadequate disposal of farm organic and inorganic water pollutes surface and groundwater bodies	Basin scale; effect on groundwaters and surface waters
	Erosion	Cultivation without proper soil conservation measures and uncontrolled grazing on steep slopes increase erosion	Basin scale; effects are very site-dependent
Draining wetlands	Seasonal flow	Upland peat bogs, groundwater fens, and African dambos have little effect in maintaining dry season flows	Basin scale; drainage or removal of wetland will not reduce, and may increase, dry season flows

Source: After Maidment [18]

volume are the infiltration rate and surface storage. Changes in interception and other factors are usually of negligible importance [17].

The effects of urbanization require special mention, as they often have a pronounced impact on the characteristics of an area's hydrology [19]. Table 11.2 summarizes the available methods of reducing or delaying runoff from urbanizing areas. Design of stormwater detention facilities is described in Section 13.7. Urbanization generally increases the volume of the runoff and peak rate of flow and decreases the watershed's time lag. Figure 11.1 illustrates the effects on lag time and hydrograph peak for hypothetical unit hydrographs [16]. The runoff volume is determined mainly by infiltration and the nature of surface storage. The land slope, soil type, nature of the vegetative cover, and degree of imperviousness of the watershed are all important factors. Figure 11.2 illustrates the combined effects of increased imperviousness and sewerage on the mean annual flood for a 1-mi^2 drainage area. An often-overlooked but potentially important effect of increased runoff is the accompanying reduction in groundwater recharge. Where urban areas are expansive, local groundwater supplies can be seriously reduced.

TABLE 11.2   Measures for Reducing and Delaying Urban Storm Runoff

Area	Reducing runoff	Delaying runoff

Large flat roof	Cistern storage
	Rooftop gardens
	Pool storage or fountain storage
	Sod roof cover
		Ponding on roof by constricted downspouts
		Increasing roof roughness
		  1. Rippled roof
		  2. Graveled roof

Parking lots	Porous pavement
	  1. Gravel parking lots
	  2. Porous or punctured asphalt
	Concrete vaults and cisterns
	  beneath parking lots in high-
	  value areas
	Vegetated ponding areas around
	  parking lots
	Gravel trenches
		Grassy strips on parking lots
		Grassed waterways draining
		  parking lot
		Ponding and detention measures
		  for impervious areas
		  1. Rippled pavement
		  2. Depressions
		  3. Basins

Residential	Cisterns for individual homes or
	  groups of homes
	Gravel driveways (porous)
	Contoured landscape
	Groundwater recharge
	  1. Perforated pipe
	  2. Gravel (sand)
	  3. Trench
	  4. Porous pipe
	  5. Dry wells
	Vegetated depressions
		Reservoir or detention basin
		Planting a high delaying grass
		  (high roughness)
		Gravel driveways
		Grassy gutters or channels
		Increased length of travel of
		  runoff by means of gutters
		  and diversions

General	Gravel alleys Porous sidewalks Mulched planters	Gravel alleys

Source: After U.S. Department of Agriculture, Soil Conservation Service, 1972.

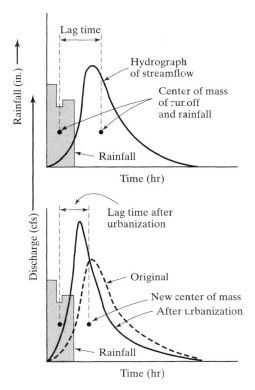

Lag time

Rainfall (in.) ——→

Hydrograph
of streamflow

Center of mass
of runoff
and rainfall

Rainfall

Time (hr)

Discharge (cfs)

Lag time after
urbanization

Original

New center of mass

After urbanization

Rainfall

Time (hr)

FIGURE 11.1
Hypothetical unit hydrographs relating the runoff to rainfall,
with definitions of significant parameters.
(U.S. Geological Survey Circular 554)

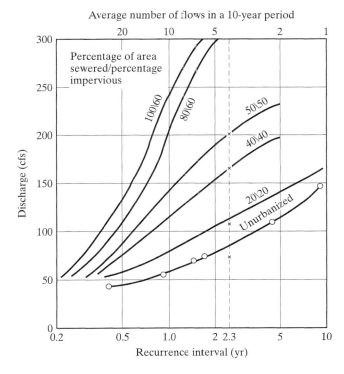

Average number of flows in a 10-year period

FIGURE 11.2

Flood frequency curves for a 1-mi^2 basin in
various states of urbanization.
(U.S. Geological Survey Circular 554)

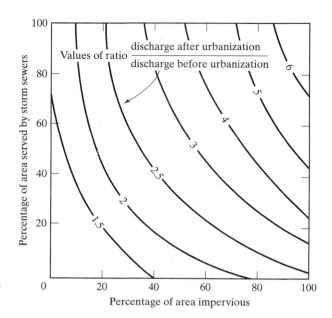

FIGURE 11.3

Effect of urbanization on the mean
annual flood for a 1-mi² drainage area.
(U.S. Geological Survey Circular 554)

In general, the peak rate of runoff will increase more rapidly than the volume of runoff as urbanization occurs. This is because of the increase in the rate of overland flow to stream channels and the resultant decrease in concentration time of the basin. Water flows more quickly from streets and roofs than from naturally vegetated areas, and conveyances such as storm sewers and lined open channels increase the flow velocities and thus decrease the lag time. The reduction in time lag (or concentration time) of the basin is extremely important as it affects the frequency or return period for a given level of flow. For example, the storm that was found to be the 50-year storm on a basin having a 6-hr lag, will no longer be the 50-year storm if the lag is reduced to 3 hr by urbanization. A study of Fig. 11.3 illustrates this point.

Water Quality

Changes in water quality due to land-use practices can be either positive or negative [20]–[24]. Table 11.1 describes some of the water quality effects of forestation and agricultural intensification. The principal effect of land-use change on water quality is the introduction of waste materials such as nutrients, road salts, various chemicals, and oil and gasoline products. An especially important water quality problem is the rapid increase in sediment load in streams owing to the exposure of bare soil to storm runoff during and after periods of development. Urbanization has caused increases in sediment yield on the order of 100–250 times that of rural areas. Such increases result from the denuding of sites and the upsetting of balances of natural drainage networks to the flows they must carry. Streams tend to construct and maintain channels that exceed the bank-full stage at a recurrence interval of about 2 years. If the number of flows above bank-full stage is increased due to urbanization, or other causes, the banks and bed of erodible channels will not remain stable but will be enlarged through erosion.

Total Urban Runoff Management

A knowledge of the manner in which land-use changes and land treatments can modify the runoff process is extremely important. Various proposed changes can be simulated and their effect evaluated before decisions to implement these practices are made. Designs can be improved and features incorporated into traditional design practices that will save funds, reduce adverse environmental impacts, and even enhance the quality of life. New uses for excess flows such as recreational ponds, artificial recharge, and urban greenway irrigation can be found. By considering the total water management instead of only the fast removal of storm water runoff, many positive impacts are obtainable.

11.3   PEAK FLOW METHODS FOR URBAN AREAS

Numerous methods are available for estimating the peak rates and volumes of runoff from urban watersheds. Some of the peak flow methods incorporate equations describing the rainfall-runoff process, whereas others predict peak runoff rates by correlating the flow rates with simple drainage basin characteristics such as total area, impervious area, slope, and other factors. This section describes several commonly used methods.

Rational Method

The rational formula for estimating peak runoff rates was introduced in the United States by Emil Kuichling in 1889 [25]. Since then it has become the most widely used method for designing drainage facilities for small urban and rural watersheds. Peak flow is found from:

$$Q_p = CIA \qquad (11.1)$$

where Q_p = the peak runoff rate (cfs)

   C = the runoff coefficient (assumed to be dimensionless)

   I = the *average* rainfall intensity (in./hr) for a storm with a duration equal to a critical period of time t_c

   t_c = the time of concentration (see Chapter 9)

   A = the size of the drainage area (acres)

   CI = the average net rain intensity (in./hr) for a storm with duration $= t_c$

The runoff coefficient can be assumed to be dimensionless because 1.0 acre-in./hr is equivalent to 1.008 ft^3/sec. Typical C values for storms of 5–10-year return periods are provided in Table 11.3. As described later, the rational equation was developed for relatively frequent storms, and the peak flow rate from Eq. 11.1 should be increased for more extreme storms.

TABLE 11.3   Typical *C* Coefficients for 5- to 10-Year Frequency Design

Description of area	Runoff coefficients
Business	
Downtown areas	0.70–0.95
Neighborhood areas	0.50–0.70
Residential	
Single-family areas	0.30–0.50
Multiunits, detached	0.40–0.60
Multiunits, attached	0.60–0.75
Residential (suburban)	0.25–0.40
Apartment dwelling areas	0.50–0.70
Industrial	
Light areas	0.50–0.80
Heavy areas	0.60–0.90
Parks, cemeteries	0.10–0.25
Playgrounds	0.20–0.35
Railroad yard areas	0.20–0.40
Unimproved areas	0.10–0.30
Streets	
Asphaltic	0.70–0.95
Concrete	0.80–0.95
Brick	0.70–0.85
Drives and walks	0.75–0.85
Roofs	0.75–0.95
Lawns; sandy soil:	
Flat, 2%	0.05–0.10
Average, 2–7%	0.10–0.15
Steep, 7%	0.15–0.20
Lawns; heavy soil:	
Flat, 2%	0.13–0.17
Average, 2–7%	0.18–0.22
Steep, 7%	0.25–0.35

Another assumption with the rational method is that the equation is most applicable to antecedent moisture conditions that exist for frequent storms, in the range of the 2- to 10-yr recurrence interval, representative of storms traditionally used for design of residential storm drain systems. Because more severe, less frequent storms often have wetter antecedent moisture conditions, the rational coefficient is increased by multiplying it by a frequency factor. The commonly used multipliers for less frequent storms are:

Return period (yrs)	Multiplier
2–10	1.0
25	1.1
50	1.2
100	1.25

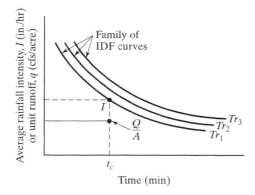

FIGURE 11.4

Rainfall-runoff relation for the rational method.

The rationale for the method lies in the concept that application of a steady, uniform rainfall intensity will cause runoff to reach its maximum rate when all parts of the watershed are contributing to the outflow at the point of design. That condition is met after the elapsed time t_c, the time of concentration, which usually is taken as the time for a wave to flow from the most remote part of the watershed. At this time, the runoff rate matches the net rain rate.

Figure 11.4 graphically illustrates the relation. The IDF curve is the rainfall intensity–duration–frequency relation for the area, and the peak intensity of the runoff is $Q/A = q$, which is proportional to the value of I defined at t_c. The constant of proportionality is thus the runoff coefficient, $C = (Q/A)/I$. Note that Q/A is a point value and that the relation, as it stands, yields nothing of the nature of the rest of the hydrograph.

The definition chosen for t_c can adversely affect a design using the rational formula. If the average channel velocity is used to estimate the travel time from the most remote part of the watershed (a common assumption), the resulting design discharge could be less than that which might actually occur during the life of the project. The reason is that wave travel time through the watershed is faster than average discharge velocity (see Section 9.5). As a result of using the slower velocity V, the peak time (t_c) is overestimated, the resulting intensity I from IDF curves is too small, and the rational flow rate Q is underestimated.

Rational Method Applications   In determining peak flow rates, most applications of the rational formula utilize the following steps:

1. Estimate the time of concentration of the drainage area.
2. Estimate the runoff coefficient; use Table 11.3.
3. Select a return period T_r and find the intensity of rain that will be equaled or exceeded, on the average, once every T_r years. To produce equilibrium flows, this design storm must have a locally derived IDF curve such as Fig. 11.4 using a rainfall duration equal to the time of concentration.
4. Determine the desired peak flow Q_p from Eq. 11.1.

Some design situations produce larger peak flows if design storm intensities for durations less than t_c are used. Substituting intensities for durations less than t_c is justified only if the contributing area term in Eq. 11.1 is also reduced to accommodate the shortened storm duration.

One of the principal assumptions of the rational method is that the predicted peak discharge has the same return period as the rainfall IDF relation used in the prediction. Another assumption, and one that has received close scrutiny by investigators [26],[27], is the constancy of the runoff coefficient during the progress of individual storms and also from storm to storm. The coefficient is usually selected from a list based on the degree of imperviousness and infiltration capacity of the drainage surface. Because $C = I_{net}/I$, the coefficient must vary if it is to account for antecedent moisture, nonuniform rainfall, and the numerous conditions that cause abstractions and attenuation of flood-producing rainfalls. In practice, a composite, area-weighted average runoff coefficient is computed for the various surface conditions. Times of concentration are determined from the hydraulic characteristics of the principal flow path, which typically is divided into two parts, overland flow and flow in defined channels; the times of flow in each segment are added to obtain t_c.

Example 11.1

Use the rational method to find the 10-year and 50-year design runoff rates for the area shown in Fig. 11.5. The IDF rainfall curves shown in Fig. 11.6 are applicable.

Solution

   1. Time of concentration:

$$t_c = t_1 + t_2 = 15 + 5 = 20\,\text{min}$$

   2. Runoff coefficient:

$$\overline{C} = [(3 \times 0.3) + (4 \times 0.7)]/(3 + 4) = 0.53 \text{ for 10-yr event}$$
$$\overline{C} = 1.2(0.53) = 0.64 \text{ for 50-yr event}$$

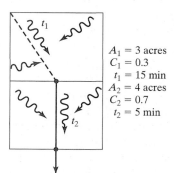

$A_1 = 3$ acres
$C_1 = 0.3$
$t_1 = 15$ min
$A_2 = 4$ acres
$C_2 = 0.7$
$t_2 = 5$ min

FIGURE 11.5

Hypothetical drainage system for Example 11.1.

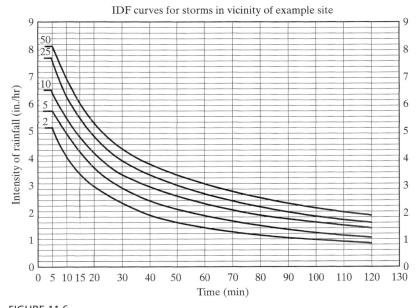

FIGURE 11.6

Intensity–duration–frequency curves used in Example 11.1.

3.   Rainfall intensity—from Fig. 11.6:

$$I_{10} = 4.2 \text{ in./hr}$$

$$I_{50} = 5.3 \text{ in./hr}$$

4.   Design peak runoff:

$$Q_{10} = CIA = 0.53 \times 4.2 \times 7 = 16 \text{ cfs}$$

$$Q_{50} = CIA = 0.64 \times 5.3 \times 7 = 24 \text{ cfs}$$

Selecting Storm Duration   The most critical (greatest peak flow) runoff event is often assumed to be caused by a storm having a duration equal to the time of concentration of the watershed. Intensity–duration–frequency (IDF) curves can be found in many local or regional drainage manuals and in state highway agency hydraulic manuals, or can be derived from rain depths for various durations and recurrence intervals published by the U.S. National Weather Service (visit www.nws.noaa.gov and click on HYDRO-35 or NOAA Atlas 2). If the rainfall IDF curve is steep in the design range, several durations should be tested for the given frequency to assure that no other storm of equal probability produces a higher peak runoff rate. Most applications of the rational method do not include this test because the assumption that the peak occurs at time t_c is commensurate with the other inherent assumptions.

The rational method is used in the design of urban storm drainage systems serving areas up to 600 acres in size. For areas larger than 1 mi^2, hydrograph or other techniques are generally warranted. Considerable judgment is required in selecting both the runoff coefficients and times of concentration. A common procedure is to select coefficients and assume that they remain constant throughout the storm. As the design proceeds from point to point downstream, a composite weighted C factor is computed for the drainage area above each point. The time of concentration is composed of an inlet time (the overland and any channel flow times to the first inlet) plus the accumulated time of flow in the system to the point of design.

Figure 11.7 is an example of a design aid for predicting overland flow times.

SCS TR-55 Graphical Peak Flow Method

The U.S. Soil Conservation Service (now known as the Natural Resources Conservation Service, NRCS) developed procedures in 1975 for estimating runoff volume and peak rates of discharge from urban areas [29] known as the TR-55 *graphical method, chart method*, and *tabular method*. The three methods adjust rural procedures

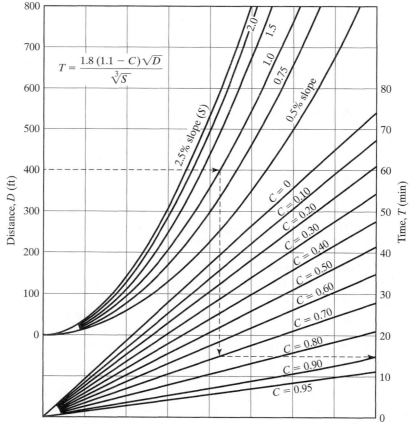

FIGURE 11.7
Surface flow time curves.
(After Federal Aviation Agency [28])

in NEH-4 [30] to urban conditions by increasing the curve number CN for impervious areas and reducing the lag time t_l for imperviousness and channel improvements. Allowances are also made for various watershed shapes, slopes, and times of concentration. The SCS designed the first two methods to be used for estimating peak flows, and the third for synthesizing complete hydrographs (see Section 11.4). The chart method (used for small watersheds up to 2,000 acres) was deleted in 1986 [31]. All three were developed for use with 24-hr storms. Use with other storm durations is not advised.

The graphical method was developed for homogeneous watersheds, up to 20 mi^2 in size, on which the land use and soil type may be represented by the runoff curve number (Chapter 7). The method uses Table 11.4 and Fig. 11.8 to provide urban runoff curve numbers for certain instances indicated in the table.

The 24-hr design rain depth is estimated, and then an initial abstraction I_a is determined from the SCS runoff equation (Chapter 7) or from Table 11.5. The peak

TABLE 11.4   Runoff Curve Numbers for Urban Areas (see Section 7.9 for other values)

Cover description		Curve numbers for hydrologic soil group[a]			
Cover type and hydrologic condition	Average percent impervious area[b]	A	B	C	D
Fully developed urban areas (vegetation established)					
Open space (lawns, parks, golf courses, cemeteries, etc.)[c]					
Poor condition (grass cover <50%)		68	79	86	89
Fair condition (grass cover 50–75%)		49	69	79	84
Good condition (grass cover >75%)		39	61	74	80
Impervious areas					
Paved parking lots, roofs, driveways, etc. (excluding right-of-way)		98	98	98	98
Streets and roads					
Paved; curbs and storm sewers (excluding right-of-way)		98	98	98	98
Paved; open ditches (including right-of-way)		83	89	92	93
Gravel (including right-of-way)		76	85	89	91
Dirt (including right-of-way)		72	82	87	89
Western desert urban areas					
Natural desert landscaping (pervious areas only)[d]		63	77	85	88
Artificial desert landscaping (impervious weed barrier, desert shrub with 1–2-in. sand or gravel mulch and basin borders)		96	96	96	96
Urban districts					
Commercial and business	85	89	92	94	95
Industrial	72	81	88	91	93
Residential districts by average lot size					
$\frac{1}{8}$ acre or less (townhouses)	65	77	85	90	92
$\frac{1}{4}$ acre	38	61	75	83	87
$\frac{1}{3}$ acre	30	57	72	81	86
$\frac{1}{2}$ acre	25	54	70	80	85
1 acre	20	51	68	79	84
2 acres	12	46	65	77	82

(*Continued*)

TABLE 11.4 (*Continued*)

Cover description		Curve numbers for hydrologic soil group[a]			
Cover type and hydrologic condition	Average percent impervious area[b]	A	B	C	D
Developing urban areas					
Newly graded areas (pervious areas only, no vegetation)[e]		77	86	91	94
Idle lands (CNs are determined using cover types similar to those in Table 7.8)					

[a] Average runoff condition, and $I_a = 0.2S$.

[b] The average percent impervious area shown was used to develop the composite CNs. Other assumptions are as follows: impervious areas are directly connected to the drainage system, impervious areas have a CN of 98, and pervious areas are considered equivalent to open space in good hydrologic condition. CNs for other combinations of conditions may be computed using Fig. 7.14 or 11.8.

[c] CNs shown are equivalent to those of pasture. Composite CNs may be computed for other combinations of open-space cover type.

[d] Composite CNs for natural desert landscaping should be computed using Fig. 11.8 based on the impervious area percentage (CN = 98) and the pervious area CN. The pervious area CNs are assumed equivalent to desert shrub in poor hydrologic condition.

[e] Composite CNs to use for the design of temporary measures during grading and construction should be computed using Fig. 11.8 based on the degree of development (impervious area percentage) and the CNs for the newly graded pervious areas.

Source: U.S. Soil Conservation Service [30]

flow is found by locating the basin in Figure 11.9 and then interpolating the curves in Figs. 11.10–11.13, depending on the rainfall distribution type. If the computed I_a/P ratio falls outside the curves, the nearest curve should be used. If the watershed contains a percentage of ponds or swampy areas, the peak flow is multiplied by a reduction coefficient from Table 11.6.

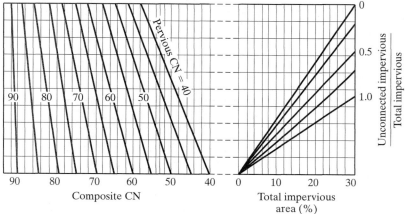

FIGURE 11.8

Graph of 1986 TR-55 composite CN with unconnected impervious area, or total impervious area, less than 30 percent.
(After U.S. Soil Conservation Service [31])

TABLE 11.5 I_a Values for Runoff Curve Numbers

Curve number	I_a (in.)	Curve number	I_a (in.)
40	3.000	70	0.857
41	2.878	71	0.817
42	2.762	72	0.778
43	2.651	73	0.740
44	2.545	74	0.703
45	2.444	75	0.667
46	2.348	76	0.632
47	2.255	77	0.597
48	2.167	78	0.564
49	2.082	79	0.532
50	2.000	80	0.500
51	1.922	81	0.469
52	1.846	82	0.439
53	1.774	83	0.410
54	1.704	84	0.381
55	1.636	85	0.353
56	1.571	86	0.326
57	1.509	87	0.299
58	1.448	88	0.273
59	1.390	89	0.247
60	1.333	90	0.222
61	1.279	91	0.198
62	1.226	92	0.174
63	1.175	93	0.151
64	1.125	94	0.128
65	1.077	95	0.105
66	1.030	96	0.083
67	0.985	97	0.062
68	0.941	98	0.041
69	0.899		

Source: U.S. Soil Conservation Service [31]

TABLE 11.6   Adjustment Factor (F_p) for Pond and Swamp Areas That Are Spread Throughout the Watershed

Percentage of pond and swamp areas	F_p
0	1.00
0.2	0.97
1.0	0.87
3.0	0.75
5.0	0.72

Source: U.S. Soil Conservation Service [31]

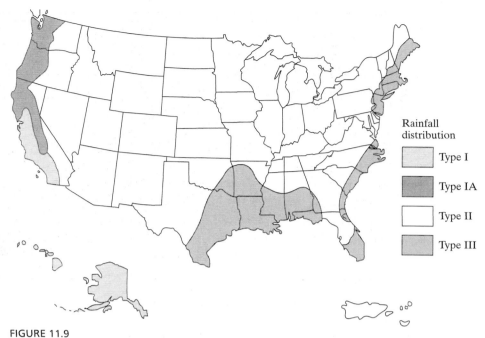

FIGURE 11.9

SCS 24-hr rainfall zones I, IA, II, and III.
(After Ref. 30)

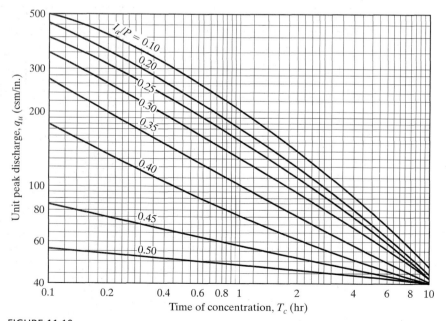

FIGURE 11.10

Unit peak discharge (q_u) for SCS Type-I rainfall distribution.
(After U.S. Soil Conservation Service [31])

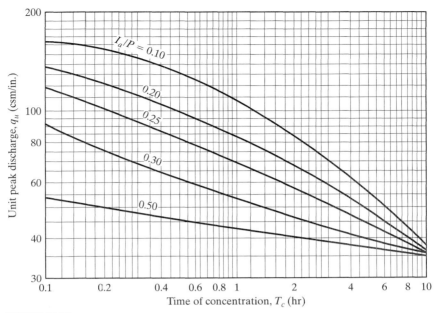

FIGURE 11.11

Unit peak discharge (q_u) for SCS Type-IA rainfall distribution.
(After U.S. Soil Conservation Service [31])

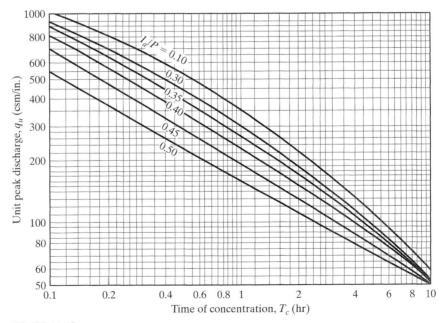

FIGURE 11.12

Unit peak discharge (q_u) for SCS Type-II rainfall distribution.
(After U.S. Soil Conservation Service [31])

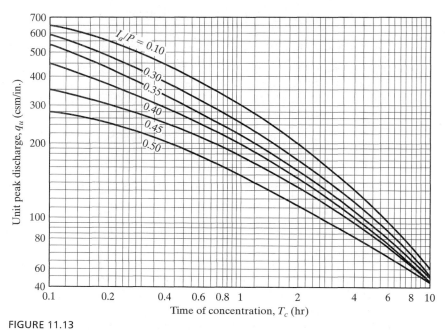

FIGURE 11.13

Unit peak discharge (q_u) for SCS Type-III rainfall distribution.
(After U.S. Soil Conservation Service [31])

Rainfall distributions for each zone of Fig. 11.9 are described in more detail in Chapter 13. Types I and IA represent the Pacific maritime climate with wet winters and dry summers. Type III represents Gulf of Mexico and Atlantic coastal areas, where tropical storms bring large 24-hr rainfall amounts. Type II represents the rest of the country. For more precise information on boundaries in a state having more than one storm type, contact the respective NRCS state conservation engineer.

Example 11.2

A 1,280-acre urban Tennessee watershed has a 6.0-hr time of concentration, a CN = 75 from Table 11.4, and 5 percent of the area is ponded. The 25-year, 24-hr rain is 6.0 in. Find the 25-year peak discharge.

Solution. From Fig. 11.9, the Type-II storm applies to Tennessee. From Table 11.5, $I_a = 0.667$. Thus $I_a/P = 0.11$ From Fig. 11.12, $q_u = 96$ csm in. (cubic ft per second per mi^2 per inch of net rain). From Chapter 7, the runoff from 6.0 in. is 3.28 in. Since 5 percent of the area is ponded, the peak flow is adjusted using Table 11.6, giving $F_p = 0.72$. Thus:

$$Q = (96 \text{ csm/in.})(3.28 \text{ in.})(2.0 \text{ mi}^2)(0.72) = 453 \text{ cfs}$$

Assumptions of the graphical method include:

The method should be used only if the weighted CN is greater than 40.

The T_c values with the method may range from 0.1 to 10 hr.

The watershed must be hydrologically homogeneous, that is, describable by one CN. Land use, soils, and cover must be distributed uniformly throughout the watershed.

The watershed may have only one main stream or, if more than one, the branches must have nearly equal times of concentration.

The method cannot perform channel or reservoir routing.

The F_p factor can be applied only for ponds or swamps that are not on the flow path.

Accuracy of peak discharge estimated by this method will be reduced if I_a/P values are used that are outside the range given. When this method is used to develop estimates of peak discharge for present and developed conditions of a watershed, use the same procedure for estimating t_c.

Several computer software packages for TR-55 hydrology have been developed [32],[33]. Caution should be applied in assuming that the commercial programs fully imitate TR-55 or other SCS handbook methods. An ideal TR-55 package would include all three methods, would carry SCS endorsement, would state all assumptions and limitations, and would incorporate all SCS adjustments for peak coefficient, percent imperviousness, percentage of channel improved, ponding or swampy areas, length/width ratio variations, and slope. Use of any TR-55 procedure should also be cautioned for other than 24-hr storms having a Type-II SCS distribution. Packages not adhering to these limitations would not be qualified as TR-55 procedures.

USGS Urban Peak Flow Regression Equations

The U.S. Geological Survey, in cooperation with the Federal Highway Administration, conducted a nationwide study of flood magnitude and frequency in urban watersheds [34]. The investigation involved 269 gauged basins at 56 cities in 31 states, including Hawaii. The locations are shown in Fig. 11.14. Basin sizes ranged from 0.2 to 100 mi^2.

Multiple linear regression (see Chapter 3) of a variety of independent parameters was conducted to develop peak flow equations that could be applied to small, ungauged urban watersheds throughout the United States. Similar USGS regression equations for large rural basins are described in Chapter 3.

The simplest form of the developed regression equations involves the three most significant variables identified. These are contributing area A (mi^2), basin development factor BDF (dimensionless), and the corresponding peak flow RQ_i (cfs) for the ith frequency from an identical rural basin in the same region as the urban watershed. The latter variable accounts for regional variations, and estimates can be developed from any of the applicable USGS flood frequency reports (see Section 3.7). The

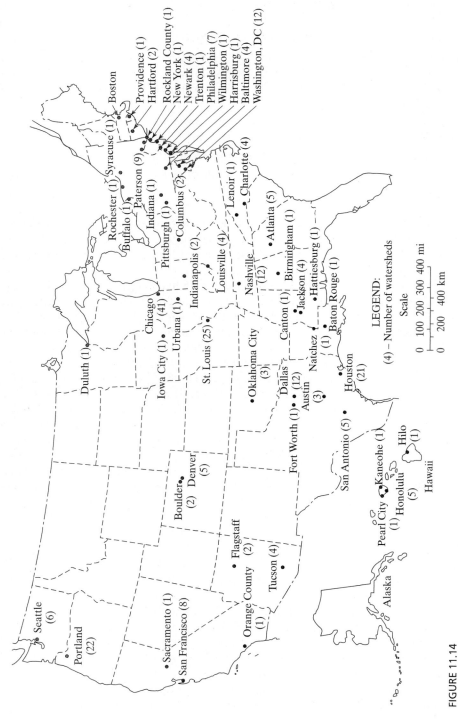

FIGURE 11.14

Location of runoff stations used in nationwide urban flood-frequency study.

(After Sauer et al. [34])

three-parameter equations for the 2-, 5-, 10-, 25-, 50-, 100-, and 500-year flows are given as [34]:

$$Q_2 = 13.2A^{0.21}(13 - \text{BDF})^{-0.43}RQ_2^{0.73} \tag{11.2}$$

$$Q_5 = 10.6A^{0.17}(13 - \text{BDF})^{-0.39}RQ_5^{0.78} \tag{11.3}$$

$$Q_{10} = 9.51A^{0.16}(13 - \text{BDF})^{-0.36}RQ_{10}^{0.79} \tag{11.4}$$

$$Q_{25} = 8.68A^{0.15}(13 - \text{BDF})^{-0.34}RQ_{25}^{0.80} \tag{11.5}$$

$$Q_{50} = 8.04A^{0.15}(13 - \text{BDF})^{-0.32}RQ_{50}^{0.81} \tag{11.6}$$

$$Q_{100} = 7.70A^{0.15}(13 - \text{BDF})^{-0.32}RQ_{100}^{0.82} \tag{11.7}$$

$$Q_{500} = 7.47A^{0.16}(13 - \text{BDF})^{-0.30}RQ_{500}^{0.82} \tag{11.8}$$

These were developed from data at 199 of the 269 original sites. The other sites were deleted because of the presence of detention storage or missing data. All these equations have coefficients of determination above 0.90.

To illustrate the accuracy of these equations, Fig. 11.15 shows the correspondence of estimated and observed values used in developing Eq. 11.4. Forty percent of the values fall within one standard deviation of the regression line. Graphs for other recurrence intervals are similar to the 10-year graph shown in Fig. 11.15.

The basin development factor BDF is an index of the prevalence of four drainage elements: storm sewers, channel improvements, impervious channel linings, and

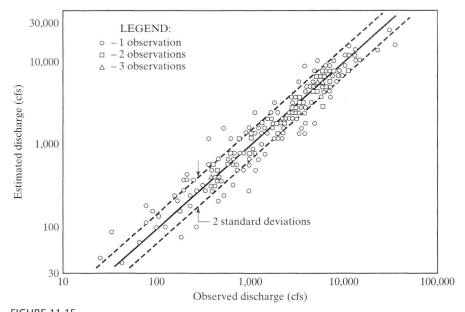

FIGURE 11.15

Comparison of observed and estimated 10-year urban peak discharges by Eq. 11.4.
(After Sauer et al. [34])

curb-and-gutter streets. A value of zero indicates the development elements are not prevalent. A maximum value of 12 indicates full development. Each of the four elements is given up to three units of BDF. This is accomplished by dividing the basin area into thirds and giving each third up to four BDF points (one for each element), depending on whether more than 50 percent of the drainage features are developed.

Example 11.3

An undeveloped basin has a $Q_{50} = 3,960$ cfs. Use the USGS regression equations to develop an estimate of the 50-year flood for the 100-mi^2 rural watershed. The upper third of the watershed has 60 percent storm sewers, 30 percent channel improvements, 74 percent impervious channel linings, and 82 percent curb-and-gutter streets.

Solution.   Because development has exceeded 50 percent in the upper third of the watershed for three of the four categories, the BDF is 3.0. From Eq. 11.6:

$$Q_{50} = 8.04(100)^{0.15}(10)^{-0.32}(3,960)^{0.81}$$
$$= 6,300 \text{ cfs for the developed condition.}$$

Other Peak Flow Methods

The literature describes numerous other urban peak flow methods [1],[6],[33],[35],[36]. Those presented here are illustrative. Because there are so many techniques available and many give different solutions to the same problem, most communities that are experiencing urban growth have selected applicable standard methods for peak flow and hydrograph calculations acceptable in the region. These standard methods are generally published in the form of local drainage design manuals. In any design, the engineer or hydrologist should attempt to determine which methods are contemporaneous for the locale.

11.4   URBAN HYDROGRAPH METHODS AND MODELS

Except for homogeneous watersheds that are suited to lumped-parameter approaches, determination of watershed runoff has transitioned from the use of graphical or empirical peak flow methods to use of computer simulation models of the rainfall-runoff process. Chapter 12 introduces the subject of simulation of hydrologic processes and provides the reader with an understanding of types and classes of models, protocols for conducting simulation modeling studies, and limitations and data needs of models. The models described in Chapter 12 are primarily used for nonurban applications. Methods and models for evaluating urban runoff are described here.

Studies of urban watershed response have resulted in the development of a class of single-event and continuous streamflow models of the unique processes in urbanized systems. These models, summarized in Table 11.8, not only simulate the rainfall-runoff

TABLE 11.8   Frequently Used Urban Storm Water Simulation Models

Code name	Model name	Agency originating	Year
CHM	Chicago Hydrograph Method	City of Chicago	1959
RRL	Road Research Laboratory Method	Road Research Lab	1962
ILLUDAS	Illinois Urban Drainage Area Simulator	Ill. Water Survey	1972
STORM	Storage, Treatment, Overflow Runoff Model	Corps of Engineers	1974
TR-55	SCS Technical Release 55	SCS	1992
DR3M	Distributed Routing Rainfall-Runoff Model	USGS	1978
HYDRA	Hydrologic Component of HYDRAIN Package	FHWA	1990
SWMM	Storm Water Management Model	EPA	1971
UCURM	U. of Cincinnati Urban Runoff Model	U. of Cincinnati	1972

process, but also allow the user to analyze an existing network of interconnected stormwater management facilities or to design new components (underground storm sewers, detention ponds, ditches, channels, street-side curbs, and storm sewer inlet sizes and locations) of an existing system. A number of the most widely adopted methods and models are described. The applications of the models to stormwater design are presented in Chapter 13.

Typical Elements of Urban Runoff Models

Many of the available software packages for modeling urban runoff incorporate *hydraulic* aspects as well as hydrologic aspects. The depth that a given flow will reach in an open channel or roadside gutter, or the capacity (maximum rate of discharge) of an underground storm sewer, is a product of hydraulic analysis. Rather than requiring users to run separate models, some storm event models incorporate both. ILLUDAS, for example, allows the user to determine the storm sewer pipe size needed to carry a frequency-based flood event. Users can also use the model to analyze whether an existing storm sewer will carry all the water generated in any given storm event. Though not limited to urban runoff applications, general-use models like HEC-1 (see Chapter 12) also allow the user to incorporate hydraulic designs along with their hydrologic analyses.

Figure 11.16 shows the typical elements modeled in urban runoff analysis. After calculating the runoff from the pervious and impervious areas, the model can route the flow in gutters, discharge it through storm sewers, and convey in into and through open channels.

Modified Rational Method

The rational method is truly "rational" in that the peak flow rate is simply set equal to the net rain rate after sufficient time occurs for the entire watershed to contribute runoff. This results for any storm equaling or exceeding the time of concentration of the watershed. The net rain rate is a fraction of the gross rain rate, I, and the fraction is C. Incorporating A in the equation simply transforms runoff from inches per hour to volume per hour. This trait allows the conceptual extension of the rational method to

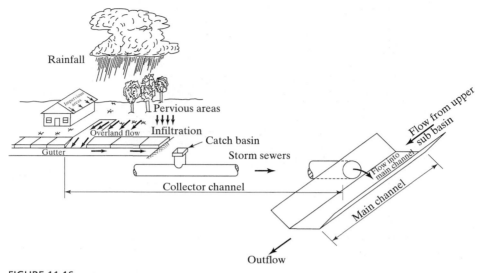

FIGURE 11.16

Typical elements of an urban runoff model.
(After Ref. 37)

problems in which the runoff hydrograph is required. This is particularly applicable in designs, such as detention basin design, requiring volume of runoff as well as peak flow rates.

In the *modified rational method*, a full hydrograph is developed rather than simply estimating the peak flow rate, using the following reasoning. If the storm duration exceeds the time of concentration, the runoff rate would rise to the rational formula peak value and then stay constant until net rain ceases. At that point, runoff rates would decrease to zero as excess rain is released from the basin. If the rainfall-excess release time (see Chapter 9) is equal to the time of concentration, the hydrograph would have an approximate trapezoid shape rising to the peak at $t = t_c$, remaining flat until $t =$ the rain duration, D, and then falling along a straight line until $t = D + t_c$. Many software packages for urban hydrology incorporate the modified rational method for hydrograph analysis. The method is approximate and should not be applied to watersheds over 50 acres in size.

Example 11.4

A 100-acre watershed has a t_c of 30 minutes and a runoff coefficient of 0.50. Use the modified rational method to develop a hydrograph for a 1-hour storm with an average intensity of 2.0 in./hr.

Solution.   Because the duration exceeds t_c, the hydrograph duration will be the rain duration plus the rainfall-excess release time, assumed equal to t_c. The hydrograph peak flow rate of $Q = CIA = 100\,\text{cfs}$ occurs at $t_c = 30$ minutes and continues at the

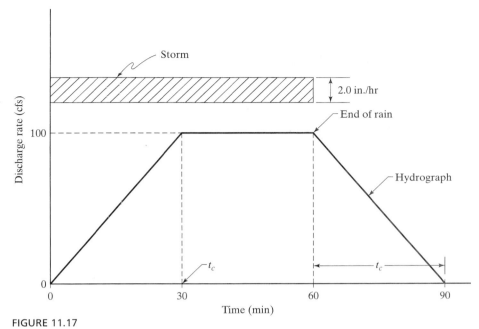

FIGURE 11.17

Modified rational method hydrograph for Example 11.4.

net rain rate until rain ceases, at which time the recession occurs. Figure 11.17 shows the resulting hydrograph.

Chicago Hydrograph Method (CHM)

Tholin's hydrograph method, known as the *Chicago hydrograph method* [38], is an example of early urban runoff models. The procedure programmed is (1) develop a design storm pattern from local intensity–duration–frequency curves and an average chronological storm pattern, (2) compute the overland flow using selected Horton-type infiltration capacity curves, the estimated depth of the rainfall retained in surface depressions, and Izzard's [39] overland flow equations, (3) route overland flow through gutters using the storage equation to obtain the runoff into catch basins, (4) synthesize hydrographs from roofs and street inlets along a typical sewer lateral to produce a lateral outflow hydrograph, and (5) route the lateral outflow hydrograph by a time-offset method along submains and the main sewer to a point of discharge. The method originally involved a graphical hand computation but was later programmed for computer solution by Keifer [40].

SCS TR-55 Tabular Hydrograph Method

The graphical method described in Section 11.3 provides peak discharges only. If a hydrograph is needed or watershed subdivision is required, the *tabular* method [31] should be used. The event simulation model TR-20 should be used if the watershed is very complex or a higher degree of accuracy is required (see Chapter 12).

Both the graphical and tabular methods are derived from TR-20 output. The use of t_c permits them to be used for any size watershed within the scope of the curves or tables. The tabular method can be used for a heterogeneous watershed that is divided into a number of homogeneous subwatersheds. Hydrographs for the subwatersheds can be routed and added.

In using the tabular method, the following steps are employed:

1. Subdivide the watershed into areas that are relatively homogeneous and have convenient routing reaches.
2. Determine the drainage area of each subarea in square miles.
3. Estimate t_c for each subarea in hours. The procedure for estimating t_c is outlined in Chapter 9.
4. Find the travel time for each routing reach in hours.
5. Develop a weighted CN for each subarea.
6. Select an appropriate rainfall distribution according to Fig. 11.9.
7. Determine the 24-hr rainfall for the selected frequency (Chapter 13).
8. Calculate the total runoff in inches computed from the CN and rainfall (Chapter 7).
9. Find I_a for each subarea from Table 11.5.
10. Using the ratio of I_a/P and T_c for each subarea, select one of the hydrographs tabulated in TR-55.
11. Multiply the hydrograph ordinates (csm/in.) by the area (mi^2) and runoff (in.) of each respective subarea.
12. Route and combine the hydrographs.

The SCS recommends that TR-20, rather than the tabular method, be used if any of the following conditions apply:

Travel time is greater than 3 hr.
T_c is greater than 2 hr.
Drainage areas of individual subareas differ by a factor of 5 or more.
An error of 25 percent or greater in predicted volume is unacceptable.
The urban watershed is very complex or a higher degree of accuracy is required.

Even though SCS documentation of TR-55 procedures for analyzing peak flows and runoff hydrographs from urbanized areas recommends manual rather than computerized applications of the procedures, several vendors have programmed the techniques for PC use and made them available through a number of outlets. A public domain version is available from the U.S. National Technical Information Service [41].

A significant problem in some of the commercial software packages is the use of a triangular-shaped unit hydrograph for convolution to produce hydrographs for storms of various durations. The SCS used a triangular shape to conceptualize the peak flow rate of a curvilinear unit hydrograph, but has never endorsed use of other than either the curvilinear shape discussed in Section 9.4 or the tabulated hydrographs given in the TR-55 manual. For further reading, the SCS released a guide [42] for the use of the 1975 TR-55 intended to clarify procedures in the original technical release.

Users of the vendor-developed renditions of TR-55 should perform initial studies using hand-checks to verify the code. An ideal TR-55 program would be one that uses SCS source code or has SCS endorsement, states all assumptions used, notifies the user of range violations, and incorporates options for making adjustments to account for all or most of the following:

1. Changes in the 484 coefficient of Eq. 9.32 for steep, average, or flat watersheds.
2. Percent imperviousness.
3. Percent of channel that is improved.
4. Ponding area.
5. Subarea length-over-width ratios that fall outside the assumed range.
6. Slope.
7. Antecedent moisture conditions.
8. Different storm distributions.
9. Proper lag time equation.
10. Recognition of the SCS recommendation that the duration for the derived unit hydrograph be about 13 percent of the subarea time of concentration.
11. Allowance for watersheds that have initial abstractions, I_a, greater than 20 percent of the potential maximum retention, S.

Road Research Laboratory (RRL) Method

An urban runoff model that utilizes the time–area runoff routing method described in Section 9.4 was developed in England and described by Watkins [43]. The technique was developed specifically for the analysis of urban runoff and ignores all pervious areas and all impervious areas that are not directly connected to the storm drain system; hence the estimates of peak flow rates and runoff volumes are likely to be low for systems that have these components.

The RRL model could be used for continuous streamflow simulation but tends to be used as an event-simulation model. It has been extensively applied in Great Britain, and moderate success has been reported by Terstriep and Stall [44] for North American applications in the Chicago, Baltimore, and Champaign, Illinois, areas. Other applications are reported in Refs. 45–49.

Illinois Urban Drainage Area Simulator (ILLUDAS)

The Illinois urban drainage area simulator (ILLUDAS) (Ref. 49) is an improved version of the RRL model that has a wider range of applications. It incorporates the impervious and pervious areas neglected by the RRL model and is a demonstrated improvement over RRL.

The hydrologic processes modeled by RRL are summarized in column 3 of Table 11.9. Also tabulated are procedures used by the storm water management model (SWMM) and the USGS distributed routing rainfall-runoff model (DR3M). SWMM and DR3M are described in subsequent sections, followed by a comparison applied to

TABLE 11.9 Comparison of Urban Runoff Model Simulation Procedures

Process (1)	SWMM (2)	ILLUDAS (3)	DR3M (4)
Simulation	Single-event or continuous	Noncontinuous	Continuous or single-event
Interception	Part of depression storage	Neglects	Part of depression storage
Evaporation	Input by user	Neglects	ET from soil zone using pan coefficient
Transpiration	Input by user	Neglects	Neglects
Depression storage	Fills before overland flow begins—part of impervious area assigned zero depression storage Depleted by infiltration	Variable, defaults to 0.2 in.	One-third of rain on directly connected previous areas Soil moisture accounted for during dry periods
Infiltration	Horton or Green–Ampt equation Satisfied by water on ground surface and rainfall	Holtan's equation Standard curves for SCS soil types *A, B, C, D* Infiltration reduced for antecedent moisture conditions	Green–Ampt equation on hourly data, percentage of rain on daily data
Overland flow	Uniform depth of detention	Time–area curve routing Defaults to linear time–area	Unsteady laminar flow by kinematic-wave methods
Hydrograph synthesis	Kinematic wave	Time–area method	Modified rational method
Reach/reservoir routing	Storage routing using Manning turbulent flow equation and continuity equation Quasisteady state	Time of entry required as input data Storage routing used in pipes and open channels	Linear reservoir or modified Puls Channels routed by kinematic-wave method Models ponding behind culverts
Gutter flow	Uniform flow storage routing	Neglects	Treats as open channel
Inlet pits and junctions	Outflow = sum of inflows	Outflow = sum of inflows	Sums, allows external input at nodes
Pipe flow	Storage routing (Manning equation based on the slope of energy line) Kinematic wave or full dynamic wave	Storage routing (Manning equation for uniform flow) Lagged using full bore velocity Quasisteady state	Kinematic routing in nonpressure pipes Unsteady, nonuniform flow; kinematic wave
Surcharge	Simulates pressure flow conditions	Increases pipe diameter in design mode, determines excess volume in analysis mode	Not permitted, uses Manning equation for free-surface flow

Source: Modified from Heeps and Mein [45]

several urban areas. Table 11.9 is reproduced in part from a study by Heeps and Mein [45]. The results of their investigation are discussed in a later section along with results of similar investigations by Marsalek et al. [46] and Lager [47].

The RRL method only simulates runoff from paved areas of the basin that are directly connected to the storm drainage system. Grassed areas and nonconnected paved areas are excluded from consideration. The ILLUDAS [49] model incorporates the directly connected paved-area technique of the RRL method but also recognizes and incorporates runoff from grassed and nonconnected paved areas.

Computation of grassed-area hydrographs for the subbasins is very similar to the approach for paved-area hydrographs. This is illustrated in Fig. 11.18. The shaded area is the contributing grassed area, which is largely the front yards of residences. Rain falling on any not-directly-connected paved area is assumed to run off instantly onto the surrounding grassed area, and grassed-area hydrology takes over. Runoff from back and side yards often drains gradually to a common back lot line and then laterally to the nearest street. The travel time required for this virtually eliminates such grassed areas from consideration during relatively short intense storms normally used for drainage design.

After the contributing grassed area has been identified, the curve in Fig. 11.18 can be constructed. Travel times across the grass strips are equivalent to the time of equilibrium from Izzard's equation [39]:

$$t_e = 0.033 K L q_e^{-0.67} \qquad (11.9)$$

which is the time when the overland flow discharge reaches 97 percent of q_e, that is:

$$q_e = 0.0000231 I L \qquad (11.10)$$

where q_e = discharge of overland flow (cfs/ft of width) at equilibrium
I = rain supply rate (in./hr), assumed to be 1.0 in ILLUDAS
L = length of overland flow (ft)

and $$K = (0.0007 I + c) S^{-0.33} \qquad (11.11)$$

where S = surface slope (ft/ft)
c = coefficient having a value of 0.046 for bluegrass turf.

The time–area curve of Fig. 11.18(b) is often assumed to be a straight line. The endpoint, as illustrated in Fig. 11.18, represents the travel time from the farthest point on the most remote contributing grassed area.

In ILLUDAS, depression storage is normally set at 0.20 in. but can be varied. Infiltration is modeled using Holtan's equation [50]:

$$f = a(S - F)^{1.4} + f_c \qquad (11.12)$$

where f = infiltration rate at time t (in./hr)
a = a vegetative factor = 1.0 for bluegrass turf
S = storage available in the soil mantle (in.) (storage at the soil porosity minus storage at the wilting point)

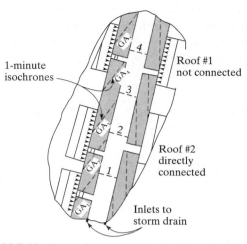

1-minute isochrones

Roof #1 not connected

Roof #2 directly connected

Inlets to storm drain

(a) Subbasin map (contributing grassed area shaded)

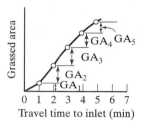

(b) Time versus grassed area curve

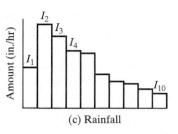

(c) Rainfall

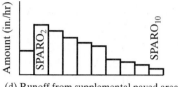

(d) Runoff from supplemental paved area

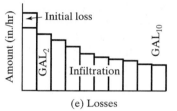

(e) Losses

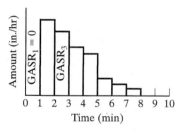

(f) Grassed area supply rate

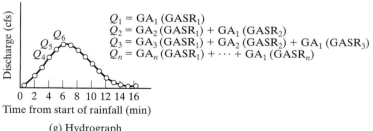

$Q_1 = GA_1 (GASR_1)$
$Q_2 = GA_2 (GASR_1) + GA_1 (GASR_2)$
$Q_3 = GA_3 (GASR_1) + GA_2 (GASR_2) + GA_1 (GASR_3)$
$Q_n = GA_n (GASR_1) + \cdots + GA_1 (GASR_n)$

(g) Hydrograph

FIGURE 11.18

Elements in the development of grassed-area hydrographs.
(After Terstriep and Stall [49])

F = water already stored in the soil at time t in excess of the wilting point (in.) (amount accumulated from infiltration prior to time t)

$S - F$ = storage space remaining in the soil mantle at time t (in.)

f_c = final constant infiltration rate (in./hr), generally equivalent to the saturated hydraulic conductivity (in./hr) of the tightest horizon present in the soil profile

If physical properties of the soil are known, the equation can be used to compute an infiltration curve. Figure 11.19 shows the general interrelation between the various infiltration rates and storage factors involved.

Table 11.10 provides an example computation of an infiltration curve for bluegrass on a silt loam soil in which soil moisture S of 6.95 in. is available. The equation is:

$$f = 1(6.95 - F)^{1.4} + 0.50 \tag{11.13}$$

Standard infiltration curves have been devised for use in ILLUDAS for soils having SCS hydrologic groups A, B, C, and D (Chapter 7). These curves were synthesized from the Horton equation:

$$f = f_c + (f_0 - f_c)e^{-kt} \tag{11.14}$$

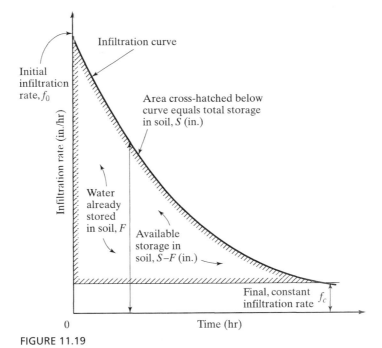

FIGURE 11.19

Diagram of infiltration relations used in ILLUDAS, Eq. 11.12.
(After Terstriep and Stall [49])

TABLE 11.10 Computation of Infiltration Curve for Silt Loam

Available storage, $S - F$ (in.)	ΔF (in.)	Water stored, F (in.)	$(S - F)^{1.4}$	Infiltration rate		Time	
				f (in./hr)	f_{avg} (in./hr)	Δt^a (hr)	t (hr)
6.95		0	15.0	15.5			0
6.00	0.95	0.95	12.3	12.8	14.1	0.07	0.07
5.0	1.0	1.95	9.5	10.0	11.4	0.09	0.16
4.0	1.0	2.95	7.0	7.5	8.7	0.11	0.27
3.0	1.0	3.95	4.65	5.15	6.3	0.16	0.43
2.0	1.0	4.95	2.64	3.14	4.2	0.24	0.67
1.0	1.0	5.95	1.0	1.50	2.3	0.43	1.10
0	1.0	6.95	0	0.50	0.7	1.43	2.53

[a] Incremental time, $\Delta t = \Delta F / f_{avg}$.

Source: Terstriep and Stall [49]

where f_0 = initial infiltration rate (in./hr)
f_c = ultimate infiltration rate
e = base of natural logarithms
k = a shape factor selected as 2
t = time from start of rainfall

This equation is solved in ILLUDAS by the Newton–Raphson technique. The curves are shown in Fig. 11.20.

To account for wet versus dry conditions, ILLUDAS divides the antecedent moisture condition (AMC) into four user-selected ranges, shown in Table 11.11. Each is based on the total 5-day precipitation prior to the storm day. Infiltration from Eq. 11.14 is varied, depending on the AMC value specified.

ILLUDAS allows the user to operate in two modes, analysis and design. For design mode, the model generates hydrographs and provides nominal storm sewer pipe diameters that are adequate, without surcharge, to pass the peak flows. In analysis mode, the model generates hydrographs throughout the basin nodes and links and then alerts the user if any input pipe diameters are too small. It also sums the volume

TABLE 11.11 Antecedent Moisture Conditions for Bluegrass Lawns

AMC number	Description	Total rainfall during 5 days preceding storm (in.)
1	Bone dry	0
2	Rather dry	0–0.5
3	Rather wet	0.5–1
4	Saturated	Over 1

Source: Terstriep and Stall [49]

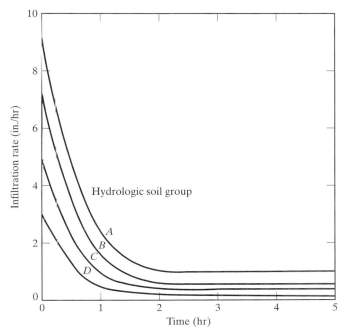

FIGURE 11.20

Standard infiltration curves for bluegrass turf on four SCS soil types
used in ILLUDAS.
(After Terstriep and Stall [49])

of runoff water backed up at inlets because it could not be accommodated by the
undersized storm sewers.

The ILLUDAS model requires estimation of several input parameters. Other
studies [51]–[53] evaluated the sensitivity of ILLUDAS to variations in parameters.

Storage, Treatment, Overflow Runoff Model (STORM)

STORM is the Corps of Engineers' continuous simulation model of the quantity and
quality of urban storm water resulting from single events or continuous daily rainfall
[54]. It also simulates dry weather flow from domestic, commercial, or industrial discharges. Wet weather hydrographs, simulated from intermittent or continuous hourly
rainfall, can be used for a variety of hydrologic study purposes.

Wet weather pollutographs (hydrographs that also provide water quality characteristics) can be predicted for individual historical or synthetic events and used in
assessments of impacts of runoff on receiving streams. The pollutographs consist of
hourly runoff rates, amounts of pollutants, and pollutant concentrations.

The model is conceptualized in Fig. 11.21. Snowmelt is simulated by the degree-
day method (Chapter 8). Statistical information is output to aid in the selection of storage capacities and treatment rates required to achieve desired control of storm water
runoff. Statistics, such as average annual runoff, average annual erosion, average

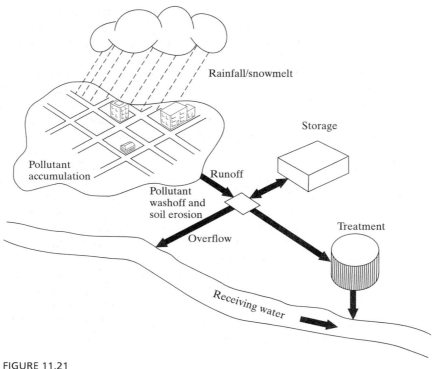

FIGURE 11.21

Conceptual view of urban system as used in STORM.
(After U.S. Army Corps of Engineers [54])

annual overflow volume from storage, and average annual pollutant overflow from storage, are all provided.

The model simulates the interaction of precipitation, air temperature (to signal snowfall), runoff, pollutant accumulation, land surface erosion, dry weather flow, storage, treatment rates, and overflows from the storage or treatment system. The program computes continuous or single-event runoff from rainfall.

Runoff is computed as a fraction of the difference between rainfall and depression storage. The fraction selected depends on land use. Runoff in excess of the specified treatment capacity is diverted into storage for subsequent treatment. Runoff in excess of both the treatment rate and storage capacity becomes overflow and is diverted directly into the receiving waters.

USGS Distributed Routing Rainfall-Runoff Model (DR3M)

The U.S. Geological Survey simulation model for urban rainfall-runoff applications originated in 1978 as a lumped-parameter, single-event model for small watersheds (see Chapter 12) and subsequently was expanded to distributed parameter status, intended primarily for urban applicability [55]. Also, a soil moisture routine was added, allowing quasicontinuous simulation.

The model can be applied to watersheds from a few acres to several square miles in size (an upper limit of $10\,\text{mi}^2$ is recommended). It does not simulate subsurface or interflow contributions to streamflow; these must be externally added if considered important to the simulation.

Routing of rainfall to channels is by unsteady overland flow hydraulics, and routing hydrographs through channel reaches is accomplished by kinematic-wave methods (refer to Chapter 9). The differential routing equations are solved by one of three optional numerical methods. The user may specify an explicit or implicit finite-difference algorithm, or the method of characteristics.

Time may be discretized by the user in as small as 1-min increments. The smallest time increment is used by the program during any days having short-time interval rainfall, called *unit days*. Other days are simulated as 24-hr intervals. Movement of surface water is simulated only during unit days. For the rain-free intervals, daily rainfall is input and used to modify the soil moisture balance leading into the next unit day(s). The format for rain data is compatible with that of the U.S. Geological Survey system, WATSTORE (Waterdata Storage and Retrieval System). Input data can also be obtained from any local National Weather Service office.

Practically any basin can be studied with DR3M by breaking the basin into several sets of four types of model segments. These include overland flow segments (must be approximately rectangular), detention storage facilities, channels, and nodes. This is illustrated in Fig. 11.22. Each segment in the figure may have inflow from either lateral or upstream sources (or both, as occurs for segment CH2).

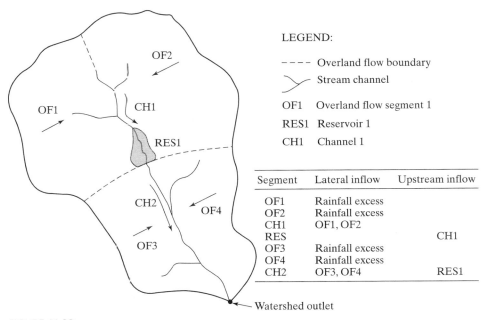

LEGEND:

---- Overland flow boundary

Stream channel

OF1   Overland flow segment 1

RES1   Reservoir 1

CH1   Channel 1

Segment	Lateral inflow	Upstream inflow
OF1	Rainfall excess	
OF2	Rainfall excess	
CH1	OF1, OF2	
RES		CH1
OF3	Rainfall excess	
OF4	Rainfall excess	
CH2	OF3, OF4	RES1

Watershed outlet

FIGURE 11.22

Segmentation of watershed for DR3M.

Rainfall is uniformly distributed over the overland flow rectangles. Each has a given length, slope, roughness, and percent imperviousness. Laminar flow is assumed to occur over these segments.

Channels in Fig. 11.22 represent either natural or artificial gutters or storm sewers (either open channels or nonpressure pipes are allowed). Inflow to channels comes from other channels, overland flow (as lateral inflow), or nodes. Nodes are used when more than three segments contribute to a channel or reservoir, or when the user wishes to specify an input or base flow hydrograph.

Channel routing is by kinematic-wave techniques [1],[18],[19],[56]. If overbank flow is possible, a second set of routing parameters can be input for all flows in excess of the channel capacity.

Reservoir inflow hydrographs are routed by either of two storage routing methods. If a linear reservoir model is appropriate, the storage coefficient K, relating outflow to storage by $O = KS$, is input. If the modified Puls method is desired, a table of outflows and corresponding storage levels must be input. The model assumes an initial reservoir level equal to that corresponding to an outflow of 0 cfs.

Ponding behind culverts can be modeled as a reservoir if an S–O relation is input. This should include data points corresponding to roadway overflow to allow simulation of this common phenomenon.

Excess rainfall (runoff) from pervious areas is developed from the precipitation input, minus several abstractions. Infiltration is simulated by:

$$S = K\left[1 + \frac{P(m - m_0)}{\text{SMS}}\right] \tag{11.15}$$

where K is the hydraulic conductivity, P is the average suction head across the capillary zone, and m_0 and m are the soil moisture contents before and after wetting. The term SMS is the soil moisture storage. The rate of excess rain is found from:

$$I_e = \frac{I^2}{2S} \quad \text{if } I \leq S \tag{11.16}$$

and

$$I_e = I - \frac{S}{2} \quad \text{if } I > S \tag{11.17}$$

where I_e is the rate of rain supplied to infiltration.

Runoff from impervious areas depends on whether the areas are directly connected to the drainage system. Those not directly connected are assumed to flow immediately onto pervious areas, where they are added as lateral inflow. One-third of the rain on directly connected areas is abstracted, and the rest is lateral inflow to the gutter or channel.

Soil moisture is accounted for in a two-layer hypothetical storage zone. The amount in storage affects the infiltration rate and allows continuous soil moisture accounting between rain events. During unit days (days with short-duration rain input) all infiltrated moisture from Eq. 11.15 is added to the upper storage zone. Between unit days, a user-specified proportion of the daily rain is added to the SMS. During rainless days, evapotranspiration occurs from the SMS, using input pan evaporation rates multiplied by a coefficient. This process continues until the next rain event, at which time infiltration is governed by the amount of soil moisture.

Applications of DR3M have been documented across the continental United States and in Alaska and Hawaii [57]. The U.S. Geological Survey maintains updates of software for performing DR3M computations. A download of the program DR3M is available at http://water.usgs.gov/software/surface_water.html. Calibration of computed and measured runoff for almost 400 storms over 37 watersheds reveals a median error in peak flow estimates and volume of 21 and 24 percent, respectively, with the best results obtained for highly impervious watersheds. Indications from verification studies to date are that the model may overestimate the peak flow rates for simulated floods from storms having magnitudes in the design range of flood control facilities, and may give better results for smaller storms typically used in runoff quality studies.

FHWA Storm Sewer Design Model, HYDRA

As part of a package of integrated design computer programs called HYDRAIN [58], the U.S. Federal Highway Administration developed the HYDRA storm drain design model for use by federal and other engineers. Like all FHWA software, the model is distributed under contract with the FHWA through McTrans Software Center at the Civil Engineering Department of the University of Florida at Gainesville. Commercial vendors have linked HYDRA to an integrated CADD/GIS system for interactive design.

The program's primary use is in analyzing the adequacy of existing storm drains or designing new storm drains and inlets by the rational method described in Section 11.3 or by the modified rational method (Section 11.4); the latter represents the hydrograph as a trapezoid having a volume equal to the calculated net rain.

In addition to the modified rational method, commercial versions of HYDRA allow design by SCS methods or the Santa Barbara hydrograph method [58],[59]. HYDRA has an uncommon but useful feature of allowing the hydraulic grade lines to be checked through underground storm sewers. Backwater calculations determine the total system energy losses and hydraulic grade line elevation, allowing an indication of whether inlets, manholes, or junction boxes will overflow. Another useful feature is that street and gutter flows that exceed the inlet capacity of the storm sewers are routed by HYDRA to the next downstream location and added to the hydrograph at that point.

Storm Water Management Model (SWMM)

A very widely accepted and applied storm runoff simulation model was jointly developed by Metcalf and Eddy, Inc., the University of Florida, and Water Resources Engineers [60],[61] for use by the U.S. Environmental Protection Agency (EPA). This model is designed to simulate the runoff of a drainage basin for any predescribed rainfall pattern. The total watershed is broken into a finite number of smaller units or subcatchments that can readily be described by their hydraulic or geometric properties. A flow chart for the process is shown in Fig. 11.23.

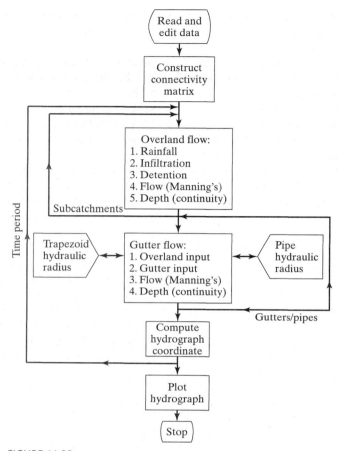

FIGURE 11.23

Flow chart for SWMM Runoff Block hydrographic computation.
(After Metcalf and Eddy, Inc. [61])

The SWMM model has the capability of determining, for short-duration storms of given intensity, the locations and magnitudes of local floods as well as the quantity and quality of storm water runoff at several locations both in the system and in the receiving waters. The original SWMM was an event-simulation model, and later versions [62] keep track of long-term water budgets.

The fine detail in the design of the model allows the simulation of both water quantity and quality aspects associated with urban runoff and combined sewer systems. Only the water quantity aspects are described here. Information obtained from SWMM would be used to design storm sewer systems for storm water runoff control. Use of the model is limited to relatively small urban watersheds in regions where seasonal differences in the quality aspects of water are adequately documented.

The simulation is facilitated by five main subroutine blocks. Each block has a specific function, and the results of each block are entered on working storage devices to be used as part of the input to other blocks.

The main calling program of the model is called the Executive Block. This block is the first and last to be used and performs all the necessary interfacing among the other blocks.

The Runoff Block uses Manning's equation to route the uniform rainfall intensity over the overland flow surfaces, through the small gutters and pipes of the sewer system into the main sewer pipes, and out of the sewer pipes into the receiving streams. This block also provides time-dependent pollutional graphs (pollutographs).

A third package of subroutines, the Transport Block, determines the quality and quantity of dry weather flow, calculates the system infiltration, and calculates the water quality of the flows in the system.

A package of subroutines for water quality determination is contained in the Storage Block. The Storage Block allows the user to specify or have the model select sizes of several treatment processes in an optional wastewater treatment facility that receives a user-selected percentage of the peak flow. If used, this block simulates the changes in the hydrographs and pollutographs of the sewage as the sewage passes through the selected sequence of unit processes.

The earlier version [61] allowed simulation of any reservoir for which the outflow could be approximated as either a weir or orifice, or if the water was pumped from the reservoir. The newer version [62] allows input of 11 points of any storage–outflow relation and routes hydrographs through natural or artificial reservoirs, including backwater areas behind culverts. Routing is by the modified Puls method, which assumes that the reservoir is small enough that the water surface is always level.

Evaporation from reservoirs is simulated by a monthly coefficient (supplied by the user) multiplied by the surface area.

The Extran Block [63] completes the hydraulic calculations for overland flows, in channels, and in pipes and culverts. It solves the complete hydrodynamic equations, assesses surcharging, performs dynamic routing, and provides all the depth, velocity, and energy grade line information requested.

Subcatchment areas, slopes, widths, and linkages must be specified by the user. Manning's roughness coefficients can be supplied for pervious and impervious parts of each subcatchment.

As indicated in Table 11.15, SWMM is the only event-simulation model listed that uses Horton's equation for calculating watershed infiltration losses. If parameters for Horton's equation are unavailable, the user can specify ASCE standard infiltration capacity curves. Infiltration amounts thus determined for each time step are compared with instantaneous amounts of water existing on the subcatchment surface plus any rainfall that occurred during the time step, and if the infiltration loss is larger, it is set equal to the amount available. Input for Horton's equation consists of the maximal and minimal infiltration rates and the recession constant k. The Green–Ampt equation is also used in SWMM.

Urban storm drainage components are modeled using Manning's equation and the continuity equation. The hydraulic radius of the trapezoidal gutters and circular pipes is calculated from component dimensions and flow depths. A pipe surcharges if it is full, provided that the inflow is greater than the outflow capacity. In this case, the surcharged amount will be computed and stored in the Runoff and Transport Blocks at the head end of the pipe. The pipe will remain full until the stored water is completely drained. Alternatively, the Extran Block can be used to conduct a dynamic simulation of the system under pressure-flow conditions.

Necessary inputs in the model are the surface area, width of subcatchment, ground slope, Manning's roughness coefficient, infiltration rate, and detention depth. Channel descriptions are the length, Manning's roughness coefficient, invert slope, diameter for pipes, or cross-sectional dimensions. General data requirements are summarized in Table 11.12. A step-by-step process accounts for all inflow, infiltration losses, and flow from upstream subcatchment areas, providing a calculated discharge hydrograph at the drainage basin outlet. A published description of the simulation process incorporated in early versions of SWMM will aid in understanding the logic of the model [64].

Three general types of output are provided by SWMM. If waste treatment processes are simulated or proposed, the capital, land, and operation and maintenance costs are printed. Plots of water quality constituents versus time form the second type of output. These pollutographs are produced for several locations in the system and in the receiving waters. Quality constituents handled by SWMM include suspended solids, settleable solids, BOD, nitrogen, phosphorus, and grease. The third type of output is hydrologic. Hydrographs at any point, for example, the end of a gutter or inlet, are printed for designated time periods. The Statistics Block provides frequency analysis of storm events from a continuous simulation.

Urban Hydrograph Time Relationships

Dimensioning of time in unit hydrograph applications can have a substantial impact on the shape of the hydrograph. Maintaining the linearity of superposition by unit hydrograph methods requires that the instantaneous unit hydrograph (IUH) have a time base of t_c. If the synthetic unit hydrograph selected for a study or design is not based on this prerequisite, results of applying it may be misleading.

SCS procedures are by far the most prevalent hydrograph methods being incorporated in computer software packages for urban hydrology. The lag time, concentration time, and time base assumptions that shape the SCS dimensionless unit

---

TABLE 11.12   General Data Requirements for the Storm Water Management Model (SWMM)

Item 1. *Define the Study Area.* Land use, topography, population distribution, census tract data, aerial photos, and area boundaries.

Item 2. *Define the System.* Plans of the collection system to define branching, sizes, and slopes; types and general locations of inlet structures.

Item 3. *Define the System Specialties.* Flow diversions, regulators, and storage basins.

Item 4. *Define the System Maintenance.* Street sweeping (description and frequency), catch basin cleaning; trouble spots (flooding).

Item 5. *Define the Base Flow (DWF).* Measured directly or through sewerage facility operating data; hourly variation and weekday versus weekend; the DWF characteristics (composited BOD and SS results); industrial flows (locations, average quantities, and quality).

Item 6. *Define the Storm Flow.* Daily rainfall totals over an extended period (6 months or longer) encompassing the study events; continuous rainfall hyetographs, continuous runoff hydrographs, and combined flow quality measurements (BOD and SS) for the study events; discrete or composited samples as available (describe fully when and how taken).

Source: After Ref. 64

hydrograph were developed for rural, undeveloped watersheds. SCS procedures such as TR-55 make adjustments to account for the urban effects, but the standard shape and time relationships remain the same.

The relationships among time parameters in SCS hydrologic methods have not been completely reconciled with time relationships known to occur in urban rainfall-runoff processes. For example, the rational method assumes that the peak flow for urban areas occurs at the watershed time of concentration, t_c, as confirmed by Izzard [39] in his classical observations of runoff hydrographs from paved areas. By defining the lag time as the time from the centroid of rain to the peak flow (Fig. 9.18), the SCS Mockus equation assumes that the peak occurs at around 60 percent of t_c, especially for short-duration storms. The time base of an SCS unit hydrograph is about 5 times the time to peak (Fig. 9.18), which becomes $3.3t_c$ if the Mockus equation is used. Because the storm duration for the rational method is t_c, and because the recession is also t_c, the time base is $2t_c$. These differences in time to peak and time base are not substantial and fall within the range of empirical observations.

The most serious difference arises from comparing the excess-rainfall release time, t_r, defined in this text as the time from end of excess rain to end of direct runoff, which is exactly the definition of the time base of the IUH. Table 11.13 shows the assumed time dimensions of the SCS and conventional urban methods, and reveals that t_r for the SCS method is 3.2 times t_c. This does not match the previously described requirement. Thus, time relationships for the SCS dimensionless unit hydrograph may result in prolonged runoff durations compared to other synthetic unit hydrographs.

Urban Runoff Models Compared

Several quantitative comparisons of urban storm water models have been reported in the literature. A qualitative comparison of several was prepared by Lager and Smith [47] and is shown in Table 11.14. Table 11.15 provides a bullet matrix showing components of most of the same models. Other comparisons involve quantitative analysis of results of the models when applied to actual gauged storm events. One of the first was an application by Heeps and Mein [45] of three models to two urban catchments in

TABLE 11.13   Comparison of SCS Time Relationships for a D-HR Unit Hydrograph with Standard Urban Runoff Methods

Rational/Izzard models	SCS dimensionless unit hydrograph
Assumed by the method:	Assumed by the method:
$\quad D = t_c$	$\quad D = 0.2 \times$ time to peak
$\quad$Time to peak $= t_c$	$\quad$Time to peak $=$ lag time $+ D/2$
$\quad$Release time $= t_c$	$\quad$Lag time $= 0.6t_c$ (Mockus)
	$\quad$Time base $= 5.0 \times$ time to peak
Solving:	Solving:
$\quad D = t_c$	$\quad D = 0.133t_c$
$\quad$Time to peak $= t_c$	$\quad$Time to peak $= 0.666t_c$
$\quad$Time base $= D + t_r = 2t_c$	$\quad$Time base $= 3.33t_c$
$\quad$Release time $= t_c$	$\quad$Release time $=$ Time base $- D = 3.20t_c$

TABLE 11.14 Comparison of Urban Runoff Models and Methods

Model	Surface routing	Pipe flow routing	Quality routing	Degree of sophistication of surface flow routing	Degree of sophistication of pipe flow routing	Accurate modeling of surcharging	Flexibility of modeling of storm drain components	Explicit modeling of in-system storage	Treatment modeling	Receiving model available	Degree of calibration/ verification required	Simulation period	Availability	Documentation	Data requirements
Rational method	Peak flows only	Peak flows only	No	Low	Low	No	Low	No	NA	No not verified	Usually storms	Individual	Nonproprietary	Good	Low
Chicago	Yes	No	No	Moderate	NA	NA	NA	NA	NA	No	Moderate	Individual storms	Nonproprietary	Fair	Moderate
Unit hydrograph	Yes	In combination with surface	No	Low	Low	No	Low	No	NA	No	High	Individual storms	Nonproprietary	Fair	Moderate
STORM	Yes	In combination with surface	Yes	Low	Low	No	Low	No	Yes	No	Low	Long-term	Nonproprietary	Good	Moderate
RRL	Yes	Yes	No	Moderate	Low-moderate	No	Low	No	NA	No	Moderate	Individual storms	Nonproprietary	Good	Moderate
MIT	Yes	No	No	High	NA	NA	NA	NA	NA	No	Moderate	Individual storms	Proprietary	Fair	Moderate
EPA-SWMM	Yes	Yes	Yes	High	Moderate	Yes	High	Yes	Yes	Yes	Moderate	Individual or continuous storms	Nonproprietary	Good	Extensive
Cincinnati (UCURM)	Yes	Yes	Yes	High	Low	No	Low	No	No	No	Moderate	Individual storms	Nonproprietary	Fair	Extensive
HSPF	Yes	Yes	Yes	Moderate	Moderate	No	Low	No	No	Yes	High	Individual storms or long-term	Nonproprietary	Fair	Extensive
ILLUDAS	Yes	Yes	No	Moderate	High	No	Low	No	NA	No	Moderate	Individual storms	Nonproprietary	Good	Extensive

Source: Modified after Lager and Smith [47]

TABLE 11.15 Comparison of Urban Runoff Model Components and Capabilities

Category	Component / Capability	British Road Research Laboratory	Chicago hydrograph method	Corps of Engineers STORM	Environmental Protection Agency SWMM	Hydrocomp HSPF	Massachusetts Institute of Technology MITCAT	University of Cincinnati UCURM	Illinois ILLUDAS
Catchment hydrology	Multiple catchment inflows	•	•		•	•		•	•
	Dry weather flow	•	•		•	•		•	•
	Input of several hyetographs	•	•		•	•		•	
	Snowmelt			•	•	•			•
	Runoff from impervious areas	•	•	•	•	•	•	•	•
	Runoff from pervious areas		•	•	•	•	•	•	•
	Water balance between storms		•	•	•	•			
Sewer hydraulics	Flow routing in sewers	•	•		•	•	•	•	•
	Upstream and downstream flow control				•	•			•
	Surcharging and pressure flow				•				
	Diversions			•	•	•			•
	Pumping stations				•				•
	Storage			•	•				•
	Prints stage				•	•	•		•
	Prints velocities				•	•	•		•
Wastewater quality	Dry weather quality				•				
	Storm water quality			•	•	•			
	Quality routing				•				
	Sedimentation and scour				•				
	Quality reactions			•	•	•			
	Wastewater treatment				•				
	Quality balance between storms			•	•	•			
	Receiving water flow simulation	•	•		•	•		•	
	Receiving water quality simulation				•	•			
Miscellaneous	Continuous simulation			•	•	•			
	Can choose time interval	•	•		•	•		•	•
	Design computations		•		•			•	•
	Real-time control				•				
	Computer program available	•	•	•	•			•	•

Source: Modified after Brandstetter [51]

443

Australia for a total of 20 storm events. A similar statistical comparison of the same three models applied to 12 storms over each of three urban watersheds was performed by Marsalek et al [46]. Another quantitative comparison is provided by Huber [65].

11.5   VENDOR-DEVELOPED URBAN STORMWATER SOFTWARE

Numerous computer software packages are available for use in urban stormwater analysis and design [66]–[68]. Surveys of available software have been conducted to help hydrologists and engineers choose among the many options [69],[70]. Federal agency software has been inventoried and summarized by Jennings [71]. The American Society of Civil Engineers' Task Committee on Microcomputer Software in Urban Hydrology inventoried the makeup, cost, and applicability of packages available from over 40 commercial vendors and public domain sources [69]. Some of the findings of the ASCE survey are summarized in Table 11.16.

Hydrologic Processes Modeled

A full range of processes is modeled by the packages [69]. Common among popular renditions are routines available in TR-20, HEC-1, and TR-55. The predominant hydrologic methods incorporated in the majority of commercial codes are SCS techniques—SCS unit hydrographs, SCS rainfall distributions, SCS curve number rainfall abstractions, and, to a lesser extent, SCS stream and channel routing procedures. This change has been progressive as documented by the comparison of three surveys over time shown in Table 11.17. This phenomenon is attributed to the ease of programming SCS methods rather than their particular pertinence to urban rainfall-runoff modeling.

TABLE 11.16   Results of ASCE Inventory of 40 Software Packages [69]

Software characteristics	No. packages
Number of public domain packages	12
Number of commercial packages with copyright	25
Number of commercial packages without copyright	3
Number offering some technical support	38
Number with full hydrograph option	35
Number with storm sewer option	29
Number with detention basin option	31
Number tested against gauged data	30
Typical costs for complete package	$150–$3500

Hydrologic abstraction methods	No. pkgs. using
SCS curve number	17
Rational C coefficient	7
Green–Ampt infiltration equation	6
Horton infiltration equation	6
Constant or uniform loss rate	5
Holtan infiltration equation	3
Other	9

TABLE 11.16   (*Continued*)

Hydrograph synthesis methods	No. pkgs. using
SCS dimensionless unit hydrograph	14
Rational method	9
Snyder's synthetic unit hydrograph	6
Kinematic-wave hydrograph synthesis	5
Clark's IUH/time–area method	5
TR-55 1986 tabular hydrograph	3
Santa Barbara urban hydrograph method	2
Other	15

Hydrograph routing methods	No. pkgs. using
Translation without attenuation, based on travel time	10
Muskingum/Muskingum–Cunge method	9
Kinematic-wave routing	7
Modified-Puls/storage-indication routing (channels and reservoirs)	6
SCS convex/att-kin method	2
Full hydrodynamic St. Venant equations	2
Other	10

As shown in Tables 11.16 and 11.17, three-fourths of the codes now use the CN method for rainfall abstraction; the rest employ infiltration by Horton's equation, the Green–Ampt method, constant losses, exponential losses, and a handful of other methods. Hydrograph synthesis (Chapter 9) is predominantly by the SCS dimensionless unit hydrograph method. Several codes employ a triangular unit hydrograph having a peak flow rate calculated from Eq. 9.32. This is described in the vendor literature as an SCS-based method; however, this practice has not been endorsed by the SCS. Hydrograph routing in channels (Chapter 9) is about evenly divided among the Muskingum, Muskingum–Cunge, kinematic-wave, and storage-indication methods. Reservoir routing is almost universally executed by the storage-indication method. Where needed, detailed reservoir routing has been accomplished by solving the hydrodynamic equations of motion [19].

Options Available in Vendor Software

In reviewing the software, one finds that some packages have unique features and options that would be highly applicable in certain circumstances. For example,

TABLE 11.17   Predominance of SCS Methods in Vendor-Developed Urban Runoff Software*

Year of survey:	1985 (Ref. 28)	1988 (Ref. 29)	1991 (Ref. 21)
SCS CN method	40%	70%	75%
SCS NEH-4 hydrograph synthesis	30%	60%	70%
SCS TR-55 methods	10%	20%	40%

*Tabulated values are percentages of responses falling in the class.

relatively flat watersheds tend to retain significantly high percentages of the initial rainfall. In comparing observed and calculated runoff, matches are sometimes impossible without revisiting the SCS assumption that the initial abstraction is 20 percent of the potential maximum, or $I_a/S = 0.2$ (see Chapter 7). Increases in the ratio up to values of 0.5 or more are sometimes justified. Among other examples of unique features, commercial packages are available for:

1. Simultaneous tracking and graphic output of hydrographs for the existing and proposed conditions.

2. Combined calculation of combined storm drain and sanitary sewer flows in systems that still incorporate both.

3. Tracking and rerouting of flow that is not able to enter underdesigned storm drains. This overflow generally passes down streets and overland until it reaches a swale or channel, or it may combine with hydrographs from other subareas before being routed to other inlets.

4. Sizing of reservoir capacity by the "bowstring" method, which is essentially the mass-curve analysis method described in Chapter 12.

SUMMARY

Many urban and small watershed drainage structures can be designed by first applying procedures presented in this chapter to develop estimates of the peak flow rate for a given design storm. Chapter 13 describes design standards and develops the techniques for synthesizing design storms that can be input to urban stormwater models.

Hundreds of thousands of culverts, bridge waterways, bridge footings, river training structures, scour control measures, stream bank erosion protection structures, storm drains, detention structures, and open channels have been designed by first calculating peak flow rates using methods in this chapter and then using hydraulic procedures to determine the water depths that allow selection of the appropriate structure size.

By far the greatest practical application of public domain and private vendor single-event rainfall-runoff models is for urbanized or urbanizing watersheds. The models described in this chapter are but a few of the many available to the urban stormwater analyst. The reader is encouraged to evaluate any chosen software package for applicability to the problem and the precision required. It is essential that the user understand the model's algorithms to be sure that they provide the appropriate method of abstracting rain losses, synthesizing hydrographs, and routing the hydrographs through streams and reservoirs.

PROBLEMS

SECTIONS 11.1–11.3: PEAK FLOW METHODS FOR URBAN AREAS

11.1 In using the rational formula, $Q = CIA$, for the design of any structure, what do the terms Q, C, I, and A represent? In selecting a T-year design storm intensity, why are the rainfall duration and time of concentration equated? Answer by noting the effect of selecting duration less than, and greater than, the time of concentration.

11.2 The time of concentration for a 6-acre parking lot is 20 min. Which of the following storms gives the greatest peak rate of runoff by the rational formula ($Q = CIA$) if $C = 0.6$? State any assumptions.

(a) 4 in./hr for 10-min duration.

(b) 1 in./hr for 40-min duration.

11.3 A 10.0-mi^2 drainage basin with a time of concentration of 100 min receives rainfall at a rate of 2.75 in./hr for a period of 200 min. If the runoff coefficient is 0.8, determine the discharge rate (in cfs) from the basin 130 min after the rain began.

11.4 A 53,200-acre area has a ϕ index of 0.10 in./hr. A storm with a constant rainfall rate of 0.7 in./hr lasts for 6 hr.

(a) What is the rational formula peak discharge in cfs if the time of concentration is 4 hr?

(b) What is the runoff rate (in cfs) at the end of the fifth hour after the rainfall begins?

11.5 A storm gutter receives drainage from both sides. On the left it drains a rectangular 600-acre area of $t_c = 60$ min. On the right it drains a relatively steep 300-acre area of $t_c = 10$ min. The ϕ index on both sides is 0.5 in./hr. Use the intensity–duration–frequency curves in Fig. 11.6 to determine the peak discharge (in cfs) with a 25-year recurrence interval for (a) the 600-acre area alone, (b) the 300-acre area alone, and (c) the combined area, assuming that the proportion of the 600-acre area contributing to runoff at any time t after rain begins is $t/60$.

11.6 A 1.0-mi^2 parking lot has a runoff coefficient of 0.8 and a time of concentration of 40 min. For the following three rainstorms, determine the peak discharge (in cfs) by the rational method: (a) 4.0 in./hr for 10 min, (b) 1.0 in./hr for 40 min, and (c) 0.5 in./hr for 60 min. State any assumptions regarding area contributing after various rainfall durations.

11.7 The concentration time actually varies with storm magnitude but is relatively constant for moderate to extreme events. From this statement, why do engineers feel confident in using the rational formula?

11.8 A 10.00-mi^2 watershed with a 100-min time of concentration receives rainfall at a rate of 2.75 in./hr for a period of 200 min. The time–area curve is shown below.

(a) Determine the peak discharge (in cfs) from the watershed if $C = 0.4$.

(b) Estimate the discharge rate (in cfs) 150 min after the beginning of rainfall.

(c) Estimate the discharge rate from the watershed 40 min after the beginning of rainfall.

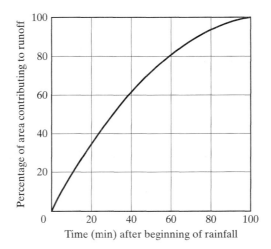

11.9 The 4-hr unit hydrograph for a 5,600-acre watershed is:

Time (hr)	0	2	4	6	8	10	12
Q (cfs)	0	400	1,000	800	400	200	0

The local 10-year IDF curve is linear with the equation $I = 5.6 - 0.2D$, where I is rain intensity in in./hr and D is the rain duration in hours. Use the rational method to determine the peak 10-year flow rate for the watershed. Compare this with the unit hydrograph estimate of the 10-year peak ($\phi = 1.0$ in./hr). Use t_c as the time from the cessation of rain to the cessation of runoff.

SECTIONS 11.4–11.5: HYDROGRAPH METHODS FOR URBAN AREAS

11.10 A rural watershed with a composite CN of 70 is being urbanized. Eventually, 36% of the area will be impervious. Use TR-55 assumptions to determine the increase in runoff that can be expected for a 6.2-in. rain.

11.11 Determine the maximum and minimum infiltration rates and the recession constant k (give units) for the infiltration capacity curve in Fig. 7.6. Compare these with corresponding values for bluegrass used in ILLUDAS.

11.12 Using the initial and ultimate infiltration rates from Table 11.13, calculate and plot Horton's equation (Eq. 11.11) infiltration rates at the intermediate times. Then plot and compare corresponding infiltration curves by Holtan's (Chapter 7) and Horton's methods. Discuss the differences.

11.13 Compare the DR3M method of infiltration in Eqs. 11.12–11.14 with the procedure employed by SWM-IV in Chapter 12.

REFERENCES

[1] W. Viessman and G. L. Lewis, *Introduction to Hydrology, 4th ed.,* New York: HarperCollins College, 1996.

[2] J. Amorocho and W. E. Hart, "A Critique of Current Methods in Hydrologic Systems Investigations," *Trans. Am. Geophys. Union* **45**(2), 307–321(June 1964).

[3] K. P. Singh, "Nonlinear Instantaneous Unit-Hydrograph Theory," *ASCE J. Hyd. Div.* **90**(HY2), Part I, 313–347(Mar. 1964).

[4] W. T. Sittner, C. E. Schauss, and J. C. Monro, "Continuous Hydrograph Synthesis with an API-Type Hydrologic Model," *Water Resources Res.* **5**(5), 1007–1022(1969).

[5] J. E. Nash, "The Form of the Instantaneous Unit Hydrograph," *Int. Assoc. Sci. Hyd.* **3**(45), 114–121(1957).

[6] V. T. Chow, D. R. Maidment, and L. W. Mays, *Applied Hydrology,* New York: McGraw-Hill, 1988.

[7] D. R. Dawdy and T. O'Donnel, "Mathematical Models of Catchment Behavior," *Proc. ASCE J. Hyd. Div.* **91**(HY4), 124–127(July 1965).

[8] S. L. S. Jacoby, "A Mathematical Model for Nonlinear Hydrologic Systems," *J. Geophy. Res.* **71**(20), 4811–4824(Oct. 1966).

[9] R. Prasad, "A Nonlinear Hydrologic System Response Model," *Proc. ASCE J. Hyd. Div.* **93**(HY4) (1967).

[10] A. L. Tholin and C. T. Keifer, "Hydrology of Urban Runoff," *J. ASCE* **85**, 47–106(Mar. 1959).

[11] N. H. Crawford and R. K. Linsley, Jr., "Digital Simulation in Hydrology: Stanford Watershed Model IV," Department of Civil Engineering, Stanford University, Stanford, CA, Tech. Rep. No. 39, July 1966.

[12] W. H. Espey, C. W. Morgan, and F. D. Masch, "Study of Some Effects of Urbanization on Storm Runoff from a Small Watershed," Texas Water Development Board, Rep. No. 23, 1966.

[13] L. D. James, "Using a Computer to Estimate the Effects of Urban Development of Flood Peaks," *Water Resources Res.* **1**(2), (1965).

[14] F. J. Keller, "The Effect of Urban Growth on Sediment Discharge," Northwest Branch Anacostia River Basin, Maryland, in "Short Papers in Geology and Hydrology," U.S. Geological Survey Professional Paper 450-C, 1962.

[15] R. L. Bras and F. E. Perkins, "Effects of Urbanization on Catchment Response," *Proc. ASCE J. Hyd. Div.* **101**(HY3) (Mar. 1975).

[16] Luna B. Leopold, "Hydrology for Urban Land Planning," USGS Circular No. 554, Washington, D.C.: U.S. Government Printing Office, 1968.

[17] Victor Mockus, "Hydrologic Effect of Land Use and Treatment," Chap. 12, *SCS National Engineering Handbook*, Sec. 4, "Hydrology," Washington, D.C.: U.S. Department of Agriculture, Soil Conservation Service, 1972.

[18] D. R. Maidment (ed.), *Handbook of Hydrology*, New York: McGraw-Hill, 1993.

[19] American Society of Civil Engineers, "Design and Construction of Urban Stormwater Management Systems," *ASCE Manuals and Reports of Engineering Practice No. 77*, New York: ASCE, 1992.

[20] G. A. Burton, *Stormwater Effects Handbook: A Toolbox for Watershed Managers, Scientists, and Engineers*, Boca Raton, FL: CRC Press, 2001.

[21] American Public Works Association, "Water Pollution Aspects of Urban Runoff," Federal Water Pollution Control Administration, 1969.

[22] W. Viessman, Jr., "Modeling of Water Quality Inputs from Urbanized Areas," Urban Water Resources Research, Study by ASCE Urban Hydrology Research Council, Sept. 1968, pp. A79–A103.

[23] Division of Water Resources, Department of Civil Engineering, University of Cincinnati, Cincinnati, OH, "Urban Runoff Characteristics," Water Pollution Control Research Series, EPA, 1970.

[24] P. N. Felton and H. W. Lull, "Suburban Hydrology Can Improve Watershed Conditions," *Public Works* **94** (1963).

[25] E. Kuichling, "The Relation Between the Rainfall and the Discharge of Sewers in Populous Districts," *Trans. ASCE*, **20**(1889).

[26] W. W. Horner, "Modern Procedure in District Sewer Design," *Eng. News* **64**, 326(1910).

[27] W. W. Horner and F. L. Flynt, "Relation Between Rainfall and Runoff from Small Urban Areas," *Trans. ASCE* **20**(140), (1936).

[28] Federal Aviation Agency, Department of Transportation, "Airport Drainage," Advisory Circular, A/C 150-5320-5B. Washington, D.C.: U.S. Government Printing Office, 1970.

[29] U.S. Soil Conservation Service, "Urban Hydrology for Small Watersheds," Tech. Release No. 55, Jan. 1975.

[30] U.S. Soil Conservation Service, "Hydrology," *National Engineering Handbook*, Sec. 4. Washington, D.C.: U.S. Government Printing Office, 1972.

[31] U.S. Soil Conservation Service, "Urban Hydrology for Small Watersheds," Tech. Release 55 (2nd ed.), June 1986.

[32] G. L. Lewis and D. P. Gilbert, "A Shopper's Guide to Urban Stormwater Micro Software," *Proceedings, ASCE Hydraulics Division Specialty Conference,* Orlando, FL, August 1985.

[33] R. H. McCuen, *Hydrologic Analysis and Design,* Englewood Cliffs, NJ: Prentice Hall, 1989.

[34] V. B. Sauer, W. D. Thomas, V. A. Stricker, and K. V. Wilson, "Flood Characteristics of Urban Watersheds in the United States," U.S. Geological Survey Water-Supply Paper 2207. Washington, D.C.: U.S. Government Printing Office, 1983.

[35] R. S. Gupta, *Hydrology and Hydraulic Systems,* Prospect Heights, IL: Waveland Press, 1995.

[36] R. H. McCuen, "Hydrologic Design of Highways," *U.S. Federal Highway Administration Hydraulic Design Series 2,* 1995.

[37] A. D. Feldman, Hydrologic Engineering Center Models for Urban Hydrologic Analysis, *Transportation Research Record 1471,* Washington, D.C.: National Academy Press, 1994.

[38] A. L. Tholin and C. T. Keifer, "Hydrology of Urban Runoff," *Proc. ASCE J. San. Eng. Div.* **85**(SA2), 47–106(Mar. 1959).

[39] C. F. Izzard, "Hydraulics of Runoff from Developed Surfaces," *Proceedings 26th Annual Meeting Highway Research Board,* **26**, 129–146(1946).

[40] C. J. Keifer, J. P. Harrison, and T. O. Hixson, "Chicago Hydrograph Method Network Analysis of Runoff Computations," Preliminary Report, City of Chicago, Bureau of Engineering, July 1970.

[41] National Technical Information Service, "TR-55, Hydrology for Small Urban Watersheds," U.S. Department of Commerce, Springfield, VA, 1986.

[42] U.S. Soil Conservation Service, "Guide for the Use of Technical Release No. 55," Albany, NY, Dec. 1977.

[43] L. H. Watkins, *The Design of Urban Sewer Systems,* Road Research Tech. Paper No. 55, Department of Scientific and Industrial Research, London: Her Majesty's Stationery Office, 1962.

[44] Michael L. Terstriep and John B. Stall, "Urban Runoff by the Road Research Laboratory Method," *Proc. ASCE J. Hyd. Div.* **95**(HY6), 1809–1834(Nov. 1969).

[45] D. P. Heeps and R. G. Mein, "Independent Comparison of Three Urban Runoff Models," *Proc. ASCE J. Hyd. Div.* **100**(HY7), 995–1010(July 1974).

[46] J. Marsalek, T. M. Dick, P. E. Wisner, and W. G. Clarke, "Comparative Evaluation of Three Urban Runoff Models," *Water Resources Bull. AWRA* **11**(2), 306–328(Apr. 1975).

[47] J. A. Lager and W. G. Smith, "Urban Stormwater Management and Technology—An Assessment," U.S. EPA Rep. EPA-670/2-74-040, Dec. 1974.

[48] John B. Stall and Michael L. Terstriep, "Storm Sewer Design—An Evaluation of the RRL Method," EPA Technology Series EPA-R2-72-068, Oct. 1972.

[49] M. L. Terstriep and J. B. Stall, "The Illinois Urban Drainage Area Simulator, ILLUDAS," *Illinois State Water Surv. Bull.* **58**, 1–30(1974).

[50] N. H. Holtan, "A Concept for Infiltration Estimates in Watershed Engineering," U.S. Department of Agriculture, Agricultural Research Service, 1961.

[51] A. Brandstetter, "Comparative Analysis of Urban Stormwater Models," Battelle Memorial Institute, Aug. 1974.

[52] H. G. Wenzel and M. L. Terstriep, "Sensitivity of Selected ILLUDAS Parameters," Illinois State Water Survey, Contract Rep. 178, Aug. 1976.

[53] J. Han and J. W. Delleur, "Development of an Extension of ILLUDAS for Continuous Simulation of Urban Runoff Quantity and Discrete Simulation of Runoff Quality," Purdue University, July 1979.

[54] U.S. Army Corps of Engineers, "Urban Storm Water Runoff, STORM," Computer Program 723-S8-L2520, Hydrologic Engineering Center, Davis, CA, Oct. 1974.

[55] W. M. Alley and P. E. Smith, "Distributed Routing Rainfall-Runoff Model—Version II, User's Manual," USGS Open-File Rep. 82-344, 1982.

[56] D. R. Maidment (ed.), *Handbook of Hydrology,* New York: McGraw-Hill, 1993.

[57] W. M. Alley, "Summary of Experience with the Distributed Routing Rainfall-Runoff Model (DR3M)," U.S. Geological Survey, Reston, VA, 1986.

[58] U.S. Federal Highway Administration, *HYDRAIN—Integrated Drainage Design Computer System,* Washington, D.C., 1990.

[59] C. T. Haan, B. J. Barfield, and J. C. Hayes, *Design Hydrology and Sedimentology for Small Catchments,* San Diego: Academic Press, 1994.

[60] Metcalf and Eddy, Inc., University of Florida, Gainesville, Water Resources Engineers, Inc., "Storm Water Management Model," Environmental Protection Agency, Vol. 1, 1971.

[61] P. B. Bedient and W. C. Huber, *Hydrology and Floodplain Analysis,* Reading, MA: Addison-Wesley, 1988.

[62] W. C. Huber et al., "Storm Water Management Model User's Manual, Version III," EPA-600/2-84-109a (NTIS PB84-198423), Nov. 1981.

[63] L. A. Roesner et al., "Storm Water Management Model User's Manual, Version III: Addendum I, Extran," EPA-600/2-84-109b (NTIS PB84-198431), Nov. 1981.

[64] R. K. Linsley, Jr., M. A. Kohler, and J. A. H. Paulhus, *Applied Hydrology,* New York: McGraw-Hill, 1949.

[65] W. C. Huber, "Modeling for Storm Water Strategies," *APWA Reporter,* May 1975.

[66] S. J. Nix, *Urban Stormwater Modeling and Simulation,* Boca Raton, FL: Lewis, 1994.

[67] D. H. Hoggan, *Computer-Assisted Floodplain Hydrology and Hydraulics,* New York: McGraw-Hill, 1989.

[68] V. P. Singh, *Hydrologic Systems,* Englewood Cliffs, NJ: Prentice Hall, 1989.

[69] D. F. Kibler, M. E. Jennings, G. L. Lewis, B. A. Tschantz, and S. G. Walesh, ASCE Task Committee on Urban Stormwater Software, "Microcomputer Software in Urban Hydrology," HYDATA, American Water Resources Association, Vol. 10, No. 5, September 1991.

[70] G. L. Lewis, "A Shopper's Guide to Urban Stormwater Software Revisited," *Proceedings, ASCE 1988 National Conference on Hydraulic Engineering,* Colorado Springs, CO, 1988.

[71] M. E. Jennings et al., "Federal Microcomputer Software for Urban Hydrology," *Proceedings, ASCE 1988 National Conference on Hydraulic Engineering,* Colorado Springs, CO, 1988.

Hydrologic Simulation and Streamflow Synthesis

OBJECTIVES

The purpose of this chapter is to:

- Introduce the types and classes of hydrologic simulation models
- Illustrate the limitations, alternatives, modeling steps, general elements, and data needs of hydrologic simulation models
- Present a proven protocol for successfully conducting modeling studies
- Describe how single-event models are structured and how they are used to simulate direct runoff hydrographs
- Describe the most widely used federal agency–developed single-event models for simple and complex watersheds
- Describe the structure and application of the most widely used continuous streamflow simulation models
- Present a detailed description of the Stanford watershed model, showing how continuous streamflow simulation models link the various physical processes
- Introduce the principles of groundwater flow simulation models
- Introduce methods of synthesizing continuous streamflow records using mass curve analysis and time–series–analysis principles from probability and statistics.

Information regarding rates and volumes of flow at any point of interest along a stream is necessary in the analysis and design of many types of water projects. Although many streams have been gauged to provide continuous records of streamflow, planners and engineers are sometimes faced with little or no available streamflow information and must rely on *synthesis* and *simulation* as tools to generate artificial flow sequences for use in rationalizing decisions regarding structure sizes, the effects of land use, flood control measures, water supplies, water quality, and the effects of natural or induced watershed or climatic changes [1],[2].

Simulation is defined as the mathematical description of the response of a hydrologic water resource system to a series of events during a selected time period. A *simulation model* is a set of equations and algorithms (e.g., operating policies for reservoirs) that describe the real system and imitate the behavior of the system. Simulation can mean calculating daily, monthly, or seasonal streamflow based on rainfall; computing the discharge hydrograph resulting from a known or hypothetical storm; or simply filling in the missing values in a streamflow record.

After providing an overview to simulation, this chapter describes the most-used single-storm event models, continuous streamflow simulation models, groundwater flow models, and time-series analysis models.

12.1   HYDROLOGIC SIMULATION OVERVIEW

In this chapter, *simulation* of all or parts of a surface, groundwater, or combined system implies the use of computers to imitate historical events or predict the future response of the physical system to a specific plan or action. Physical, analog, hybrid, or other models for simulating the behavior of hydraulic and hydrologic systems and system components have had, and will continue to have, application in imitating prototype behavior but are not discussed here.

A few of the numerous event, continuous, and urban runoff computer models for simulating rainfall-runoff components of the hydrologic cycle are compared in Table 12.1. As shown in the table, most of the models were developed for, or by, universities or federal agencies. All have moderate-to-extensive input data requirements, and all have from 1 to 10 percent of inputs that are judgment parameters. These are normally validated by repeated trials with the models. The urban runoff models are primarily event simulation models but have been isolated in Table 12.1 because the descriptions of urban models were given in Chapter 11. Additional details for the models listed in Table 12.1 are provided in several of the references [3]–[7].

Types of Mathematical Models

The American Society of Civil Engineers' Task Committee on Hydraulic Modeling Terms developed a full glossary of terms used in simulation by researchers and practitioners [8]. The types of models identified by the committee are:

1. *Analytical model.* Mathematical model in which the solution of the governing equations is obtained by mathematical analysis, as opposed to numerical manipulation.
2. *Deterministic model.* Mathematical model in which the behavior of every variable is completely determined by the governing equations.
3. *Dynamic model.* Mathematical model in which time is included as an independent variable.
4. *Empirical model.* Representation of a real system by a mathematical description based on experimental data rather than on general physical laws.
5. *Heuristic model.* Representation of a real system by a mathematical description based on an unreasoned, but unproven argument.

TABLE 12.1   Digital Simulation Models of Hydrologic Processes

Code name	Model name	Agency or organization	Percentage of inputs by judgment[a]	Date of original development
	Rainfall-runoff event-simulation models — Section 12.2			
HEC-1	HEC-1 Flood Hydrograph Package	Corps	1	1973
HEC-HMS	Hydrologic Modeling System	Corps	1	1998
TR-20	Computer Program for Project Hydrology	SCS	5	1965
USGS	USGS Rainfall–Runoff Model	USGS	10	1972
HYMO	Hydrologic Model Computer Language	ARS	1	1972
SWMM	Storm Water Management Model	EPA	5	1971
	Continuous streamflow simulation models — Section 12.3			
API	Antecedent Precipitation Index Model	Private	1	1969
USDAHL	1970, 1973, 1974 Revised Watershed Hydrology	ARS	1	1970
SWM-IV	Stanford Watershed Model IV	Stanford University	10	1959
HSPF	Hydrocomp Simulation Program —FORTRAN	EPA	10	1967
NWSRFS	National Weather Service Runoff Forecast System		10	1972
SSARR	Streamflow Synthesis and Reservoir Regulation	Corps	3	1958
PRMS	Precipitation-Runoff Modeling System	USGS	5	1982
SWRRB	Simulator for Water Resources in Rural Basins	USDA	10	1990
	Urban runoff simulation models — Chapter 11			
UCUR	University of Cincinnati Urban Runoff Model	University of Cincinnati	2	1972
STORM	Quantity and Quality of Urban Runoff	Corps	3	1974
MITCAT	MIT Catchment Model	MIT	5	1970
SWMM	Storm Water Management Model	EPA	5	1971
ILLUDAS	Illinois Urban Drainage Area Simulator	Illinois State Survey	1	1972
DR3M	Distributed Routing Rainfall– Runoff Model	USGS	5	1978
PSURM	Pennsylvania State Urban Runoff Model	Pennsylvania State University	5	1979
TR-55	Technical Release 55	SCS	5	1986

[a] Judgment percentages are from U.S. Army Waterways Experiment Station [4].

6. *Interactive model.* Numerical model that allows interaction by the modeler during computation.

7. *Linear model.* Mathematical model based entirely on linear equations.

8. *Nonlinear model.* Mathematical model using one or more nonlinear equations.

9. *Numerical model.* Mathematical model in which the governing equations are not solved analytically. Using discrete numerical values to represent the variables involved and using arithmetic operations, the governing equations are solved approximately.

10. *Probabilistic model.* Mathematical model in which the behavior of one or more of the variables is either completely or partially described by equations of probability.

11. *Semiempirical model.* Representation of a real system by a mathematical description based on general physical laws but containing coefficients determined from experimental data.

12. *Simulation model.* Mathematical model that is used with actual or synthetic data, or both, to produce long-term time series or predictions.

13. *Stochastic model.* Another name for a probabilistic model.

14. *Theoretical model.* Representation of a real system by a mathematical description.

Most of these terms are used in the remainder of this chapter.

Classification of Hydrologic Simulation Models

In recent decades the science of computer simulation of groundwater and surface water resource systems has passed from scattered academic interests to a practical engineering procedure. The varied nature of developed and applied simulation models has caused a proliferation of categorization attempts. A few of the most descriptive classifications are presented [3].

Physical vs. Mathematical Models   Physical models include analog technologies and principles of similitude applied to small-scale models. In contrast, mathematical models rely on mathematical statements to represent the system. A laboratory flume may be a 1 : 10 physical model of a stream, while the unit hydrograph theory of Chapter 9 is a mathematical model of the response of a watershed to various effective rain hyetographs.

Lumped- vs. Distributed-Parameter Models   Models that ignore spatial variations in parameters throughout an entire system are classified as *lumped-parameter* models. An example is the use of a unit hydrograph for predicting time distributions of surface runoff for different storms over a homogeneous drainage area. The lumped parameter is the X-hour unit hydrograph used for convolution with rain to give the storm hydrograph. The time from end of rain to end of runoff is also a lumped parameter as it is held constant for all storms. *Distributed-parameter* models account for behavior variations from point to point throughout the system. Most modern groundwater simulation models are distributed in that they allow variations in storage and transmissivity parameters over a grid or lattice system superimposed over the plan of an aquifer. More recently, surface water systems are being analyzed through use of distributed-parameter geographical information system (GIS) technologies [9].

Stochastic vs. Deterministic Models   Many stochastic processes are approximated by deterministic approaches if they exclude all consideration of random parameters or inputs. For example, the simulation of a reservoir system operating policy for water supply would properly include considerations of uncertainties in natural inflows, yet many water supply systems are designed on a deterministic basis by mass curve analyses, which assume that sequences of historical inflows are repetitive.

Deterministic methods of modeling hydrologic behavior of a watershed have become popular. *Deterministic simulation* describes the behavior of the hydrologic cycle in terms of mathematical relations outlining the interactions of various phases of the hydrologic cycle. Frequently, the models are structured to simulate a streamflow value, hourly or daily, from given rainfall amounts within the watershed boundaries. The model is verified or calibrated by comparing results of the simulation with existing records. Once the model is adjusted to fit the known period of data, additional periods of streamflow can be generated.

Event-Based vs. Continuous Models   Hydrologic systems can be investigated in greater detail if the time frame of simulation is shortened. Many short-term hydrologic models could be classified as *single-event simulation* models as contrasted with *sequential* or *continuous* models. An example of the former is the Corps of Engineers' single-event model, HEC-1 [10], and an example of the latter is the Stanford watershed model developed by Crawford and Linsley [7], which is normally operated to simulate 3, 4, 5, or more years of streamflow. A typical single-event simulation model might use a time increment of 1 hr or perhaps even 1 min.

Water Budget vs. Predictive Models   Several model classifications have arisen that distinguish between the purposes of the model types. One important comparison is whether the model proposes to predict future conditions using synthesized precipitation and watershed conditions or by verifying historical events. A *water budget model* is defined as a model or set of relationships that confirm the historical balance of inflows, outflows, and changes in storage for a system under study. It is advised that simulation model studies begin by structuring a water budget model; then use the parameters that affirm the balance in any predictive simulation of the watershed response. For example, meteorologic data, streamflow records, crop patterns, and water application amounts might be known for a given agricultural watershed. A water budget model would be used to determine the correct evapotranspiration (ET) formula parameters by testing a range of values until a balance in the continuity equation occurs for all time increments. This is often performed on a day-by-day or month-by-month basis.

Once the ET parameters are derived from the water budget model, predictive simulations of different crop patterns, meteorologic conditions, or farming practices could be performed with the satisfaction that the relationships in the model corroborate historical water budgets. Because only *primary* hydrologic inputs and outputs are measured (precipitation, temperature, runoff, land use), normal modeling studies require the simultaneous development of water budget parameters for secondary processes such as ET, infiltration, groundwater storage, and temporal and spatial distribution of water applications.

Components of Hydrologic Simulation Models

Numerous mathematical models have been developed for the purpose of simulating various hydrologic phenomena and systems. A general conceptual model including most of the important components is shown in Fig. 12.1; several others are described subsequently. Imported water in the lower left could be input to reservoir or groundwater storage or channel flow, or it might be guided directly to water allocations on the far right if either storage or distribution were deemed unnecessary. The routing of channel flow or overland flow could be accomplished by simple lumped-parameter techniques, or solutions of the unsteady-state flow equations for discrete segments of the channel could be used. The selection of techniques and algorithms to represent each component depends on the degree of refinement desired as output and also on

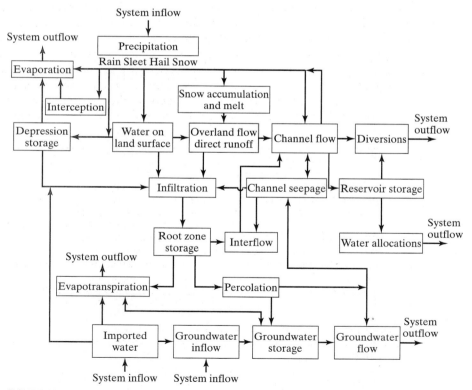

FIGURE 12.1
Components of a surface and subsurface water resource system.

knowledge of the system. A distributed-parameter approach is justified only when available information is adequate.

Steps in Digital Simulation

A fundamental first step in preparing for a simulation model study involves a detailed analysis of all existing and proposed components of the system and the collection of pertinent data. This step is called the system *identification* or inventory phase. Included items of interest are site locations, reservoir characteristics, rainfall and streamflow histories, water and power demands, and so forth. Most model input data requirements (90 percent or more) are map or field available, or can be empirically determined or obtained from engineering handbooks and equations. A general list of data inventory topics that encompasses most hydrologic–economic modeling needs is presented in Table 12.2. Examining and evaluating the basic data are essential. An annotated, bibliographic record of the data sources should be maintained. It is always good practice to program models that output (echo) data values as they are read in. Verification of the numerical values and proper entry of the data can be established from the echo.

The second phase is model *conceptualization*, which involves identifying system components that are important to the behavior of the system. This step includes (1) selecting a technique or techniques that are to be used to represent the system elements, (2) formulating the comprehensive mathematical description of the techniques, and (3) translating the proposed formulation into a working computer program that interconnects all the subsystems and algorithms.

Assumptions used in the conceptualization are also important to the success of a simulation. Assumptions were made by the programmer when developing the model, and additional assumptions are made by users. No computer program should be used prior to reading and understanding the assumptions made by the programmer and becoming aware of the assumptions implicit in the hydrologic process that was programmed.

TABLE 12.2   Data Checklist for Hydrologic Simulation Modeling

---

A. Basin and subbasin characteristics
 1. Lag times, travel times in reaches, times of concentration.
 2. Contributing areas, depressions, mean overland flow distances and slopes.
 3. Design storm abstractions: evapotranspiration, infiltration, depression, and interception losses. Composite curve numbers, infiltration capacities and parameters, ϕ indexes.
 4. Land-use practices, soil types, surface and subsurface divides.
 5. Water-use sites for recreation, irrigation, flood damage reduction, diversions, flow augmentation, and pumping.
 6. Numbering system for junctions, subareas, gauging and precipitation stations.
 7. Impervious areas, forested areas, areas between isochrones, irrigable acreages.

B. Channel characteristics
 1. Channel bed and valley floor profiles and slopes.
 2. Manning or Chézy coefficients for various reaches, or hydraulic or field data from which these coefficients could be estimated.
 3. Channel and valley cross-sectional data for each river reach.
 4. Seepage information; channel losses and base flows.
 5. Channel and overbank storage characteristics, existing or proposed channelization and levee data.
 6. Sediment loads, bank stability, and vegetative growth.

(Continued)

TABLE 12.2   (*Continued*)

C. Meteorologic data
   1. Hourly and daily precipitation for gauges in or near the watershed.
   2. Temperature, relative humidity, and solar radiation data.
   3. Data on wind speed and direction.
   4. Evaporation pan data.
D. Water use data
   1. Flows returned to streams from treatment plants or industries.
   2. Diversions from streams and reservoirs.
   3. Transbasin diversions from and to the basin.
   4. Stream and ditch geometric properties and seepage characteristics.
   5. Irrigated acreages and irrigation practices, including water use efficiencies.
   6. Crop types and water consumption requirements.
   7. Past conservation practices such as terracing, installation of irrigation return pits, and conservation tillage.
   8. Stock watering practices.
   9. Presence and types of phreatophytes in stream valleys and along ditch banks.
E. Streamflow data
   1. Hourly, daily, monthly, annual streamflow data at all gauging stations, including statistical analyses.
   2. Flood frequency data and curves at gauging stations, or regional curves for ungauged sites, preferably on an annual and seasonal basis.
   3. Flow duration data and curves at gauging stations (also any synthesized data for ungauged areas).
   4. Rating curves; stage–discharge, velocity–discharge, depth–discharge curves for certain reaches.
   5. Flooded area curves.
   6. Stage versus area flooded.
   7. Stage versus frequency curves.
   8. Stage versus flood damage curves, preferably on a seasonal basis.
   9. Hydraulic radius versus discharge curves.
   10. Streamflows at ungauged sites as fractions of gauged values.
   11. Return flows as fractions of water-use allocations diverted for consumptive use.
   12. Seasonal distributions of allocations to users.
   13. Minimal streamflow to be maintained at each site.
   14. Mass curves and storage–yield analyses at gauged sites.
F. Design floods and flood routing
   1. Design storm and flood determination; temporal and spatial distribution and intensity.
   2. Maximum regional storms and floods.
   3. Selection and verification of flood routing techniques to be used and necessary routing parameters.
   4. Base flow estimates during design floods.
   5. Available records of historic floods.
G. Reservoir information
   1. List of potential sites and location data.
   2. Elevation–storage curves.
   3. Elevation–area curves.
   4. Normal, minimal, and other pool levels.
   5. Evaporation and seepage loss data or estimates.
   6. Sediment, dead storage requirements.
   7. Reservoir economic life.
   8. Flood control operating policies and rule curves.
   9. Outflow characteristics, weir and outlet equations, controls.
   10. Reservoir-based recreation benefit functions.
   11. Costs versus reservoir storage capacities.
   12. Purposes of each reservoir and beneficiaries and benefits.

Following the system identification and conceptualization phases are several steps of the *implementation* phase. These include (1) validating the model, preferably by demonstrating that the model reproduces any available observed behavior for the actual or a similar system; (2) modifying the algorithms as necessary to improve the accuracy of the model; and (3) putting the model to work by carrying out the simulation experiments.

A study plan can be developed with approximate time and monetary limits to use as a guide during a simulation project. Combining several investigations in a single run is another way to conduct an efficient simulation. Some of the models available allow this. For example, TR-20 (see Section 12.2) allows the generation of flood hydrographs from several storms at once. It is often desirable to generate the 2-, 5-, 10-, 25-, 50-, 100-, and 500-year flood discharge at a single watershed location.

The computer is able to generate far more output than the hydrologist can analyze. Most models incorporate options allowing the user to specify output quantity. In addition to controlling output, a predetermination should be made of which specific analyses will be performed. A tabulation of key output data can be developed to compile and evaluate trends (and make course corrections) after each run. Because deterministic hydrology is about 80 percent accounting, many opportunities exist in simulation for assessing *water budget* balances. For example, if the total recharge to an aquifer is less than the total outflow and withdrawals, but simulated water tables are rising, a check of input and model parameters should be made. Writing important conclusions on the printed output of simulation runs helps document the study and guide revisions in future runs.

Documentation of simulation studies is sometimes neglected in practice. The record should communicate the findings in a way that provides a later reviewer general understanding of the work plan followed, decisions made, and reasons for each run. The documentation should state assumptions made, provide samples of the input and output, explain input preparation requirements, state how sensitive the results are to parameter changes and assumptions, and document reasons out-of-range parameters were accepted. Documentation is an ongoing and continual task. It is especially crucial if the model will be employed in regulatory procedures or litigation.

Quality Assurance in Computer Simulation Studies

Any hydrologic study or design has the potential of being sufficiently complex that modern computer simulation techniques could enhance the work product and efficiently decrease uncertainty about the solution. Balancing the cost of the study with value received is strongly governed by the choices made in planning and implementing the study. Overkill of data collection, mapping, discretization, or calibration and validation runs may cause excessive, unwarranted costs. Knowledgeable and professional care in model selection, calibration, and use is therefore beneficial.

No professional organization has published standards for the decisions regarding appropriate processes and level of detail, though some useful guidelines have been promulgated by a number of organizations [11],[12]. James and Robinson [13] surveyed the literature and summarized the following list of quality assurance guidelines for consideration in any computer-based study:

- *Problem review*. A problem review should identify all significant study area elements so that the model selected includes all relevant processes and so that it can be disaggregated to the appropriate level.

- *Objectives*. The study objectives should be quantified so that any alternative solution can be shown to impact the objectives.

- *Performance criteria*. The model performance criteria need to be preset and correctly identified so that model alternatives can be compared and the simplest modeling that provides the desired performance results can be selected.

- *Data availability and accuracy*. The field measurements to be used for calibration, validation, and prediction should be reviewed to assure that the data are acceptable and that they are not so extensive as to make model runs inordinately expensive. Possible sources of both systematic and random errors should be identified.

- *Model options*. Several models should be reviewed, including physical process models and object-oriented models, and the sequence of model uses should be determined.

- *Study resources*. Schedule, human resource availability, budget, data availability, and other study resources should be assessed to guide the model selection.

- *Model selection*. Model selection should be based on the performance criteria and study resources available for the application.

- *Verification*. Most models contain tutorials or example problems, which should be reviewed and replicated to assure that the model is performing as intended on the particular hardware being used.

- *Calibration and validation*. The data sets should be divided approximately in half, with one half being used for calibration. In this step, parameters are specified and altered within applicable ranges to provide the best match of output with measured values. Validation tests then proceed with the second half of the data, either fixing the parameters or using appropriate changes in parameters that reflect any differences in the validation data set.

- *Discretization*. The smallest number of subareas required for discretizing and modeling the system should be commensurate with the study objectives and resources (time, money, and data) available.

- *Sensitivity analysis*. Sensitivity tests should be conducted on key parameters to identify which are of most significance, and effort put into their estimation must be commensurate with their importance.

- *Data preparation*. All input and output for the design or predictive runs should be interpreted to demonstrate that the model is performing logically.

- *Documentation*. The study should be thoroughly documented, including the version of model used, assumptions made, and parameters selected and why, and a machine-readable input and output file should be archived for future use.

Delivery of solutions and services by hydrologists will be improved for studies and designs that follow these guidelines. Among other advantages, they reduce the possibility and occurrence of using inappropriate models, making inappropriate uses of appropriate models, conducting erroneous data preparations or output interpretations, and applying and accepting models and output rather than under-standing the physical processes involved.

Corps of Engineers HEC Simulation Models

In 1964, the U.S. Army Corps of Engineers developed a specialty branch located at the Hydrologic Engineering Center (HEC) in Davis, California. The facility provides a center for applying academic research results to practical needs of the Corps field offices. In addition, the center provides training and technical assistance to govern-ment agencies in advanced hydrology, hydraulics, and reservoir operations.

HEC Models   Over the years, a large number of analytical tools were developed at HEC. Table 12.3 summarizes the computer programs in categories of hydrology, river/reservoir hydraulics, reservoir operations, stochastic hydrology, river/reservoir

TABLE 12.3   HEC Water Resource Computer Programs

Name	Date of inception	Purpose
	HYDROLOGY MODELS	
HEC-1, Flood Hydrograph Package	September 1980	Simulates the precipitation runoff process in any complex river basin. Optimizes parameters. Computes expected annual flood damage. Optimizes size of flood control system components.
HEC-HMS, Hydrologic Modeling System [15]	1995	Replaces HEC-1 with additional hydrologic routines and a complete GIS-based framework.
Basin Rainfall and Snowmelt Computation	July 1966	Computes area-average precipitation and snowmelt for many subbasins of a river basin using gauge data and weightings (included in HEC-1).
Unit Graph and Hydrograph Computation	July 1966	Computes subbasin interception/ infiltration, unit hydrograph base flow, and runoff hydrograph (included in HEC-1).
Unit Graph Loss Rate Optimization	August 1966	Estimates best-fit values for unit graph and loss rate parameters from given precipitation and subbasin runoff (included in HEC-1).
Hydrograph Combining and Routing	August 1966	Combines runoff from subbasins at confluences and routes hydrographs through a river network using hydrologic routing methods (included in HEC-1).
Streamflow Routing Optimization	July 1966	Estimates best-fit values for hydrologic streamflow routing parameters with given upstream, downstream, and local inflow hydrographs (included in HEC-1).

(Continued)

TABLE 12.3 (*Continued*)

Name	Date of inception	Purpose
	HYDROLOGY MODELS (continued)	
HEC-IFH, Interior Flood Hydrology Package [16]	February 1992	Computes seepage, gravity and pressure flow, pumping and overtopping discharges for pond areas behind levees or other flow obstructions. Main river elevation and ponding area elevation–area–capacity data are used in computing discharges.
Storage, Treatment, Overflow, Runoff Model (STORM)	July 1976	Simulates the precipitation runoff process for a single, usually urban, basin for many years of hourly precipitation data. Simulates quality of urban runoff and dry weather sewage flow. Evaluates quantity and quality of overflow for combinations of sewage treatment plant storage and treatment rate.
	RIVER/RESERVOIR HYDRAULICS	
HEC-2, Water Surface Profiles	August 1979	Computes water surface profiles for steady, gradually varied flow in rivers and tributaries using natural or artificial cross sections. Flow may be sub- or supercritical. Analyzes allowable encroachment for a given rise in water surface.
Gradually Varied Unsteady Flow Profiles	January 1976	Simulates one-dimensional, unsteady, free surface flows in a branching river network. Natural and artificial cross sections may be used. Uses an explicit centered difference computational scheme.
Geometric Elements from Cross Section Coordinates (GEDA)	June 1976	Computes tables of hydraulic elements for use by the Gradually Varied Unsteady Flow Profiles or other programs. Interpolates values for area, top width, n value, and hydraulic radius at evenly spaced locations along a reach.
Stream Hydraulics Package (SHP)	October 1978	Performs dynamic streamflow routing in a complex river network using full St. Venant, kinematic-wave, modified Puls, or Muskingum routines. Can optimize the storage-discharge function for modified Puls given inflow and outflow hydrographs from a reach.

TABLE 12.3   (*Continued*)

Name	Date of inception	Purpose
RIVER/RESERVOIR HYDRAULICS (continued)		
Spillway Rating and Flood Routing	October 1966	Computes a spillway rating curve for concrete ogee or broadcrested weir spillway with or without discharge from a conduit or sluice. Routes the spillway design flood using a gated or uncontrolled spillway to determine maximum water surface elevation.
Spillway Rating—Partial Tainter Gate Opening	July 1966	Computes discharge rating curve for ogee-type weirs with Tainter (radial-type) gates at any size opening.
Spillway Gate Regulation Curve	February 1966	Computes gate regulation schedule curves for a reservoir knowing the area–capacity curves, induced surcharge envelope curve, and the slope of the recession portion of an inflow hydrograph.
Reservoir Area–Capacity Tables by Conic Method	July 1966	Computes surface area and storage volume for various reservoir water surface elevations using a conic method.
RESERVOIR OPERATION		
HEC-3, Reservoir System Analysis for Conservation	July 1973	Simulates operation of reservoirs for a complex system of reservoirs and purposes for multiyear sequences of monthly streamflows without streamflow routing. Demands for hydropower, low-flow augmentation, and various water supplies are satisfied at the reservoir or at downstream control points within constraints for reservoir storage release capacities, balance between reservoirs, and various hydrothermal power systems.
HEC-5, Simulation of Flood Control and Conservation Systems	June 1979	Simulates operation of reservoirs for flood control, low-flow augmentation, hydropower, and water supply throughout a complex river system. Reservoir operation may be accomplished at variable time intervals during a multiyear simulation. Downstream flood routing is performed for flood control operation and expected annual damages may be computed.

(*Continued*)

TABLE 12.3   (*Continued*)

Name	Date of inception	Purpose
RESERVOIR OPERATION (continued)		
Reservoir Yield	August 1966	Simulates the operation of a single reservoir and one downstream control point for hydropower, water supply, and water quality. Operates on a monthly time interval without streamflow routing.
STOCHASTIC HYDROLOGY		
HEC-4, Monthly Streamflow Simulation	January 1971	Analyzes monthly streamflows at a number of interrelated stations to determine their statistical characteristics. Can generate multiyear synthetic monthly streamflow series using the log–Pearson Type III distribution representation of the historical data.
Flood Flow Frequency Analysis	June 1976	Computes log–Pearson Type III frequency statistics of annual maximum flood peaks according to the United States Water Resource Council Guidelines, Bulletin 17a. Program automatically adjusts for zero flood years, incomplete records, low and high outliers, and historical information. Computes expected probability.
Regional Frequency Computation	July 1972	Performs log–Pearson Type III frequency computations of annual maximum hydrologic events at several stations. Used in regional frequency analysis, preserving intercorrelation between stations and regional skew. Computes statistics of flow at each station for several durations if desired.
RIVER/RESERVOIR WATER QUALITY		
Water Quality for River-Reservoir Systems (WQRRS)	October 1978	Simulates multiparameter water quality in rivers (horizontal segments) and in reservoirs (vertical segments) for steady-state or dynamic flow conditions. Complete aquatic-biologic system is simulated to obtain resulting water quality.
Statistical and Graphical Analysis of Water Quality Data	October 1978	Performs statistical analyses of water quality time series and station results obtained from Water Quality for River-Reservoir Systems or other models. Plots up to 11 parameters versus time at desired stations.

TABLE 12.3   (*Continued*)

Name	Date of inception	Purpose
RIVER/RESERVOIR WATER QUALITY (continued)		
HEC-5Q, Reservoir System Operation including Water Quality Control	September 1979	Simulates the operation of a single reservoir in the same manner as HEC-5 but also includes operation for control of water temperature in the reservoir and at downstream control points.
Heat Exchange Program	December 1972	Computes equilibrium temperatures and surface heat exchange coefficients for use in estimating net heat exchange between a water surface and the atmosphere. Requires daily meteorologic data at locations of interest.
Thermal Simulation Program	June 1973	Determines the annual temperature cycle of a water body by heat balance analysis of inflows, outflows, and surface heat transfer.
Reservoir Temperature Stratification	September 1969	Simulates monthly temperature variations in horizontal strata within a reservoir using inflow, outflow, and meteorologic data, reservoir and dam configuration, and downstream release requirements.
RIVER/RESERVOIR SEDIMENTATION		
HEC-6, Scour and Deposition in Rivers and Reservoirs	March 1976	Simulates one-dimensional sediment transport, scour, and deposition in a river system that may have reservoirs. Steady-state water surface profiles are computed; and at each cross section, discharge, inflowing sediment load, gradation of bed material, and armoring are considered in computing the scour or deposition. Dredging may be analyzed.
Suspended Sediment Yield	March 1968	Computes annual suspended sediment load corresponding to observed daily water discharge and suspended sediment loads.
Reservoir Delta Sedimentation	July 1967	Computes the expected ultimate profile of sediment deposits forming the delta at a reservoir inflow point.
Deposit of Suspended Sediment	June 1967	Computes the distribution and location of sediments in a reservoir. Sediment inflow, trap efficiency, and size distribution of passing sediments are calculated. Sedimentation is calculated in main body of reservoir as well as in tributary arms.

(*Continued*)

TABLE 12.3   (*Continued*)

Name	Date of inception	Purpose
GROUNDWATER		
Finite Element Solution of Steady State Potential Flow Problems	July 1970	Computes groundwater or seepage flows for steady, two-dimensional or axisymmetric flow through heterogeneous, anisotropic porous media of virtually any geometry.
FLOOD DAMAGE		
Expected Annual Flood Damage Computation (EAD)	June 1977	Computes flood damage for the economic evaluation of flood control and floodplain management plans. Damages may be computed for a specific event, expected annual, or equivalent annual for a given interest rate and time period. Analyzes multiple damage reaches and types of damage for several flood management plans.
Interactive Nonstructural Analysis Package	February 1980	Analyzes flood damage reduction benefits for various nonstructural measures such as floodproofing, raising structures, and relocating structures.
GEOGRAPHIC INFORMATION ANALYSIS		
Resource Information and Analysis (RIA)	September 1978	Analyzes geographic data stored in grid cell data banks. Analyzes, tabulates, and displays (printer plots) map analyses for locational attractiveness, impact assessment, distance determination, and coincidence of geographic features.
Hydrologic Parameters (HYDPAR)	November 1978	Computes interception/infiltration and unit-hydrograph parameters from grid cell data for use in HEC-1 watershed model. Computes parameters for subbasins in a river basin and stores results for access by HEC-1.
Damage Calculation (DAMCAL)	November 1978	Computes stage versus damage functions for index locations along the floodplain. Uses grid cell data bank representation of land use and topography and aggregates damages to index location for a range of water surface elevations.

Source: A. D. Feldman, "HEC Models for Water Resources Systems Simulation: Theory and Experience," in Advances in Hydroscience, *Vol. 12. Orlando, FL: Academic Press, 1981 [14].*

water quality, river/reservoir sedimentation, groundwater, flood damage, and geographic information analysis [14]. Figure 12.2 shows the relations among, and potential use of, three of the earliest HEC models. This figure illustrates the concept that many of the HEC models can be combined in a single investigation to provide useful tools for total system analysis. A listing of private outlets for the software and vendors who provide technical support can be obtained from the HEC.

HEC-HMS Hydrologic Modeling System   The Hydrologic Engineering Center is developing next-generation computer models to replace those in current use. This

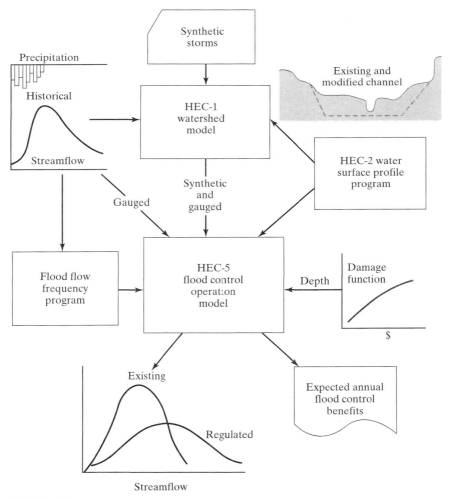

FIGURE 12.2

Example of the use of three HEC models to study flood control alternatives.
(After A. D. Feldman, "HEC Models for Water Resources System Simulation: Theory and Experience,"
Advances in Hydroscience, *Vol. 12. Orlando, FL: Academic Press, 1981 [14])*

package, called the HEC-HMS, is targeted to include object-oriented versus traditional procedural-oriented technology. A new graphical user interface allows the user to edit, execute, and view model data in a window environment. The new approach is designed to provide a logical way to express problems, breaking them down into individual understandable entities and defining interactions among the entities. The technology is described by Pabst [15].

Optimization Models in Hydrology

Planners are continually required to anticipate the future and ask "What if?" and "What's best?" questions. Quantitative planning techniques, such as simulation can provide detailed information about more planning alternatives for less cost than any other approach available; however, they are not the most efficient tools for determining the best plan of action.

Screening models, or optimization models as they are often called, are designed to utilize limited system information to select a best plan among many alternatives for a specified objective or set of objectives. Hence screening models are oriented toward *plan formulation* in contrast to the *plan evaluation* function of simulation models. Simulation models are suited to detailed analyses of specific alternatives and yield reliable information on which to base final designs or operating policies. Screening models address the question, "If our goal is . . . , how can we best proceed?" Simulation models, on the other hand, ask, "If our plan is to . . . , will the plan work and what will be the system response after we are finished?" Used together to take advantage of the special merits of each, these two tools become a powerful adjunct to traditional planning technologies. Complete descriptions of techniques and case studies of screening, optimization, and simulation models are presented in Refs. 17–21.

Final design values should be determined by assigning the optimization model results to the system elements and operating a detailed simulation model over time using a sequence of known or synthesized precipitation amounts and/or streamflows, while at the same time accumulating benefits over time for flood control, reservoir and stream-side recreation, water yields, streamflow augmentation, sediment and erosion control, and any other factors not considered or approximated in the preliminary screening model. Several simulation runs with slight adjustments in decision variables should result in a plan that best meets the objectives and is a significant improvement in plans generated by conventional methods.

12.2 SINGLE-EVENT RAINFALL-RUNOFF MODELS

Many severe floods are caused by short-duration, high-intensity rainfall events. A single-event watershed model simulates runoff during and shortly following these discrete rain events. Users of single-event models are normally interested in the peak flow rate, or the entire direct runoff hydrograph if timing or volume of runoff is needed. Single-event models simulate the rainfall-runoff process and make no special effort to account for the rest of the hydrologic cycle. Few, if any, simulate soil moisture, evapotranspiration, interflow, base flow, or other processes occurring between discrete rainfall events.

Models described in this chapter are applicable to studies of watersheds that are primarily rural in makeup. For watersheds that are principally urbanized, the single-event models described in Chapter 11 are more applicable. Coastal flooding that is induced by surges created by wind action on the ocean surface is modeled by a different class of single-event models, described in Ref. 3.

Single-Event Model Structure

Event simulation model structures closely imitate the rainfall and runoff processes described in earlier chapters. Lumped-parameter approaches, such as unit-hydrograph methods, are generally incorporated even though some use distributed-parameter techniques. Preparation for implementing most single-event simulation studies begins with a watershed subdivision into homogeneous subbasins as illustrated in Fig. 12.3. Computations proceed from the most remote upstream subbasin in a downstream direction.

In any single-event model for a typical basin (Fig. 12.3), the runoff hydrographs for each of subbasins $A, B, \ldots, E$ are computed independently, and then routed and

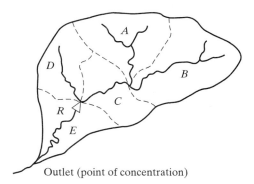

Outlet (point of concentration)

1. Subdivide basin to accommodate reservoir sites, damage centers, diversion points, surface and subsurface divides, gauging stations, precipitation stations, land uses, soil types, and geomorphologic features.

2. Computation sequence in event simulation models:
   a. Compute hydrograph for subbasin B.
   b. Compute hydrograph for subbasin A.
   c. Add hydrographs for A and B.
   d. Route combined hydrograph to upstream end of reservoir R.
   e. Compute hydrograph for subbasin C.
   f. Compute hydrograph for subbasin D.
   g. Combine three hydrographs at R.
   h. Route combined hydrograph through reservoir R.
   i. Route reservoir outflow hydrograph to outlet.
   j. Compute hydrograph for subbasin E.
   k. Combine two hydrographs at outlet.

FIGURE 12.3

Typical watershed subdivision and computation sequence for event-simulation models.

combined at appropriate points (called *nodes*) to obtain design hydrographs throughout the basin. The model reads input parameters for the storm; applies the storm to the first upstream subbasin, *B*; computes the hydrograph resulting from the storm event; repeats the hydrograph computation for subbasin *A*; combines the two computed hydrographs into a single hydrograph; routes the hydrograph by conventional techniques through reach *C* to the upstream end of reservoir *R*, where it is combined with the computed hydrograph for subbasin *C*; and so on through the procedure detailed in Fig. 12.3.

Hydrograph computations for subbasins are most often determined using unit-hydrograph procedures as illustrated in Fig. 12.4. The precipitation hyetograph is input uniformly over the subbasin area, and precipitation losses are abstracted, leaving an excess precipitation hyetograph that is convoluted (see Chapter 9) with the prescribed unit hydrograph to produce a surface runoff hydrograph for the subbasin. The abstracted losses are divided among the loss components on the basis of prescribed parameters. Subsurface flows and waters derived from groundwater storage are transformed into a subsurface runoff hydrograph, which when combined with the surface runoff hydrograph forms the total streamflow hydrograph at the subbasin outlet. This hydrograph can then be routed downstream, combined with another contributing hydrograph, or simply output if this subbasin is the only, or the final, subbasin being considered.

The rainfall-runoff processes depicted in Figs. 12.3 and 12.4 are recognized by most of the event simulation models named in Table 12.1. Specific computation techniques for losses, unit hydrographs, river routing, reservoir routing, and base flow are compared in Table 12.4 for five of the major federal agency rainfall-runoff event simulation models. All the models allow selection among available techniques. Brief descriptions of each of these models are followed by an illustrative example of an application of the HEC-1 model to a single storm occurring over a 250-mi^2 watershed near Lincoln, Nebraska. More detailed descriptions are available in other references [3],[22]–[26].

Calibration Alternatives

Calibration is defined as the process of modifying algorithms, parameters, connectivity, and sequencing of hydrologic processes, within acceptable ranges of each, to cause the model to replicate known conditions. This is necessary in order to assure that the model adequately represents the system being modeled, allowing use of the model to forecast outputs for other input conditions. To illustrate, if the user wanted to predict the 100-year flood from a watershed, a model could be constructed and calibrated to a known rainfall-runoff event of lesser severity, and then a 100-year design storm (see Section 13.4) could be input to meet the objective.

Calibration of single-event rainfall-runoff models is normally accomplished using one or more of the following procedures:

- Comparison of model hydrograph output with gauged runoff for the same storm event.
- Comparison of model hydraulic output with measured or observed high-water marks for historical storm events.

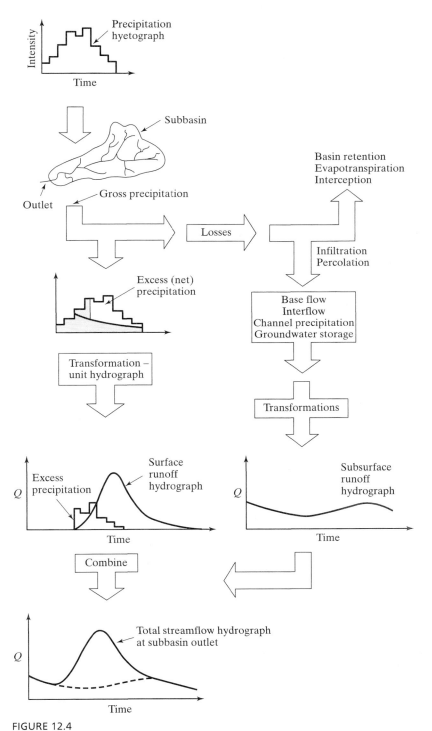

FIGURE 12.4

Typical lumped-parameter event-simulation model of the rainfall-runoff process.

TABLE 12.4   Hydrology Processes and Options Used by Several Agency Rainfall-Runoff Event Simulation Models

Modeled components	HEC-1 (Corps)	TR-20 (SCS)	USGS (USGS)	HYMO (ARS)	SWMM (EPA)
Infiltration and losses					
Holtan's equation	X				
Horton's equation					X
Green–Ampt equation	X				X
Phillip's equation			X		
SCS curve number method	X	X		X	
Exponential loss rate	X				
Standard capacity curves					X
Unit hydrograph					
Input	X	X			
Clark's equation	X		X		
Snyder's equation	X				
Two-parameter gamma response				X	
SCS dimensionless unit hydrograph	X	X			
River routing					
Kinematic wave equation	X				X
Full dynamic wave equation					X
Muskingum method	X				
Muskingum–Cunge method	X				
Modified Puls equation	X				
Normal depth	X				
Variable storage coefficient	X			X	
Att-kin method		X			
Translation only			X		
Reservoir routing					
Storage-indication (Puls) method	X	X		X	
Base flow					
Input	X		X		X
Constant value		X	X		X
Recession equation	X				
Snowmelt routine	Yes	No	No	No	Yes

- Comparison of model output of flooded areas with measured or observed flooded areas for historical storm events.
- Comparison of peak flow rates with those described in previous studies of similar watersheds in the region.
- Transposition of model parameters from nearby studies or other studies of similar topography.

Validation of runoff models is accomplished by applying the calibrated model to storms other than those used in the calibration.

Adjustments in parameters for calibration are achieved by making comparisons of measured values with the simulated volume of runoff, simulated peak flow rates, timing of changes in the hydrograph, and overall shape of the simulated hydrograph. Adjustments in initial abstraction and infiltration are usually made to match the overall runoff volume. For subwatersheds near the stream gauges, these parameters can significantly affect the rising limb of the simulated hydrograph. Adjustments in infiltration, synthetic unit hydrograph parameters, or channel routing parameters affect the rising and falling limb shape, timing, and magnitudes. Watershed lag times, reservoir or roadway detention, and lag parameters have the greatest effects on the timing of arrival of runoff from remote subwatersheds and on the simulated recession of the hydrograph.

SCS Project Formulation Hydrology (TR-20)

One of the earliest simulation models of hydrologic processes and water surface profiles was developed by CEIR, Inc. [27] and is known by the code name TR-20, which is an acronym for the U.S. Soil Conservation Service Technical Release Number 20. The model is a computer program of methods used by the Soil Conservation Service as presented in the *National Engineering Handbook* [28].

The program is recognized as an engineer-oriented rather than computer-oriented package, having been developed with ease of use as a purpose. Input data sheets and output data are designed for ease of use and interpretation by field engineers, and the program contains a liberal number of operations that are user-accommodating, even at the expense of machine time.

The TR-20 was designed to use soil and land-use information to determine runoff hydrographs for known storms and to perform reservoir and channel routing of the generated hydrographs. It is a single-event model, with no provision for additional losses or infiltration between discrete storm events. The program has been used in all 50 states by engineers for flood insurance and flood hazard studies, for the design of reservoir and channel projects, and for urban and rural watershed planning.

Surface runoff is computed from an historical or synthetic storm using the SCS curve number approach described in Chapter 7 to abstract losses. The standard dimensionless hydrograph shown in Fig. 9.18 is used to develop unit hydrographs for each subarea in the watershed. The excess rainfall hyetograph is constructed using the effective rain and a given rainfall distribution and is then applied incrementally to the unit hydrograph to obtain the subarea runoff hydrograph for the storm.

As shown in Table 12.4, TR-20 uses the storage-indication method to route hydrographs through reservoirs (see Section 9.5). The base flow is added to the direct runoff hydrographs at any time to produce the total flow rates. The program uses the logic depicted in Fig. 12.3 by computing the total flow hydrographs, routing the flows through stream channels and reservoirs, combining the routed hydrographs with those from other tributaries, and routing the combined hydrographs to the watershed outlet. Prior to 1983, the model routed stream inflow hydrographs by the convex method (see Ref. 3), which has since been replaced by the modified att-kin method (see Section 9.5). As many as 200 channel reaches and 99 reservoirs or floodwater-retarding structures can be accommodated in any single application of the model. To add to this capability, the program allows the concurrent input of up to nine different storms over the watershed area.

ARS Problem-Oriented Hydrologic Modeling (HYMO)

A unique computer language designed for use by hydrologists who have no conventional computer programming experience was developed by the Agricultural Research Service (Williams and Hann) [29]. Once the program has been compiled, the user forms a sequence of commands that synthesize, route, store, plot, or add hydrographs for subareas of any watershed. Seventeen commands are available to use in any sequence to transform rainfall data into runoff hydrographs and to route these hydrographs through streams and reservoirs. The HYMO model also computes the sediment yield in tons for individual storms on the watershed.

Watershed runoff hydrographs are computed by HYMO using unit-hydrograph techniques. Unit hydrographs can either be input or synthesized according to the dimensionless unit hydrograph shown in Fig. 12.5. Terms in the equations are

q = flow rate (cfs) at time t

q_p = peak flow rate (cfs)

t_p = time to peak (hr)

n = dimensionless shape parameter

q_0 = flow rate at the inflection point (cfs)

t_0 = time at the inflection point (hr)

K = recession constant (hr)

Once K, t_p, and q_p are known, the entire hydrograph can be computed from the three segment equations shown in Fig. 12.5. The peak flow rate is computed by the equation:

$$q_p = \frac{BAQ}{t_p} \tag{12.1}$$

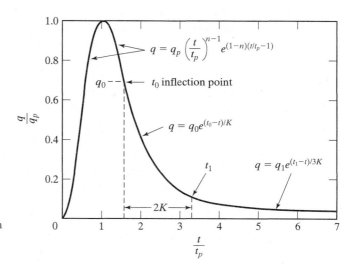

FIGURE 12.5

Dimensionless unit hydrograph used in HYMO.

(After Williams and Hann [29])

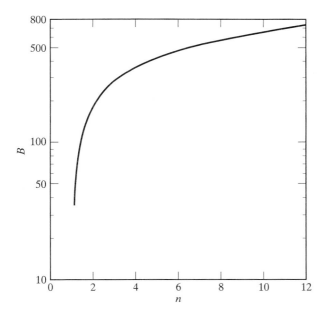

FIGURE 12.6

Relation between dimensionless shape parameter n and watershed parameter B. *(After Williams and Hann [29])*

where B = a watershed parameter, related to n as shown in Fig. 12.6

A = watershed area (mi^2)

Q = volume of runoff (in.), determined by HYMO from the SCS rainfall-runoff equation described in Chapter 7

The duration of the unit hydrograph is equated with the selected time increment. The runoff Q for the unit hydrograph would of course be 1.0 in. The parameter n in Fig. 12.6 is obtained from Fig. 12.7. Parameters K and t_p for ungauged watersheds are determined from regional regression equations based on 34 watersheds located in Texas, Oklahoma, Arkansas, Louisiana, Mississippi, and Tennessee, ranging in size from 0.5 to 25 mi^2, or:

$$K = 27.0A^{0.231}\text{SLP}^{-0.777}\left(\frac{L}{W}\right)^{0.124} \tag{12.2}$$

and

$$t_p = 4.63A^{0.422}\text{SLP}^{-0.46}\left(\frac{L}{W}\right)^{0.133} \tag{12.3}$$

where   SLP = the difference in elevation (ft), divided by floodplain distance (mi), between the basin outlet and the most distant point on the divide

L/W = the basin length/width ratio

River routing is accomplished in HYMO by a revised *variable storage coefficient* (VSC) method [30]. The continuity equation, $I - O = dS/dt$, and the storage equation,

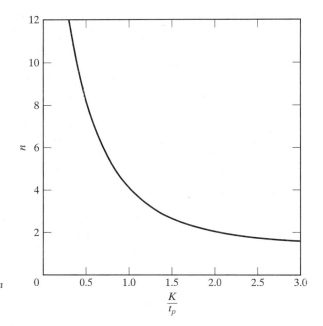

FIGURE 12.7

Relation between dimensionless shape parameter n and ratio of recession constant to time to peak. *(After Williams and Hann [29])*

$S = KO$, are combined and discretized according to the methods outlined in Chapter 9. The VSC method recognizes the variability in K as the flow leaves the confines of the stream channel and inundates the floodplain and valley area. Relations between K and O are determined by HYMO from the input cross-sectional data, or HYMO will calculate the relation using Manning's equation if the floodplain and channel roughness coefficients are specified. The bed slope and reach length are also part of the required input.

The widely adopted storage-indication method (see Chapter 9) is used to route inflow hydrographs through reservoirs. The storage-outflow curve must be determined externally by the user and is input to the program as a table containing paired storages and outflows, using storage defined as zero whenever the outflow is zero.

Output from HYMO includes the synthesized or user-provided unit hydrographs, the storm runoff hydrographs, the river- and reservoir-routed hydrographs, and the sediment yield for individual storms on each subwatershed. Hydrographs computed by HYMO compared closely with measured hydrographs from the 34 test watersheds.

Storm Water Management Model (SWMM)

The Environmental Protection Agency model, SWMM [31], is listed in Table 12.1 in two locations corresponding to rainfall-runoff event simulation and urban runoff simulation. The model is primarily an urban runoff simulation model and was described in Chapter 11.

Like most other models, SWMM has undergone numerous modifications and improvements since its first release in 1972. The initial version [31] was a single-event model, and newer versions [32],[33] allow its use in continuous modeling of urban storm water flows and water quality parameters. The latest release includes a new snowmelt routine, a new storm water storage and treatment package, a sediment scour and deposition routine, and a revised infiltration simulation.

SWMM's hydrograph and routing routines are hydraulic rather than hydrologic. A distributed-parameter approach is used for subcatchments consisiting of single parking lots, city lots, and so on. Accumulated rainfall on these plots is first routed as overland flow to gutter or storm drain inlets, where it is then routed as open or closed channel flow to the receiving waters or to some type of treatment facility. Of the five event-simulation models compared in Table 12.4, SWMM gives the greatest detail in simulation, but is not used in large rural watershed simulations.

Overland flow depths and flow rates are computed for each time step using Manning's equation along with the continuity equation. The water depth over a sub-catchment will increase without inducing an outflow until the depth reaches a specified detention requirement. If and whenever the resulting depth over the subcatchment, D_r, is larger than the specified detention requirement, D_d, an outflow rate is computed using a modified Manning's equation:

$$V = \frac{1.49}{n} (D_r - D_d)^{2/3} S^{1/2} \qquad (12.4)$$

and

$$Q_w = VW(D_r - D_d) \qquad (12.5)$$

where V = the velocity
n = Manning's coefficient
S = the ground slope
W = the width of the overland flow
Q_w = the outflow discharge rate

After flow depths and rates from all subcatchments have been computed, they are combined along with the flow from the immediate upstream gutter to form the total flow in each successive gutter.

The gutter and pipe flows are routed by the Manning and continuity equations to any points of interest in the network, where they are added to produce hydrograph ordinates for each time step in the routing process. The time step is advanced in incre-ments until the runoff from the storm is no longer being produced. The parameters of the gutter shape, slope, and length must be supplied by the user. Manning's roughness coefficients for the pipes or channels must also be supplied and are available in most hydraulics textbooks.

Other input required for a typical simulation with SWMM include the following:

1. Watershed characteristics such as the infiltration parameters, percent impervious area, slope, area, detention storage depth, and Manning's coefficients for over-land flow.
2. The rainfall hyetograph for the storm to be simulated.

3. The land-use data, average market values of dwellings in subareas, and populations of subareas.

4. Characteristics of gutters such as the geometry, slope, roughness coefficients, maximum allowable depths, and linkages with other connecting inlets or gutters.

5. Street-cleaning frequency.

6. Treatment devices selected and their sizes.

7. Indexes for costs of facilities.

8. Boundary conditions in the receiving waters.

9. Storage facilities, location, and volume.

10. Inlet characteristics such as surface elevations and invert elevations.

11. Characteristics of pipes such as type, geometry, slope, Manning's n, and downstream and upstream junction data.

HEC-1 Flood Hydrograph Package

The U.S. Army Corps of Engineers' Hydrologic Engineering Center developed the flood hydrograph package, HEC-1 [34]. The HEC-1 model consists of a calling program and six subroutines. Two of these subroutines determine the optimal unit hydrograph, loss rate, or streamflow routing parameters by matching recorded and simulated hydrograph values. The other subroutines perform snowmelt computations, unit-hydrograph computations, hydrograph routing and combining computations, and hydrograph balancing computations. In addition to being capable of simulating the usual rainfall-runoff event processes, HEC-1 will also simulate multiple floods for multiple basin development plans and perform the economic analysis of flood damages by numerically integrating areas under damage–frequency curves for existing and postdevelopment conditions.

HEC-1 underwent revisions in the early 1970s and again in the 1990s. Several features were added (e.g., SCS curve number method, hydraulic routing), and a PC version was developed in 1984. The 1985 release expanded earlier versions to include kinematic hydrograph routing, simulation of urban runoff, hydrograph analysis for flow over a dam or spillway, analysis of downstream impacts of dam failures, multistage pumping plants for interior drainage, and flood control system economics. The 1990 version of HEC-1, available for PCs or Harris minicomputers, incorporates yet other improvements. It adds report-quality graphic and table capability, storage and retrieval of data from other programs, and new hydrologic procedures including the popular Green–Ampt infiltration equation (Chapter 7) and the Muskingum–Cunge flood routing method (Chapter 9).

In addition to the unit-hydrograph techniques of the earlier versions, the modified HEC-1 allows hydrograph syntheses by kinematic-wave overland runoff techniques, similar to those developed for use in the SWMM. The runoff can either be concentrated at the outlet of the subarea or uniformly added along the watercourse length through the subarea, distributing the inflow to the channel or gutter in linearly increasing amounts in the downstream direction. The 1990 version allows the use of the Muskingum–Cunge routing method in a land surface runoff calculation mode.

Precipitation can be directly input, or one of three synthetic storms (refer to Chapter 13) can be selected. The model allows simulation of any storm duration from

5 min to 10 days. The user need only specify the desired duration and depth, and the program balances the depth around the central portion of the duration using the blocked IDF method of Section 13.4.

The later versions of HEC-1 include all the precipitation loss, synthetic unit hydrograph, and routing functions developed for earlier versions. Additional loss methods include both the SCS curve number method and Holtan's loss rate equation (an exponential decay function).

Because of the popularity of SCS techniques, the HEC-1 now includes TR-20 procedures for losses and hydrograph synthesis. The duration of the SCS dimensionless unit hydrograph is interpreted by HEC-1 as approximately 0.2 times the time to peak, but not exceeding 0.25 times the time to peak (this converts to 0.29 times the lag time).

For routing through streams and reservoirs, the newest version of HEC-1 includes all previous methods, and additionally performs kinematic-wave channel routing for several standard geometric cross section shapes.

In comparison to other event-simulation models, HEC-1 is relatively compact and still able to execute a variety of computational procedures in a single computer run. The model is applicable only to single-storm analysis because there is no provision for precipitation loss rate recovery during periods of little or no precipitation.

The program logic for HEC-1 is shown in Fig. 12.8. Hydrologic processes such as the subarea runoff computation, routing computation, hydrograph combining, subtracting diverted flow, balancing, comparing, or summarizing are specified in the input using the sequence illustrated in Fig. 12.3.

One loss rate in the HEC-1 model is an exponential decay function that depends on the rainfall intensity and the antecedent losses. The instantaneous loss rate, in in./hr, is:

$$L_t = K' P_t^E \qquad (12.6)$$

where
L_t = the instantaneous loss rate (in./hr)
P_t = the intensity of the rain (in./hr)
E = the exponent of recession (range of 0.5–0.9)
K' = a coefficient, decreasing with time as losses accumulate

K' is defined as follows:

$$K' = K_0 C^{-\text{CUML}/10} + \Delta K \qquad (12.7)$$

where
K_0 = the loss coefficient at the beginning of the storm (when CUML = 0), an average value of 0.6
CUML = the accumulated loss (in.) from the beginning of the storm to time t
C = a coefficient, an average of 3.0

If ΔK is zero, the loss rate coefficient K' becomes a parabolic function of the accumulated loss, CUML, and would thus plot as a straight-line function of CUML

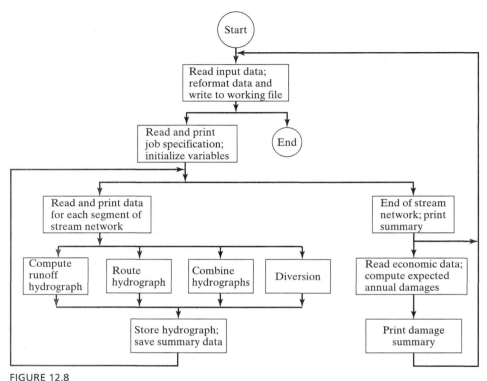

FIGURE 12.8

HEC-1 program operations overview [34].

on semilogarithmic graph paper if K were plotted on the logarithmic scale. The straight-line relation is depicted in Fig. 12.9, showing the decrease in the loss rate coefficient as the losses accumulate during any storm. Because loss rates typically decrease much more rapidly during the initial minutes of a storm, the loss rate K' is increased above the straight-line amount by an amount equal to ΔK, which in turn is made a function of the amount of losses that will accumulate before the K' value is again equal to the straight-line value, K. The initial accumulated loss, $CUML_1$, is user-specified. It is related to ΔK in such a fashion that the initial loss rate K' is $K_0 + 0.2\,CUML_1$ (see Fig. 12.9). Initial loss coefficients K_0 are difficult to estimate, and standard curves in Chapters 7 and 11 are available to determine initial infiltration rates, L_0. For gauged basins, HEC-1 allows the user to input rainfall and runoff data from which the loss rate parameters are optimized to give a best fit to the information provided. Estimates of parameters for ungauged basins fall in the judgment realm noted in Table 12.1. An alternative to the described loss rate function is available in HEC-1, which is an initial abstraction followed by a constant loss rate, similar to a ϕ index.

The HEC-1 model provides separate computations of snowmelt in up to 10 elevation zones. The precipitation in any zone is considered to be snow if the zone temperature is less than a base temperature, usually 32°F, plus 2°. The snowmelt is

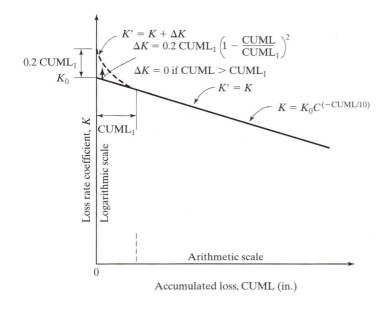

FIGURE 12.9

Variation of the loss rate coefficient K' with the accumulated loss amount CUML.

computed by the degree-day or energy budget methods whenever the temperature is equal to or greater than the base temperature. The elevation zones are usually considered in increments of 1,000 ft although any equal increments can be used.

Unit hydrographs for each subarea can be provided by the user, or Clark's method [35] of synthesizing an instantaneous unit hydrograph (IUH) can be used. Clark's method is more commonly recognized as the time–area curve method of hydrograph synthesis described in Section 9.4. The time–area histogram, determined from an isochronal map of the watershed, is convoluted with a unit design-storm hyetograph using Eq. 9.44, as illustrated in Fig. 9.20. The methods described in Section 9.4 are then used to route the resulting hydrograph through linear reservoir storage using Eq. 9.46 with a watershed storage coefficient K. Input data for Clark's method consists of the time–area curve ordinates, the time of concentration for the Clark unit graph, and the watershed storage coefficient K. If the time–area curve for the watershed under consideration is not available, the model provides a synthetic time–area curve at the user's request.

Because the Corps of Engineers commonly uses Snyder's method (see Chapter 9) in unit-hydrograph synthesis for large basins, the Snyder time lag from Eq. 9.24 and Snyder's peaking coefficient C_p from Eq. 9.28 can be input, and Clark's parameters will be determined by HEC-1 from the Snyder coefficients. The actual or synthetic time–area curve is still required.

Base flow is treated by HEC-1 as an exponential recession using an exponent of 0.1 in the following equation:

$$Q_2 = \frac{Q_1}{R^{0.1}} \tag{12.8}$$

where Q_1 = the flow rate at the beginning of the time increment

Q_2 = the flow rate at the end of the time increment

R = the ratio of the base flow to the base flow 10 time increments later

The base flow determined from this equation is added to the direct runoff hydrograph ordinates determined from unit-hydrograph techniques. The starting point for the entire computation is the user-prescribed base flow rate at the beginning of the simulation, which is normally the flow several time increments prior to any direct runoff. If the initial base flow rate is specified as zero, the computer program output contains only direct runoff rates.

The HEC-1 package allows the user a choice of several hydrologic or "storage-routing" techniques for routing floods through river reaches and reservoirs. All use the continuity equation and some form of the storage–outflow relation; some are described in more detail in Chapter 9. The five routing procedures included in the program are the following:

1. Modified Puls. This method is also called the storage or storage-indication method and is a level-pool-routing technique normally reserved for use with reservoirs or flat streams. The technique was described in detail in Section 9.5.
2. Muskingum. Described in detail in Section 9.5.
3. Muskingum–Cunge. A blended hydrologic and hydraulic routing method detailed in Section 9.5.
4. Kinematic wave. Described in Ref. 3.
5. Straddle-stagger. Also known as the progressive average lag method. The technique simply averages a subset of consecutive inflow rates, and the averaged inflow value is lagged a specified number of time increments to form the outflow rate [34].

Input to HEC-1 is facilitated by arranging three categories of data in a sequence compatible with the desired computation sequence, summarized in Table 12.5. The

TABLE 12.5   Subdivisions of Input Data for HEC-1

Job control	Hydrology and hydraulics	Economics and end of job
I_, Job initialization	K_, Job step control	E_, Economics data
V_, Variable output summary	H_, Hydrograph transformation	ZZ, End of Job
O_, Optimization	Q_, Hydrograph data	
J_, Job type	B_, Basin data	
	P_, Precipitation data	
	L_, Loss (infiltration) data	
	U_, Unit graph data	
	M_, Melt data	
	R_, Routing data	
	S_, Storage data	
	D_, Diversion data	
	W_, Pump withdrawal data	

Source: After Ref. 34

individual input records are preceded by a two-character code. The first character indicates the type of hydrologic or program operation (P_ for precipitation, U_ for unit hydrograph, etc.), and the second character is reserved for the suboption for the operation (PM for probable maximum storm, PI for IDF balanced storm, etc.).

Output from HEC-1 is both complete and descriptive. Options are available for graphical and tabular report-quality displays of intermediate or summary hydrographs or precipitation hyetographs [36]. The extent of output from HEC-1 is further illustrated in Example 12.1.

Example 12.1

In June 1963 the Oak Creek watershed shown in Fig. 12.10 experienced a severe flood-producing storm in a 6-hr period. Average excess rain depths over each of the nine sub-areas $A, B, \ldots, I$ ranged from 1.0 in. in the southwest to 7.8 in. in the north. Crest-stage flood records show that the storm produced peak flows of 27,500 cfs at Agnew (point 3) and 21,600 cfs at the watershed outlet (point 8). The map in Fig. 12.10 gives the sub-basin divides, reach lengths, and stream bed elevations (underlined).

A reservoir located at point 9 will store runoff from virtually any probable storm over subarea I. For the remaining elongated watershed area A–H within the boldface border, use the June storm to simulate hydrographs at each of points 1–8 using a single run of HEC-1, and compare peak flows with recorded values at points 3 and 8.

Solution.   The net storm depths shown in the table of Fig. 12.10 are applied uniformly over each subarea. The records provide the time distribution for the 12 successive half-hour periods of the thunderstorm in percent: 3, 5, 6, 9, 37, 10, 8, 6, 4, 5, 3, and 4 percent. The net rain was determined from the measured depths using the SCS runoff equation [28] and a basin-wide composite curve number of 73 (see Chapter 7).

The computation logic to simulate runoff for this storm consists of the following 22 steps:

1. Compute the hydrograph for area A at point 1.
2. Route the A hydrograph from point 1 to point 2.
3. Compute the hydrograph for area B at point 2.
4. Combine the two hydrographs at point 2.
5. Route the combined hydrograph to point 3.
6. Compute the hydrograph for area C at point 3.
7. Combine the two hydrographs at point 3.
8. Route the combined hydrograph to point 4.
9. Compute the hydrograph for area D at point 4.
10. Combine the two hydrographs at point 4.
11. Route the combined hydrograph to point 5.
12. Compute the hydrograph for area E at point 5.
13. Combine the two hydrographs at point 5.
14. Route the combined hydrograph to point 6.
15. Compute the hydrograph for area H at point 6.

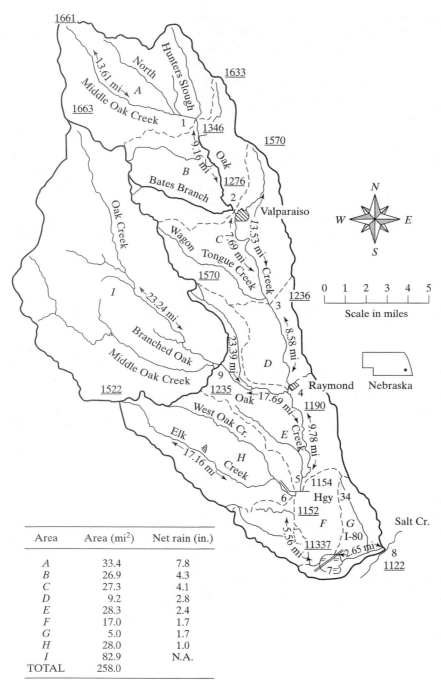

Area	Area (mi²)	Net rain (in.)
A	33.4	7.8
B	26.9	4.3
C	27.3	4.1
D	9.2	2.8
E	28.3	2.4
F	17.0	1.7
G	5.0	1.7
H	28.0	1.0
I	82.9	N.A.
TOTAL	258.0	

FIGURE 12.10

Oak Creek watershed subarea map and data sheet, June 1963.

16. Combine the two hydrographs at point 6.
17. Route the combined hydrograph to point 7.
18. Compute the hydrograph for area F at point 7.
19. Combine the two hydrographs at point 7.
20. Route the combined hydrograph to point 8.
21. Compute the hydrograph for area G at point 8.
22. Combine the two hydrographs at point 8.

Runoff hydrographs for subareas are simulated by convoluting the net storm hyetograph with unit hydrographs synthesized by Clark's method using Snyder's coefficients (see Chapter 9). A C_p value of 0.8 is applied for Oak Creek because of the moderately high retention capacity of the watershed. Subarea time lag values for each subarea are found from Eq. 9.24 using a C_t value of 2.0.

Hydrograph stream routing is performed using the Muskingum technique (Chapter 9) with $x = 0.15$ and K = the approximate reach travel time, using length divided by the average velocity. A Chézy average velocity determined as 100 times the square root of the average reach slope is used. If K exceeds three routing increments, the reach is further subdivided by HEC-1 into shorter lengths to ensure computational resolution.

A sample of the input and output for this job is shown as Table 12.6. Each of the 22 computational steps are separated in sequence. Only Steps 1–5 are included in the sample. Note that the HEC-1 loss rate function was not used so that the end-of-period excess and rain depths are equal. Note also that hydrograph routing of the A hydrograph from point 1 to point 2 was facilitated using three equal reach lengths each with a K-value (AMSKK) of 1.2 hr.

A summary of HEC-1 peak and time-averaged flow rates for each of Steps 1–22 is given in Table 12.7. Note that the simulated peak at point 3 is 27,539 cfs, which agrees very well with the recorded value of 27,500. The corresponding simulated and observed peak flows at point 8 are 22,453 and 21,600 cfs, respectively.

HEC-HMS Hydrologic Modeling System

The U.S. Army Corps of Engineers' Hydrologic Modeling System (HEC-HMS) [37], [38] supersedes HEC-1. Because it is primarily a graphical user interface (GUI) linked to an HEC-1 engine (with some improvements described as follows), the descriptions of the previously described HEC-1 hydrologic options and algorithms contained within HEC-HMS are not repeated here.

HEC-HMS is public domain software, and the most recent version can be downloaded from http://wrc-hec.usace.army.mil/software/software_distrib/index.html. Training in its use is provided at most universities and on a regular schedule of 2-day workshops sponsored by the American Society of Civil Engineers. The model uses object-oriented technology, which provides a graphical, building-block method of decomposing the complexity of a problem prior to implementing the code to solve the problem, stressing the objective of the study versus the procedures of solutions. The program provides a variety of options for simulating precipitation-runoff processes,

TABLE 12.6 HEC-1 Input and Partial Output Listing for Oak Creek Problem in Example 12.1

FLOOD HYDROGRAPH PACKAGE (HEC-1)
SEPTEMBER 1990
VERSION 4.0

U.S. ARMY CORPS OF ENGINEERS
HYDROLOGIC ENGINEERING CENTER
609 SECOND STREET
DAVIS, CALIFORNIA 95616
(916) 756-1104

RUN DATE 02/07/1994   TIME 12:10:01

PARTIAL HEC-1 INPUT FOR UPPER THIRD OF WATERSHED

LINE	CARD	VALUES	DESCRIPTION
1	ID		OAK CREEK WATERSHED STUDY
2	ID		TEST HEC1 USING STORM OF RECORD ON JUNE 23–24, 1963
3	ID		USE ACTUAL RAIN, SNYDER'S CP = 0.8, VEL = 100 ROOT S, AVG LAG EQ, SNYD, X = .15
4	IT	30 23 JUN 63 1400 100	
5	IO	0	
6	KK	1	COMPUTE HYDROGRAPH FOR AREA A
7	BA	33.4	
8	PB	7.8	
9	PI	.234 .234 .468 .702 2.886 .780 .624 .390	
10	PI	.312 .390	
11	US	2.93 0.8	
12	KK	2	ROUTE HYDROGRAPH FROM AREA A TO POINT 2
13	RM	3 0.15 3.6	
14	KK	2	COMPUTE HYDROGRAPH FOR AREA B
15	BA	26.9	
16	PB	4.3	
17	PI	.129 .129 .258 .387 1.591 .430 .344 .215	
18	PI	.172 .215	
19	US	2.51 0.8	
20	KK	2	ADD HYDROGRAPHS AT 2
21	HC	2	
22	KK	3	ROUTE COMBINED HYDROGRAPH TO POINT 3
23	RM	3 0.15 3.6	
24	KK	3	COMPUTE HYDROGRAPH FOR AREA C
25	BA	27.3	
26	PB	4.1	
27	PI	.123 .123 .246 .369 1.517 .410 .328 .205	
28	PI	.164 .205	
29	US	2.76 0.8	
30	KK	3	ADD HYDROGRAPHS AT 3
31	HC	2	

TABLE 12.6 (Continued)

6 KK | 1 | COMPUTE HYDROGRAPH FOR AREA A

7 BA   SUBBASIN RUNOFF DATA
            SUBBASIN CHARACTERISTICS
                TAREA   33.40   SUBBASIN AREA
            PRECIPITATION DATA

8 PB   STORM   7.80   BASIN TOTAL PRECIPITATION
9 PI   INCREMENTAL PRECIPITATION PATTERN
      .23   .39   .47   .70   .78   2.89   .62   .47   .31   .39
      .23   .31

11 US   SNYDER UNITGRAPH
                    TP   2.93   LAG
                    CP   .80   PEAKING COEFFICIENT
            SYNTHETIC ACCUMULATED-AREA VS. TIME CURVE WILL BE USED
                                        ***

APPROXIMATE CLARK COEFFICIENTS FROM GIVEN SNYDER CP AND TP ARE TC = 3.74 AND R = 1.46 INTERVALS

                    UNIT HYDROGRAPH PARAMETERS

                CLARK   TC = 3.74 HR,   R = 1.46 HR
                SNYDER TP = 2.92 HR,   CP = .79

                            UNIT HYDROGRAPH
                      20 END-OF-PERIOD ORDINATES
         437.   1546.   2928.   4319.   5397.   5864.   5704.   4815.   3546.   2507.
        1772.   1253.   885.   826.   442.   313.   221.   156.   110.   78.

                                                                            (Continued)

TABLE 12.6 (Continued)

HYDROGRAPH AT STATION 1

DA	MON	HRMN	ORD	RAIN	LOSS	EXCESS	COMP Q	DA	MON	HRMN	ORD	RAIN	LOSS	EXCESS	COMP Q
23	JUN	1400	1	.00	.00	.00	0.	24	JUN	1500	51	.00	.00	.00	0.
23	JUN	1430	2	.23	.00	.23	102.	24	JUN	1530	52	.00	.00	.00	0.
23	JUN	1500	3	.39	.00	.39	532.	24	JUN	1600	53	.00	.00	.00	0.
23	JUN	1530	4	.47	.00	.47	1493.	24	JUN	1630	54	.00	.00	.00	0.
23	JUN	1600	5	.70	.00	.70	3183.	24	JUN	1700	55	.00	.00	.00	0.
23	JUN	1630	6	2.89	.00	2.89	6665.	24	JUN	1730	56	.00	.00	.00	0.
23	JUN	1700	7	.78	.00	.78	12356.	24	JUN	1800	57	.00	.00	.00	0.
23	JUN	1730	8	.62	.00	.62	19107.	24	JUN	1830	58	.00	.00	.00	0.
23	JUN	1800	9	.47	.00	.47	25802.	24	JUN	1900	59	.00	.00	.00	0.
23	JUN	1830	10	.31	.00	.31	31125.	24	JUN	1930	60	.00	.00	.00	0.
23	JUN	1900	11	.39	.00	.39	34078.	24	JUN	2000	61	.00	.00	.00	0.
23	JUN	1930	12	.23	.00	.23	34475.	24	JUN	2030	62	.00	.00	.00	0.
23	JUN	2000	13	.31	.00	.31	32163.	24	JUN	2100	63	.00	.00	.00	0.
23	JUN	2030	14	.00	.00	.00	28114.	24	JUN	2130	64	.00	.00	.00	0.
23	JUN	2100	15	.00	.00	.00	23855.	24	JUN	2200	65	.00	.00	.00	0.
23	JUN	2130	16	.00	.00	.00	19854.	24	JUN	2230	66	.00	.00	.00	0.
23	JUN	2200	17	.00	.00	.00	16164.	24	JUN	2300	67	.00	.00	.00	0.
23	JUN	2230	18	.00	.00	.00	12780.	24	JUN	2330	68	.00	.00	.00	0.
23	JUN	2300	19	.00	.00	.00	9759.	25	JUN	0000	69	.00	.00	.00	0.
23	JUN	2330	20	.00	.00	.00	7176.	25	JUN	0030	70	.00	.00	.00	0.
24	JUN	0000	21	.00	.00	.00	5117.	25	JUN	0100	71	.00	.00	.00	0.
24	JUN	0030	22	.00	.00	.00	3605.	25	JUN	0130	72	.00	.00	.00	0.
24	JUN	0100	23	.00	.00	.00	2527.	25	JUN	0200	73	.00	.00	.00	0.
24	JUN	0130	24	.00	.00	.00	1760.	25	JUN	0230	74	.00	.00	.00	0.
24	JUN	0200	25	.00	.00	.00	1206.	25	JUN	0300	75	.00	.00	.00	0.
24	JUN	0230	26	.00	.00	.00	693.	25	JUN	0330	76	.00	.00	.00	0.
24	JUN	0300	27	.00	.00	.00	447.	25	JUN	0400	77	.00	.00	.00	0.
24	JUN	0330	28	.00	.00	.00	281.	25	JUN	0430	78	.00	.00	.00	0.
24	JUN	0400	29	.00	.00	.00	173.	25	JUN	0500	79	.00	.00	.00	0.
24	JUN	0430	30	.00	.00	.00	105.	25	JUN	0530	80	.00	.00	.00	0.
24	JUN	0500	31	.00	.00	.00	53.	25	JUN	0600	81	.00	.00	.00	0.
24	JUN	0530	32	.00	.00	.00	24.	25	JUN	0630	82	.00	.00	.00	0.
24	JUN	0600	33	.00	.00	.00	0.	25	JUN	0700	83	.00	.00	.00	0.

TOTAL RAINFALL = 7.80, TOTAL LOSS = .00, TOTAL EXCESS = 7.80

PEAK FLOW (CFS)	TIME (HR)
34475.	5.50

MAXIMUM AVERAGE FLOW

	6-HR	24-HR	72-HR	49.50-HR
(CFS)	24048.	6974.	3382.	3382.
(INCHES)	6.694	7.766	7.766	7.766
(AC-FT)	11925.	13834.	13834.	13834.

TABLE 12.6 (Continued)

| 12 | KK | 2 | | | ROUTE HYDROGRAPH FROM AREA A TO POINT 2 |
| 13 | RM | HYDROGRAPH ROUTING DATA | | | |

MUSKINGUM ROUTING

NSTPS	3	NUMBER OF SUBREACHES
AMSKK	3.60	MUSKINGUM K
X	.15	MUSKINGUM X

HYDROGRAPH AT STATION 2

DA	MON	HRMN	ORD	FLOW
23	JUN	1400	1	0.
23	JUN	1430	2	0.
23	JUN	1500	3	0.
23	JUN	1530	4	5.
23	JUN	1600	5	27.
23	JUN	1630	6	104.
23	JUN	1700	7	302.
23	JUN	1730	8	731.
23	JUN	1800	9	1576.
23	JUN	1830	10	3055.
23	JUN	1900	11	5313.
23	JUN	1930	12	8350.
23	JUN	2000	13	11974.
23	JUN	2030	14	15836.
23	JUN	2100	15	19499.
23	JUN	2130	16	22529.
23	JUN	2200	17	24612.
23	JUN	2230	18	25622.
23	JUN	2300	19	25601.
23	JUN	2330	20	24696.
24	JUN	0000	21	23101.
24	JUN	0030	22	21021.
24	JUN	0100	23	18651.
24	JUN	0130	24	16166.
24	JUN	0200	25	13714.
24	JUN	0230	26	11409.
24	JUN	0300	27	9324.
24	JUN	0330	28	7496.
24	JUN	0400	29	5931.
24	JUN	0430	30	4622.
24	JUN	0500	31	3552.
24	JUN	0530	32	2694.
24	JUN	0600	33	2019.
24	JUN	0630	34	1495.
24	JUN	0700	35	1094.
24	JUN	0730	36	791.
24	JUN	0800	37	566.
24	JUN	0830	38	401.
24	JUN	0900	39	281.
24	JUN	0930	40	195.
24	JUN	1000	41	135.
24	JUN	1030	42	92.
24	JUN	1100	43	63.
24	JUN	1130	44	43.
24	JUN	1200	45	29.
24	JUN	1230	46	19.
24	JUN	1300	47	13.
24	JUN	1330	48	8.
24	JUN	1400	49	6.
24	JUN	1430	50	4.
24	JUN	1500	51	2.
24	JUN	1530	52	2.
24	JUN	1600	53	1.
24	JUN	1630	54	1.
24	JUN	1700	55	0.
24	JUN	1730	56	0.
24	JUN	1800	57	0.
24	JUN	1830	58	0.
24	JUN	1900	59	0.
24	JUN	1930	60	0.
24	JUN	2000	61	0.
24	JUN	2030	62	0.
24	JUN	2100	63	0.
24	JUN	2130	64	0.
24	JUN	2200	65	0.
24	JUN	2230	66	0.
24	JUN	2300	67	0.
24	JUN	2330	68	0.
25	JUN	0000	69	0.
25	JUN	0030	70	0.
25	JUN	0100	71	0.
25	JUN	0130	72	0.
25	JUN	0200	73	0.
25	JUN	0230	74	0.
25	JUN	0300	75	0.
25	JUN	0330	76	0.
25	JUN	0400	77	0.
25	JUN	0430	78	0.
25	JUN	0500	79	0.
25	JUN	0530	80	0.
25	JUN	0600	81	0.
25	JUN	0630	82	0.
25	JUN	0700	83	0.
25	JUN	0730	84	0.
25	JUN	0800	85	0.
25	JUN	0830	86	0.
25	JUN	0900	87	0.
25	JUN	0930	88	0.
25	JUN	1000	89	0.
25	JUN	1030	90	0.
25	JUN	1100	91	0.
25	JUN	1130	92	0.
25	JUN	1200	93	0.
25	JUN	1230	94	0.
25	JUN	1300	95	0.
25	JUN	1330	96	0.
25	JUN	1400	97	0.
25	JUN	1430	98	0.
25	JUN	1500	99	0.
25	JUN	1530	100	0.

PEAK FLOW (CFS)	TIME (HR)
25622.	8.50

MAXIMUM AVERAGE FLOW

	6-HR	24-HR	72-HR	49.50-HR
+ (CFS)	20848.	6974.	3382.	3382.
+ (INCHES)	5.803	7.766	7.766	7.766
(AC-FT)	10338.	13833.	13834.	13834.

(Continued)

TABLE 12.6 (Continued)

14 KK | 2 | COMPUTE HYDROGRAPH FOR AREA B

15 BA SUBBASIN RUNOFF DATA
        SUBBASIN CHARACTERISTICS
        TAREA 26.90 SUBBASIN AREA
        PRECIPITATION DATA
16 PB STORM 4.30 BASIN TOTAL PRECIPITATION
17 PI INCREMENTAL PRECIPITATION PATTERN
        .13 .21 .26 .39 1.59 .43 .34 .26 .17 .21
        .13 .17

19 US SNYDER UNITGRAPH
        TP 2.51 LAG
        CP .80 PEAKING COEFFICIENT
        SYNTHETIC ACCUMULATED-AREA VS. TIME CURVE WILL BE USED
                                ***

APPROXIMATE CLARK COEFFICIENTS FROM GIVEN SNYDER CP AND TP ARE TC = 3.40 AND R = 1.06 INTERVALS
                        UNIT HYDROGRAPH PARAMETERS
                CLARK   TC = 3.40 HR,   R = 1.06 HR
                SNYDER   TP = 2.49 HR,   CP = .79

                        UNIT HYDROGRAPH
                    16 END-OF-PERIOD ORDINATES
        528. 1820. 3343. 4722. 5523. 5512. 4681. 3271. 2026. 1254.
        777.   481.   298.   184.   114.   71.

HYDROGRAPH AT STATION 2

DA	MON	HRMN	ORD	RAIN	LOSS	EXCESS	COMP Q
23	JUN	1400	1	.00	.00	.00	0.
23	JUN	1430	2	.13	.00	.13	68.
23	JUN	1500	3	.22	.00	.22	348.
23	JUN	1530	4	.26	.00	.26	959.
23	JUN	1600	5	.39	.00	.39	2002.
23	JUN	1630	6	1.59	.00	1.59	4135.
23	JUN	1700	7	.43	.00	.43	7534.
23	JUN	1730	8	.34	.00	.34	11324.
23	JUN	1800	9	.26	.00	.26	14700.
23	JUN	1830	10	.17	.00	.17	16833.
23	JUN	1900	11	.21	.00	.21	17310.
23	JUN	1930	12	.13	.00	.13	16128.
23	JUN	2000	13	.17	.00	.17	13732.
23	JUN	2030	14	.00	.00	.00	11199.
23	JUN	2100	15	.00	.00	.00	9031.
23	JUN	2130	16	.00	.00	.00	7148.
23	JUN	2200	17	.00	.00	.00	5479.
23	JUN	2230	18	.00	.00	.00	3991.
23	JUN	2300	19	.00	.00	.00	2728.
23	JUN	2330	20	.00	.00	.00	1742.
23	JUN	0000	21	.00	.00	.00	1062.
24	JUN	0030	22	.00	.00	.00	588.
24	JUN	0100	23	.00	.00	.00	345.
24	JUN	0130	24	.00	.00	.00	199.
24	JUN	0200	25	.00	.00	.00	112.
24	JUN	0230	26	.00	.00	.00	62.
24	JUN	0300	27	.00	.00	.00	29.
24	JUN	0330	28	.00	.00	.00	12.
24	JUN	0400	29	.00	.00	.00	0.
24	JUN	0430	30	.00	.00	.00	0.
24	JUN	0500	31	.00	.00	.00	0.
24	JUN	0530	32	.00	.00	.00	0.
24	JUN	0600	33	.00	.00	.00	0.

DA	MON	HRMN	ORD	RAIN	LOSS	EXCESS	COMP Q
24	JUN	1500	51	.00	.00	.00	0.
24	JUN	1530	52	.00	.00	.00	0.
24	JUN	1600	53	.00	.00	.00	0.
24	JUN	1630	54	.00	.00	.00	0.
24	JUN	1700	55	.00	.00	.00	0.
24	JUN	1730	56	.00	.00	.00	0.
24	JUN	1800	57	.00	.00	.00	0.
24	JUN	1830	58	.00	.00	.00	0.
24	JUN	1900	59	.00	.00	.00	0.
24	JUN	1930	60	.00	.00	.00	0.
24	JUN	2000	61	.00	.00	.00	0.
24	JUN	2030	62	.00	.00	.00	0.
24	JUN	2100	63	.00	.00	.00	0.
24	JUN	2130	64	.00	.00	.00	0.
24	JUN	2200	65	.00	.00	.00	0.
24	JUN	2230	66	.00	.00	.00	0.
24	JUN	2300	67	.00	.00	.00	0.
24	JUN	2330	68	.00	.00	.00	0.
24	JUN	0000	69	.00	.00	.00	0.
25	JUN	0030	70	.00	.00	.00	0.
25	JUN	0100	71	.00	.00	.00	0.
25	JUN	0130	72	.00	.00	.00	0.
25	JUN	0200	73	.00	.00	.00	0.
25	JUN	0230	74	.00	.00	.00	0.
25	JUN	0300	75	.00	.00	.00	0.
25	JUN	0330	76	.00	.00	.00	0.
25	JUN	0400	77	.00	.00	.00	0.
25	JUN	0430	78	.00	.00	.00	0.
25	JUN	0500	79	.00	.00	.00	0.
25	JUN	0530	80	.00	.00	.00	0.
25	JUN	0600	81	.00	.00	.00	0.
25	JUN	0630	82	.00	.00	.00	0.
25	JUN	0700	83	.00	.00	.00	0.

TOTAL RAINFALL = 4.30, TOTAL LOSS = .00, TOTAL EXCESS = 4.30

PEAK FLOW (CFS)	TIME (HR)	MAXIMUM AVERAGE FLOW			
			6-HR	24-HR	72-HR	49.50-HR

		6-HR	24-HR	72-HR	49.50-HR
17310.	5.00	(CFS)	11207.	3100.	1503.	1503.
		(INCHES)	3.873	4.286	4.286	4.286
		(AC-FT)	5557.	6149.	6149.	6149.

+

+

(Continued)

493

TABLE 12.6 (*Continued*)

```
20  KK          2            ADD HYDROGRAPHS AT 2
21  HC       HYDROGRAPH COMBINATION
                  ICOMP          2   NUMBER OF HYDROGRAPHS TO COMBINE
```

HYDROGRAPH AT STATION 2
SUM OF 2 HYDROGRAPHS

DA	MON	HRMN	ORD	FLOW	DA	MON	HRMN	ORD	FLOW	DA	MON	HRMN	ORD	FLOW	DA	MON	HRMN	ORD	FLOW.
23	JUN	1400	1	0.	24	JUN	0230	26	11471.	24	JUN	1500	51	2.	25	JUN	0330	76	0.
23	JUN	1430	2	68.	24	JUN	0300	27	9353.	24	JUN	1530	52	2.	25	JUN	0400	77	0.
23	JUN	1500	3	349.	24	JUN	0330	28	7508.	24	JUN	1600	53	1.	25	JUN	0430	78	0.
23	JUN	1530	4	963.	24	JUN	0400	29	5931.	24	JUN	1630	54	1.	25	JUN	0500	79	0.
23	JUN	1600	5	2029.	24	JUN	0430	30	4622.	24	JUN	1700	55	0.	25	JUN	0530	80	0.
23	JUN	1630	6	4239.	24	JUN	0500	31	3552.	24	JUN	1730	56	0.	25	JUN	0600	81	0.
23	JUN	1700	7	7835.	24	JUN	0530	32	2694.	24	JUN	1800	57	0.	25	JUN	0630	82	0.
23	JUN	1730	8	12055.	24	JUN	0600	33	2019.	24	JUN	1830	58	0.	25	JUN	0700	83	0.
23	JUN	1800	9	16277.	24	JUN	0630	34	1495.	24	JUN	1900	59	0.	25	JUN	0730	84	0.
23	JUN	1830	10	19888.	24	JUN	0700	35	1094.	24	JUN	1930	60	0.	25	JUN	0800	85	0.
23	JUN	1900	11	22624.	24	JUN	0730	36	791.	24	JUN	2000	61	0.	25	JUN	0830	86	0.
23	JUN	1930	12	24478.	24	JUN	0800	37	566.	24	JUN	2030	62	0.	25	JUN	0900	87	0.
23	JUN	2000	13	25706.	24	JUN	0830	38	401.	24	JUN	2100	63	0.	25	JUN	0930	88	0.
23	JUN	2030	14	27035.	24	JUN	0900	39	281.	24	JUN	2130	64	0.	25	JUN	1000	89	0.
23	JUN	2100	15	28530.	24	JUN	0930	40	195.	24	JUN	2200	65	0.	25	JUN	1030	90	0.
23	JUN	2130	16	29677.	24	JUN	1000	41	135.	24	JUN	2230	66	0.	25	JUN	1100	91	0.
23	JUN	2200	17	30092.	24	JUN	1030	42	92.	24	JUN	2300	67	0.	25	JUN	1130	92	0.
23	JUN	2230	18	29613.	24	JUN	1100	43	63.	24	JUN	2330	68	0.	25	JUN	1200	93	0.
23	JUN	2300	19	28328.	24	JUN	1130	44	43.	25	JUN	0000	69	0.	25	JUN	1230	94	0.
23	JUN	2330	20	26438.	24	JUN	1200	45	29.	25	JUN	0030	70	0.	25	JUN	1300	95	0.
24	JUN	0000	21	24163.	24	JUN	1230	46	19.	25	JUN	0100	71	0.	25	JUN	1330	96	0.
24	JUN	0030	22	21609.	24	JUN	1300	47	13.	25	JUN	0130	72	0.	25	JUN	1400	97	0.
24	JUN	0100	23	18996.	24	JUN	1330	48	8.	25	JUN	0200	73	0.	25	JUN	1430	98	0.
24	JUN	0130	24	16364.	24	JUN	1400	49	6.	25	JUN	0230	74	0.	25	JUN	1500	99	0.
24	JUN	0200	25	13826.	24	JUN	1430	50	4.	25	JUN	0300	75	0.	25	JUN	1530	100	0.

	PEAK FLOW	TIME		MAXIMUM AVERAGE FLOW			
+	(CFS)	(HR)		6-HR	24-HR	72-HR	49.50-HR
			(CFS)				
+	30092.	8.00		26453.	10074.	4885.	4885.
			(INCHES)	4.079	6.213	6.213	6.213
			(AC-FT)	13117.	19982.	19982.	19982.

TABLE 12.6 (*Continued*)

22	KK	*	3	*

ROUTE COMBINED HYDROGRAPH TO POINT 3

23	RM	HYDROGRAPH ROUTING DATA

MUSKINGUM ROUTING

NSTPS	3	NUMBER OF SUBREACHES
AMSKK	3.60	MUSKINGUM K
X	.15	MUSKINGUM X

HYDROGRAPH AT STATION 3

DA	MON	HRMN	ORD	FLOW	DA	MON	HRMN	ORD	FLOW	DA	MON	HRMN	ORD	FLOW	DA	MON	HRMN	ORD	FLOW
23	JUN	1400	1	0.	24	JUN	0230	26	25320.	24	JUN	1500	51	151.	25	JUN	0330	76	0.
23	JUN	1430	2	0.	24	JUN	0300	27	23988.	24	JUN	1530	52	110.	25	JUN	0400	77	0.
23	JUN	1500	3	0.	24	JUN	0330	28	22346.	24	JUN	1600	53	79.	25	JUN	0430	78	0.
23	JUN	1530	4	3.	24	JUN	0400	29	20482.	24	JUN	1630	54	57.	25	JUN	0500	79	0.
23	JUN	1600	5	18.	24	JUN	0430	30	18482.	24	JUN	1700	55	41.	25	JUN	0530	80	0.
23	JUN	1630	6	68.	24	JUN	0500	31	16428.	24	JUN	1730	56	29.	25	JUN	0600	81	0.
23	JUN	1700	7	195.	24	JUN	0530	32	14394.	24	JUN	1800	57	21.	25	JUN	0630	82	0.
23	JUN	1730	8	469.	24	JUN	0600	33	12438.	24	JUN	1830	58	15.	25	JUN	0700	83	0.
23	JUN	1800	9	1006.	24	JUN	0630	34	10608.	24	JUN	1900	59	10.	25	JUN	0730	84	0.
23	JUN	1830	10	1942.	24	JUN	0700	35	8933.	24	JUN	1930	60	7.	25	JUN	0800	85	0.
23	JUN	1900	11	3369.	24	JUN	0730	36	7432.	24	JUN	2000	61	5.	25	JUN	0830	86	0.
23	JUN	1930	12	5294.	24	JUN	0800	37	6113.	24	JUN	2030	62	3.	25	JUN	0900	87	0.
23	JUN	2000	13	7626.	24	JUN	0830	38	4973.	24	JUN	2100	63	2.	25	JUN	0930	88	0.
23	JUN	2030	14	10211.	24	JUN	0900	39	4003.	24	JUN	2130	64	2.	25	JUN	1000	89	0.
23	JUN	2100	15	12867.	24	JUN	0930	40	3189.	24	JUN	2200	65	1.	25	JUN	1030	90	0.
23	JUN	2130	16	15447.	24	JUN	1000	41	2517.	24	JUN	2230	66	1.	25	JUN	1100	91	0.
23	JUN	2200	17	17869.	24	JUN	1030	42	1968.	24	JUN	2300	67	1.	25	JUN	1130	92	0.
23	JUN	2230	18	20102.	24	JUN	1100	43	1525.	24	JUN	2330	68	0.	25	JUN	1200	93	0.
23	JUN	2300	19	22120.	24	JUN	1130	44	1172.	24	JUN	0000	69	0.	25	JUN	1230	94	0.
23	JUN	2330	20	23868.	24	JUN	1200	45	893.	24	JUN	0030	70	0.	25	JUN	1300	95	0.
24	JUN	0000	21	25273.	24	JUN	1230	46	676.	24	JUN	0100	71	0.	25	JUN	1330	96	0.
24	JUN	0030	22	26255.	24	JUN	1300	47	507.	24	JUN	0130	72	0.	25	JUN	1400	97	0.
24	JUN	0100	23	26759.	24	JUN	1330	48	378.	24	JUN	0200	73	0.	25	JUN	1430	98	0.
24	JUN	0130	24	26759.	24	JUN	1400	49	280.	24	JUN	0230	74	0.	25	JUN	1500	99	0.
24	JUN	0200	25	26265.	24	JUN	1430	50	206.	24	JUN	0300	75	0.	25	JUN	1530	100	0.

	PEAK FLOW	TIME			
	(CFS)	(HR)			
+	26759.	11.00			

MAXIMUM AVERAGE FLOW

		6-HR	24-HR	72-HR	49.50-HR
+	(CFS)	24061.	10070.	4885.	4885.
	(INCHES)	3.710	6.211	6.213	6.213
	(AC-FT)	11931.	19973.	19982.	19982.

(*Continued*)

TABLE 12.6 (Continued)

```
24 KK    3              COMPUTE HYDROGRAPH FOR AREA C

25 BA           SUBBASIN RUNOFF DATA
                SUBBASIN CHARACTERISTICS
                     TAREA      27.30      SUBBASIN AREA
                PRECIPITATION DATA
26 PB           STORM      4.10      BASIN TOTAL PRECIPITATION
27 PI           INCREMENTAL PRECIPITATION PATTERN
                  .12    .21    .25    .37   1.52    .41    .33    .25    .16    .20
                  .12    .16

29 US           SNYDER UNITGRAPH
                     TP      2.76      LAG
                     CP       .80      PEAKING COEFFICIENT
                SYNTHETIC ACCUMULATED-AREA VS. TIME CURVE WILL BE USED

                                        ***

APPROXIMATE CLARK COEFFICIENTS FROM GIVEN SNYDER CP AND TP ARE TC = 3.65 AND R = 1.30 INTERVALS
                        UNIT HYDROGRAPH PARAMETERS
              CLARK     TC = 3.65 HR,      R = 1.30  HR
              SNYDER    TP = 2.77 HR,      CP = .80

                            UNIT HYDROGRAPH
                     18 END-OF-PERIOD ORDINATES
       406.   1424.   2670.   3891.   4777.   5075.   4795.   3858.   2685.   1821.
      1234.    837.    567.    385.    261.    177.    120.     81.
```

TABLE 12.6 (*Continued*)

DA	MON	HRMN	ORD	RAIN	LOSS	EXCESS	COMP Q	DA	MON	HRMN	ORD	RAIN	LOSS	EXCESS	COMP Q
23	JUN	1400	1	.00	.00	.00	0.	24	JUN	1500	51	.00	.00	.00	0.
23	JUN	1430	2	.12	.00	.12	50.	24	JUN	1530	52	.00	.00	.00	0.
23	JUN	1500	3	.21	.00	.21	258.	24	JUN	1600	53	.00	.00	.00	0.
23	JUN	1530	4	.25	.00	.25	720.	24	JUN	1630	54	.00	.00	.00	0.
23	JUN	1600	5	.37	.00	.37	1526.	24	JUN	1700	55	.00	.00	.00	0.
23	JUN	1630	6	1.52	.00	1.52	3184.	24	JUN	1730	56	.00	.00	.00	0.
23	JUN	1700	7	.41	.00	.41	5873.	24	JUN	1800	57	.00	.00	.00	0.
23	JUN	1730	8	.33	.00	.33	9008.	24	JUN	1830	58	.00	.00	.00	0.
23	JUN	1800	9	.25	.00	.25	12033.	24	JUN	1900	59	.00	.00	.00	0.
23	JUN	1830	10	.16	.00	.16	14308.	24	JUN	1930	60	.00	.00	.00	0.
23	JUN	1900	11	.20	.00	.20	15400.	24	JUN	2000	61	.00	.00	.00	0.
23	JUN	1930	12	.12	.00	.12	15268.	24	JUN	2030	62	.00	.00	.00	0.
23	JUN	2000	13	.16	.00	.16	13880.	24	JUN	2100	63	.00	.00	.00	0.
23	JUN	2030	14	.00	.00	.00	11836.	24	JUN	2130	64	.00	.00	.00	0.
23	JUN	2100	15	.00	.00	.00	9861.	24	JUN	2200	65	.00	.00	.00	0.
23	JUN	2130	16	.00	.00	.00	8061.	24	JUN	2230	66	.00	.00	.00	0.
23	JUN	2200	17	.00	.00	.00	6436.	24	JUN	2300	67	.00	.00	.00	0.
23	JUN	2230	18	.00	.00	.00	4967.	24	JUN	2330	68	.00	.00	.00	0.
23	JUN	2300	19	.00	.00	.00	3679.	25	JUN	0000	69	.00	.00	.00	0.
23	JUN	2330	20	.00	.00	.00	2596.	25	JUN	0030	70	.00	.00	.00	0.
24	JUN	0000	21	.00	.00	.00	1760.	25	JUN	0100	71	.00	.00	.00	0.
24	JUN	0030	22	.00	.00	.00	1180.	25	JUN	0130	72	.00	.00	.00	0.
24	JUN	0100	23	.00	.00	.00	780.	25	JUN	0200	73	.00	.00	.00	0.
24	JUN	0130	24	.00	.00	.00	445.	25	JUN	0230	74	.00	.00	.00	0.
24	JUN	0200	25	.00	.00	.00	279.	25	JUN	0300	75	.00	.00	.00	0.
24	JUN	0230	26	.00	.00	.00	171.	25	JUN	0330	76	.00	.00	.00	0.
24	JUN	0300	27	.00	.00	.00	102.	25	JUN	0400	77	.00	.00	.00	0.
24	JUN	0330	28	.00	.00	.00	60.	25	JUN	0430	78	.00	.00	.00	0.
24	JUN	0400	29	.00	.00	.00	30.	25	JUN	0500	79	.00	.00	.00	0.
24	JUN	0430	30	.00	.00	.00	13.	25	JUN	0530	80	.00	.00	.00	0.
24	JUN	0500	31	.00	.00	.00	0.	25	JUN	0600	81	.00	.00	.00	0.
24	JUN	0530	32	.00	.00	.00	0.	25	JUN	0630	82	.00	.00	.00	0.
24	JUN	0600	33	.00	.00	.00	0.	25	JUN	0700	83	.00	.00	.00	0.

TOTAL RAINFALL = 4.10, TOTAL LOSS = .00, TOTAL EXCESS = 4.10

	PEAK FLOW (CFS)	TIME (HR)		MAXIMUM AVERAGE FLOW			
				6-HR	24-HR	72-HR	49.50-HR
+	15400.	5.00	(CFS)	10503.	2995.	1452.	1452.
			(INCHES)	3.577	4.080	4.080	4.080
+			(AC-FT)	5208.	5941.	5941.	5941.

(*Continued*)

TABLE 12.6 (Continued)

```
30 KK   [3]        ADD HYDROGRAPHS AT 3
31 HC        HYDROGRAPH COMBINATION
             ICOMP      2    NUMBER OF HYDROGRAPHS TO COMBINE

                    SUMMED HYDROGRAPH AT STATION   3
```

DA	MON	HRMN	ORD	FLOW	DA	MON	HRMN	ORD	FLOW	DA	MON	HRMN	ORD	FLOW	DA	MON	HRMN	ORD	FLOW
23	JUN	1400	1	0.	24	JUN	0230	26	25492.	24	JUN	1500	51	151.	25	JUN	0330	76	0.
23	JUN	1430	2	50.	24	JUN	0300	27	24090.	24	JUN	1530	52	110.	25	JUN	0400	77	0.
23	JUN	1500	3	259.	24	JUN	0330	28	22407.	24	JUN	1600	53	79.	25	JUN	0430	78	0.
23	JUN	1530	4	723.	24	JUN	0400	29	20512.	24	JUN	1630	54	57.	25	JUN	0500	79	0.
23	JUN	1600	5	1544.	24	JUN	0430	30	18495.	24	JUN	1700	55	41.	25	JUN	0530	80	0.
23	JUN	1630	6	3251.	24	JUN	0500	31	16428.	24	JUN	1730	56	29.	25	JUN	0600	81	0.
23	JUN	1700	7	6067.	24	JUN	0530	32	14394.	24	JUN	1800	57	21.	25	JUN	0630	82	0.
23	JUN	1730	8	9477.	24	JUN	0600	33	12438.	24	JUN	1830	58	15.	25	JUN	0700	83	0.
23	JUN	1800	9	13039.	24	JUN	0630	34	10608.	24	JUN	1900	59	10.	25	JUN	0730	84	0.
23	JUN	1830	10	16250.	24	JUN	0700	35	8933.	24	JUN	1930	60	7.	25	JUN	0800	85	0.
23	JUN	1900	11	18769.	24	JUN	0730	36	7432.	24	JUN	2000	61	5.	25	JUN	0830	86	0.
23	JUN	1930	12	20561.	24	JUN	0800	37	6113.	24	JUN	2030	62	3.	25	JUN	0900	87	0.
23	JUN	2000	13	21507.	24	JUN	0830	38	4973.	24	JUN	2100	63	2.	25	JUN	0930	88	0.
23	JUN	2030	14	22047.	24	JUN	0900	39	4003.	24	JUN	2130	64	2.	25	JUN	1000	89	0.
23	JUN	2100	15	22728.	24	JUN	0930	40	3189.	24	JUN	2200	65	1.	25	JUN	1030	90	0.
23	JUN	2130	16	23508.	24	JUN	1000	41	2517.	24	JUN	2230	66	1.	25	JUN	1100	91	0.
23	JUN	2200	17	24305.	24	JUN	1030	42	1968.	24	JUN	2300	67	1.	25	JUN	1130	92	0.
23	JUN	2230	18	25070.	24	JUN	1100	43	1525.	24	JUN	2330	68	0.	25	JUN	1200	93	0.
23	JUN	2300	19	25798.	24	JUN	1130	44	1172.	25	JUN	0000	69	0.	25	JUN	1230	94	0.
23	JUN	2330	20	26464.	24	JUN	1200	45	893.	25	JUN	0030	70	0.	25	JUN	1300	95	0.
24	JUN	0000	21	27033.	24	JUN	1230	46	676.	25	JUN	0100	71	0.	25	JUN	1330	96	0.
24	JUN	0030	22	27435.	24	JUN	1300	47	507.	25	JUN	0130	72	0.	25	JUN	1400	97	0.
24	JUN	0100	23	27539.	24	JUN	1330	48	378.	25	JUN	0200	73	0.	25	JUN	1430	98	0.
24	JUN	0130	24	27204.	24	JUN	1400	49	280.	25	JUN	0230	74	0.	25	JUN	1500	99	0.
24	JUN	0200	25	26545.	24	JUN	1430	50	206.	25	JUN	0300	75	0.	25	JUN	1530	100	0.

PEAK FLOW	TIME
(CFS)	(HR)
27539.	11.00

	MAXIMUM AVERAGE FLOW			
	6-HR	24-HR	72-HR	49.50-HR
+ (CFS)	25828.	13056.	6337.	6337.
+ (INCHES)	2.741	5.543	5.549	5.549
(AC-FT)	12807.	25897.	25923.	25923.

TABLE 12.7   Runoff Summary of Simulated Peak and Average Flows at Points 1 through 8 for the June 1963 Storm

Simulation	Node	Peak	6-hr	24-hr	72-hr	Area
Hydrograph at	1	34475.	24048.	6974.	3382.	33.40
Routed to	2	25622.	20848.	6974.	3382.	33.40
Hydrograph at	2	17310.	11207.	3100.	1503.	26.90
2 Combined	2	30092.	26453.	10074.	4885.	60.30
Routed to	3	26759.	24061.	10070.	4885.	60.30
Hydrograph at	3	15400.	10503.	2995.	1452.	27.30
2 Combined	3	27539.	25828.	13056.	6337.	87.60
Routed to	4	25912.	24551.	13052.	8409.	87.60
Hydrograph at	4	3362.	2382.	689.	441.	9.20
2 Combined	4	25925.	24606.	13702.	8850.	96.80
Routed to	5	23911.	22858.	13523.	8875.	96.80
Hydrograph at	5	7100.	5604.	1816.	1162.	28.30
2 Combined	5	23911.	22874.	14717.	10038.	125.10
Routed to	6	23911.	22874.	14717.	10039.	125.10
Hydrograph at	6	3349.	2478.	750.	480.	28.00
2 Combined	6	23911.	22874.	15268.	10518.	153.10
Routed to	7	22949.	22024.	14984.	10482.	153.10
Hydrograph at	7	4113.	2764.	773.	495.	17.00
2 Combined	7	22949.	22924.	15112.	10977.	170.10
Routed to	8	22453.	21595.	14994.	10908.	170.10
Hydrograph at	8	1684.	868.	228.	146.	5.00
2 Combined	8	22453.	21595.	14995.	11054.	175.10

including most of those available in previous Corps and SCS models. Table 12.8 lists the options available in HEC-HMS.

An interactive, multitasking, multiuser platform is operational in either X-Windows or Microsoft Windows environments. Components include a graphical user interface, integrated hydrologic analysis algorithms, data storage and management controls, and full graphics and reporting capabilities. The program has an interface to the Corps' Hydrologic Engineering Center decision-support system, HEC-DSS for storage of time series, paired function data, and gridded data.

Key new capabilities included are continuous hydrograph simulation over extended periods, a spatially distributed runoff computation algorithm using a grid-cell depiction of the watershed, a versatile parameter optimization routine, and graphical user interfaces linking the hydrologic process algorithms to GIS coverages [37]. Additional discussion is provided by Maidment [22].

The software uses a mix of programming languages and was designed for multi-platform usages, primarily PCs and workstations. Flood-damage assessment routines in HEC-1 have not been included in HEC-HMS, and instead are incorporated in a new, separate flood damage analysis package, HEC-FDA.

Though HEC-HMS allows continuous hydrograph simulation through a soil-moisture accounting routine that continues to be under development, its primary use is in single-event simulation. All computations are performed in metric units, but input and output can be in either English or metric units.

TABLE 12.8   Hydrologic Element Options Available in HEC-HMS

Hydrologic elements	Options in HEC-HMS
Losses	Initial/constant
	Deficit/constant (soil moisture accounting)
	Green–Ampt
	SCS curve no.
Transformation of rainfall to runoff	Modified Clark
	Kinematic wave gridded
	Snyder unit hydrograph
	Clark unit hydrograph
	SCS dimensionless unit hydrograph
	Input unit hydrograph
Base flow	Exponential recession
Routing	Lag
	Muskingum
	Modified Puls
	Muskingum–Cunge
Precipitation	Grid-based (radar)
	Average grid-based
	Import hyetograph
	Gauge data with weights
	Inverse distance gauge weighting
	Frequency-based blocked IDF design storm
	Standard project storm (eastern U.S. only)

The river basin network is configured in a dendritic pattern by selecting and connecting icons that represent the elements and by graphically selecting their interconnections. System setup and editing are accomplished by clicking and dragging icons that represent various hydrologic elements, including subbasins, routing reaches, junctions, uncontrolled reservoirs, diversions, and sources and sinks. Digital quadrangle maps, digital orthoquadrangle maps, digital elevation models (DEMs), and standard GIS procedures can be used to provide background mapping and automatically develop subbasin, hydrograph, and abstraction input values such as drainage areas, channel lengths and slopes, centroidal distances, times of concentration, curve numbers, and a number of other standard measures.

Data from an existing HEC-1 file can be imported, or pop-up menus can be invoked by clicking on the elements, allowing the user to select the hydrologic method and input parameters and data enabling simulation for that element. The GUI also allows global input of data of a given type for all applicable elements. Computations proceed in an upstream-to-downstream sequence.

The execution of a run of HEC-HMS is accomplished through construction of three data sets. The *basin model* contains connectivity, parameters, and data for the hydrologic elements (subbasins, routing reaches, reservoirs, and junctions). The *precipitation model* incorporates meteorological data and parameters for processing the data. The third set, the *control specifications,* provides temporal information. Several sets of data sets can be input, allowing more than one run.

Basin Model   Runoff is computed by either lumped (subbasin) methods similar to HEC-1, or in a new gridded approach. Rainfall and losses are uniform within each grid cell or subbasin, but can vary across cells or subbasins. Options for losses include the initial/constant loss rate, SCS CN methods, gridded SCS CNs, and the Green–Ampt equation (Chapter 7). Unit hydrographs for the subbasin approach can be in tabular form, or can be constructed by inputting parameters for the Clark, Snyder, or SCS methods (Chapter 9). Runoff for the gridded approach is computed from each cell by the kinematic-wave method, where the user breaks each cell into two rectangular over-land flow planes and the model routes the runoff through a central collector channel by either the kinematic-wave or Muskingum–Cunge method.

Alternatively, a new modified Clark method can be selected for runoff calculations within either a subbasin or a gridded approach. In either case, cells are superimposed on the basin and rainfall excess computed from each cell is translated to the basin outlet on a distance-to-outlet and time-of-travel basis and then combined and routed through a linear reservoir to reflect the storage effect. Selections of parameters for time-of-travel and storage effect are described in Section 9.4.

Channel routing can be by the Muskingum, modified Puls, kinematic-wave, Muskingum–Cunge, or dynamic-wave, unsteady-flow routing methods, described in Section 9.5. Standard geometric cross-sectional shapes may be used, or an 8-point cross section can be input representing irregular-shaped sections. If streamflow spills over the subbasin divide or is intentionally diverted, these diversions are modeled by inputting the relationship between inflow rate and diversion rate. These diversions can be added back to the system at any downstream point.

Precipitation Model   The temporal distribution of either a historical or design storm (Section 13.4) is input as either a total rain depth and distribution histogram or directly as the desired distribution of the precipitation amounts in each time step. Standard agency distributions (SCS, Corps) can be specified if only the depth is known. Frequency-based design storms are input using the blocked IDF method described in Section 13.4. Spatial distribution of the precipitation can be cell-by-cell, subbasin-by-subbasin, by rain gauge location and weighting (such as Thiessen polygons), or by an automated inverse distance-squared weighting method spread among points of known values.

Control Specifications   This segment incorporates user-supplied control information such as the clock time and starting date of the rainfall, actual hydrographs for comparisons with simulated graphs, time interval for computation, and input/output specifications. If measured hydrographs are available, the parameter optimization routine can be selected to generate best-fit estimates of the subbasin and routing-reach parameters.

Popularity of Single-Event Models

Hagen completed a survey of the current use of single-event hydrologic models in estimating flood flow rates and flood hydrographs for small (less than 30 mi^2) watershed analysis and design [24]. The results, shown in Table 12.9, reveal that over 60 percent of the reported 21,000 applications surveyed involved two SCS models, and the rational method was used in almost 20 percent of the applications. As noted earlier, the ease of use

TABLE 12.9   1994 Survey of U.S. Uses of Hydrologic Methods in Urban Hydrology

Hydrologic method	No. of studies	Percent
TR-55	10,763	51.3
Rational	4,054	19.5
TR-20	1,954	9.3
USGS rural regression equations	1,265	6.0
USGS urban regression equations	529	2.5
FHWA small rural watersheds	500	2.4
HEC-1	485	2.3
log–Pearson Type III	360	1.7
SCS hand methods	219	1.0
Synthetic flood frequency	155	0.7
Other runoff hydrograph models	138	0.7
Miscellaneous	553	2.6
TOTALS:	20,975	100.0

Source: After Hagen [24]

and widespread availability of SCS storm-event simulation models make them the most popular and most utilized procedures, especially in urban applications. Though these models are popular, users of any vendor-developed software that emulates SCS procedures are advised to review the SCS *National Engineering Handbook* [28] and other references [3],[27] in order to accurately apply the procedures to urban or rural watersheds.

Hagen [24] also presents data regarding uses by state highway departments in sizing bridge and culvert openings, and by the Federal Emergency Management Agency (FEMA) contractors in assessing the 100-yr flood flow rates required for completing flood insurance studies. The state highway department survey results are shown in Table 12.10, and the FEMA study survey is summarized in Table 12.11. FEMA's Web site (www.fema.gov) lists techniques that have been approved for hydrologic and hydraulic analysis of the nation's floodways and floodplains, showing that a number of other techniques are accepted. Both tables reveal that practitioners dealing with larger, rural watersheds normally encountered in bridge design and floodplain analysis do not adopt the SCS procedures as readily as analysts working in small, urban watersheds.

TABLE 12.10   1990 Survey of Uses of Hydrologic Methods by State Highway Departments

Hydrologic method	No. of states	Percent of states
Regional regression equations (USGS)	36	80
Rational	19	42
Bulletin 17B	8	18
Other (used by at least one state)	7	16
TR-55	6	13
TR-20	4	9
HEC-1	3	7
FHWA Hydrology Manual	3	7

Source: After Hagen [24]

Source: After Hagen [24]

TABLE 12.11   1993 Survey of Hydrologic Methods Used by FEMA
Contractors

Hydrologic method	No. of studies	Percent
HEC-1	34	25
Regional regression equations (USGS)	28	21
Bulletin 17B	25	18
TR-20	23	17
Rational	10	7
TR-55	8	6
Other (used by at least one contractor)	8	6
TOTALS:	136	100

12.3   CONTINUOUS SIMULATION MODELS

Simulation models described in this section provide hydrologists with tools for estimating streamflow by *continuously* accounting in time for precipitation, direct runoff,
infiltration, evapotranspiration, interflow, deep percolation, base flow, and streamflow.
During rain-free intervals between storms, continuous simulation models track the
storage of water and its depletion to evaporation, deep percolation, and base flow, until
the next rain or snow event occurs.

The models are based on the physical processes described in Chapters 1–11. As
such, they are classified as deterministic tools. Stochastic modeling procedures
described in Section 12.5 can be used to generate synthetic streamflows without simulating the physical processes involved in converting rain and snow into runoff and
streamflow. Alternatively, the methods in Section 12.5, or other similar procedures, can
be used to synthesize precipitation sequences, which are then input to continuous simulation models and converted to streamflow.

The Stanford watershed model, Version IV (SWM-IV), is presented in detail in
the following section as typical of the other models. Many of the others are, in fact,
based on SWM-IV, and several simulate various components of the hydrologic cycle in
the same manner. Reference 3 describes all of the models and presents and compares
two independent case studies of Stanford model studies, showing how the parameters
were determined and how the models were calibrated and applied to the problems
being assessed.

Stanford Watershed Model IV (SWM-IV)

Crawford and Linsley designed this digital computer program to simulate portions
(the land phase) of the hydrologic cycle for an entire watershed [39]. The model has
undergone much development since its conception and is currently available from the
U.S. Environmental Protection Agency under the name HSPF, which is a public
domain FORTRAN version (discussed subsequently) of the original program. The
SWM-IV has been widely accepted as a tool to synthesize a continuous hydrograph of
hourly or daily streamflows at a watershed outlet. A lumped-parameter approach is

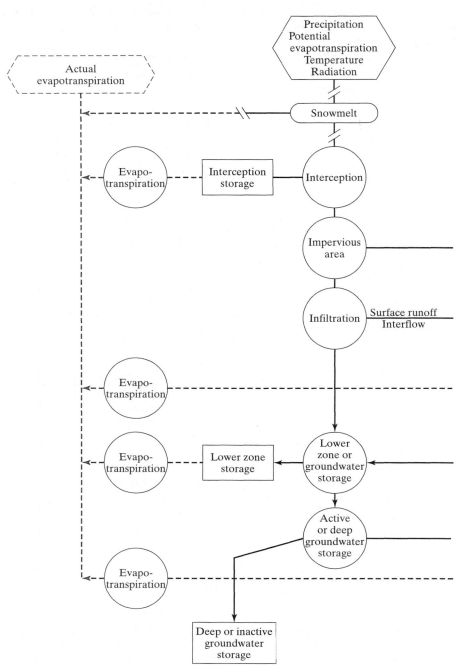

FIGURE 12.11

Stanford watershed model IV flow chart.
(After Crawford and Linsley [39])

used, and data requirements are much less than for alternative distributed models. Hourly and daily precipitation data, daily evaporation data, and a variety of watershed parameters are input.

The relations and linkage of the various components of SWM-IV are shown in Fig. 12.11. Hydrologic fundamentals are used at each point to transform the input data

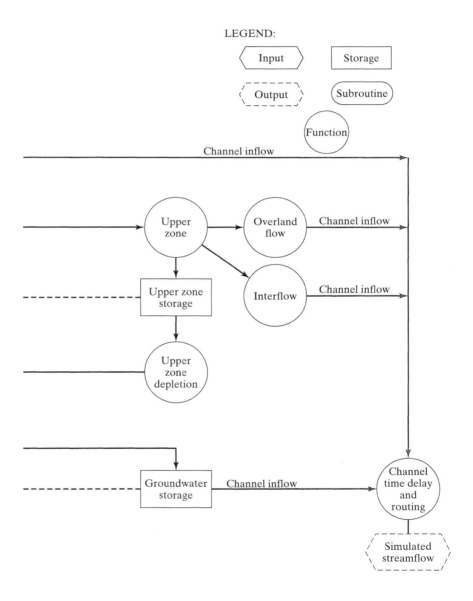

FIGURE 12.11   (*Continued*)

into a hydrograph of streamflow at the basin outlet. Rainfall and evaporation data are first entered into the program. Incoming rainfall is distributed, as shown in Fig. 12.11, among interception, impervious areas such as lakes and streams, and water destined to be infiltrated or to appear in the upper zone as surface runoff or interflow, both of which contribute to the channel inflow. The infiltration and upper zone storage eventually percolate to lower zone storage and to active and inactive groundwater storage. User-assigned parameters govern the rate of water movement between the storage zones shown in Fig. 12.11.

Three zones of moisture regulate soil moisture profiles and groundwater conditions. The rapid runoff response encountered in smaller watersheds is accounted for in the upper zone, while both upper and lower zones control such factors as overland flow, infiltration, and groundwater storage. The lower zone is responsible for longer-term infiltration and groundwater storage that is later released as base flow to the stream. The total streamflow is a combination of overland flow, groundwater flow, and interflow.

Model Structure   The SWM-IV is made up of a sequence of computation routines for each process in the hydrologic cycle (interception, infiltration, routing, and so on). Separate discussions of each component are provided in the following paragraphs. Calculations proceed from process to process as illustrated by the arrows in Fig. 12.11. All the moisture that was originally stored in the watershed or was input as precipitation during any time period is balanced in the continuity equation:

$$P = E + R + \Delta S \tag{12.9}$$

where P = precipitation
E = evapotranspiration
R = runoff
ΔS = the total change in storage in the upper, lower, and groundwater storage zones

The change in storage for each zone is calculated as the difference between the volumes of inflow and outflow. Furthermore, all hydrologic activity in a time interval is simulated and balanced before the program proceeds to the next time interval. The simulation terminates when no additional data are input.

Interception   Interception is the first of several abstractions modeled by the SWM-IV. All incoming precipitation is intercepted unless the precipitation intensity exceeds the interception rate or if the interception storage fills. Interception rates depend on the precipitation rate and on the watershed cover. Typical values of interception maximums are provided in Table 12.12.

Evapotranspiration   In SWM-IV evapotranspiration (ET) is assumed to occur at the potential rate from interception storage and the "upper" storage zone. The upper zone simulates the depressions and highly permeable surface soils. The lower soil zone simulates the linkage to the groundwater storage zone.

TABLE 12.12   Typical Maximum Interception Rates

Watershed cover	Interception rate (in./hr)
Grassland	0.10
Moderate forest cover	0.15
Heavy forest cover	0.20

Source: After Crawford and Linsley [39]

   Evapotranspiration from the lower zone is set equal to the *ET opportunity*, defined in Fig. 12.12. ET opportunity is defined as the maximum amount of water available for ET at a particular location during a prescribed time interval. In the modeling logic, ET occurs from several locations (see Fig. 12.11), including the interception storage, upper zone storage, lower zone storage, stream and lake surfaces, and groundwater storage. Evapotranspiration from interception and upper zone storage is set equal to the potential rate, E_p, which is assumed to be the lake evaporation rate, calculated as the product of a pan coefficient times the input values of the evaporation pan data. The evaporation of any intercepted water is assumed to occur at a rate equal to the potential evapotranspiration rate and ceases when the interception storage has been depleted.

   Evaporation from stream and lake surfaces also occurs at the potential rate. The total volume is governed by the total surface area of streams and lakes (ETL), defined as the ratio of the total stream and lake area in the watershed to the total watershed area. Evapotranspiration from groundwater storage also occurs at the potential rate and is calculated in a similar fashion using a surface area equal to a factor K24EL multiplied by the watershed area. Thus the parameter K24EL represents the fraction of the total watershed area over which evapotranspiration from the groundwater storage will occur. Most investigators set this parameter at a value equal to the fraction of the watershed area covered by phreatophytes. Its value is normally small but can be large, for example, in an agricultural area that has many acres of subirrigated alfalfa.

   If interception storage is depleted, the model will attempt to satisfy the potential for ET by drawing from the upper zone storage at the potential rate. Once the upper zone storage is depleted, ET occurs from the lower zone but not at the potential rate; the ET rate from the lower zone is always less than E_p. When interception and the

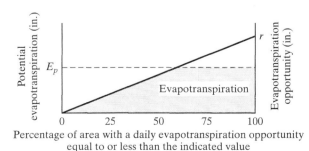

FIGURE 12.12

Evapotranspiration relation used in the Stanford watershed model.
(After Crawford and Linsley [39])

upper zone storage do not satisfy the potential, any excess enters as E_p in Fig. 12.12, and the rate of evapotranspiration from the lower zone is determined from the shaded area, or:

$$E = E_p - \frac{E_p^2}{2r} \tag{12.10}$$

The variable r is the evapotranspiration opportunity, defined as the maximum water amount available for ET at a particular location during a prescribed time period. This factor varies from point to point over any watershed from zero to a maximum value of:

$$r = K3 \frac{\text{LZS}}{\text{LZSN}} \tag{12.11}$$

where   LZS = the current soil moisture storage in the lower zone (in.)

   LZSN = a nominal storage level, normally set equal to the median value of the lower zone storage (in.)

   K3 = an input parameter that is a function of watershed cover as shown in Table 12.13

The ratio LZS/LZSN is known as the lower zone soil moisture ratio and is used to compare the actual lower zone storage with the nominal value at any time. Values of ET opportunity are assumed to vary over a watershed from zero to r along the straight line shown in Fig. 12.12. This assumed linear cumulative distribution of the parameter over an area is also used in evaluating areal distributions of infiltration rates.

Infiltration   Like the evapotranspiration opportunity, the infiltration capacity of a watershed is highly variable from point to point and is assumed to be distributed according to a linear cumulative distribution function shown as a line from the origin to point b in Fig. 12.13.

Infiltration into the lower and groundwater storage zones is determined as a function of the moisture supply $\bar{x}$ available for infiltration. Steps to determine infiltration for a given moisture supply $\bar{x}$ are:

1. The net infiltration is determined from the area labeled *infiltration* in Fig. 12.13. This water is assumed to infiltrate into the lower and groundwater storage zones. The area enclosed by the trapezoid is given by the equations in the first

TABLE 12.13   Typical Lower Zone
                Evapotranspiration Parameters

Watershed cover	K3
Open land	0.20
Grassland	0.23
Light forest	0.28
Heavy forest	0.30

Source: Crawford and Linsley [39]

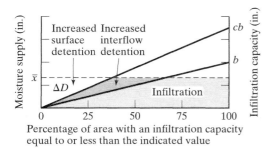

FIGURE 12.13

Assumed linear areal variation of infiltration capacity over a watershed.

(After Crawford and Linsley [39])

row of Table 12.14. If the moisture supply $\bar{x}$ exceeds the maximum infiltration capacity b, the maximum allowed net infiltration is $b/2$, which is the median infiltration capacity.

2. Some of the moisture supply contributes to an increase in the interflow detention during any time increment and is calculated as the region indicated by an arrow in Fig. 12.13. Equations for this area using various ranges of $\bar{x}$ are provided in the second row of Table 12.14. The volume of water in a state of being transported as interflow at any instant is called the *interflow detention* or *detained interflow*.

3. Any remaining moisture supplied, ΔD in Fig. 12.13, contributes to increasing the surface detention during the time increment. Equations for this triangular-shaped area are included in Table 12.14 for various values of $\bar{x}$.

The quantity of net infiltration is largely controlled by the maximum infiltration capacity b, while the parameter c significantly affects hydrograph shapes because this parameter controls the amount of water detained during the time increment. The values of b and c for any time interval depend on the soil moisture ratio, LZS/LZSN, and on the input parameters CB and CC. CB is an index that controls the rate of infiltration and depends on the soil permeability and the volume of moisture that can be stored in the soil. Values in the range from 0.3 to 1.2 are common. The parameter CC is

TABLE 12.14   Equations for the Shaded Areas in Fig. 12.13

Component	$\bar{x} < b$	$b < \bar{x} < cb$	$\bar{x} > cb$
Net infiltration	$\bar{x} - \dfrac{\bar{x}^2}{2b}$	$\dfrac{b}{2}$	$\dfrac{b}{2}$
Increase in interflow detention	$\dfrac{\bar{x}^2}{2b}\left(1 - \dfrac{1}{c}\right)$	$\bar{x} - \dfrac{b}{2} - \dfrac{\bar{x}^2}{2cb}$	$\dfrac{b}{2}(c - 1)$
Increase in surface detention	$\dfrac{\bar{x}^2}{2cb}$	$\dfrac{\bar{x}^2}{2cb}$	$\bar{x} - \dfrac{cb}{2}$
Percentage of increased detention assigned to interflow	$100\left(1 - \dfrac{1}{c}\right)$	$100\left(1 - \dfrac{\bar{x}^2}{2cb(\bar{x} - b/2)}\right)$	$100\,\dfrac{c - 1}{2\bar{x}/b - 1}$

Source: After Crawford and Linsley [39]

an input value that fixes the level of interflow relative to the overland flow. Values of CC range from 1.0 to 5.0.

If the soil moisture ratio is less than 1.0, the variable b is found from:

$$b = \frac{CB}{2^{(4LZS/LZSN)}} \tag{12.12}$$

and when LZS/LZSN is greater than 1.0, the equation for b is:

$$b = \frac{CB}{2^{\{4.0+2[(LZS/LZSN)-1.0]\}}} \tag{12.13}$$

These equations were developed by Crawford and Linsley from numerous trials using SWM-IV in many different watersheds. When the soil moisture ratio reaches a value of 2.0, the variable b reaches its minimum value of $\frac{1}{64}$ of CB. The parameter c is determined from:

$$c = (CC)2^{(LZS/LZSN)} \tag{12.14}$$

Variations in parameters b and c with changes in LZS/LZSN are shown in Figs. 12.14 and 12.15. Midrange values of CB = 1.0 and CC = 1.0 were used in developing these curves.

Figure 12.16 is a graph of distribution of water among infiltration, interflow, and overland flow for various values of the moisture supply $\bar{x}$. Different values of b and c would produce a different set of curves.

Water stored as overland flow surface detention will either contribute to streamflow or enter the upper zone storage as depicted in Fig. 12.11. The portion that enters the upper zone storage is called *delayed infiltration* and is a function of the upper zone soil moisture ratio, UZS/UZSN, as shown in Fig. 12.17. The inflection point occurs at a

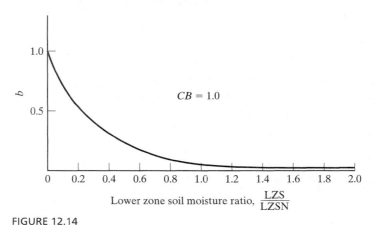

FIGURE 12.14

Variation in parameter b for various values of the soil moisture ratio. *(After Crawford and Linsley [39])*

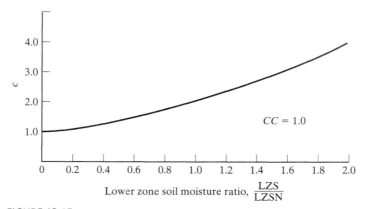

FIGURE 12.15

Variation in parameter c for various values of the soil moisture ratio.
(After Crawford and Linsley [39])

soil moisture ratio of 2.0. If the ratio is less than 2.0, the percentage retained by the upper zone is given by:

$$P_r = 100\left[1.0 - \left(\frac{\text{UZS}}{2\text{UZSN}}\right)\left(\frac{1.0}{1.0 + \text{UZI1}}\right)^{\text{UZI1}}\right] \tag{12.15}$$

where UZI1 is determined from:

$$\text{UZI1} = 2.0\left[\frac{\text{UZS}}{2\text{UZSN}} - 1.0\right] + 1.0 \tag{12.16}$$

The curve is defined to the right of the inflection point by:

$$P_r = 100\left[\left(\frac{1.0}{1.0 + \text{UZI2}}\right)^{\text{UZI2}}\right] \tag{12.17}$$

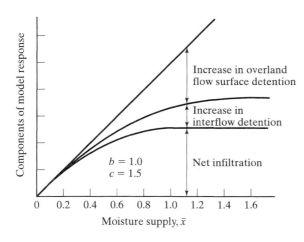

FIGURE 12.16

Typical SWM-IV response to moisture supply variations.
(After Crawford and Linsley [39])

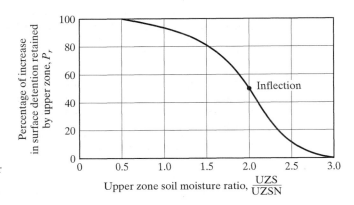

FIGURE 12.17

Delayed infiltration as a function of upper
zone soil moisture ratio.
(After Crawford and Linsley [39])

where UZI2 is determined from:

$$UZI2 = 2.0\left[\frac{UZS}{UZSN} - 2.0\right] + 1.0 \tag{12.18}$$

Upper Zone Storage   The upper storage zone, as shown in Fig. 12.11, receives a large
portion of the rain during the first few hours of the storm, while the lower and groundwater storage zones may or may not receive any moisture. The portion of the upper
zone storage that is not evaporated or transpired is proportioned to the surface runoff,
interflow, and percolation. Percolation (upper zone depletion) from the upper zone to
the lower zone in Fig. 12.11 ocurs only when UZS/UZSN exceeds LZS/LZSN. When
this occurs, the percolation rate in in./hr is determined from:

$$PERC = 0.003(CB)(UZSN)\left(\frac{UZS}{UZSN} - \frac{LZS}{LZSN}\right)^3 \tag{12.19}$$

where CB is an index that controls the rate of infiltration. This index ranges from 0.3 to
1.2 depending on the soil permeability and on the volume of moisture that can be
stored in the soil. The variables UZS and UZSN are defined as the actual and nominal
soil moisture storage amounts in the upper zone. The nominal value of UZSN is
approximately a function of watershed topography and cover and is always considered
to be much smaller than the nominal LZSN value. The initial estimates of UZSN relative to LZSN are found from Table 12.15.

The parameters LZSN and CB must also be estimated at the beginning of a simulation study. The combination that will most satisfactorily reproduce both long- and
short-term historical responses to hydrologic inputs can be determined by the following procedure [39]:

1. Assume an initial value for LZSN equal to one-quarter of the mean annual rainfall plus 4 in. (used in arid and semiarid regions), or one-eighth of the annual
   mean rainfall plus 4 in. (used in coastal, humid, or subhumid climates).
2. Determine the initial value of UZSN from Table 12.15.

TABLE 12.15   Values of UZSN as a Function of LZSN for Initial Estimates in
              Simulation with SWM-IV

Watershed	UZSN
Steep slopes, limited vegetation, low depression storage	0.06LZSN
Moderate slopes, moderate vegetation, moderate depression storage	0.08LZSN
Heavy vegetal or forest cover, soils subject to cracking, high depression storage, very mild slopes	0.14LZSN

Source: After Crawford and Linsley [39]

3. Assume a value for CB in the normal range from 0.3 to 1.2.

4. Simulate a period of record using the streamflow, rainfall, and evaporation data and systematically adjust LZSN, UZSN, CB, and other parameters until agreement between synthesized and recorded streamflows is satisfactory. If the annual water budgets do not balance, LZSN is adjusted; CB is adjusted on the basis of comparisons between synthesized and recorded flow rates for individual storms.

Lower Zone Storage and Groundwater   The lower groundwater storage zone in Fig. 12.11 receives water from the net infiltration and from percolation. The percolation rate is determined from Eq. 12.19. The percentage of net infiltration that reaches groundwater storage depends on the soil moisture ratio LZS/LZSN, as shown in Fig. 12.18. If this ratio is less than 1.0, the percentage P_g is found from:

$$P_g = 100\left[\frac{LZS}{LZSN}\left(\frac{1.0}{1.0 + LZI}\right)^{LZI}\right]$$ (12.20)

and if LZS/LZSN is greater than 1.0, the percentage is:

$$P_g = 100\left[1.0 - \left(\frac{1.0}{1.0 + LZI}\right)^{LZI}\right]$$ (12.21)

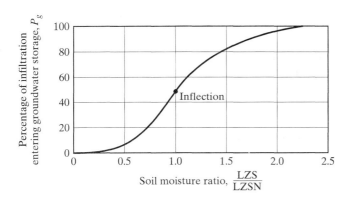

FIGURE 12.18

Percentage of infiltrated water that reaches groundwater storage.
(After Crawford and Linsley [39])

In both equations, the variable LZI is defined as:

$$LZI = 1.5\left[\frac{LZS}{LZSN} - 1.0\right] + 1.0 \tag{12.22}$$

Note from Fig. 12.18 that the nominal storage LZSN equals the lower zone storage LZS when 50 percent of all the incoming moisture enters groundwater storage.

The outflow from the groundwater storage, GWF, at any time is based on the commonly used linear semilogarithmic plot of base flow discharge versus time. This technique was described in Section 9.1 and illustrated in Fig. 9.7. In modified form the base flow equation is:

$$GWF = (LKK4)[1.0 + KV(GWS)](SGW) \tag{12.23}$$

where LKK4 is defined by:

$$LKK4 = 1.0 - (KK24)^{1/96} \tag{12.24}$$

in which KK24 is the minimum of all the observed daily recession constants (see Section 9.1), where each constant is the ratio of the groundwater discharge rate to the groundwater discharge rate 24 hr earlier. Thus the recession constant KK24 (K in Eq. 9.1) is determined using $t = 1$ day. The variable GWS in Eq. 12.23 has values that depend on the long-term inflows to groundwater storage. Its value on any given day (e.g., the ith day) is calculated as 97 percent of the previous day's value, adjusted for any inflow to groundwater storage, or $GWS_i = 0.97$ (GWS_{i-1} + inflow to groundwater storage during day i).

In Eq. 12.23, SGW is a groundwater storage parameter that reflects the fluctuations in the volume of water stored and ranges from 0.10 to 3.90 in. The term KV in Eq. 12.23 allows for changes that are known to exist in the groundwater recession rates as time passes. When KV is zero, Eq. 12.23 reduces to Eq. 9.1, and the groundwater recession follows the linear semilog relation. If the usual dry season recession rate KK24 is too large for wet periods (when groundwater storages are being recharged by seepage from the streams), the parameter KV is hand-adjusted so that the term $1.0 + KV(GWS)$ will reduce the effective rate to some desired value during recharge periods. Table 12.16 illustrates this computation by showing effective recession rates for various combinations of KK24 and GWS when KV is set equal to 1.0.

The fraction of active or deep groundwater storage that is either lost to deep or inactive groundwater storage (Fig. 12.11) or is diverted as flow across the drainage basin boundary is input as parameter K24L. This fraction is the total inflow to groundwater and represents all the active groundwater storage that does not contribute to streamflow.

Overland Flow   The overland flow process has been studied by many investigators. A wide range of methods for estimating the velocities and depths of sheet flow over a land surface has been applied and falls in the hydraulic-hydrologic categories of Chapter 9. Hydraulic overland flow methods involve finite-difference and other numerical techniques to solve at various points the partial differential equations of

TABLE 12.16   Effective Recession Rates for
Various Combinations of KK24
and GWS when KV = 1.0

	GWS			
KK24	0	0.5	1.0	2.0
0.99	0.99	0.985	0.98	0.97
0.98	0.98	0.970	0.96	0.94
0.97	0.97	0.955	0.94	0.91
0.96	0.96	0.940	0.92	0.88

Source: After Crawford and Linsley [39]

continuity and momentum for unsteady overland flow. The hydrologic methods, including those adopted in SWM-IV, approximate the velocities and depths for unsteady overland flows by a lumped-parameter approach that requires much less data than the hydraulic techniques.

Average values of lengths, slopes, and roughnesses of overland flow in the Manning and continuity equations are used in SWM-IV to continuously calculate the surface detention storage D_e. The overland flow discharge rate q is then related to D_e.

As the rain supply rate continues in time, the amount of water detained on the surface increases until an equilibrium depth is established. The amount of surface detention at equilibrium estimated by SWM-IV is:

$$D_e = \frac{0.000818 i^{0.6} n^{0.6} L^{1.6}}{S^{0.3}} \tag{12.25}$$

where D_e = the surface detention at equilibrium (ft^3/ft of overland flow width)
   i = the rain rate (in./hr)
   S = the slope (ft/ft)
   L = the length of overland flow (ft)
   n = Manning's roughness coefficient

The overland flow discharge rate is next determined as a function of detention storage from:

$$q = \frac{1.486}{n} S^{1/2} \left(\frac{D}{L}\right)^{5/3} \left[1.0 + 0.6\left(\frac{D}{D_e}\right)^3\right]^{5/3} \tag{12.26}$$

where q = the overland flow discharge rate (cfs per ft of width)
   D = the average detention storage during the time interval (ft^3/ft)

The equation also applies during the recession that occurs after rain ceases, but the ratio D/D_e is assumed to be 1.0. Typical overland flow roughness coefficients are provided in Table 12.17.

TABLE 12.17   Typical Manning Equation Overland Flow Roughness
Parameters

Watershed cover	Manning's n for overland flow
Smooth asphalt	0.012
Asphalt or concrete paving	0.014
Packed clay	0.03
Light turf	0.20
Dense turf	0.35
Dense shrubbery and forest litter	0.40

Source: After Crawford and Linsley [39]

The time at which detention storage reaches an equilibrium is determined from:

$$t_e = \frac{0.94 L^{3/5} n^{3/5}}{i^{2/5} S^{3/10}}$$
(12.27)

where t_e is the time to equilibrium (min). Crawford and Linsley show that these equations very accurately reproduce measured overland flow hydrographs [39].

For each time interval Δt, an end-of-interval surface detention D_2 is calculated from the initial value D_1 plus any water added ΔD (see Fig. 12.13) to surface detention storage during the time interval, less any overland flow discharge $\overline{q}$ that escapes from detention storage during the time interval. This is simply an expression of continuity, or:

$$D_2 = D_1 + \Delta D - \overline{q}\,\Delta t$$
(12.28)

The discharge $\overline{q}$ is found from Eq. 12.26 using a value of $D = (D_1 + D_2)/2$. Equations 12.25–12.28 allow the complete determination of overland flow using easily found basin-wide values of the average length, slope, and roughness overland flow.

Interflow   The water temporarily detained as interflow storage is treated in the same fashion as overland flow detention storage. The inflow to interflow detention was defined in Fig. 12.13. The outflow is simulated using a daily recession constant similar to that defined for groundwater discharge. The interflow recession constant IRC is the average ratio of the interflow discharge at any time to the interflow discharge 24 hr earlier. For each 15-min time interval modeled, the outflow from detention storage is:

$$\text{INTF} = \text{LIFC4(SRGX)}$$
(12.29)

where

$$\text{LIFC4} = 1.0 - (\text{IRC})^{1/96}$$
(12.30)

The variable SRGX is the water stored in the interflow detention at any time. Its value continuously changes when the continuity equation is applied to each time interval. The end-of-interval value of SRGX depends, according to continuity, on the value at the beginning of the interval and any inflow to or discharge from the interflow detention during the interval.

Channel Translation and Routing   The Stanford watershed model utilizes a hydrologic watershed routing technique to translate the channel inflow to the watershed outlet. Clark's IUH time–area method described in Section 9.3 is adopted almost as presented in Chapter 9. In place of the net rain hyetograph, the Stanford model views the sum of all channel inflow components as an "inflow" hyetograph. This inflow is then translated in time through the channel to the basin outlet, where it is next routed through an equivalent storage system to account for the attenuation caused by storage in the channel system. Routing through the linear reservoir (linear in the sense that storage is assumed to be directly proportional to the outflow, Eq. 9.44) is accomplished from:

$$O_2 = \bar{I} - KS1(\bar{I} - O_1) \tag{12.31}$$

where O_2 = the outflow rate at the end of the time interval
   O_1 = the outflow rate at the beginning of the time interval
   $\bar{I}$ = the average inflow rate during the time interval

Also:

$$KS1 = \frac{K - \Delta t/2}{K + \Delta t/2} \tag{12.32}$$

Examples of the determination of K and other necessary parameters from watershed data are included in Section 9.2.

Applications of the SWM-IV   Applications of this model typically begin with data for a 3- to 6-year calibration period for which rainfall and runoff data are available. These data are used to allow successive adjustments of several parameters until the simulated and recorded hydrographs of the streamflow agree. If sufficient data are available, a second period of record may be reserved for use as a control to check the accuracy of the parameters derived from a calibration with the first half of the data.

Hydrocomp Simulation Program—FORTRAN (HSPF)

The Stanford watershed model was originally developed in 1959 and has undergone several modifications since that time. James [40] translated the Crawford and Linsley model from ALGOL to FORTRAN. Several modifications of the FORTRAN version have evolved from a variety of investigations. Included among these are the Kentucky watershed model (KWM) [40],[41]; the Kentucky self-calibrating version (OPSET) [42]; the Ohio State University version; the Texas version [43]; the Hydrocomp Simulation Program (HSP) written in PL/1; the EPA-produced, nonproprietary FORTRAN version of HSP called HSPF; and the National Weather Service runoff forecasting model. Brief descriptions of several of these are included in the following paragraphs. The U.S. Geological Survey maintains updates of the HSPF model. A download of the program HSPF is available at http://water.usgs.gov/software/surface_water.html.

HSPF has been used in hundreds of applications ranging in size from a few acres to over 60,000 square miles. The physical process algorithms are similar to SWM-IV but were enhanced in the 1980s, and improvements continue to be made, including user-friendly pre- and postprocessing routines. In addition to continuous streamflow simulation, HSPF simulates water quality constituents including dissolved oxygen, biochemical oxygen demand (BOD), fecal coliforms, sediment detachment and transport, temperature, pesticides, pH, ammonia, organic nitrogen, organic phosphorus, phytoplankton, zooplankton, and nitrite-nitrate.

The model is often used to simulate effects of land-use change, snowmelt runoff and forecasting, reservoir operations, effects of point or nonpoint source treatment options, and soil erosion. Time steps of any increment up to 1 day can be specified. To simulate streamflow, the model requires two time series, precipitation and estimates of potential evapotranspiration. For snowmelt simulation, additional inputs include air temperature, dew point temperature, wind speed, and solar radiation. Water-quality simulation requires the streamflow and snowmelt simulation inputs plus humidity, cloud cover, tillage practices, point sources, and areal pesticide applications. Training in HSPF is offered annually at various universities, by some private consulting firms, and by the EPA and USGS.

Expert Systems

Recent trends in systems analysis are leading to development of *expert system* (ES) techniques that rely on *artificial intelligence* for use in planning and design of water resources projects [44],[45]. A computer can be given information obtained from extensive interviews of one or more experts in some field. The computer can then make decisions in much the same way as the experts, applying their judgment and experience and making these available to others through the expert system model.

Streamflow models, especially those that perform continuous simulation, incorporate input parameters that require considerable judgment. Developers and users of the watershed models have accumulated decades of experience in assigning coefficients and parameters. Their experience and judgment can be extracted by an interview process involving hundreds of questions to build an expert system model. The model not only incorporates direct answers but also addresses uncertainties about each. Early applications with this modeling technique show considerable promise [45].

In addition to streamflow simulation, expert systems have the potential to be useful in the design and management of complex river basin systems of dams, reservoirs, power plants, diversion canals, and flood control structures. Operations for such systems involve independent and collective decisions by dozens of professionals. These experts are normally in radio or telephone contact with numerous other controllers and decision-makers. If ES data could be developed from these teams, the potential for improved management exists. A prime incentive of implementing expert systems in water resources systems involves capturing insights of experienced professionals before they retire or move into other positions.

The U.S. Geological Survey maintains updates of expert system software for calibrating HSPF using information compiled from interviews with the developer of SWM-IV. A download of the program HSPEXP is available at http://water.usgs.gov/software/surface_water.html.

12.4 GROUNDWATER FLOW SIMULATION MODELS

Several popular public domain computer codes for solving various types of groundwater flow problems are listed in Table 12.18. The codes become models when the system being studied is described to the code by inputting the system geometry and known internal operandi (aquifer and flow field parameters, initial and boundary conditions, and water use and flow stresses applied in time to all or parts of the system). Codes have emerged in four general categories: *groundwater flow* codes, *solute transport* codes, *particle tracking* codes, and *aquifer test data analysis* programs [46].

Groundwater flow codes provide the user with the distribution of heads in an aquifer that would result from a simulated set of distributed recharge-discharge stresses at cells or line segments. From Darcy's law, the flow passing any two points can

TABLE 12.18 Groundwater Modeling Codes

Acronym for code	Description	Source	Year
Groundwater flow models			
PLASM	Two-dimensional finite-difference	Ill. SWS	1971
MODFLOW	Three-dimensional finite-difference	USGS	1988
AQUIFEM-1	Two- and three-dimensional finite-element	MIT	1979
GWFLOW	Package of 7 analytical solutions	IGWMC	1975
GWSIM-II	Storage and movement model	TDWR	1981
GWFL3D	Three-dimensional finite-difference	TDWR	1991
MODRET	Seepage from retention ponds	USGS	1992
Solute transport models			
SUTRA	Dissolved substance transport model	USGS	1980
RANDOMWALK	Two-dimensional transient model	Ill. SWS	1981
MT3D	Three-dimensional solute transport	EPA	1990
AT123D	Analytical solution package	DOE	1981
MOC	Two-dimensional solute transport	USGS	1978
HST3D	3-D heat and solute transport model	USGS	1992
Particle tracking models			
FLOWPATH	Two-dimensional steady-state	SSG	1990
PATH3D	Three-dimensional transient solutions	Wisc GS	1989
MODPATH	Three-dimensional transient solutions	USGS	1991
WHPA	Analytical solution package	EPA	1990
Aquifer test analyses			
TECTYPE	Pump and slug test by curve matching	SSG	1988
PUMPTEST	Pumping and slug test	IGWMC	1980
THCVFIT	Pumping and slug test	IGWMC	1989
TGUESS	Specific capacity determination	IGWMC	1990

Note: IGWMC = International Groundwater Modeling Center; Ill. SWS = Illinois State Water Survey; SSG = Scientific Software Group; EPA = Environmental Protection Agency; USGS = U.S. Geological Survey; Wisc. GS = Wisconsin Geological Survey; MIT = Massachusetts Institute of Technology; TDWR = Texas Department of Water Resources; DOE = Department of Energy.

be calculated from the head differential. The codes are used to model both confined and unconfined aquifers. Each can be structured to model regional flow, or flow in proximity of a single well or well field. Steady-state and transient conditions can be evaluated. Boundaries can be barriers, full or partially penetrating streams and lakes, leaky zones, or constant head or constant gradient perimeters. By application of Darcy's law, the seepage velocities of groundwater can be determined after solving for the head differentials.

When groundwater seepage velocities are known, the advection, dispersion, and changes in concentration of solutes can be modeled. Solute transport models build on groundwater flow models by the addition of advection, dispersion, and/or chemical reaction equations. If the chemical, dispersion, or dilution concentration changes due to groundwater flow are not important, particle tracking codes model transport by advection and provide an easier method than solute transport models to track the path and travel times of solutes that move under the influence of head differentials. Aquifer test data programs provide users with computer solutions to many of the hand calculations (see Chapter 10) needed to graph and interpret aquifer test data for determining aquifer and well parameters.

Solution Techniques

With few exceptions, the hydrodynamic equations for groundwater flow have no analytical solutions, and groundwater modeling relies on *finite-difference* and *finite-element* methods to provide approximate solutions to a wide variety of groundwater problems. The choice of method is normally driven by the system to be modeled. Other numerical methods include *boundary integral* methods, *integrated finite-difference* methods, and *analytic element* methods.

These solutions, as with streamflow simulation models, are facilitated by first subdividing the region to be modeled into subareas. Groundwater system subdivision depends more on geometric criteria and less on topographic criteria in the sense that the region is overlaid by a regular or semiregular pattern of node points at which (or between which) specific measures of aquifer and water system parameters are input and other parameters are calculated. Approximate solutions of simultaneous linear and nonlinear equations are found by making initial estimates of the solution values, testing the estimates in the equations of motion and continuity, adjusting the values, and finally accepting minor violations in the basic principles or making further adjustments of the parameters in an orderly and converging fashion.

The orderly solution of finite-difference analogs of the steady-state or unsteady-state partial differential equation of motion for flow of groundwater in a confined aquifer or an unconfined aquifer is obtained by *relaxation* methods. An early relaxation solution of the equation is discussed by Jacob [47]. For two-dimensional problems, the iterative alternating-direction-implicit (ADI) method developed by Peaceman and Rachford [48] is often adopted.

Prickett and Lonnquist [49] used the ADI technique to calculate fluctuations in water table elevations at all nodes in an aquifer model by proceeding through time in small increments from a known initial state. Their model is computationally efficient and readily applied and is particularly attractive for use with problems involving time variables and numerous nodes. The primary aquifer parameters are the permeability and storage coefficient, which, if assumed constant over the aquifer plan, result in a homogeneous and isotropic condition. For those familiar with relaxation methods, the Gauss–Seidel and the successive overrelaxation (SOR) methods have had application in solving difference equations.

Data Requirements

Input to groundwater system models may be classified as spatial and temporal. Spatial input includes initial or projected water table maps, saturated thickness data over the region, land surface contour maps, transmissivity maps, regional variations in storage coefficients, locations and types of wells and canals, locations and types of aquifer boundaries both lateral and vertical, a node coordinate system, actual or net pumpage rates, percolation and recharge rates for precipitation and other applied waters, logs of drilled wells, geologic stratigraphy, and soil types and cropping patterns.

Time-dependent data requirements for aquifer models principally involve the formulation of time schedules, using a range of time increments for such variables as pumping rates, precipitation hyetographs, canal and streamflow hydrographs, ground-water evapotranspiration rates, and development variables such as the timing of added wells or other system components. Because each temporal schedule can apply only to a particular subset of node positions, the time-dependent requirements are also spatial.

In addition to the listed input parameters, aquifer models require reliable estimates of the percentages of waters in the land phase that actually percolate to the aquifer being modeled. These estimates can be based on knowledge of the physical processes involved in unsaturated flow through porous medium but are most often obtained as judgment parameters that are modified during the calibration phase of the simulation. Simply stated, the lateral movement and the changes in piezometer or water table levels are easily modeled if the node-by-node stresses (withdrawal rates or recharge rates) are known. The latter parameters are governed by the complex movement of water in the unsaturated soil zone and by the random precipitation and consumptive use patterns of the region. The art of modeling groundwater systems lies in the ability to evaluate these parameters.

12.5   STREAMFLOW SYNTHESIS

Time-series analysis of hydrologic variables has become a practical methodology for generating synthetic sequences of precipitation or steamflow values that can be used for a range of applications from filling in missing data in a gauged record to extending monthly streamflow records [50], and from analyzing long-term reliability of yields of watersheds [51] or reservoirs [52],[53], to forecasting floods or snowmelt runoff quantities from synthetic precipitation sequences [54]. These *synthetic hydrology* techniques augment the simulation tools described in Sections 12.1–12.4. Both have experienced widespread use by hydrologists and engineers [55]. Synthesis involves the generation of a sequence of values for some hydrologic variable (daily, monthly, seasonal, or annual). The techniques are most often applied to produce streamflow sequences for use in reservoir design or operation studies but can also be used to generate rainfall sequences that can subsequently be input to simulation models.

If historical flows could be considered to be representative of all possible future variations that some project will experience during its lifetime, there would be little need for synthetic hydrology. The historical record is seldom adequate for predicting future events with certainty. The exact historical pattern is unlikely to recur, sequences of dry years (or wet years) may not have been as severe as they may become, and the single historical record gives the planner limited knowledge of the magnitude of risks involved.

Hydrologic synthesis techniques are classified as (1) *historical repetition methods*, such as mass curve analyses, which assume that historical records will repeat themselves in as many end-to-end repetitions as required to bracket the planning period; (2) *random generation techniques*, such as Monte Carlo techniques, which assume that the historical records are a number of random, independent events, any of which could occur within a defined probability distribution; and (3) *persistence methods*, such as Markov generation techniques, which assume that flows in sequence are dependent and that the next flow in sequence is influenced by some subset of the previous flows. Historical repetition or random generation techniques are normally applied only to annual or seasonal flows. Successive flows for shorter time intervals are usually correlated, necessitating analysis by the Markov generation method.

As with most subfields of hydrology, a number of computer programs for time-series analysis and hydrologic data synthesis have been developed. One of the first, and one of the most widely applied, was the U.S. Army Corps of Engineers' model HEC-4 (see Section 12.1), published in 1971 [50]. Its use is limited, though, to synthesizing sequences of serially dependent monthly streamflows in a river reach. Other codes [56],[57] are available to the hydrologist. Additional models and descriptions of theory and applications of time-series analysis of precipitation and streamflow are detailed in a number of available texts and publications [3],[25],[58]–[61].

Mass Curve Analysis

One of the earliest and simplest synthesis techniques was devised by Rippl [62] to investigate reservoir storage capacity requirements. His analysis assumes that the future inflows to a reservoir will be a duplicate of the historical record repeated in its entirety as many times end to end as is necessary to span the useful life of the reservoir. Sufficient storage is then selected to hold surplus waters for release during critical periods when inflows fall short of demands. Reservoir size selection is easily accomplished from an analysis of peaks and troughs in the mass curve of accumulated synthetic inflow versus time [63]–[65]. Future flows can be similar, but are unlikely to be identical to past flows. Random generation and Markov modeling techniques produce sequences that are different from, although still representative of, historical flows.

Example 12.2

Streamflows past a proposed reservoir site during a 5-year period of record were, respectively, in each year 14,000, 10,000, 6,000, 8,000, and 12,000 acre-ft. Use Rippl's mass curve method to determine the size of reservoir needed to provide a yield of 9,000 acre-ft in each of the next 10 years.

TABLE 12.19   Streamflows for Example 12.2

				Flows (thousands of acre-ft)						
Year	1	2	3	4	5	6	7	8	9	10
Inflow	14	10	6	8	12	14	10	6	8	12
Cumulative inflow	14	24	30	38	50	64	74	80	88	100

Solution.   A 10-year sequence of synthetic flows, using Rippl's assumptions, is shown in Table 12.19. Inflows are set equal to the historical record repeated twice.

When cumulative inflow and cumulative draft are plotted, the maximum deficiency shown in Fig. 12.19 is 4,000 acre-ft. Thus a reservoir with a 4,000-acre-ft capacity should be placed in the stream. Starting with a full reservoir at the beginning of year 1, the reader should verify the adequacy of the reservoir by "simulating" a draft of 9,000 acre-ft per year for 10 years.

Random Generation

One method of generating sequences of future flows is a simple random rearrangement of past records. If the stream is ungauged and records are not available, a probability distribution can be selected and a sequence of future flows that follow the distribution and have prescribed statistical moments is generated.

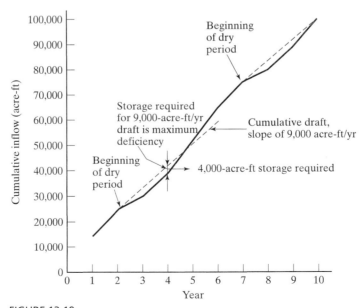

FIGURE 12.19

Mass curve for Example 12.2: — cumulative inflow; --- cumulative draft.

Whenever historical flows are available, a reasonable sequence of future flows can be synthesized by first consulting any set of random numbers, selecting a number, matching this with the rank-in-file number of a past flow, and listing the corresponding flow as the first value in the new sequence. The next random number would be used in a similar fashion to generate the next flow, and so on. Random numbers having no corresponding flows are neglected and the next random number is selected. Any random number generator can be used, or the reader can use Table B.3 in Appendix B, a table of uniformly distributed random numbers (each successive number has an equal probability of taking on any of the possible values). To illustrate the use of Table B.3 in the random generation process, the first three years of a synthetic flow sequence could be generated by selecting the 53rd, 74th, and 23rd from the list of past flows. Alternatively, the flows in 1953, 1974, and 1923 could also be selected as the new random sequence.

Most PCs have random number generation capabilities in their system libraries. Rather than storing large tables of numbers such as Table B.3, successive random integers are usually generated by the computer.

Example 12.3

Annual flows in Crooked Creek were 19,000, 14,000, 21,000, 8,000, 11,000, 23,000, 10,000, and 9,000 acre-ft, respectively, for years 1, 2, 3, 4, 5, 6, 7, and 8. Generate a 5-year sequence of annual flows, Q_j, by matching five random numbers with year numbers.

Solution.   Random integers between 0 and 9 are generated by the computer. The Q_j values in Table 12.20 are selected from the eight given flows by matching the respective year number with the random number. The digit 9 has no corresponding flow in the 8-year sequence, so the next random number, 2, places the 14,000-cfs flow in year 2 in the first position of the synthetic 5-year sequence.

TABLE 12.20   Development of 5-Year Synthetic Sequence

j	Random digit	Q_j (acre-ft)
1	9	Skip
2	2	14,000
3	5	11,000
4	8	9,000
5	1	19,000
6	4	8,000

Streamflow synthesis for ungauged streams requires either a regional frequency curve or an equation of the cumulative distribution function. The synthetic sequence can be generated using an assumed CDF shape (see Chapter 3) with regional estimates of distribution parameters such as the mean and standard deviation.

The random generation process for gauged or ungauged basins begins with the selection of a sequence of random numbers p_i from Table B.3 or from computers. Each p_i is made fractional by placing a decimal point in front, and the resulting decimal fraction is treated either as $G(x)$ or $F(x)$ (see Chapter 3) in solving for streamflow x.

Example 12.4

Generate a 5-year sequence of annual flows in Crooked Creek (see Example 12.3) having a Pearson Type III CDF (see Chapter 3) with a mean of 14,000 acre-ft, a standard deviation of 5,000 acre-ft, and a skew coefficient of 0.2.

Solution. Successive random values of p_i between 0 and 1,000 are generated from a random number generator. These are treated as exceedance probabilities $G(Q)$, which in turn are entered into Table B.2 to find the interpolated Pearson Type III frequency factor K (see Table 12.21). Corresponding flows are found from (see Chapter 3):

$$Q_j = \overline{Q} + K_j(s) \tag{12.33}$$

or

$$Q_j = 14{,}000 + K_j(5{,}000) \tag{12.34}$$

TABLE 12.21 Development of 5-Year Pearson Type III Sequence

i	p_i	$G(Q)$ (%)	j	K_i (Table B.2)	Q_i [Eq. 12.34] (acre-ft)
1	123	12.3	1	1.17	19,850
2	61	6.1	2	1.56	21,800
3	529	52.9	3	−0.11	13,450
4	716	71.6	4	−0.62	10,900
5	361	36.1	5	0.36	15,800

Markov Generation

The synthesis of sequences of streamflows by random generation ignores the existence of *persistence* present to some extent in most hydrologic sequences. Persistence is the tendency for high flows to be followed by high flows, and low flows to be followed by low flows. Markov models assume that each flow is dependent on one or more of the most recent flows.

 Single-period and *multiperiod* Markov models view hydrologic series as chains of serially dependent values, where each value has a deterministic part and a random error part. To illustrate, the *lag-one single-period* Markov chain assumes that the next flow in sequence Q_{i+1} is the sum of the mean $\overline{Q}_{j+1}$ plus a dependent fractional part of the deviation of the previous flow Q_i from its mean, $\overline{Q}_j$, plus a random component e.

Expressed as an equation, the lag-one Markov model [66] is:

$$Q_{i+1} = \overline{Q}_{j+1} + r_j(Q_i - \overline{Q}_j) + e \qquad (12.35)$$

where Q_{i+1} = the generated streamflow in period $i + 1$ (day, month, year)
Q_{j+1} = the mean of observed flows in period $j + 1$
Q_i = the generated flow in period i
Q_j = the mean of observed flows in period j
r_j = the correlation coefficient for the relation of flows for period $j + 1$ to flows for period j
e = the simulation error due to unexplained variance

Now since e represents the error in the relation, rewriting Eq. 12.35, we obtain:

$$Q_{i+1} = \overline{Q}_{j+1} + r_j(Q_i - \overline{Q}_j) + t_i\sigma_{j+1}(1 - r_j^2)^{1/2} \qquad (12.36)$$

where t_i = a random number selected from a normal distribution having a zero mean and a unit variance
σ_{j+1} = the standard deviation of observed flows for the period $j + 1$

To apply Eq. 12.36, it is necessary to calculate the required means, variances, and correlations and then to initiate the synthetic sequence of flows with some value Q_i with $i = 1$. The mean of observed flows is often selected as the initial flow in the random synthetic runoff sequence, but the choice affects all future flows in the sequence. The first 50 flows in the sequence should be discarded as a means of providing a warmup period to eliminate starting condition bias [65]. Standard normal random deviates, t_i, are generated using the random generation procedures described earlier.

Whenever the single-period Markov generator is used for annual flows, the period $j + 1$ flows and the period j flows have equal means, or $\overline{Q}_{j+1} = \overline{Q}_j$. This assumption is not valid for smaller periods, such as monthly or daily subdivisions. For example, October and November flows seldom have equal means. To reflect different seasonal or monthly means, the *multiperiod* Markov model is used. This requires a double indexing subscript in Eq. 12.36 or:

$$Q_{i,j} = \overline{Q}_j + b_j(Q_{i-1, j-1} - \overline{Q}_{j-1}) + t_i\sigma_j(1 - r_j^2)^{1/2} \qquad (12.37)$$

where j is the number of seasonal periods in the year and the other terms have been defined except for $b_j = r_j(\sigma_{j+1}/\sigma_j)$, which must be employed since $\overline{Q}_{j+1} \neq \overline{Q}_j$. This model has been used extensively in streamflow analysis. The single-period and multiperiod Markov generation procedures sometimes result in negative flows. These flows must be retained for generating the next flows in sequence, and then they may be discarded.

Example 12.5

The parameters describing quarterly flows in millions of gallons per day for the Patapsco River are listed in Table 12.22. Apply the information in conjunction with a table of random numbers and create a 2-year record of quarterly flows. The solution is shown in Table 12.23.

TABLE 12.22   Quarterly Streamflow Parameters, Patapsco River

j	b_j	σ_j	r_j	$(1 - r_j^2)^{1/2}$	$\overline{Q}_j$
1	0.66	73	0.57	0.82	68
2	0.43	85	0.41	0.91	137
3	0.14	96	0.43	0.90	183
4	0.91	29	0.36	0.93	107

TABLE 12.23   Tabular Generation of Synthetic Quarterly Streamflows

i (1)	j (2)	$\overline{Q}_j$ (3)	$\overline{Q}_{j-1}$ (4)	$Q_{i-1, j-1}$ (5)	k (6)	t_j (7)	$b_j(Q_{i-1, j-1} - \overline{Q}_{j-1})$ (8)	$t_j\sigma_j(1 - r_j^2)^{1/2}$ (9)	$Q_{i,j}$ (10) [(3) + (8) + (9)]
1	1	68	107	107	0.5374	0.094	0	5.6	73.6
2	2	137	68	73.6	0.6338	0.342	2.4	26.5	165.9
3	3	183	137	165.9	0.3530	−0.377	4.0	−30.6	156.4
4	4	107	183	156.4	0.6343	0.343	−24.2	9.3	92.1
5	1	68	107	92.1	0.0263	−1.94	−9.8	−116.4	0
6	2	137	68	0	0.6455	0.373	−29.2	28.9	136.7
7	3	183	137	136.7	0.8507	1.04	−0.04	84.5	267.5
8	4	107	183	267.5	0.3485	0.39	76.8	10.6	194.5

Column 1. Period index, i = 1 to 8 quarters for a 2-year run.
Column 2. Quarterly index, j = 1 to 4, two cycles.
Column 3. Mean quarterly flow.
Column 4. Preceding mean quarterly flow.
Column 5. The generated flow for the preceding quarter. Note that the first entry is the same as column 4, that is, generation begins using the mean flow from the preceding quarter.
Column 6. Uniform random digits from a random number table. Could be produced simply by drawing randomly from a card deck, with replacement.
Column 7. Random normal deviates from Appendix B, Table B.1. If column 6 is less than 0.5, find z for $(0.5 - k)$ and set $t = -z$; if $k > 0.5$, find z for $(k - 0.5)$ and set $t = z$.
Column 8. Regression component of Eq. 12.37.
Column 9. Random component of Eq. 12.37.
Column 10. Generated flow. Note that a negative flow occurred in row i = 5 and was then set to 0. In practice, the negative value should not be canceled to 0 until all subsequent flows have been calculated.

SUMMARY

The models introduced in this chapter have come into popular practice over the past three decades. By far the largest number of hydrologic model applications involves the use of single-event models presented in Chapter 11. Data requirements for single-event models are nominal, far less than for continuous modeling studies. In the majority of cases, the data required for any subarea are easily obtained from readily available topographic, land-use, and soils maps. GIS procedures in models such as HEC-HMS allow automatic determination of basin area, slope, and channel length parameters needed by the model algorithms. The same GIS procedures allow overlays of the soil types, land use and topography to develop distributed or composite infiltration parameters. Given the inputs, estimates of peak flow rates and shapes of runoff hydrographs at several locations in the watershed can be obtained for given storms within a few hours' time using these models. These procedures continue to be the primary tools used by engineers in analysis and design of storm water–handling facilities.

If actual or synthesized precipitation records are available, one of the most effective means of analyzing historical flows or evaluating future possible flows under changing land-use patterns is through any of the continuous streamflow simulation models described in this chapter. Hydrologic problems and applications that can be analyzed using continuous modeling include watershed yield studies; reservoir design and operation studies; sediment yield estimating for reservoir design or analysis of impacts of erosion controls on water quality; water supply studies for municipal, industrial, or agricultural demands; litigation over impacts of well fields or direct diversions; determination of hydropower production potential; evaluations of flows that will pass through critical in-stream or riparian habitat reaches of a stream; identification of flows that will be available for recreational uses of a stream; and, among numerous other applications, analysis of water quantity and quality impacts of removing dams or making other major upstream changes.

The models described in this chapter, and other similar continuous simulation codes, are available from the federal or state agencies that originated the code, or from numerous public outlets or vendors who have been authorized to distribute the software.

Hydrologic modeling is often viewed only in terms of the deterministic models of the rainfall-runoff processes described in Sections 12.1–12.4. A complete view encompasses these tools plus the streamflow synthesis methods described in Section 12.5. A growing number of projects are analyzed, designed, constructed or operated each year on the basis of process modeling and analysis of synthesized time series of values of random variables.

PROBLEMS

SECTION 12.1: HYDROLOGIC SIMULATION OVERVIEW

12.1 Six numbered subareas for a river basin are as shown in the following sketch. Prepare a schematic diagram for a model study using boxes as subbasin runoff components, connecting lines as channel routing links, circles as hydrograph combination nodes and triangles as reservoir routing nodes. Then describe the computation sequence for this basin in the same manner shown in Fig. 12.3.

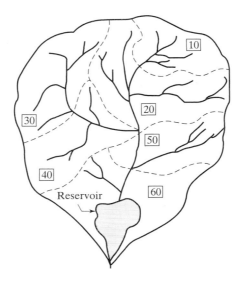

12.2 Review the description of Corps of Engineers' models in Table 12.3 and identify which ones primarily simulate hydrologic processes (as contrasted with hydraulic, economic, river system, optimization, or other types of processes).

12.3 Repeat Problem 12.2 but identify which of the models most likely fall in the following categories:

    a. Analytical models

    b. Probabilistic models

    c. Physical models

    d. Mathematical models

    e. Lumped-parameter models

    f. Distributed-parameter models

    g. Stochastic models

    h. Deterministic models

    i. Single-event models

    j. Continuous models

SECTION 12.2: SINGLE-EVENT RAINFALL-RUNOFF MODELS

12.4 Synthesize a unit hydrograph for a watershed in your locale using the HYMO model equations. Compare with corresponding unit hydrographs from Snyder's method and the SCS method in Chapter 9.

12.5 Use the HYMO model equations to synthesize a unit hydrograph for the entire 258 -mi^2 Oak Creek watershed in Fig. 12.10.

12.6 A watershed experiences a 12-hr rainstorm having a uniform intensity of 0.1 in./hr. Using $E = 0.7$, $K_0 = 0.6$, $C = 3.0$, and $\Delta K = 0$, calculate the hourly loss rates L_t as determined by the HEC-1 event-simulation model. Determine the total and percent losses for the storm.

12.7 Repeat Problem 12.6 using $CUML_1 = 0.5$ in.

12.8 Route the inflow hydrograph in Problem 9.44 to the outlet of the 30-mi reach using the HEC-1 straddle-stagger method by lagging averaged pairs of flows two time increments (12 hr). Compare the routed and measured outflow rates.

12.9 Route the inflow hydrograph in Problem 9.44 through the reach by dividing the 30-mi reach into three subreaches, and treat the outflow from each as inflow to the next in line. Lag flows one time increment in each subreach and compare the final outflows with the measured values.

12.10 Study Table 12.6 and Fig. 12.10, and then define the following terms from Table 12.6: AMSKK, X, TAREA, NP, STORM, TP, CP, TC, R, RAIN, EXCESS, and COMP Q.

12.11 Search the HEC-1 printout in Table 12.6 to determine values (give units) of the following:
   a. The time increment used in the model run.
   b. Snyder's C_p, Eq. 9.28 input for subarea B.
   c. The peak flow rate for the synthesized subarea-A unit hydrograph.
   d. The total runoff (in.) from subarea A.
   e. The peak outflow rate from subarea A.
   f. The peak-to-peak time lag in routing the outflow hydrograph from point 1 to point 2, Fig. 12.10.
   g. The percent attenuation caused by the reach between points 1 and 2.
   h. The subarea-B peak outflow rate if subarea A is neglected.
   i. The simulated subarea-B peak outflow rate.

12.12 Refer to the HEC-1 output in Tables 12.6 and 12.7 to answer the following questions:
   a. Are the values labeled PRECIPITATION PATTERN actual rainfall depths, or are they fractions of the rainfall?
   b. Name the method used to synthesize the unit hydrograph.
   c. What is the correct value of X for the X-hr unit hydrograph?
   d. How does the storm over subarea A differ from the storm over subarea B?
   e. Determine the total Muskingum K and the value of X for the river reach between points 1 and 2 of Oak Creek.

12.13 Describe how, for a given storm, you would determine the effectiveness of Branched Oak reservoir at point 9 in Fig. 12.10 to reduce flooding at point 8. Answer by enumerating your computation logic as illustrated. You are allowed a maximum of two runs with HEC-1, and any number of subareas may be used as long as you describe your subareas.

12.14 A hydrologist wishes to model a watershed using the SCS curve number method to determine net rain, and to route the watershed runoff hydrograph through a reservoir using the storage-indication method. Base flow is to be incorporated as a constant value throughout the storm duration. Which event-simulation model or models would accomplish the task?

SECTION 12.3: CONTINUOUS SIMULATION MODELS

12.15 Assume that a 30-mi^2 rural watershed in your locale receives a 3-in. rain in a 10-day period. Make a copy of Fig. 12.11 and plot approximate percentages to show, for average conditions, how the rain would be distributed (a) initially and (b) after 30 days.

12.16 A sloping, concrete parking lot experiences rain at a rate of 3.0 in./hr for 60 min. The lot is 500 ft deep and has a slope of 0.0001 ft/ft. If the water detention on the lot is zero at the start of the storm, calculate the complete overland flow hydrograph for 1 ft of width using the SWM-IV equations. Use a 5-min routing interval and continue computations until all the detained water is discharged.

12.17 Calculate the SWM-IV overland flow time to equilibrium for the lot of Problem 12.16 and compare it with the Kirpich time of concentration for the lot. Should these be equal?

12.18 Compare, by listing traits and capabilities of each, the SWM-IV with its more sophisticated offspring, HSP and HSPF.

12.19 Discuss the primary differences among the four versions of the Stanford watershed model described in this chapter.

12.20 Discuss the watershed behavior that is depicted in Fig. 12.16. Is this a typical watershed?

SECTION 12.4: GROUNDWATER FLOW SIMULATION MODELS

12.21 For the groundwater models listed in Table 12.18, state which would be most applicable to the following studies:
   a. Estimating the hydraulic conductivity of an aquifer using measurements of drawdown at a pumped well and at a remote water level observation well.
   b. Establishing which of four underground gasoline storage tanks may have caused pollution at a water supply well.
   c. Estimating the long-term potential of water supply from proposed wells on an island in a river.
   d. Estimating the lateral extent that seepage from a landfill might influence water quality in a deep aquifer.

SECTION 12.5: STREAMFLOW SYNTHESIS

12.22 Plot cumulative inflows versus time for the 8-year record in Example 12.3 and determine by mass curve analysis the size of the reservoir needed to provide a yield of 12,000 acre-ft in each of the next 24 years. What is the maximum yield possible?

12.23 Use the annual rain depths from Table 3.1 to generate a 46-year sequence of synthetic annual rain depths for Richmond using Rippl's mass curve assumption.

12.24 Repeat Problem 12.23 using random generation rather than mass curve methods. Use two-digit random numbers from Table B.3 and match these with the last two digits of the year numbers in Table 3.1.

12.25 Repeat Problem 12.23 using random generation to generate a 46-year synthetic sequence of annual rain depths that has a normal CDF curve with a mean and standard deviation equal to those of the annual rain data for Richmond from Table 3.1.

12.26 Repeat Problem 12.25 assuming that the annual rain depths follow a log–Pearson Type III distribution. For statistics use the mean, standard deviation, and skew of the logarithms of annual rain at Richmond.

12.27 Select a gauged stream in your geographic location and prepare a quarterly model using (a) normal distribution, (b) lognormal distribution, and (c) Pearson Type III distribution.

12.28 Can the simulation problem in Example 12.5 be converted to a lognormal distribution simulation? What difficulties are encountered with the data given in the example?

12.29 Select a month of thunderstorm activity in your region. From published NOAA hourly rainfall data, fit a distribution to the time between storms, and duration of storms, for 20 years of recorded data covering the selected month. Prepare a computer program to randomly generate the times between storms and the durations of storms.

12.30 Flows during 6 years of record were used in synthesizing the mass curve shown on the following sketch.

    a. Use Rippl's assumption and the graph to determine the missing magnitude of the flow for the 12th year.

    b. Determine the reservoir capacity required to allow an annual yield of 2,000 acre-ft/yr. Repeat for 500 acre-ft/yr.

    c. Determine the maximum yield possible at the site. How does this value relate statistically to the flows?

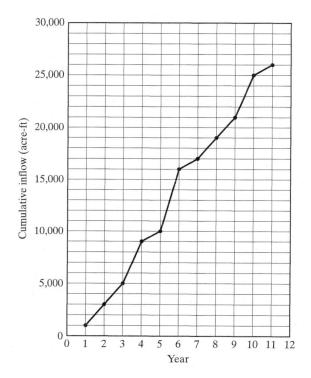

12.31 Describe with words and equations how you would develop a table of random precipitation depths that follow a normal distribution and have a mean of 4 in. and a standard deviation of 3 in. Use your method to calculate the first three depths.

12.32 Uniformly distributed random numbers are 20, 01, 90, 03, and 80. Use random generation to generate a 5-year sequence of annual rain depths that will follow a Pearson Type III distribution and will have a mean of 25.8 in., a standard deviation of 4.0 in., and a skew coefficient of −2.20.

12.33 Total July runoff from a basin is randomly distributed according to a Pearson Type III distribution. The mean is 10,000 acre-ft, the standard deviation is 1,000 acre-ft, the skew is −0.6, and the lag-one serial correlation coefficient is 0.50. Start with $Q_1 = 10,000$ and find five more Markov-generated flows if a sequence of randomly selected return periods gives 2, 100, 10, 2, and 50 years.

12.34 Review the differences between *water budget* and *simulation* models discussed in Section 12.1 and determine which of the simulation models described in this chapter could be used to perform water budget calculations.

12.35 Repeat Problem 12.34, contrasting between lumped-parameter vs. distributed-parameter models in this chapter.

12.36 Repeat Problem 12.34 for stochastic vs. deterministic models described in this chapter.

12.37 Repeat Problem 12.34 for single-event vs. continuous models described in this chapter.

12.38 Simulation and synthesis are treated separately in this chapter. List the most distinguishing characteristics of each method and give an example of each.

12.39 You are asked to determine a design inflow hydrograph to a reservoir at a site where no records of streamflow are available. List the general steps you would take as a hydrologist in developing the entire design inflow hydrograph.

REFERENCES

[1] G. A. Burton, *Stormwater Effects Handbook, A Toolbox for Watershed Managers, Scientists, and Engineers,* Boca Raton, FL: CRC Press, 2001.

[2] I. Watson and A. Burnett, *Hydrology—An Environmental Approach,* Boca Raton, FL: Lewis, 1995.

[3] W. Viessman and G. L. Lewis, *Introduction to Hydrology, 4th ed.,* New York: HarperCollins College, 1996.

[4] U.S. Army Waterways Experiment Station, "Models and Methods Applicable to Corps of Engineers Urban Studies," Miscellaneous Paper H-74-8, National Technical Information Service, Aug. 1974.

[5] George Fleming, *Computer Simulation Techniques in Hydrology*, New York: American Elsevier, 1975.

[6] D. R. Maidment (ed.), *Handbook of Hydrology*, New York: McGraw-Hill, 1993.

[7] N. H. Crawford and R. K. Linsley, Jr., "Digital Simulation in Hydrology: Stanford Watershed Model IV," Department of Civil Engineering, Stanford University, Stanford, CA, Tech. Rep. No. 39, July 1966.

[8] ASCE Task Committee on Glossary of Hydraulic Modeling Terms, "Modeling Hydraulic Phenomena: A Glossary of Terms," *Proc. ASCE J. Hyd. Div.,* Vol. 108, No. HY7, July 1982.

[9] Hydrologic Engineering Center, *HEC-HMS Hydrologic Modeling System and User's Manual,* U.S. Army Corps of Engineers: Davis, CA, March 1998.

[10] U.S. Army Corps of Engineers, "HEC-1 Flood Hydrograph Package," Users and Programmers Manuals, HEC Program 723-X6-L2010, Jan. 1973.

[11] Association of Professional Engineers of Canada, *Guidelines to Standards of Practice for the Use of Computer Programs in Engineering,* April 1977.

[12] W. James, "Developing Simulation Models," *J. Water Resources Research,* Vol. 8, No. 6, Dec. 1972.

[13] W. James and M. A. Robinson, "Standards for Computer-Based Design Studies," *Proc. ASCE J. Hyd. Div.,* Vol. 107, No. HYY, July 1981.

[14] A. D. Feldman, "HEC Models for Water Resources Systems Simulation: Theory and Experience," in *Advances in Hydroscience,* Vol. 12. Orlando, FL: Academic Press, 1981.

[15] W. Charley, A. Pabst, and J. Peters, "The Hydrologic Modeling System (HEC_HMS): Design and Development Issues," *Proc. ASCE,* Second Congress on Computing in Civil Engineering, Atlanta, 1995.

[16] Hydrologic Engineering Center, *HEC-IFH, Interior Flood Hydrology Package, User's Manual,* U.S. Army Corps of Engineers, Davis, CA, 1992.

[17] W. K. Johnson, "Use of Systems Analysis in Water Resource Planning," *Proc. ASCE J. Hyd. Div.* (1974).

[18] R. deNeufville and D. H. Marks, *Systems Planning and Design Case Studies in Modeling Optimization and Evaluation,* Englewood Cliffs, NJ: Prentice Hall, 1974.

[19] D. P. Loucks, "Stochastic Methods for Analyzing River Basin Systems," Cornell University Water Resources and Marine Sciences Center, Ithaca, NY, Aug. 1969.

[20] A. Maass (ed.), *Design of Water Resources Systems*, Cambridge, MA: Harvard University Press, 1962.

[21] D. P. Loucks, J. R. Stedinger, and D. A. Haith, *Water Resource Systems Planning and Analysis,* Englewood Cliffs, NJ: Prentice Hall, 1981.

[22] D. R. Maidment (ed.), *Handbook of Hydrology,* New York: McGraw-Hill, 1993.

[23] V. P. Singh, *Hydrologic Systems,* Englewood Cliffs, NJ: Prentice Hall, 1989.

[24] V. K. Hagen, "Small Urban Watershed Use of Hydrologic Procedures," National Research Council, Transportation Research Record No. 1471, Washington, D.C., 1994.

[25] R. S. Gupta, *Hydrology and Hydraulic Systems,* Prospect Heights, IL: Waveland Press, 1995.

[26] V. T. Chow, D. R. Maidment, and L. W. Mays, *Applied Hydrology,* New York: McGraw-Hill, 1988.

[27] U.S. Department of Agriculture Soil Conservation Service, "Computer Program for Project Formulation Hydrology," Tech. Release No. 20, June 1973.

[28] Hydrology, Section 4, *National Engineering Handbook*, Washington, D.C.: U.S. Department of Agriculture, Soil Conservation Service, 1985.

[29] J. R. Williams and R. W. Hann, "HYMO: Problem-Oriented Computer Language for Hydrologic Modeling," U.S. Department of Agriculture, Agriculture Research Service, May 1973.

[30] J. R. Williams, "Flood Routing with Variable Travel Time or Variable Storage Coefficients," *Am. Soc. Agric. Eng. Trans.* **12**(1), 100–103(1969).

[31] "SWMM, Volume No. 1, Final Report," Report for U.S. Environmental Protection Agency, Water Resources Engineers, and Metcalf and Eddy Inc., July 1971.

[32] W. C. Huber, J. P. Heaney, S. J. Nix, R. E. Dickinson, and D. J. Polmann, "Storm Water Management Model User's Manual, Version III," EPA-600/2-84-109a (NTIS PB84-198423), Environmental Protection Agency, Cincinnati, OH, Nov. 1981.

[33] L. A. Roesner, R. P. Shubinski, and J. A. Aldrich, "Storm Water Management Model User's Manual, Version III: Addendum I, Extran," EPA-600/2-84-109b (NTIS PB84-198431), Environmental Protection Agency, Cincinnati, OH, Nov. 1981.

[34] U.S. Army Corps of Engineers, "HEC-1 Flood Hydrograph Package," Users Manual, Hydrologic Engineering Center, Sept. 1990.

[35] C. O. Clark, "Storage and the Unit Hydrograph," *Trans. ASCE* **110**, 1419–1488(1945).

[36] D. H. Hoggan, *Computer-Assisted Floodplain Hydrology and Hydraulics,* New York: McGraw-Hill, 1989.

[37] W. Charley, A. Pabst, and J. Peters, "The Hydrologic Modeling System (HEC_HMS): Design and Development Issues," *Proc. ASCE,* Second Congress on Computing in Civil Engineering, Atlanta, 1995.

[38] K. C. Patra, *Hydrology and Water Resources Engineering,* Boca Raton, FL: CRC Press, 2000.

[39] N. H. Crawford and R. K. Linsley, Jr., "Digital Simulation in Hydrology: Stanford Watershed Model IV," Department of Civil Engineering, Stanford University, Tech. Rep. No. 39, July 1966.

[40] L. D. James, "An Evaluation of Relationship Between Streamflow Patterns and Watershed Characteristics Through the Use of OPSET," Research Rep. No. 36, Water Resources Institute, University of Kentucky, Lexington, 1970.

[41] L. D. James, "Hydrologic Modeling, Parameter Estimation, and Watershed Characteristics," *J. Hydrology* **17**, 283–307(1972).

[42] E. Y. Liou, "OPSET: Program for Computerized Selection of Watershed Parameter Values for the Stanford Watershed Model," Research Rep. No. 34, Water Resources Institute, University of Kentucky, Lexington, 1970.

[43] B. J. Claborn and W. Moore, "Numerical Simulation in Watershed Hydrology," Hydraulic Engineering Laboratory, University of Texas Rep. No. HYD 14-7001, 1970.

[44] C. T. Haan, B. J. Barfield, and J. C. Hayes, *Design Hydrology and Sedimentology for Small Catchments,* San Diego: Academic Press, 1994.

[45] "Expert Systems in Water Resources," American Society of Civil Engineering, *Civil Eng.* 46(Oct. 1984).

[46] National Research Council, *Ground Water Models—Scientific and Regulatory Applications.* Water Science and Technology Board, Commission on Physical Sciences, Mathematics, and Resources, National Academy Press, Washington, D.C., 1990.

[47] C. E. Jacob, "Flow of Groundwater," in *Engineering Hydraulics* (Hunter Rouse, ed.), New York: Wiley, 1950.

[48] D. W. Peaceman and N. H. Rachford, "The Numerical Solution of Parabolic and Elliptic Differential Equations," *J. Soc. Indust. Appl. Math.* **3**, (1955).

[49] T. A. Prickett and C. G. Lonnquist, "Selected Digital Computer Techniques for Groundwater Resource Evaluation," Illinois State Water Survey Bull. No. 55, 1971.

[50] U.S. Army Corps of Engineers, "HEC-4 Monthly Streamflow Simulation," Hydrologic Engineering Center, 1971.

[51] R. M. Hirsch, "Synthetic Hydrology and Water Supply Reliability," *Water Resources Research,* v. 15, no. 6, 1979.

[52] R. M. Vogel and J. R. Stedinger, "The Value of Stochastic Streamflow Models in Overyear Reservoir Design Applications," *Water Resources Research,* v. 25, no. 9, 1988.

[53] D. K. Frevert, et al., "Use of Stochastic Hydrology in Reservoir Operation," *J. Irrigation and Drainage Engineering,* ASCE, v. 115, no. 3, 1989.

[54] J. W. Delleur, et al., "An Evaluation of the Practicality and Complexity of Some Rainfall and Runoff Time Series Models," *Water Resources Research,* v. 12, no. 5, 1976.

[55] J. D. Salas, et al., "Applied Modeling of Hydrologic Time Series," Water Resources Publications, Littleton, CO, 1980.

[56] U.S. Bureau of Reclamation, "Applied Stochastic Techniques, Personal Computer Version 5.2, User's Manual," Earth Sciences Division, Denver, CO, 1990.

[57] J. C. Grygier and J. R. Stedinger, "SPIGOT, A Synthetic Streamflow Generation Software Package," School of Civil and Environmental Engineering, Cornell University, Ithaca, NY, 1990.

[58] V. Yevjevich, "Structural Analysis of Hydrologic Time Series," Hydrology Paper No. 56, Colorado State University Hydrology Papers, Ft. Collins, CO, 1972.

[59] P. J. Brockwell, and R. A. Davis, *Time Series: Theory and Methods,* (2nd ed.), Springer-Verlag, New York, 1991.

[60] M. B. Fiering, *Streamflow Synthesis,* Harvard University Press, Cambridge, MA, 1967.

[61] J. W. Delleur, and M. L. Kavvas, "Stochastic Models for Monthly Rainfall Forecasting and Synthetic Generation," *J. Applied Meteorology,* v. 17, no. 10, 1978.

[62] W. Rippl, "The Capacity of Storage Reservoirs for Water Supply," *Proc. Inst. Civil Eng. London* **71**, 270(1883).

[63] H. E. Hurst, "Long-Term Storage Capacities of Reservoirs," *Trans. ASCE* **116**, 776(1951).

[64] H. E. Hurst, R. P. Black, and Y. M. Sunaika, *Long-Term Storage,* London: Constable & Company Ltd., 1965.

[65] M. B. Fiering and B. B. Jackson, "Synthetic Streamflows," American Geophysical Union Water Resources Monograph No. 1, 1971.

[66] H. A. Thomas and M. B. Fiering, "Mathematical Synthesis of Streamflow Sequences for the Analysis of River Basins by Simulation," in *Design of Water Resources Systems* (A. Maass et al., eds), Cambridge, MA: Harvard University Press, 1962.

CHAPTER 13

Hydrology in Design

OBJECTIVES

The purpose of this chapter is to:

- Introduce the hydrologist to procedures used in the United States for designing structures for safe and effective passage of flood flows
- Give sufficient information for the designer to select the applicable criteria for designing hydraulic structures
- Provide a discussion of design storm hyetographs and provide methods for selecting the duration, depth, and distribution of precipitation for design
- Demonstrate how design floods can be developed without using precipitation data
- Discuss particular design methods including airport drainage, urban storm sewer design, detention basin design, and flood control reservoir and spillway design.

Hydrologic methods for designing minor and major structures are described in this chapter. Predicting peak discharge rates and synthesizing complete discharge hydrographs for use in designing minor and major structures are two of the more challenging aspects of engineering hydrology. Minor types of hydraulic structures range from small crossroad culverts, levees, drainage ditches, urban storm drain systems, and airport drainage structures to the spillway appurtenances of small dams. When lumped together with major structures, all require varying amounts of hydrologic design information. Generally, a hydrologist is required to provide peak rates of discharge for a design frequency or a complete discharge hydrograph for a design storm [1]. Other information, such as sedimentation rates, low-flow frequency analysis, groundwater analysis, and reservoir yield studies, is often required as part of a design project [2]–[5].

Most designs involving hydrologic analyses use a design flood that simulates some severe future event or imitates some historical event. If streamflow records are unavailable, design flood hydrographs are synthesized from available storm records using the rainfall-runoff procedures of Chapters 8 and 9. Only in rare cases are streamflow records adequate for complex designs, particularly in small watersheds. Regional analyses and the empiric-correlative methods discussed in Chapter 3 are useful for determining peak flow rates at ungauged sites.

13.1 HYDRAULIC STRUCTURE DESIGN METHODS

Procedures for estimating design flood flows (interest can be in either the peak flow rate or the entire hydrograph) include methods that examine historical or projected flood flows to arrive at a suitable estimate (*flow-based* methods), and methods that evaluate the storms that produce floods, and then convert the storms to flood flow rates (*precipitation-based* methods). In each case, the analysis can be based on selecting a design frequency and determining the associated flood (called *frequency-based* methods), developing designs for a range of flood frequencies and narrowing the final choice on the basis of long-term costs and benefits (called *risk-based* methods), or designing on the basis of an estimate of the probable maximum storm or maximum flood that could occur at the site (called *critical-event* methods).

Most information and techniques presented in this chapter are directed toward the flood protection aspect of small drainage structures and spillways and flood control pools for small and large dams. Most major dam structures are designed for more than just flood protection; they are multipurpose and may provide storage for irrigation, hydropower, water supply, navigation, and low-flow augmentation. Additional guidelines for hydraulic structure design are presented in Reference 1.

Flow-Based Methods

For design locations where records of stream flows are available, or where flows from another basin can be transposed to the design location, a design flood magnitude can be estimated directly from the stream flows by any of the following methods:

1. Frequency analysis of flood flows at the design location or from a similar basin in the region.
2. Use of regional flood frequency equations, normally developed from regression analysis (see Chapter 3) of gauged flood data.
3. Examination of the stream and floodplain for signs of highest historical floods and estimation of the flow rates using measurements of the cross section and slope of the stream.

Precipitation-Based Methods

Where stream-gauging records are unavailable or inadequate for streamflow estimation, design floods can be estimated by evaluating the precipitation that would produce the flood, and then converting the precipitation into runoff by any of the rainfall-runoff methods described in Chapters 9, 11, and 12. Typical methods include:

1. Design using the greatest storm of record at the site, by converting the precipitation to runoff.

2. Transposition of a severe historical storm from another similar watershed in the region.
3. Frequency analysis of precipitation and conversion of the design storm to runoff.
4. Use of a theoretical probable maximum precipitation (PMP), or fraction of PMP, based on meteorological analyses.

Because the flood flow rate is desired in all cases, the flow-based methods are preferred over conversion of precipitation to runoff. Due to the relatively longer period of time and greater number of locations at which precipitation amounts have been recorded, precipitation-based methods are used in the majority of designs, especially with small and very large basins. Flow-based methods are typically used in the midrange of basin sizes.

Frequency-Based Methods

Regardless of whether flow or precipitation data are used, designs most often proceed by selecting a minimum acceptable recurrence interval and using procedures from Chapter 3 to determine the corresponding worst-condition storm or flood that could be equaled or exceeded during the selected recurrence interval. Criteria for selecting design recurrence intervals are summarized in Section 13.3. Results from frequency analysis of flood flow data normally provide reliable estimates of 2-, 5-, 10-, and 25-year flows. Extrapolation beyond the range of the period of flow records is allowed, but is less reliable.

Risk-Based Methods

Recent trends in design of minor (and major) structures are toward the use of *economic risk analyses* rather than frequency-based designs. The risk method selects the structure size as that which minimizes total expected costs. These costs are made up of the structure costs plus the potential flood losses associated with the particular structure [6]. The procedure is illustrated in Fig. 13.1. The total expected cost curve is the sum of the other two curves. Risk costs (flood damages, structure damages, road and bridge losses, and traffic interruptions) and structure costs are estimated for each of several sizes. The optimal size is that with the smallest sum. Structures selected by risk analysis are normally constrained to sizes equal to or larger than those resulting from traditional frequency-based methods. Reference 7 provides a comparison of risk-based and frequency-based methods.

Critical-Event Methods

Because of the high risk to lives or property below major structures, the design of these structures generally includes provisions for a flood caused by a combination of the most severe meteorologic and hydrologic conditions that are possible. Instead of designing for some frequency or least expected total cost, flood-handling facilities for the structures are sized to safely store or pass the most critical storm or flood possible. Methods for designing by critical event techniques include:

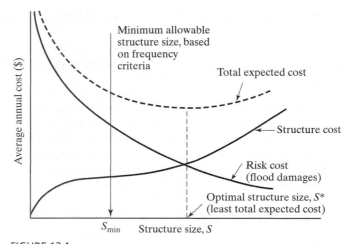

FIGURE 13.1

Principles of economic risk analysis for structure size selection.
(U.S. Federal Highway Administration, Hydraulic Engineering Circular No. 17 [6])

1. Estimating the probable maximum precipitation (PMP) and determining the associated flood flow rates and volumes by transforming the precipitation to runoff.

2. Determining the probable maximum flood (PMF) by determining the PMP and converting it to a flood by application of a rainfall-runoff model, including snowmelt runoff if pertinent.

3. Examining the floodplain and stream to identify palaeo-flood evidences such as high-water marks, boulder marks on trees or banks, debris lines, historical accounts by local residents, or geologic or geomorphologic evidences.

4. In some cases, the critical-event method involves estimating the magnitude of the 500-yr event by various frequency or approximate methods. Often, such as in mapping floodplains, the 500-yr flood is estimated as a multiple of the 100-yr event, ranging from 1.5 to 2.5, with 1.7 in common use. Due to lack of longer-term records, frequency-based estimates are seldom attempted for recurrence intervals exceeding 500 years.

13.2   HYDROLOGIC DESIGN DATA

The design of any structure requires a certain amount of data, even if only a field estimate of the drainage area and a description of terrain type and cover. The following material identifies some general data types and sources.

Physiographic Data

The hydrologic study for any structure requires a reliable topographic map. U.S. Geological Survey topographic maps usually are available. The mapping of the United

States is almost complete with 15-minute quadrangles, and many of these areas are mapped by 7.5-minute quadrangles. County maps and aerial photos can also be used to advantage in making preliminary studies of the watershed.

Based on an area map, a careful investigation of the watershed's drainage behavior must be made. Additional information can be obtained from USGS maps that depict predominant rock formations. Soil types and the infiltration and erosive characteristics of soils can be secured from U.S. Soil Conservation districts or university extension divisions.

The drainage areas contributing to large dams require stricter analysis of an area's hydrology than is necessary in designing minor structures. The possibility of a uniformly intense rainfall over the entire basin is an unrealistic assumption for large watersheds. The influence of temporal and spatial variations of the rainfall should thus be considered. For major dams, the estimated worst possible rainfall values are generally converted to a design discharge hydrograph, which is then used in reservoir routing calculations to proportion reservoir and spillway size, surcharge storage, and any additional outlets needed to maintain power requirements or sustained downstream flow for navigation, irrigation, or water supply. The basic concern in hydrologic design of a large dam is to protect downstream interests using a realistic estimate for the design storm hydrograph.

Topographic map detail necessarily shifts with the type and purpose of the structure being designed. Field reconnaissance always increases the understanding of an area's hydrology, no matter how insignificant the structure might be.

Hydrologic Data

One difficulty in hydrologic design is that of getting adequate data for the region under study. Considerable data can be acquired from previously published reports issued by governmental agencies and/or universities. The following is a list of federal agencies that publish hydrologic data:

> Agricultural Research Service
> Natural Resources Conservation Service
> Forest Service
> U.S. Army Corps of Engineers
> National Oceanic and Atmospheric Administration
> Bureau of Reclamation
> Federal Highway Administration
> U.S. Geological Survey, Topographic Division
> U.S. Geological Survey, Water Resources Division

Additional data can often be procured from departments of state governments, interstate commissions, and regional and local agencies.

Meteorologic Data

The National Weather Service (NWS), couched in the National Oceanic and Atmospheric Administration, is the primary source of meteorologic data published in

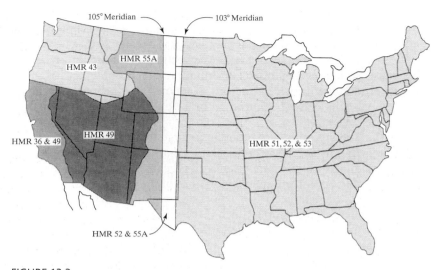

FIGURE 13.2

Hydrometeorological report series coverage of the conterminous United States.
(U.S. Bureau of Reclamation [9])

a variety of forms; these data include the NWS' Hydrometeorologic Report (HMR) series [8]. Figure 13.2 shows the applicable reports for various geographic and topographic regions of the United States [9]. Numerous other federal, state, and local agencies collect and analyze precipitation information—especially those who design, inspect, or regulate large hydraulic structures.

Current practice for estimating design storm hyetographs requires knowledge of the meteorologic characteristics of storms in the region, maximum amount of precipitable moisture in the atmosphere, causes of precipitation, frequencies of total storm depths for various durations of storms, and influence of snowmelt for storms over the region. In some areas, such as foothill regions of major mountain chains, topography has a very distinct impact on precipitation.

13.3   HYDROLOGIC DESIGN STANDARDS AND CRITERIA

Selection of frequency for the design of a structure is most often based on potential damage to property, danger to life, and economic losses such as interruption of commerce. A standard of practice involves selecting a frequency, and then designing for the worst condition expected to occur for that frequency. Where danger to life is involved, a great amount of controversy exists over appropriate design standards.

All projects involve some risks to property and life, but where direct danger to human life is absent, the design can proceed through selection of an accepted frequency level and design of the least-cost structure that provides this protection. As an alternative to least structure cost, economic risk analysis can be used. For this method, the final design frequency is optimized rather than preselected. Structure sizes that would accommodate storms for several frequencies are tested, and the one with the least total expected cost is used. These costs include not only the actual construction costs but also the flood damage risks and costs due to interruption of services and commerce. Either annual or present worth economic analyses can be used.

Meaning of the 100-year Storm

Use of the terms *100-year storm* or *100-year flood* is common, but these terms are also commonly misunderstood. A relatively correct definition of the 100-year storm is the maximum point rainfall over a specified duration, X, with a 1.0 percent chance of being equaled or exceeded in any given year. As noted, the definitions involve a probability, storm duration X, and the qualifier that it is a *point* rainfall. The reason for defining it as a point value is that the rain depths are measured at points, and frequency analyses of measured rain depths for various durations are conducted using the point data.

As shown in this section, rain depths for given durations and probabilities over areas such as watersheds or regions are less than the point values for the same durations and probabilities. Though extreme values of rainfall occur at points within a storm, the average depth over an area is smaller than the maximum point value.

Duration is also an important aspect of specifying the 100-year storm. There is no single 100-year storm at any location. When defining this storm, it is necessary to specify the duration X, where X is the duration of generally continuous rain. As durations increase, 100-year rain depths increase. To illustrate, the following are all 100-year storms at Limon, Colorado (from NOAA Atlas 2, 1973) [10]:

Duration	100-year storm depth, in.
30 minutes	2.16
1 hour	2.73
6 hours	3.60
24 hours	4.50

Setting X equal to 6 hours, the 100-year, 6-hour storm is 3.60 inches, having an average intensity of 0.6 inches per hour. Though 3.60 inches could occur at a point, it would be incorrect to state that this depth would occur over an entire county with the same recurrence interval. Similarly, the 100-year, 24-hour storm is 4.50 inches. The important point is that the 3.60- and 4.50-inch depths are *both* 100-year storms, as are the other two.

The third aspect, probability, is the most frequently misunderstood component. The X-hour, 100-year storm is the maximum point rain depth for duration X that has a 1 percent chance of being equaled or exceeded in any given year. No other definition should be used, and the 100-year event should never be identified as the maximum storm depth that will occur in any 100-year period. Though confusing, a location may experience several 100-year storms within a relatively short number of years. This could happen in a relatively wet cycle, but in any event, the frequency analysis of the data should be updated because the previous 100-year rain depths may have been underestimated.

The 100-year flood is most commonly defined as the maximum instantaneous flow rate at any point in a stream that has a 1 percent chance of being equaled or exceeded in any given year. Others define it as the entire hydrograph of a storm having this maximum instantaneous flow rate. Common definition errors include equating this with the maximum daily discharge rate or with the maximum flow that will occur in any 100-year period. As with the 100-year storm, several 100-year floods may occur within a relatively short period of time.

Logic and common practice would tend to suggest that a 100-year storm causes a 100-year flood, but due to the complexities of watersheds, these may not be related. Many areas requiring hydrologic design have relatively ample precipitation gauge data but no streamflow gauge data. For these areas, it is necessary to define the 100-year flood as the simulated flow rate from a watershed experiencing a 100-year storm. In evaluating the 100-year flood for any watershed, the analyst should test all locally possible 100-year storm durations to assess the worst possible case.

For the example above, a 30-minute storm for a small watershed at Limon, Colorado, may produce a larger 100-year flood than a 1-hour or 6-hour, 100-year storm. All three storms would produce estimates of the 100-year flood, so all possible cases should be considered in any 100-year flood design. Rather than always calculating flood rates from all types of 100-year storms, it is commonly accepted for small watersheds that a storm having a duration approximately equal to the time of concentration for the watershed (see Chapter 9) will produce the worst-case peak flow rate. For larger watersheds with times of concentration over 1 hour, common practice is to evaluate only the 6-hour or 24-hour storms because these tend to represent, respectively, the durations of convective thunderstorms and heaviest-rainfall portions of frontal storms common to many parts of the United States.

Minor Structures

The design frequencies shown in Table 13.1 are typical of levels generally encountered in minor structure design. An example of variations that do occur is the design frequency of a culvert, which under cases of excessive backwater, could effectively halt traffic.

TABLE 13.1 Minor Structure Design Frequencies

Type of minor structure		Return period, T_r	Frequency = $1/T_r$
Highway crossroad drainage[a]			
0–400	ADT[a]	10 yr	0.10
400–1700	ADT	10–25 yr	0.10–0.04
1700–5000	ADT	25 yr	0.04
5000+	ADT	50 yr	0.02
Airfields		5 yr	0.20
Railroads		25–50 yr	0.04–0.02
Storm drainage		2–10 yr	0.50–0.10
Levees		2–50 yr	0.50–0.02
Drainage ditches		5–50 yr	0.20–0.02

[a] ADT = average daily traffic.

(After Ref. 11)

The Soil Conservation Service recommends the use of a 25-year frequency for minor urban drainage design if there is no potential loss of life or risk of extensive damage, such as for first-floor elevations of homes. A 100-year frequency is commonly recommended when extensive property damage may occur [11].

Minor structure design is largely based on frequency-based or sometimes risk-based methods. Several steps in the hydrologic approach to minor structure design are common to most design handbooks and adopted techniques. The general steps (each is illustrated subsequently) are:

1. Determine the duration of the critical storm, usually equated to the time of concentration of the watershed.
2. Choose the design frequency.
3. Obtain the storm depth based on the selected frequency and duration.
4. Compute the net direct runoff (several methods were presented in Chapter 7).
5. Select the time distribution of the rainfall excess.
6. Synthesize the unit hydrograph for the watershed (see Chapter 9).
7. Apply the derived rainfall excess pattern to the synthetic unit hydrograph to get the runoff hydrograph.
8. Establish the frequency of the calculated flood (usually assumed equal to the design storm frequency).

Design Criteria for Small Dams

Small dams customarily are designed using two or more levels of frequency to provide an emergency spillway and ensure an adequate allowable freeboard. Figure 13.3 shows a typical small dam with normal freeboard (NF) and minimal freeboard (MF). The freeboard values for earth dams with riprap protection on the upstream slope, shown in Table 13.2, are based on wave runup caused by storms with 100-mph wind velocities. Minimal freeboard pertains to wind velocities of 50 mph. The *fetch* is defined as the perpendicular distance from the structure to the windward shore. If smooth concrete rather than riprap is used on the upstream face, the freeboard values shown should be increased 50 percent [12].

TABLE 13.2   USBR Recommended Normal (NF) and Minimum Freeboard (MF) Values, ft

Fetch (mi)	NF	MF
<1	4	3
1	5	4
2.5	6	5
5	8	6
10	10	7

Source: After Ref. 12

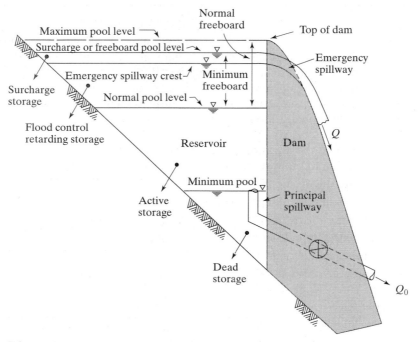

FIGURE 13.3

Multipurpose reservoir pool levels and storage zones.

The U.S. Soil Conservation Service design criteria for principal spillways of small dams are given in Table 13.3. The SCS Technical Release No. 60 should be consulted for full interpretation of this table [13]. Design frequency requirements are selected to fit the planned or foreseeable use of the structures. The SCS classifies structures into three groups [14]:

Class a. Structures located in rural or agricultural areas where failure might damage farm buildings, agricultural land, or township or country roads.

Class b. Structures located in predominantly rural or agricultural areas where failure might damage isolated homes, main highways or minor railroads, or cause interruption of use or service of relatively important public utilities.

Class c. Structures located where failure might cause loss of life or serious damage of homes, industrial and commercial buildings, important public utilities, main highways, or railroads.

The size of a small dam can range to over 100 ft in height but generally is restricted to structures retarding less than 25,000 acre-ft of storage at the emergency spillway crest.

The design frequencies for emergency spillways for small SCS Class a, b, or c dams are provided in Table 13.4. In this case, designs are for the 100-yr flood plus a fraction of the difference between the probable maximum precipitation (PMP, see

TABLE 13.3 SCS Design Criteria for Principal Spillways of Small Dams

Class of dam	Purpose of dam	$V_s H_e^1$	Existing or planned upstream dams	Precipitation data for maximum frequency[2] of use of emergency spillway type:	
				Earth	Vegetated
(a)	Single[3] irrigation only	Less than 30,000	None	$0.5DL^4$	$0.5DL$
		Greater than 30,000	None	$0.75DL$	$0.75DL$
	Single or multiple[5]	Less than 30,000	None	P_{50}	P_{25}^6
		Greater than 30,000	None	$0.5(P_{50} + P_{100})$	$0.5(P_{25} + P_{50})$
		All	Any[7]	P_{100}	P_{50}
(b)	Single or multiple	All	None or any	P_{100}	P_{50}
(c)	Single or multiple	All	None or any	P_{100}	P_{100}

[1] Product of reservoir storage volume V_s (acre-feet) times effective height of dam H_e (feet).
[2] Precipitation depths for indicated return periods (years).
[3] Applies to irrigation dams on ephemeral streams in areas where mean annual rainfall is less than 25 in.
[4] DL = design life (years).
[5] Class (a) dams involving industrial or municipal water are to use minimum criteria equivalent to that of Class (b).
[6] In the case of a ramp spillway, the minimum criteria should be increased from P_{25} to P_{100}.
[7] Applies when the failure of the upstream dam may endanger the lower dam.

Source: Soil Conservation Service [13]

TABLE 13.4 SCS Design Criteria for Emergency Spillways of Small Dams

Class of dam	$V_s H_e^1$	Existing or planned upstream dams	Precipitation data[2] for	
			Emergency spillway hydrograph	Freeboard hydrograph
(a)[3]	Less than 30,000	None	P_{100}	$P_{100} + 0.12(PMP - P_{100})$
	Greater than 30,000	None	$P_{100} + 0.06(PMP - P_{100})$	$P_{100} + 0.26(PMP - P_{100})$
	All	Any[4]	$P_{100} + 0.12(PMP - P_{100})$	$P_{100} + 0.40(PMP - P_{100})$
(b)	All	None or any	$P_{100} + 0.12(PMP - P_{100})$	$P_{100} + 0.40(PMP - P_{100})$
(c)	All	None or any	$P_{100} + 0.26(PMP - P_{100})$	PMP

[1] Product of reservoir storage volume V_s (acre-feet) times effective height of dam H_e (feet).
[2] Precipitation depths for either 100-yr return period (P_{100}) or PMP.
[3] Class (a) dams involving industrial or municipal water are to use minimum criteria equivalent to that of Class (b).
[4] Applies when the failure of the upstream dam may endanger the lower dam.
Source: Soil Conservation Service [13]

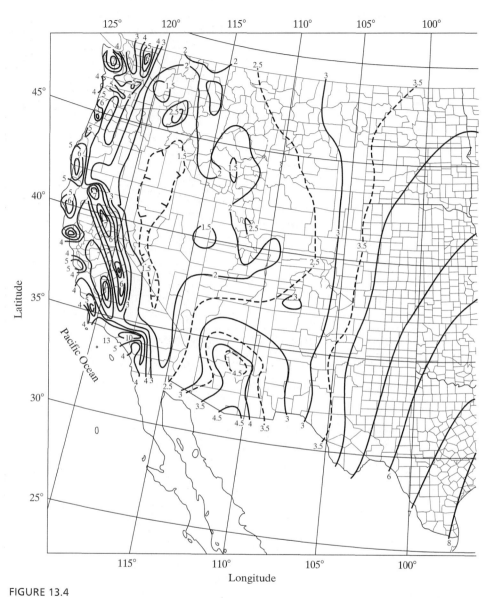

FIGURE 13.4

The 100-year, 6-hr precipitation (in.) for 10 mi^2 or less.
(U.S. Weather Bureau, NOAA. [10])

subsequent section) and the 100-yr rain. Figure 13.4 shows the 6-hr, 100-yr rain depths for the United States [10]. Design storm depths for all watersheds having a time of concentration less than 6-hr are established in Table 13.4. For those watersheds with a greater time of concentration, adjustments are made to the 6-hr storm depth to account for the greater amounts of direct runoff in a longer period of time. These adjustments are discussed in Section 13.4.

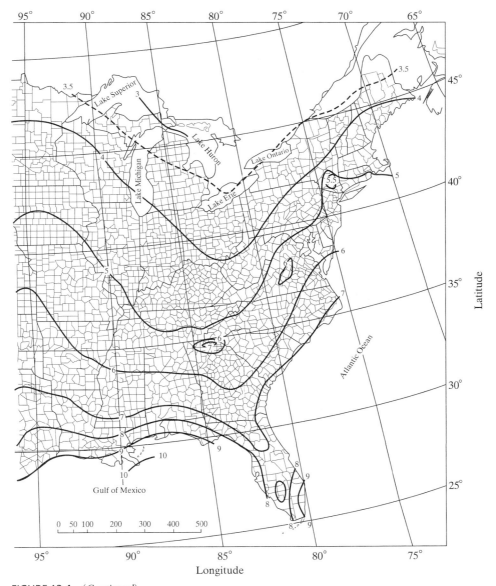

FIGURE 13.4   (*Continued*)

Design Storms for Major Structures

Four general terms are employed to designate design floods for major structures: (1) the probable maximum flood (PMF), (2) the standard project flood (SPF), (3) the frequency-based flood, and (4) the paleoflood (see Section 8.3). The concept of *flood* in this section is described by an entire discharge hydrograph that is generally synthesized from rainfall estimates. Corresponding to the first three flood designations are the storm values, that is, the depth of rainfall, referred to in terms of (1) the probable maximum precipitation, PMP; (2) the standard project storm, SPS; and (3) the frequency-based storm. Design of the reservoir system components is commonly based on one of these four representative terms.

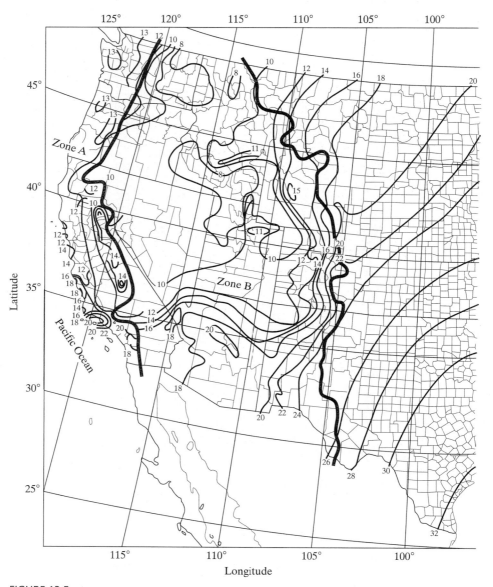

FIGURE 13.5

The 10 mi^2 or less PMP for a 6-hr duration (in.).
(U.S. Weather Bureau, NOAA. [10])

Probable Maximum Precipitation (PMP)

All dams generally receive special attention if they are constructed in populated areas where dam failure could cause the loss of life. Many flood deaths have been caused by dam or levee failure. When this possibility exists, the design storm for dams is established by use of the probable maximum precipitation, PMP. The PMP is generally defined as the reasonable maximization of the meteorological factors that operate to

FIGURE 13.5 (*Continued*)

produce a maximum storm for any given duration and aerial extent. Other definitions have been proposed [15], including

1. The PMP is the maximum amount and duration of precipitation that can be expected to occur on a drainage basin.
2. The PMP is the flood that may be expected from the most severe combination of critical meteorologic and hydrologic conditions that are reasonably possible in the region. The PMP has a low, but unknown, probability of occurrence. It is neither the maximum observed depth at the design location or region nor a value that is completely immune to exceedance.

Estimates of PMP are based on an investigation by the National Weather Service (formerly the U.S. Weather Bureau) conducted to establish the maximum possible amount of precipitable water that could be achieved throughout the United States [16],[17]. Figure 13.5 provides PMP estimates for 6-hr storms. These and similar published charts for other durations are helpful in selecting the PMP for any region in the United States [8].

Critical-Event Design for Major Structures

Hydrologic design aspects of major structures are considerably more complex than those of a small dam, crossroad culvert, or urban drainage system. A design storm hydrograph for a large dam still is required but it is put to greater use. The design storm hydrograph is routed to determine the adequacy of spillways and outlets operated in conjunction with reservoir storage. The economic selection of the spillway size from the various possibilities dictates the final design and is a function of the degree of protection provided for downstream life and property, project economy, agency policy and construction standards, and reservoir operational requirements. Major structure design is largely based on critical-event methods.

For some structures such as large dams, design of the flood-handling facilities by frequency-based methods is not appropriate. For any fixed frequency, a flood exceeding the design level is possible, and may have catastrophic consequences. Rather than selecting a frequency level for design, the design for large water control structures more often is an attempt to find the worst storm or flood of history or to calculate the worst possible future event, and then to design accordingly. These methods include the use of the PMP, SPS, record high storm depths, record high floods, multiples of frequency-based floods, and paleohydrology. The *probable maximum flood*, PMF, is defined as the flood resulting from the PMP.

The SPS differs from a PMP estimate and is patterned after a storm of record that causes the most severe rainfall depth–area–duration relation. Appropriate allowances should be made for inclusion of snowmelt in calculating design storm hydrographs from the standard project storm. Generally, the standard project storm rainfall is approximately 50 percent of the PMP. Records of the four or five largest storms should be critically examined to find a suitable composite for use in calculating the SPS. When these data are not available, a reasonable percentage of the PMP can be substituted.

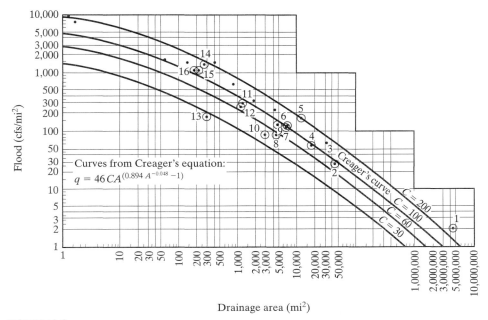

FIGURE 13.6

Creager envelope curves: ⊙ peak inflow for Harza Projects; · recorded unusual flood discharges. (1) Congo at Inga, Congo; (2) Tigris at Samarra, Iraq; (3) Caroni at Guri, Venezuela; (4) Tigris at Eski Mosul, Iraq; (5) Jhelum at Mangla, Pakistan; (6) Diyala at Derbendi Khan, Iraq; (7) Greater Zab at Bekhme, Iraq; (8) Suriname at Brokopondo, Suriname; (9) Lesser Zab at Doken Dam, Iraq; (10) Pearl River, U.S.A.; (11) Cowlitz at Mayfield, U.S.A.; (12) Cowlitz at Mossyrock, U.S.A.; (13) Karadj, Iran; (14) Agno at Ambuklao, Philippines; (15) Angat, Philippines; (16) Tachien, Formosa. *(Note: Curves taken from* Hydroelectric Handbook *by Creager and Justin, New York: Wiley, 1950.)*

Where frequency-based methods of PMP/SPS/PMF studies are unwarranted, design for critical events can be based on the greatest recorded rain or flood flow for the location. Tables or curves of flood data can be developed to give the maximum floods of record in the region under study, such as the Creager flood envelope curves in Fig. 13.6. In cases where estimates of PMP have not been made, volumes of rainfall to be expected can also be approximated from Creager rainfall envelope curves of the world record rainfalls as depicted in Fig. 13.7. Maximum flood flow data for 883 sites up to 25,900 km^2 formed the basis for the Crippen and Bue envelope equation [12] given by:

$$q_p = 10^{[C_1 + C_2 \log A + C_3(\log A)^2 + C_4(\log A)^3]} \tag{13.1}$$

where q_p is the maximum flow (m^3/sec), A is the drainage area (km^2), and the coefficients are from Table 13.5 using Figure 13.8.

Design Criteria for Large Dams

Dams require hydrologic analysis during the design of the original structure and during periodic safety evaluations. Significant economic and human losses are possible when large quantities of water are rapidly released from storage.

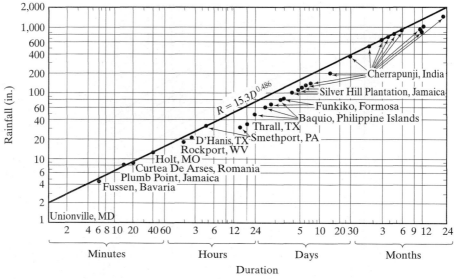

FIGURE 13.7

Creager curves of world's greatest rainfalls.
(*After* Hydroelectric Handbook *by Creager and Justin, New York: Wiley, 1950*)

TABLE 13.5   Coefficients for Crippen and Bue Peak Discharge Envelope Curves

Fig. 13.8 region	Upper limit (km²)	Coefficient			
		C_1	C_2	C_3	C_4
1	26,000	3.203865	.8049163	−.0394382	−.0029757
2	7,800	3.470923	.7472908	−.0551780	−.0000965
3	26,000	3.330746	.8443124	−.0642062	−.0021362
4	26,000	3.258400	.8906783	−.0870959	.0022803
5	26,000	3.726412	.7964721	−.0899000	.0022744
6	26,000	3.500489	.9123848	−.1013380	.0049614
7	26,000	3.326333	.8503960	−.0998747	.0042129
8	26,000	3.236183	.9193289	−.0947436	.0029486
9	26,000	3.503734	.8054884	−.0890172	.0018961
10	2,600	3.314692	1.0386350	−.0597463	−.0042542
11	26,000	3.231389	.8867450	−.1020535	.0045531
12	18,100	3.596209	.8806263	−.0747598	.0000138
13	26,000	3.461373	.8519276	−.1094456	.0058948
14	26,000	3.073497	.6472710	−.0252243	−.0038285
15	50	3.451746	.9718339	−.0617496	−.0057110
16	2,600	3.565536	.9699340	−.0649503	−.0034776
17	26,000	3.389030	.9445212	−.0678131	−.0027647
Nationwide	2,600	3.743026	.7918884	.0244991	−.0192899

Source: After J. R. Crippen and C. D. Bue, "Maximum Flood Flows in The Coterminous United States," U.S.G.S. Water Supply Paper 1887, 1977.

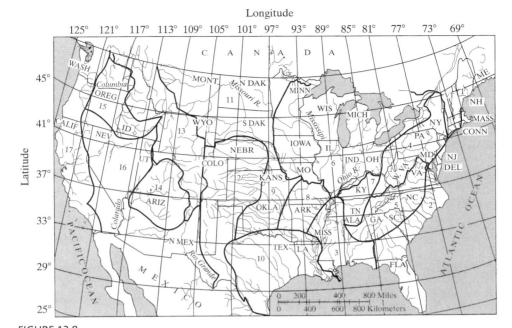

FIGURE 13.8

Map of the coterminous United States showing Crippen and Bue flood region boundaries.
(Source: After USGS Water Supply Paper 1887)

Initial heights of retarded water behind the dam, disregarding the total volume of stored water, can produce destructive flood waves for a considerable distance downstream. Based on two criteria, the Task Force on Spillway Design Floods recommended the classification of large dams as listed in Table 13.6. The type of construction has not been included in this grouping, although it affects the extent of failure resulting from overtopping.

Frequency-Based Criteria for High-Hazard Dams

The federal agency standard for designing and evaluating spillway capacities for dams of high hazard potential has been the PMF. The use of such a conservative standard has been challenged, especially with concern over the need and costs to modify existing federally licensed dams so that they safely pass the probable maximum flood.

A committee of the National Research Council recommended the continued use of the PMF as the general design standard for proposed high-hazard dams [20]. For existing high-hazard dams the committee recommended that design should take into account estimated flood probabilities, expected project performances, and incremental damages that would result from dam failure for a range of floods up to and including the probable maximum flood.

TABLE 13.6   Design Criteria for Large Dams

| Category (1) | Impoundment danger potential | | Failure damage potential[a] | | Spillway design flood (6) |
	Storage (acre-ft)[b] (2)	Height (ft) (3)	Loss of life (4)	Damage (5)	
Major; failure cannot be tolerated	>50,000	>60	Considerable	Excessive or as matter of policy	Probable maximum; most severe flood considered reasonably possible on the basin
Intermediate	1,000–50,000	40–100	Possible but small	Within financial capability of owner	Standard project; based on most severe storm or meteorological conditions considered reasonably characteristic of the specific region
Minor	<1,000	<50	None	Of same magnitude as cost of the dam	Frequency basis; 50–100-year recurrence interval

[a] Damage potential is based on consideration of height of dam above tailwater, storage volume, and length of damage reach, present and future potential population, and economic development of floodplain.
[b] Storage at design spillway pool level.
Source: After Snyder [19]

This latter recommendation of a frequency-based approach signals a significant departure from a long tradition of no-risk design based on the PMF. The U.S. Army Corps of Engineers, a long-time supporter of the PMF standard, is investigating the potential of a frequency-based design approach.

Frequency curves (see Chapter 3) can be plotted and used in major and minor structure design for projects along streams for which lengthy records are available [21]. Most often, the location of the dam is not the gauging site, and stream-routing techniques can effectively transfer the flood peaks. Regionalized flood frequency data may also be employed for small structures. For example, log–Pearson Type III estimates of peak flows for assigned return periods are readily found from Eq. 3.42 if the mean, standard deviation, and skew of logarithms of annual peaks can be estimated. The regional mean and standard deviation of logarithms often correlate well with drainage area and can be determined from nearby gauged stations. Because large samples are required for the determination of skew coefficients, regional skews such as those in Fig. 13.9 are preferred. Customarily, frequency-based floods are not a part of the design criteria for major structures.

13.4   SYNTHESIZING DESIGN STORMS

Once the design frequency has been established, the next step in a structure design is the determination of six storm parameters: the storm duration, the duration of rainfall

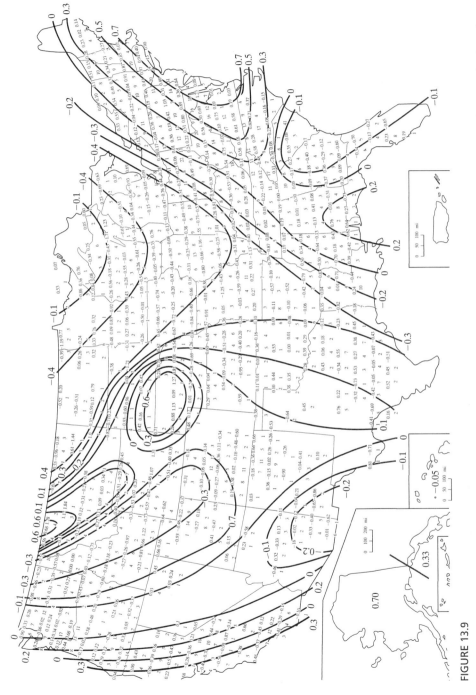

FIGURE 13.9

Generalized skew coefficients of logarithms of annual maximum streamflow: average skew coefficient by 1° quadrangles. The lower number in each quadrangle is the number of stream-gauging stations for which the average shown above it was computed.
(After Ref. 21)

excess, the point depth, any areal depth adjustment, the storm intensity and time distribution, and the areal distribution pattern.

Storm Duration

The length of storm used by the SCS in designing emergency and freeboard hydrographs for small dams is of 6-hr duration or t_c, whichever is greater. Often, the minor structure being designed cannot be justified economically on the basis of this length of storm. For many minor structures, particularly urban drainage structures, a design flood hydrograph is based on a storm duration equal to the time of concentration of the watershed. This procedure uses the rational method of Chapter 11 or the synthetic unit hydrographs of Chapter 9 along with a critical storm pattern produced by arranging the rainfall excess pattern into the most critical sequence. The SCS uses 24-hr durations for all urban watershed studies.

Durations of approximately 6 hr or less are satisfactory for small watersheds, but the lengths of storms in large areas require storm depths for periods of up to 10 days. Frequency-based values are available for durations of from 2 to 10 days for locations within the United States [22]. Similar data are also available for other selected areas outside the United States. Generally, however, design criteria for large dams require estimates of storm depths that do not have frequency levels assigned.

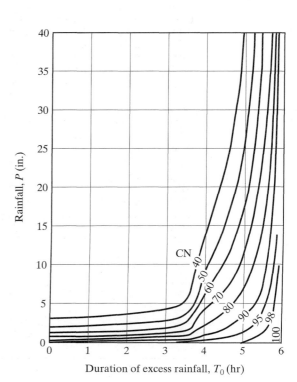

FIGURE 13.10

Duration of excess rainfall for SCS 6-hr design storms.
(After Ref. 23)

Duration of Rainfall Excess

Initial rainfall during most storms infiltrates or is otherwise abstracted, and the duration of excess rain T_0 is less than the actual rain duration by an amount equal to the time that initial abstractions occur. Excess rain duration T_0 can be estimated for a 6-hr storm as a function of the curve number CN and precipitation P from Fig. 13.10. This family of curves was developed by the SCS, [23], where P is the storm depth and CN is a loss parameter defined in Chapter 4. A CN of 100 represents zero losses so that $T_0 = 6$ hr for CN = 100. Table 13.7 is used to find the duration of excess rain for any

TABLE 13.7   Rainfall and Time Ratios for Determining T_0 When Storm Duration Is Greater Than 6 Hours

Rainfall ratio	Time ratio	Rainfall ratio	Time ratio	Rainfall ratio	Time ratio	Rainfall ratio	Time ratio
0	1.00	0.070	0.852	0.140	0.746	0.210	0.684
0.002	0.995	0.072	0.848	0.142	0.744	0.212	0.682
0.004	0.990	0.074	0.844	0.144	0.742	0.214	0.680
0.006	0.985	0.076	0.841	0.146	0.740	0.216	0.679
0.008	0.981	0.078	0.837	0.148	0.739	0.218	0.677
0.010	0.976	0.080	0.833	0.150	0.737	0.220	0.675
0.012	0.971	0.082	0.830	0.152	0.735	0.222	0.673
0.014	0.967	0.084	0.827	0.154	0.733	0.224	0.672
0.016	0.962	0.086	0.824	0.156	0.732	0.226	0.670
0.018	0.957	0.088	0.821	0.158	0.730	0.228	0.668
0.020	0.952	0.090	0.818	0.160	0.728	0.230	0.667
0.022	0.948	0.092	0.815	0.162	0.726	0.232	0.666
0.024	0.943	0.094	0.812	0.164	0.724	0.234	0.666
0.026	0.938	0.096	0.809	0.166	0.723	0.236	0.665
0.028	0.933	0.098	0.806	0.168	0.721	0.238	0.665
0.030	0.929	0.100	0.803	0.170	0.719	0.240	0.664
0.032	0.924	0.102	0.800	0.172	0.717		
0.034	0.919	0.104	0.797	0.174	0.716	(change in	
0.036	0.915	0.106	0.794	0.176	0.714	tabulation	
0.038	0.911	0.108	0.791	0.178	0.712	increment)	
0.040	0.908	0.110	0.788	0.180	0.710	0.250	0.662
0.042	0.904	0.112	0.785	0.182	0.709	0.300	0.651
0.044	0.900	0.114	0.782	0.184	0.707	0.350	0.640
0.046	0.896	0.116	0.779	0.186	0.705	0.400	0.628
0.048	0.893	0.118	0.776	0.188	0.703	0.450	0.617
0.050	0.889	0.120	0.773	0.190	0.702	0.500	0.606
0.052	0.885	0.122	0.770	0.192	0.700	0.550	0.595
0.054	0.882	0.124	0.767	0.194	0.698	0.600	0.583
0.056	0.878	0.126	0.764	0.196	0.696	0.650	0.542
0.058	0.874	0.128	0.761	0.198	0.695	0.700	0.500
0.060	0.870	0.130	0.758	0.200	0.693	0.750	0.447
0.062	0.867	0.132	0.755	0.202	0.691	0.800	0.386
0.064	0.863	0.134	0.751	0.204	0.689	0.850	0.310
0.066	0.859	0.136	0.749	0.206	0.687	0.900	0.220
0.068	0.856	0.138	0.747	0.208	0.686	0.950	0.116

Source: After Ref. 23

TABLE 13.8   Rainfall Prior to Excess Rainfall

CN	$P*$ (in.)	CN	$P*$ (in.)	CN	$P*$ (in.)	CN	$P*$ (in.)	CN	$P*$ (in.)
100	0	86	0.33	72	0.78	58	1.45	44	2.54
99	0.02	85	0.35	71	0.82	57	1.51	43	2.64
98	0.04	84	0.38	70	0.86	56	1.57	42	2.76
97	0.06	83	0.41	69	0.90	55	1.64	41	2.88
96	0.08	82	0.44	68	0.94	54	1.70	40	3.00
95	0.11	81	0.47	67	0.98	53	1.77	39	3.12
94	0.13	80	0.50	66	1.03	52	1.85	38	3.26
93	0.15	79	0.53	65	1.08	51	1.92	37	3.40
92	0.17	78	0.56	64	1.12	50	2.00	36	3.56
91	0.20	77	0.60	63	1.17	49	2.08	35	3.72
90	0.22	76	0.63	62	1.23	48	2.16	34	3.88
89	0.25	75	0.67	61	1.28	47	2.26	33	4.06
88	0.27	74	0.70	60	1.33	46	2.34	32	4.24
87	0.30	73	0.74	59	1.39	45	2.44	31	4.44

Source: After Ref. 23

storm duration greater than 6 hr. The rainfall ratio is the abstraction $P*$ lost before runoff (see Table 13.8) divided by the total precipitation amount P. The time ratio from Table 13.7 is multiplied by the rainfall duration to obtain T_0.

Storm Depth

The probable maximum precipitation or frequency-based 100-year, 6-hr storm depths at any point can be determined from Figs. 13.7 and 13.8. A convenient means of obtaining storm depths for durations other than 6 hr is to use a table or graph of multipliers for various durations. The U.S. Bureau of Reclamation [9] applies the multipliers in Table 13.9 to the PMP from Fig. 13.7 to determine other-duration PMP depths for areas west of the 105° meridian. Similar USBR data east of the meridian are not available.

TABLE 13.9   Constants for Extending 6-hr PMP Design Storms in Areas West of the 105° Meridian to Longer-Duration Periods

Duration (hr)	Constant[a]	Duration (hr)	Constant[a]
8	1.16	22	1.74
10	1.31	24	1.80
12	1.43	30	1.95
14	1.50	36	2.10
16	1.56	42	2.25
18	1.62	48	2.38
20	1.68		

[a] Multiply 6-hr point rainfall from Fig. 13.5 by the indicated constant.
Source: After Ref. 9

TABLE 13.10   TVA Ratios for Adjusting 6-hr Storm Depths for
             Other Durations

Duration	1	2	3	6	12	24
Ratio	0.51	0.68	0.80	1.00	1.13	1.24

The Tennessee Valley Authority (TVA) has regional regulatory and resource development authorities for much of the Tennessee River basin. Over the years the TVA has developed dam design criteria, including its own definition of PMP. It recommends the use of Table 13.10 for adjusting 6-hr storm depths to other-duration storms [18].

The U.S. Soil Conservation Service curve [23] of Fig. 13.11 is available for use in adjusting the PMP and 100-year, 6-hr point rainfall depths from Figs. 13.4 and 13.5. Taken together, Figs. 13.4, 13.5, and 13.11 allow the determination for minor structure design of all values in Table 13.4 for any storm duration.

Using IDF Relationships

For short-duration storms over small areas, the most convenient method of determining storm depth is to acquire the intensity–duration–frequency (IDF) curve for the locale, enter the graph with the selected duration and frequency, and convert the resulting intensity to depth of rain over the selected duration. Area adjustments in the depth may be necessary for basins larger than about one square mile. The IDF curves are available from several sources, including NOAA, the National Weather Service, and more often from the city, county or parish, or state engineer. One such set

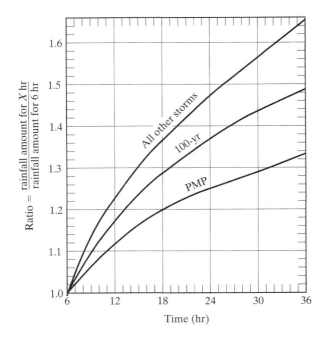

FIGURE 13.11

Relative increase in rainfall amount for storm durations over 6 hr for SCS dam designs.

(After Ref. 23)

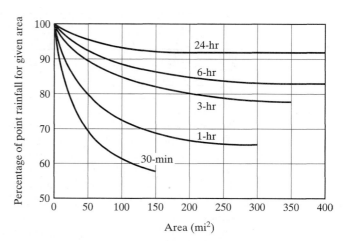

FIGURE 13.12

Area–depth curves for use with duration frequency values.
(After Ref. 25)

of curves was provided in Fig. 4.8 for use in the vicinity of Baltimore, Maryland. Equations that describe the shapes of IDF curves have been developed for a number of major U.S. cities [25]. For small structure designs, the distribution of the selected design storm from an IDF curve is often assumed to be uniform. Alternatively, other distributions described subsequently may be applied.

Area Adjustment

The rainfall depths shown in Figs. 13.4 and 13.5 were derived from frequency analyses (Chapter 3) of *point* measurements and are considered to be applicable only for areas up to 10 mi². For larger watersheds the areal depths are less; adjustment must be made to account for smaller rainfall depths over larger areas.

The U.S. Weather Bureau [25] developed Fig. 13.12 as a guide in reducing point depths to areal depths for areas up to 400 mi². For small watersheds, the SCS applies the ratios from Fig. 13.13 to 6-hr map values from Figs. 13.4 and 13.5. Any PMP value

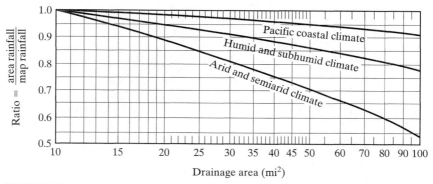

FIGURE 13.13

Rainfall ratios for 10–100 mi² for SCS dams.
(After Ref. 23)

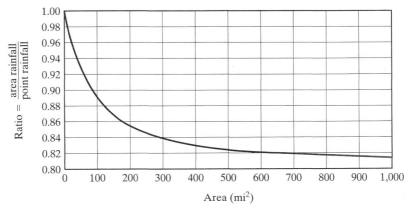

FIGURE 13.14

Conversion ratio from 6-hr point PMP rainfall to 6-hr area rainfall for area west of 105°
meridian.
(After Ref. 9)

from Fig. 13.5 for major designs is modified according to Fig. 13.14. This curve is used
by the U.S. Bureau of Reclamation in areas west of the 105° meridian. For drainage
areas up to 100 mi^2, the TVA recommends use of Fig. 13.15 for adjusting the
expected rainfall over 1 mi^2 (approximately equal to the point rainfall) to larger
areas [18].

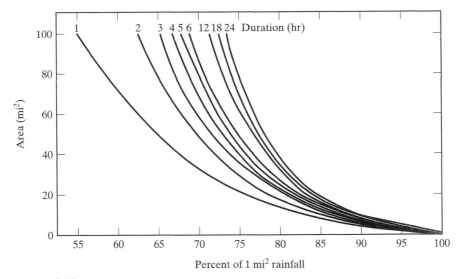

FIGURE 13.15

TVA graph for adjusting 1 mi^2 rain depth to rain depth for basin areas up to 100 mi^2.
(After Ref. 24)

Time Distributions

After the storm depth and duration have been established, the designer must select a representative hyetograph. The choice will significantly affect the shape and peak value of the resulting runoff hydrograph. Any decision must be based on either the worst possible storm pattern or on an analysis of recorded storm distribution patterns.

Huff Time Distributions   Huff [26] divided recorded storm distribution patterns from small midwestern watersheds into four equal-probability groups from the most severe (first quartile) to the mildest (fourth quartile). The median curve for the first-quartile storms is given in Fig. 13.16, which is used, for example, in the RRL and ILLUDAS simulation models of Chapter 11. A 90 percent level is the distribution occurring in 10 percent or less of the storms. Eighty percent of the total rainfall occurs in the first 20 percent of storm time for the 10 percent level in the first-quartile storm. The passage of intense, prefrontal squall lines, typical of thunderstorms, will produce this particular rainfall distribution. On the other hand, the 90 percent level is more indicative of steady rain or a series of rain showers. The 50 percent or median curve is recommended for most applications.

Figure 13.17 shows the 10, 50, and 90 percent histograms for first-quartile storms. Using these storm distributions permits the construction of hyetographs for the design rainfall.

SCS Time Distributions   Time distributions for critical storms for small dam or other minor structure designs are usually assumed to be uniform. The SCS uses a uniform distribution for short-duration storms. Alternatively, Fig. 13.18 is the SCS distribution of the 6-hr storm used in developing emergency spillway and freeboard hydrographs [23]. This curve is very similar to Huff's 50 percent (median) second-quartile curve.

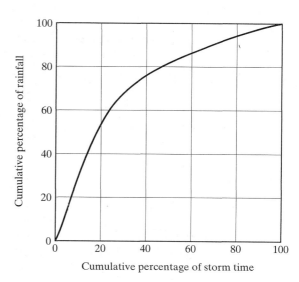

FIGURE 13.16

Time distribution of storm rainfall, median first-quartile curve for point rainfall. *(After Huff [26])*

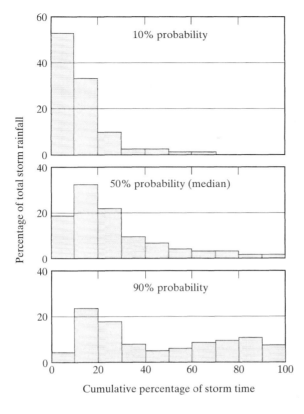

FIGURE 13.17

Selected histograms for first-quartile storms.
(After Huff [26])

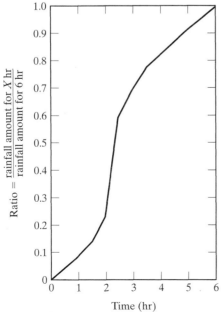

FIGURE 13.18

A 6-hr design storm distribution for SCS dam design.
(After Ref. 23)

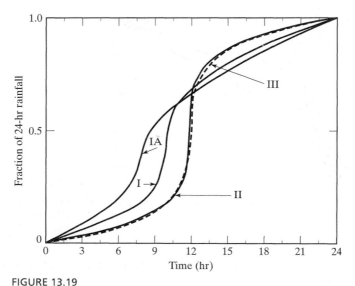

FIGURE 13.19

SCS 24-hr rainfall distributions for zones I, IA, II, and III in Fig. 11.9.
(After Ref. 23)

The SCS also developed 24-hr storm distributions to represent the critical rainfall and runoff volume for peak discharges from watershed sizes normally studied by its engineers. A set of four rainfall distributions were developed and are shown in Fig. 13.19. They are applicable to the various regions shown in Fig. 11.9 and incorporate brief central periods of intense rain within a longer-duration storm. Numerical values for plotting these curves can be found in SCS publications [27].

The greatest peak flows from small basins are usually caused by intense, brief rains. These can occur as distinct events or as portions of a longer storm. The 24-hr storm duration is longer than needed to determine peaks from small watersheds but is appropriate for determining runoff volumes. In light of this, the SCS uses 24-hr storms to study peak flows, volumes of runoff, and direct runoff hydrographs from watersheds normally studied by the agency.

Time distributions for PMP and other storms used in major structure design can be constructed from Fig. 13.20. This family of curves is used by the U.S. Bureau of Reclamation [14] in the three geographical zones shown in Fig. 13.5. The Corps of Engineers uses a distribution curve similar to Fig. 13.18 for 6-hr SPS analyses.

Triangular Distributions   The simplest design storm distribution is a triangular shape. Because the depth, P, and duration, D, of rain are already established, the peak intensity, i_{max}, is $2P/D$, found by solving for the height of the triangular hyetograph as shown in Fig. 13.21. The only remaining decision is the time to the peak, t_p. The ratio t_p/D has been investigated for a large number of storms at locations in California, Illinois, Massachusetts, New Jersey, and North Carolina. Values range from about 0.3 to 0.5 [28]. Once the triangle is constructed, the intensities at regular intervals may be

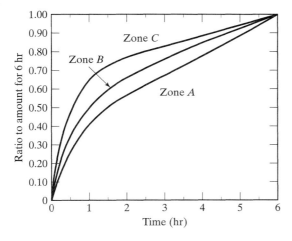

FIGURE 13.20

Distribution of 6-hr PMP for any area west of the 105° meridian.

(After Ref. 14)

graphically or analytically determined for input to the rainfall-runoff model being used for design.

Blocked IDF Distributions   A frequently used procedure for developing a design storm distribution for short-duration storms (up to about 2 hr) is to successively construct blocks of a design storm histogram by using the appropriate intensity–duration–frequency curve to find the rain intensities for Δt, $2\Delta t$, $3\Delta t$, etc., increments of time and then to organize these blocks of rain intensities in some pattern, usually symmetrical, around the center of the storm, making sure that the area under the hyetograph is equal to the design storm depth, P, spread over the design storm duration, D.

To apply the procedure, successive depths of equal-probability storms with durations of Δt, $2\Delta t$, $3\Delta t$, $4\Delta t$, etc., are determined from the IDF curve and tabulated. Next, any of a variety of procedures, such as the *alternating block method*, the *Chicago method*, or the *balanced method*, are available for distributing these blocks and

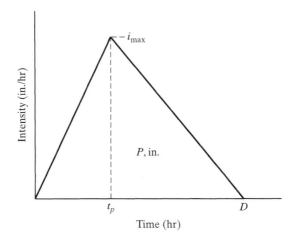

FIGURE 13.21

Triangular design hyetograph.

assuring that the total rain depth equals *P*. Most assume that the highest intensity occurs in the middle of the storm, the second highest occurs next, and so on, working out in both directions from the center block. The balanced method, for example, assumes that a Δ*t*-hr storm with intensity $i_{\Delta t}$ from the IDF curve could occur, with equal probability, during the middle of the *D*-hr design storm. This intensity is plotted as the middle block of the design storm hyetograph. Next, the rain depth for duration 2Δ*t* is obtained from the IDF curve. Its distribution is assumed to be a two-bar histogram with the first half matching the intensity of the Δ*t*-hr storm; the second-half intensity is calculated by spreading the rest of the rain depth for the 2Δ*t*-hr duration uniformly over the second Δ*t* interval. The process is repeated for rain depths for storms with durations of 3Δ*t*, 4Δ*t*, . . . , up to *D*. The goal is to develop a storm hyetograph such that a storm of any duration, centered at the middle of the blocked IDF hyetograph, will have a total rain depth matching the rain depth from the IDF curve for the given duration.

Alternate Alternating Blocked Design Storm   An alternative method of developing a blocked IDF design storm is to place the incremental storm intensity blocks described previously on alternating sides of the most intense storm increment. This alternative also assures that the model user will determine the worst-case *X*-year flood. The procedure is illustrated in Example 13.1 [29].

Example 13.1

Find the 10-year, 1-hour alternating block design storm for a watershed in St. Louis, Missouri. Assume that the storm will be modeled in 5-minute time steps. The 10-year IDF relationship for St. Louis is approximated by $i = 104.7/(t_d^{0.89} + 9.4)$, where *i* is the rainfall intensity in inches per hour and t_d is the duration in minutes.

Solution. The calculations are summarized in Table 13.11, and the resulting design storm hyetograph is shown in Fig. 13.22. Average rainfall intensities are computed for

TABLE 13.11   Development of Alternating Block Design Storm for Example 13.1

Duration (min)	Average intensity (in./hr)	Cumulative depth (in.)	Incremental depth (in.)	Incremental intensity (in./hr)
5	7.68	0.640	0.640	7.68
10	6.09	1.015	0.375	4.49
15	5.09	1.272	0.257	3.09
20	4.39	1.463	0.191	2.31
25	3.88	1.617	0.154	1.82
30	3.48	1.740	0.123	1.49
35	3.16	1.843	0.103	1.25
40	2.90	1.933	0.090	1.07
45	2.68	2.010	0.077	0.93
50	2.50	2.083	0.073	0.82
55	2.34	2.145	0.062	0.73
60	2.20	2.200	0.055	0.66

Source: ASCE [29]

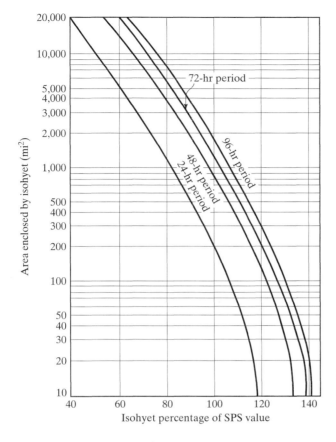

FIGURE 13.25

SPS depth–area–duration curves by 24-hr storm increments.
(*After Ref. 30*)

are multiplied by the 96-hr SPS depth to give an elliptical pattern with the desired average depth. Similar maps for 24-, 48-, or 72-hr storms can be obtained simply by modifying the 96-hr percentages of Fig. 13.25. This is accomplished using the depth–area–duration curves in Fig. 13.24. For example, if a 24-hr storm is used, first note that the A isohyet of Fig. 13.24 encloses an area of 16 mi^2. From Fig. 13.25 the corresponding SPS percentage for a 24-hr storm is 116 percent rather than the 140 percent value used with a 96-hr storm. Therefore the pattern percentages vary with the selected design storm duration.

13.5   URBAN STORM DRAINAGE DESIGN

Storm drainage from urban areas starts as sheet flow on paved areas. These flows combine with flows from rooftops and hydrographs from pervious zones forming inflow to surface swales or open street gutters. For floods up to the design frequency, surface runoff enters storm sewers through street inlets or outlets from detention or retention storage areas. Most modern storm drain design incorporates these types of facilities to keep runoff from minor storms off streets and parking lots. For major storms, good system designs incorporate facilities to handle the overflows at storm sewer inlets (flows in excess of the design flows), as well as regional storm water detention and retention facilities for 25-, 50-, or even 100-yr events.

Most urban storm systems were designed by peak flow methods such as the rational method (see Section 11.3) or through the use of simplified hydrograph methods. For a given design storm, either peak flow rates or entire hydrographs are calculated at the inlet to each storm drain (pipe, open channel, street gutter, gutter inlet, etc.), and the drain is then sized by hydraulic methods to convey the flow safely. Many software packages combine hydrologic and hydraulic algorithms to allow full design capability.

The calculation of individual hydrographs to the inlet of each drain is influenced by the infiltration capacity of the pervious areas, overland flow delays, depression storage, detention in gutters, house drains, catch basins, storm sewer systems, and interception in extensively landscaped locations.

Two items normally accounted for in urban storm drain design are:

1. *Infiltration.* The ability of the soil to infiltrate water depends on many characteristics, as noted in Chapter 7. The range of values given in Table 13.12 is typical of various bare soils after 1 hr of continuous rainfall.

TABLE 13.12   Typical Infiltration Rates for Bare Soils

Soil group	Infiltration (in./hr)
High (sandy, open-structured)	0.50–1.00
Intermediate (loam)	0.10–0.50
Low (clay, close-structured)	0.01–0.10

The influence of grass cover increases these values 3 to 7.5 times.

2. *Retention.* This is usually assumed to be 0.10 in. for pervious surfaces such as lawns and normal urban pervious surfaces.

Calculation of flow time in storm drains can readily be estimated knowing the type of pipe, slope, size, and discharge [21]. Generally, the pipe is assumed to flow full for this calculation. Nomographs like the one shown in Fig. 13.26 are also available to solve the Manning equation for flow in ditches and gutters. The estimation of inlet time is frequently based solely on judgment; reported values vary from 5 to 30 min. Densely developed areas with impervious tracts immediately adjacent to the inlet might be assigned inlet periods of 5 min, but a minimum value of 10–20 min is more usual.

Rational Method

Section 11.3 described the rational method, which allows the peak flow rate at any inlet to be calculated for any recurrence interval. Inputs are the time of concentration at the inlet, contributing drainage area, IDF curves, runoff coefficient, and a frequency

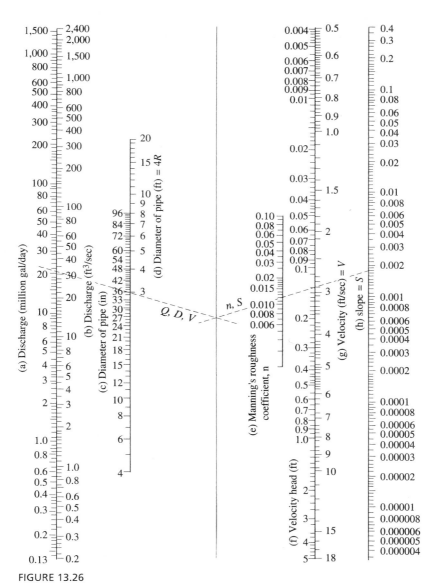

FIGURE 13.26

Flow in pipes (Manning's formula).
(After Ref. 31)

factor for more extreme events. The application of the rational formula to a single sub-area is simple, but differences exist in applying it to a multiple-component drainage system. Texts that illustrate the method aren't consistent in explaining the procedure and can result in confusion [32]–[37]. The American Society of Civil Engineers' manual for design of urban structures is recommended [29].

The key to correct applications for multi-subarea systems is to evaluate each point of design (inlets, channel outfalls, and detention sites) as a single application of Eq. 11.1, using the entire drainage area above the point and the travel time from the most remote point upstream in setting the time of concentration, and storm duration, for that point. Because a storm occurring over the contributing area with this duration is possible, this storm may produce the highest peak flow for the selected recurrence interval. Example 13.2 illustrates the procedure.

Example 13.2

Based on the storm sewer arrangement of Fig.13.27a, determine the outfall discharge. Assume that $C = 0.3$ for residential areas and $C = 0.6$ for business tracts. Use a 5-year frequency rainfall from Fig. 13.27b and assume a minimum 20-min inlet time.

Solution. The tabular solution is listed in Table 13.13. The column headings in Table 13.13 are explained in Table 13.14. Additional columns can be provided to list elevations of manhole inverts, sewer inverts, and ground elevations. This information is helpful in checking designs and for subsequent use in drawing final design plans.

Most designers applying the rational method do not use the time of concentration in its strictest sense; rather, the largest sum of inlet time plus travel time in the storm drain system is taken as the time of concentration at each point. Peak discharge is not the summation of the individual discharges, because peaks from subareas occur at different times and for different points on the IDF curve. The runoff from subareas should be rechecked for each area under consideration. The average intensity I is that for the time of concentration of the total area drained. While I decreases as the design proceeds downstream, the size of the contributing area increases and Q normally increases continuously. It should be reiterated that the design at each point downstream is a new solution of the rational method. The only direct relation from point to point derives from the means for determining an increment of time to be added for a new time of concentration. The discharge is always a new calculation, and never a sum of upstream peak flows. The effect is to provide an equal level of protection (i.e., an equal frequency) at all points in the system.

Storm Water Design Software

Sections 11.4 and 11.5 described many of the available urban hydrology software models developed by public and private vendors [38]–[41]. Computer packages for urban stormwater modeling fall into two categories: those codes intended primarily for the *analysis* of an existing or proposed system, and *design* packages that select storm sewer

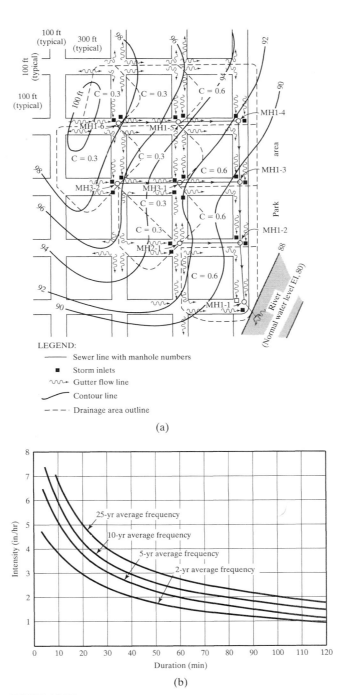

(a)

(b)

FIGURE 13.27

Sample storm drainage problem: (a) typical storm sewer design plan and (b) intensity–duration–frequency rainfall curves for Davenport, Iowa. *(After Ref. 31)*

TABLE 13.13 Typical Storm Sewer Computations for the Rational Method

	Manhole number			Area (acres)		Flow time (min)										Design flow							Sewer invert		Ground elevation	
Line	From	To	Length (ft)	Increment	Total	To upper end	In section	Average runoff coefficient	Rainfall (in./hr)	Runoff (cfs/acre)	Total runoff (cfs)	Slope of sewer (%)	Diameter (in.)	Capacity, full	Velocity, full	Velocity (fps)	Velocity head (ft)	Depth of flow (in.)	Total energy (ft)	Manhole losses (ft)	Manhole invert drop (ft)	Fall in sewer (ft)	Upper end	Lower end	Upper end	Lower end
(1)	(2)	(3)	(4)	(5)	(6)	(7)	(8)	(9)	(10)	(11)	(12)	(13)	(14)	(15)	(16)	(17)	(18)	(19)	(20)	(21)	(22)	(23)	(24)	(25)	(26)	(27)
1	1-6	1-5	400	2.64	2.64	20.0	1.4	0.30	3.7	1.11	2.93	0.85	12	3.3	4.0	4.6		9			...	3.40	93.00	89.60	98.4	94.9
1	1-5	1-4	400	3.61	6.25	21.4	1.2	0.30	3.6	1.08	6.75	0.75	18	9.2	5.1	5.6		11			0.40	3.00	89.20	86.20	94.9	91.8
1	1-4	1-3	400	3.88	10.13	22.6	1.2	0.42	3.4	1.43	14.50	0.45	24	15.2	4.8	5.6		18			0.40	1.80	85.80	84.00	91.8	89.7
3	3-2	3-1	400	5.55	5.55	20	1.1	0.30	3.7	1.11	6.16	1.00	15	6.4	5.1	5.9		12			...	4.00	91.00	87.00	96.2	92.3
3	3-1	1-3	400	6.43	11.98	21.1	1.1	0.30	3.6	1.08	12.92	0.60	24	17.5	5.5	6.1		15			0.60	2.40	86.40	84.00	92.3	89.7
1	1-3	1-2	400	3.92	26.03	23.8	1.1	0.39	3.3	1.29	33.60	0.30	36	37.0	5.1	5.9		26			0.80	1.20	83.20	82.00	89.7	89.5
2	2-1	1-2	400	2.52	2.52	20	1.4	0.30	3.7	1.11	2.80	0.90	12	3.2	4.1	4.7		9			...	3.60	87.50	83.90	92.7	89.5
1	1-2	1-1	400	3.86	32.41	24.9	1.1	0.41	3.2	1.31	42.50	0.24	42	50.0	5.2	5.9		29			0.40	0.96	81.60	80.64	89.5	88.5
1	1-1	Out-fall	125	5.44	37.85	26.0	...	0.44	3.2	1.41	53.20	0.30	42	56.0	5.7	6.6		33			0.10	0.38	80.54	80.16	88.5	...

Source: Ref. 31

576

TABLE 13.14   Definition of Column Headings in Table 13.13

Column	Comment
1	Line being investigated
2, 3	Inlet or manhole being investigated
4	Length of the line
5	Subarea of the inlet
6	Accumulated subareas
7	Value of the concentration time for the area draining into the inlet
8	Travel time in the pipeline
9	Weighted C for the area being drained
10	Rainfall intensity based on time of concentration and a 5-year frequency curve
11	Unit runoff $q = CI$
12	Accumulated runoff that must be carried by the line
13	Slope of line
14	Size of pipe
15	Pipe capacity
16	Velocity in full pipe
17	Actual velocity in pipe

diameters, sizes of channels or ponds, or other features of storm water management facilities. About 80 percent of the software packages surveyed (see Section 11.5) allow the design of detention basin size, and over 70 percent have routines for design of storm sewers or inlets. All but a few (about 10 percent) allow the user to generate full direct runoff hydrographs.

13.6   AIRPORT DRAINAGE DESIGN

Airport drainage is required to dispose of surface water and minimize the interruption of traffic into and out of the area. The total drainage system has several functions [42]: (1) collect and carry off surface water, (2) remove excess groundwater, (3) lower the water table, and (4) protect all slopes from erosion. Only the problem of collecting and removing the surface water is discussed here.

The design of most airport drains relies on the rational method, although some efforts are being made to evaluate these systems by means of other techniques such as ILLUDAS (see Section 11.4). Drainage calculations are usually based on collected rainfall data and a 1-ft contour topographic map of the proposed finished site. The entire drainage system is outlined on this map with the proper identification of subareas, main and lateral storm drains, direction of flow, gradients, inlets, and surface channels. The final design is attained by calculating the most reasonable cost to provide satisfactory drainage.

A step-by-step procedure to design a portion of the surface drainage facilities of an airport is outlined in Example 13.3 as a straightforward application of the rational method. Each subarea size is outlined and a weighted C is adopted, based on $C = 0.90$ for the pavement and $C = 0.30$ for the turf areas. The time of concentration in the system is composed of inlet time and duration of travel in the conduit. The design of the

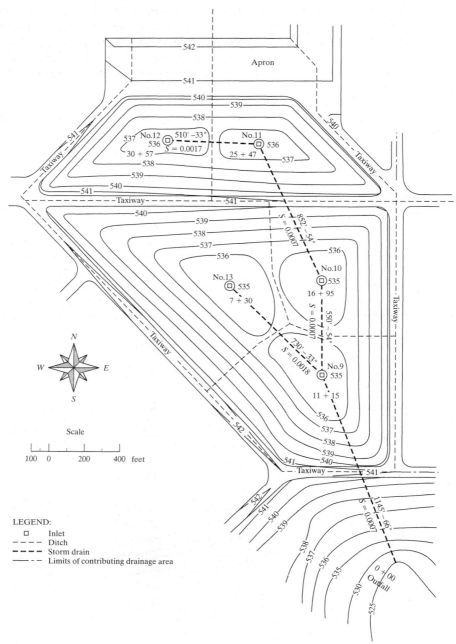

542

Apron

541

540
539
538
537 No.12 510' -33" No.11
536 □ S = 0.0017 □ 536
30 + 57 25 + 47 537
538
539
540
541 Taxiway 541
540
539
538
537 852' - 54" 536
536 S = 0.0007
No.10
No.13 □ 535
□ 535 16 + 95
7 + 30 550' - 54"
S = 0.0007
730' -33" No.9
S = 0.0018 □ 535
11 + 15
536
537
538
539
540
542 541
Taxiway 541
542
541
540
1145' - 66"
539 S = 0.0007
538
537
536
535
530 0 + 00
525 Outfall

Taxiway

N
W E
S

Scale

100 0 200 400 feet

LEGEND:
□ Inlet
- - - Ditch
▪ ▪ ▪ ▪ Storm drain
— ‐ — Limits of contributing drainage area

FIGURE 13.28

Portion of airport for Example 13.3.
(After Ref. 42)

drainage system should be adequate to ensure that ponding will not be excessive in areas adjacent to the runways. The general criterion is that these areas should be at least 75 ft from the bordering pavement. To prevent saturation of the nearby ground, rough calculations to ensure adequate ponding volumes for the design rainfall are desirable. If ponding occurs, routing techniques such as those described in Chapter 9 facilitate calculations of ponding depth from the known storage and outflow characteristics of the system.

Example 13.3

Prepare a surface drainage design for the portion of an airfield shown in Fig. 13.28 for the 5-year frequency rainfall in Fig. 13.29.

Solution. The solution is given in Table 13.15.

Computation of subarea sizes, values of weighted C, and inlet times used in Table 13.15 are listed in Table 13.16, and descriptions of the computations are given in Table 13.17. Calculation of final pipe sizes necessary to drain the system will vary with slope and type of pipe selected. The slope of the pipe is usually controlled by the outlet elevation that must be maintained to allow the system to drain freely. Designs in this example are based on use of a concrete pipe with $n = 0.015$. A nomograph for solution

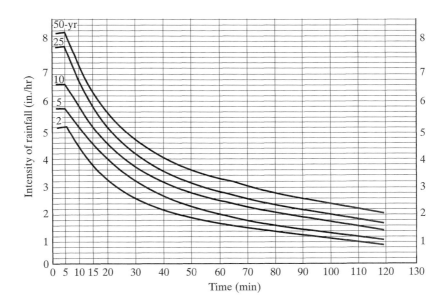

FIGURE 13.29

Intensity–duration–frequency curve for Example 13.3.
(After Ref. 42)

TABLE 13.15   Drainage System Design Data for Example 13.3

Inlet (1)	Line segment (2)	Length of segment (ft) (3)	Inlet time (min) (4)	Flow time (min) (5)	Time of concentration (min) (6)	Runoff coefficient, C (7)	Rainfall intensity (in./hr) (8)	Tributary area, A (acres) (9)	Runoff, Q (cfs) (10)	Accumulated runoff (cfs) (11)	Velocity of drain (ft/sec)[a] (12)	Size of pipe (in.)[b] (13)	Slope of pipe (ft/ft) (14)	Capacity of pipe (cfs) (15)	Invert elevation (16)	Remarks (17)
12	12–11	510	41	2.7	41	0.49	2.40	14.69	17.28	17.28	3.18	33	0.0017	18.90	530.65	(n = 0.015)
11	11–10	852	40	5.0	43.7	0.53	2.31	14.72	18.02	35.30	2.84	54	0.0007	45.00	528.03	See note below
10	10–9	550	34.8	3.3	48.7	0.35	2.15	11.97	9.01	44.31	2.84	54	0.0007	45.00	527.44	See note below
13	13–9	730	48.6	3.7	48.6	0.35	2.16	21.50	16.25	16.25	3.27	33	0.0018	19.40	530.11	
9	9–out	1,145	36.3	5.9	52.3	0.35	2.03	16.05	11.40	71.96	3.24	66	0.0007	77.00	526.05	
Out															525.25	

[a] Minimum velocity is 2.5 fps.

[b] Minimum pipe size is 12-in. diameter for maintenance purposes.

Note: The time of concentration for inlet 11 is 43.7 min (41 + 2.7 = 43.7), which is the most time-remote point for the inlet. Also, the time of concentration for inlet 10 is 48.7 min (41 + 2.7 + 5.0 = 48.7).

Source: Ref. 42

TABLE 13.16 Design Data for Drainage Example 13.3

Inlet number	Tributary area to inlets (acres)				C^a	Distance from remote point to inlet (ft)			Time for overland flow (min)		
	Pavement	Turf	Both	Subtotal		Pavement	Turf	Total	Pavement	Turf	Total
12	4.78	9.91	14.69	14.69	0.49	100	790	890	4	37	41
11	5.48	9.24	14.72	29.41	0.53	90	750	840	4	36	40
10	1.02	10.95	11.97	41.38	0.35	65	565	630	3.5	31.3	34.8
13	1.99	19.51	21.50	21.50	0.35	110	1,140	1,250	4.3	44.3	48.6
9	1.46	14.59	16.05	78.93	0.35	85	612	697	3.9	32.4	36.3
Totals	14.73	64.20	78.93								

[a] Weighted C based on C = 0.9 for pavement and C = 0.3 for turf.

Source: Ref. 42

TABLE 13.17   Descriptions of the Computations in Table 13.15

Column	Comment
1	Inlet being investigated
2	Line segment
3	Length of line
4	Inlet time
5	Flow time in line obtained by dividing the length of the line by the velocity of the drain
6	Time of concentration
7	Weighted value of C
8	Rainfall intensity based on the time of concentration as the duration and 5 years as the frequency of design
9	Acreage of subarea immediately tributary to the inlet
10	$Q = CIA$
11	Accumulated runoff that must be carried by the next line being computed
12	Velocity of flow for full pipe flow. Velocities for actual flow rates can also be used
13	Pipe size
14	Slope of the line
15	Pipe capacity that must be larger than the estimated flow
16	Invert elevation of the pipeline
17	Pertinent remarks relative to the design

of the discharge as a function of the size and slope was provided in Fig. 13.26. The minimum velocity allowed in the pipe is 2.5 fps to prevent excessive settling of sediment.

13.7   DETENTION STORAGE DESIGN

An almost universal philosophy in storm water management is that the increase in peak flow rate due to urbanization must be mitigated by reducing the peak flow rates to attain predevelopment values. Often the property developer is required to carry all or most of the economic burden for assuring that peak flows leaving the property do not exceed the predevelopment conditions.

Figure 13.30 presents the effect that detention storage can have in meeting this goal. Urbanization causes the predevelopment-condition hydrograph A to be transformed to hydrograph B (see Section 11.2 for a complete discussion of the causes).

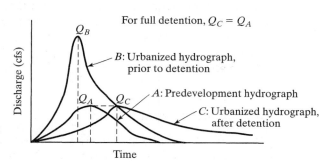

FIGURE 13.30

Effect of urbanization on predevelopment hydrograph, and effect of detention on urbanized hydrograph.

Both the peak rate and the total volume of runoff increase for any storm. Detention storage, applied to the urbanized area, has the effect of transforming hydrograph B to hydrograph C, which has a different shape but the same peak flow rate as the predevelopment hydrograph A. If no permanent retention of water is planned, hydrograph C also has the same runoff volume as hydrograph B. This section describes four methods of assessing the needed size and configuration of detention facilities.

Storm water detention involves construction of holding facilities that temporarily detain and attenuate the incoming runoff from an urbanized area. The reservoir storage routing procedure described in Section 9.5 is used to analyze the abatement effect. A depression or small dam is normally constructed with outlet facilities that force sufficient storage of the inflow hydrograph to reduce the peak outflow to the desired rate. Thus, design of the storage volume and outlet works requires knowledge of the predevelopment peak flow rate plus an estimate of the inflow hydrograph for urbanized conditions. Hydraulic analysis of the spillways is also required, but only the hydrologic aspects are described here.

Locating sufficient undeveloped land for construction of the facility is often difficult, and many cities are implementing underground disposal in gravel or perforated pipe infiltration galleries. Reduction of the increased runoff volume is generally not feasible, and the common objective in detention design is to reduce the peak flow rate for some design storm to its predevelopment value. Safety of the facility for reasonably anticipated storms exceeding the design storm should also be evaluated.

The obligation for planning and construction of detention facilities has rested with developers in most cases, although many communities choose to mitigate development by constructing regional facilities. In this case, the costs are prorated on the basis of contribution to the increases, which are usually assessed as site-development fees when the owner applies for a permit to develop the tract.

Types of Detention Facilities

Both structural and nonstructural methods have been effective in detaining urban storm runoff. Structural detention involves the construction of surface or underground storage areas that accomplish the attenuation goal. Nonstructural methods take advantage of the detention properties of facilities constructed with some other primary purpose. Table 11.2 listed some of the latter.

Detention basins or ponds are the most common structural measures. They are classified as *wet* or *dry*, depending on whether the principal outlet is constructed above the lowest point in the depression. Wet structures are designed to retain some of the inflow in a pool, and are known as *retention basins*. Inventories of constructed detention basins reveal that about 80 percent of the facilities in the United States and Canada are dry. Equal percentages are privately versus publicly owned [45].

Determining Detention Basin Size

Four of several available hydrologic methods for estimating detention and retention basin volumes and configurations are described, including the natural storage loss method, rational triangular hydrograph method, SCS TR-55 procedure, and storage indication method. Some give the required detention storage size directly, while others require repeated routing of the inflow hydrograph through a trial basin configuration until the desired result is achieved. In rare instances, hydrodynamic routing methods [1] are used to simulate the unsteady, nonuniform hydraulics within the basin and at the outlet.

Natural Storage Loss Method   An approximation of the detention volume needed can be obtained by calculating the volume difference between the predevelopment and urbanized hydrographs. This is the difference in areas under hydrographs A and B in Fig. 13.30 and represents the natural storage lost by the urbanization, so it is logical to provide an equivalent storage replacement. The presumption is that a pond built to this size could store the entire excess caused by the development, and the outlet works could simultaneously be designed to attenuate the peak outflow rate so that $Q_C = Q_A$ (see Fig. 13.30).

When full hydrographs are not available, an even simpler approximation of detention storage size, using the natural storage loss method, can be obtained by assuming that the predevelopment and urbanized hydrographs each have time bases equal to $2t_c$ (see Section 9.2). The times of concentration would be different, accounting for the natural and urbanized conditions (equations for both are provided in Section 9.2). Once peak flows for each condition are calculated by the rational formula, the area under each curve can be estimated from the known triangular shape.

Example 13.4

The 7-acre tract in Fig. 11.5 is to be fully developed (new $C_1 = C_2 = 0.9$, new $t_c = 10$ min), and the effect of development needs to be attenuated by installing detention storage at the outlet to reduce the 50-yr peak to the current value. From Example 11.1, the predevelopment Q_{50} was 24 cfs and the t_c was 20 minutes; hence, the volume of predevelopment runoff is 480 cfs-minutes.

Solution.   For the fully developed condition with $t_c = 10$ minutes, Fig. 11.6 gives $I_{50} = 7.0$ in./hr, and

$$Q_{50} = CIA = 1.2 \times 0.9 \times 7.0 \times 7 = 52.9 \text{ cfs}$$

With the urbanized $t_c = 10$ minutes, the 50-yr urbanized runoff volume is 529 cfs-minutes. The difference of 49 cfs-minutes is the required detention storage by this method. This converts to 2,940 ft^3 or 0.067 acre-ft. An area of about 1,000 ft^2 could be provided with a depth of about 3 ft.

Modified Rational Triangular Hydrograph Method   A slightly modified procedure can be applied when the target outflow rate is known, and when the inflow and outflow hydrographs can be approximated as triangles. The preliminary estimate of the storage volume can be obtained by the modified rational analysis shown in Fig. 13.31.

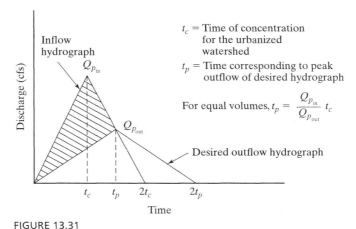

FIGURE 13.31

Rational triangular hydrograph method of estimating detention storage volume.

From the discussion in Section 9.2, the time base of an urban hydrograph is approximately the storm duration, D, plus the time of concentration, t_c. For the rational method, D equals the time of concentration, so the hydrograph time base is $2t_c$. Because the outflow hydrograph from any linear storage reservoir peaks at the intersection of the falling limb of the inflow hydrograph (see Problem 9.59), the peak time and time base of the outflow hydrograph can be found from geometry by setting the volumes under each hydrograph equal to each other. The trial value for detention storage volume is the difference, shown as the shaded area in Fig. 13.31.

SCS TR-55 Procedure   After examining a number of TR-20 runs, the SCS identified a power-series relationship between the urban detention storage volume V_s, volume of runoff V_r, and peak rate of outflow Q_p. The equation is:

$$V_s/V_r = C_0 + C_1 Q_p + C_2 Q_p^2 + C_3 Q_p^3 \qquad (13.2)$$

where $C_0, C_1, C_2,$ and C_3 are coefficients from Table 13.18. For a given design, the detention storage volume can be determined by entering known runoff inflow volumes and desired peak outflow rates into Eq. 13.2.

Storage Indication (Modified Puls) Method   Once an initial detention storage volume is estimated, the most common method of affirming its function is to route the

TABLE 13.18   Coefficients for Use in Equation 13.2

Rainfall distribution type (Fig. 11.9)	C_0	C_1	C_2	C_3
I or IA	0.660	−1.76	1.96	−0.730
II or III	0.682	−1.43	1.64	−0.804

Source: After U.S. Soil Conservation Service [27]

inflow hydrograph through the basin by the storage indication method described in Section 9.5. The method is resident and easily implemented in most storm water design software packages. Steps in designing the detention basin are:

1. Compute the inflow hydrograph for the urbanized condition at the detention basin location.
2. Determine the desired peak outflow rate, usually the rate for undeveloped conditions, or it could be the downstream channel capacity.
3. From topographic data, find the storage volume in the basin at several depths, resulting in a storage–elevation curve.
4. Set the initial crest of the overflow level of the basin at an elevation corresponding with the storage volume from Step 2, adding some freeboard height for wave action if appropriate.
5. Perform a hydraulic analysis of the proposed principal spillway, usually constructed as a drop inlet to a discharge pipe or as an overflow weir, and develop a stage–outflow curve for a range of possible stages. For higher stages, the outflow may be a combination of principal spillway plus emergency spillway discharge.
6. Select a routing period, Δt, such that there will be 5 to 10 points on the rising portion of the inflow hydrograph. If possible, have one point fall under or near the peak.
7. Develop the storage indication curve (see Fig. 9.24).
8. Route the inflow hydrograph, and then revise the crest elevation, outlet pipe size, or basin dimensions, and repeat the above steps until a satisfactory result is found.

Other Urban Design Resources

Development of hydrologic parameters for design of storm sewer pipes, street gutters, or detention basins is by the rational method or modified rational hydrograph method (see Section 11.4) when peak flow rates and approximate hydrographs are adequate, or by unit hydrograph and kinematic-wave hydrograph synthesis methods when greater detail is needed. The latter usually involve use of public domain or vendor-developed storm water design software. Chapter 12 details the hydrologic aspects of computerized hydrologic design tools. In addition to the material presented in this text, numerous urban storm water design texts and handbooks detail uses of the rational method, modified rational method, ILLUDAS, TR-55, SWMM, DR3M, or other tools in designing urban storm drainage facilities. Additionally, many state departments of transportation or city and county engineers' offices have developed locally applicable drainage design manuals. As well, the American Society of Civil Engineers has developed a manual of practice [29] for storm water design. The discussion of urban models in Section 11.5 includes a useful "shopper's guide" to urban drainage analysis and design software.

SUMMARY

Much of the material presented in hydrology textbooks is relevant to the design of hydraulic structures. This chapter summarizes the main tenets of modern hydrologic design of major and minor structures that are subjected to high flow rates.

PROBLEMS

SECTION 13.1: HYDRAULIC STRUCTURE DESIGN METHODS

13.1 Plot the frequency data from Example 8.4 on log probability paper and extrapolate the curve to estimate the 500-yr event. Compare this with the range of 500-yr events described in Section 13.1 as the fourth critical event method.

13.2 Indicate which of the five methods described in Section 13.1 would be most applicable in designing each of the following:

(a) A large dam upstream of a populated floodplain.

(b) A low-level bridge where the owners wish to minimize long-term costs, including the costs of repairs expected because the bridge will have frequent flood damage.

(c) An urban storm drain pipe that will carry runoff from a 35-acre subdivision.

SECTION 13.2: HYDROLOGIC DESIGN DATA

13.3 A design inflow hydrograph to a reservoir is needed at a site where no records of stream-flow are available. List the general steps you would take as a hydrologist in developing the entire design inflow hydrograph.

SECTION 13.3: HYDROLOGIC DESIGN STANDARDS AND CRITERIA

For Problems 13.4–13.8, refer also to Chapter 3.

13.4 What return period (in years) must an engineer use in the design of a bridge opening if it is acceptable to have only a 19 percent risk that flooding will occur at least once in two consecutive years?

13.5 A temporary spillway for a dam has a capacity of 3,000 cfs. Any discharge equaling or exceeding 3,000 cfs will result in damage or destruction of the spillway structure. If the frequency of a 3,000-cfs flood is 0.10, what is the probability that the capacity of the spillway will be equaled or exceeded at least once in a 3-year period required for construction of a permanent spillway?

13.6 A building site near a small natural stream was flooded 20 times in the last 50 years.

(a) During the next 3 years, what is the risk to any construction on the site?

(b) What is the probability of flooding next year?

(c) What is the probability of flooding in at least 1 of the next 3 years?

(d) What is the probability of three consecutive years of safe construction?

13.7 The 75-year record of peak annual earthquake magnitudes at Crete, California, reveals that the values follow an extreme-value distribution with a mean of 5.2 and a standard deviation of 2.0 on the Richter scale. Determine the probability of completing the construction of a nuclear power plant without having an earthquake magnitude exceeding 12.0 during the 10-year construction period.

13.8 Historical records of power failures in a buried power cable near a stream reveal that power failure occurs for a variety of reasons on the average of once every 5 years. Records also show that whenever the stream floods, the chances of a power failure are increased to 40 percent. The river reaches flood stage once every 10 years.

(a) Demonstrate that a flood in the stream and a power failure in the cable are dependent events.

(b) Are the two events mutually exclusive? Why?

(c) Find the probability of the joint occurrence of a flood and a power failure in any year.

(d) Find the probability of the occurrence of either event in any year.

(e) Find the probability of a power failure in both of two consecutive years.

SECTION 13.4: SYNTHESIZING DESIGN STORMS

13.9 Compare the TVA storm depth adjustments in Table 13.10 with the USBR values from Table 13.9 for durations of 6, 12, and 24 hr. Discuss.

13.10 Compare the TVA storm depth adjustments in Fig. 13.15 with the SCS values from Fig. 13.13 for areas of 20, 40, 60, 80, and 100 mi^2. Discuss.

13.11 Using the IDF curves in Fig. 4.7 for Baltimore, MD, develop a triangular 10-yr design storm hyetograph for a 300-acre watershed having a 1.2-hr time of concentration. Use $t_p/D = 0.4$.

13.12 Repeat Problem 13.11 but develop a blocked IDF 10-yr design storm hyetograph using the balanced method. Discuss the differences between the resulting triangular and blocked hyetographs, especially noting the different effects they would have on runoff hydrographs.

13.13 Calculate a 100-year minor structure design storm hyetograph for subarea I in the Oak Creek watershed, Fig. 12.10.

13.14 Calculate a major structure design storm for subarea I in the Oak Creek watershed, Fig. 12.10. Assume that the storm will have a uniform areal distribution.

13.15 Construct a major structure design storm pattern for the entire Oak Creek watershed, Fig. 12.10, using Figs. 13.23 and 13.24. Describe how you would orient the pattern to create the most severe conditions at point 8.

SECTION 13.5: URBAN STORM DRAINAGE DESIGN

13.16 The 2-hr unit hydrograph for a 5,600-acre urban watershed is:

Time (hr)	0	1	2	3	4	5	6	7
Q (cfs)	0	400	1,000	800	300	200	100	0

The local 10-year IDF curve is linear with the equation $I = 5.6 - 0.2D$, where I is the rain intensity in inches per hour and D is the rain duration in hours. Use unit-hydrograph concepts to determine the direct runoff hydrograph for a 10-year design storm. ($\phi = 0.6$ in./hr.)

13.17 Except for soil type, two drainage basins are otherwise identical. For the same storm, which basin would produce the most direct runoff, one containing SCS Group A soils or one containing Group D soils?

SECTION 13.6: AIRPORT DRAINAGE DESIGN

13.18 Using the data in Table 13.16 and Figs. 13.28 and 13.29, repeat Example 13.3 using a 10-yr design frequency.

13.19 Develop a spreadsheet to perform the calculations in Table 13.15, and verify it by reconstructing the results in the table.

SECTION 13.7: DETENTION STORAGE DESIGN

13.20 Given the following watershed data, route a storm hydrograph through a reservoir at the watershed outlet and plot the resulting outflow hydrograph. Show all the work in a neat and logical order. Area = $3.75 \, mi^2$; length = $5.80 \, mi$; elevation of the bottom of the spillway = 1,160.0 ft; spillway width B = 500 ft; spillway coefficient C = 3.0.

   (a) Develop a curve relating water elevation and storage volume.

   (b) Develop a storage–indication curve (Fig. 9.24) using Δt = 0.2 hr.

   (c) Route the storm inflow hydrograph from Problem 9.56 through the reservoir. Assume that at the start of rainfall, the elevation of the water surface in the reservoir is 1,158.0 ft.

   (d) Find the elevation for the top of the dam. Discuss.

Elevation–Area Data for Problem 13.20

Elevation (ft)	Outlet distance (ft)	Area $\times \, 10^{-6} \, ft^2$ (of respective contours)
1,110	0	0
1,120	1,400	0.85
1,140	4,600	3.75
1,160	8,000	10.80
1,180	11,500	25.00
1,200	14,700	
1,220	17,400	
1,240	20,000	
1,260	22,600	
1,280	24,800	
1,300	26,600	
1,320	28,000	
1,340	29,200	
1,360	30,000	
1,376	30,624	

13.21 Repeat Problem 13.20 for a Class (b) structure in an area selected by the instructor.

REFERENCES

[1] W. Viessman and G. L. Lewis, *Introduction to Hydrology, 4th ed.*, New York: HarperCollins College, 1996.

[2] V. P. Singh, *Hydrologic Systems,* Englewood Cliffs, NJ: Prentice-Hall, 1989.

[3] I. Watson and A. Burnett, *Hydrology—An Environmental Approach,* Boca Raton, FL: Lewis, 1995.

[4] A. D. Ward and W. J. Elliot, *Environmental Hydrology,* Boca Raton, FL: Lewis, 1995.

[5] K. C. Patra, *Hydrology and Water Resources Engineering,* Boca Raton, FL: CRC Press, 2000.

[6] "HEC-17: The Design of Encroachment on Flood Plains Using Risk Analysis," U.S. Federal Highway Administration, Washington, D.C., 1980.

[7] G. L. Lewis, "Jury Verdict: Frequency versus Risk-Based Culvert Design," *J. Water Resources Planning and Management Division,* ASCE, Vol. 118, No. 2, March–April 1992.

[8] U.S. National Weather Service, "Hydrometeorological Reports," Washington, D.C.

[9] U.S. Bureau of Reclamation, *Flood Hydrology Manual, A Water Resources Technical Publication,* U.S. Department of the Interior, Denver Office, 1989.

[10] National Oceanic & Atmospheric Administration, *NOAA Rainfall Atlas 2,* 1973.

[11] U.S. Soil Conservation Service, "Earth Dams," Engineering Memorandum 27 (rev.), Sec. A, Washington, D.C., March 1965.

[12] *Design of Small Dams,* Bureau of Reclamation, U.S. Department of the Interior, Washington, D.C.: U.S. Government Printing Office, 1965.

[13] U.S. Soil Conservation Service, "Earth Dams and Reservoirs," *Technical Release No. 60,* U.S. Department of Agriculture, Oct. 1985.

[14] H. O. Ogrosky, "Hydrology of Spillway Design: Small Structures—Limited Data," *Proc. ASCE J. Hyd. Div.* **90**(HY3), 295–310(May 1964).

[15] Federal Emergency Management Agency, *Proceedings, Probable Maximum Precipitation and Probable Maximum Flood Workshop, Proceedings, FEMA Workshop,* Berkeley Springs, W. Virginia, May 1990.

[16] "Generalized Estimates of Maximum Probable Precipitation Over the United States East of the 105th Meridian," U.S. Weather Bureau Hydrometeorological Report No. 23, June 1947.

[17] "Generalized Estimates of Probable Maximum Precipitation for the United States West of the 105th Meridian for Areas to 400 Square Miles and Durations to 24 Hours," U.S. Weather Bureau Tech. Paper 38, 1960.

[18] "Probable Maximum TVA Precipitation for Tennessee River Basins Up to 3000 Square Miles in Area and Durations to 72 Hours," U.S. Department of Commerce, ESSA, Weather Bureau Hydrometeorological Rep. No. 45, 1969.

[19] F. F. Snyder, "Hydrology of Spillway Design: Large Structures—Adequate Data," *Proc. ASCE J. Hyd. Div.* **90**(HY3), 239–259(May 1964).

[20] National Research Council, "Safety of Dams—Floods and Earthquake Criteria," Committee on Safety Criteria for Dams, National Academy Press, 1985.

[21] "Guidelines for Determining Flood Flow Frequencies," U.S. Water Resources Council, Hydrology Committee Bulletin 17B, Revised, Washington, D.C., Sept. 1981.

[22] "Two-to-Ten-Day Precipitation for Return Periods of 2 to 100 Years in the Contiguous United States," U.S. Department of Commerce, Tech. Paper No. 49, 1964.

[23] "Hydrology," Suppl. A to Sec. 4, *Engineering Handbook,* U.S. Department of Agriculture, Soil Conservation Service, 1968.

[24] National Oceanic & Atmospheric Administration, *Hydrometeorological Report No. 56,* "Probable Maximum and TVA Precipitation Estimates with Areal Distributions for Tennessee River Drainages Less Than 3000 Square Miles in Area," 1986.

[25] "Rainfall Frequency Atlas of the United States for Durations from 30 Minutes to 24 Hours and Returns Periods from 1 to 100 Years," U.S. Weather Bureau Tech. Paper 40, 1963.

[26] F. A. Huff, "Time Distribution of Rainfall in Heavy Storms," *Water Resources Res.* **3**(4), 1007–1019(1967).

[27] Soil Conservation Service, "Urban Hydrology for Small Watersheds," U.S. Department of Agriculture, Technical Release 55, June 1986.

[28] B. C. Yen, and V. T. Chow, "Design Hyetographs for Small Drainage Structures," Proceedings, ASCE, JHD, V. 106, HY6, 1980.

[29] American Society of Civil Engineers, "Design and Construction of Urban Stormwater Management Systems," *ASCE Manuals and Reports of Engineering Practice No. 77,* New York: ASCE, 1992.

[30] "Standard Project Flood Determinations," Civil Engineer Bulletin No. 52-8, EM 1110-2-1411, U.S. Department of the Army, Office of the Chief of Engineers, Washington, D.C., revised 1965, pp. 1–19.

[31] "Design and Construction of Sanitary Storm Sewers," ASCE Manuals and Reports on Engineering Practice, No. 37.

[32] G. A. Burton, *Stormwater Effects Handbook: A Toolbox for Watershed Managers, Scientists, and Engineers,* Boca Raton, FL: CRC Press, 2001.

[33] K. C. Patra, *Hydrology and Water Resources Engineering,* Boca Raton, FL: CRC Press, 2000.

[34] V. T. Chow, D. R. Maidment, and L. W. Mays, *Applied Hydrology,* New York: McGraw-Hill, 1988.

[35] R. S. Gupta, *Hydrology and Hydraulic Systems,* Prospect Heights, IL: Waveland Press, 1995.

[36] C. T. Haan, B. J. Barfield, and J. C. Hayes, *Design Hydrology and Sedimentology for Small Catchments,* San Diego: Academic Press, 1994.

[37] R. H. McCuen, "Hydrologic Design of Highways," *U.S. Federal Highway Administration Hydraulic Design Series 2,* 1995.

[38] D. H. Hoggan, *Computer-Assisted Floodplain Hydrology and Hydraulics,* New York: McGraw-Hill, 1989.

[39] S. J. Nix, *Urban Stormwater Modeling and Simulation,* Boca Raton, FL: Lewis, 1994.

[40] G. L. Lewis and D. P. Gilbert, "A Shopper's Guide to Urban Stormwater Micro Software," *Proceedings, ASCE Hydraulics Division Specialty Conference,* Orlando, FL, August 1985.

[41] D. F. Kibler, M. E. Jennings, G. L. Lewis, B. A. Tschantz, and S. G. Walesh, ASCE Task Committee on Urban Stormwater Software, "Microcomputer Software in Urban Hydrology," *HYDATA,* American Water Resources Assoc., Vol. 10, No. 5, Sept. 1991.

[42] "Airport Drainage," Federal Aviation Agency, Advisory Circular, AC No. AC150/5320-5A. Washington, D.C.: U.S. Government Printing Office, 1966.

[43] R. H. McCuen, *Hydrologic Analysis and Design,* Englewood Cliffs, NJ: Prentice Hall, 1989.

[44] D. R. Maidment (ed.), *Handbook of Hydrology,* New York: McGraw-Hill, 1993.

[45] M. Wanielista, *Hydrology and Water Quality Control,* Wiley, 1990.

[46] APWA, "Urban Stormwater Management," Special Report No. 49, American Public Works Association, 1981.

[47] David Kibler (ed.), *Water Resources Monograph 7, Urban Storm Water Hydrology,* American Geophysical Union, 1982.

Appendixes

APPENDIX: A

TABLE A.1 Water Properties, Constants, and Conversion Factors

Gas constants (R)
   $R = 0.0821$ (atm)(liter)/(g-mol)(K)
   $R = 1.987$ g-cal/(g-mol)(K)
   $R = 1.987$ Btu/(lb-mol)(°R)
Acceleration of gravity (standard)
   $g = 32.17$ ft/sec^2 $= 980.6$ cm/sec^2
Heat of fusion of water
   79.7 cal/g $= 144$ Btu/Ib

Heat of vaporization of water at 1.0 atm
   540 cal/g $= 970$ Btu/Ib

Specific heat of air
   $C_p = 0.238$ cal/(g)(°C)
Density of dry air at 0°C and 760 mm Hg $= 0.001293$ g/cm^3

Conversion factors

1 second-foot-day per square mile $= 0.03719$ inch
1 inch of runoff per square mile $= 26.9$ second-foot-days
$= 53.3$ acre-feet
$= 2,323,200$ cubic feet
1 cubic foot per second $= 0.9917$ acre-inch per hour
$= 1$ sec-ft $= 1$ cusec
1 horsepower $= 0.746$ kilowatt
$= 550$ foot-pounds per second
$e = 2.71828$
$\log e = 0.43429$
$\ln 10 = 2.30259$

Metric equivalents

1 foot $= 0.3048$ meter
1 mile $= 1.609$ kilometers
1 acre $= 0.4047$ hectare
$= 4047$ square meters
1 square mile (mi^2) $= 259$ hectares
$= 2.59$ square kilometers (km^2)
1 acre foot (acre-ft) $= 1233$ cubic meters
1 million cubic feet (mcf) $= 28,320$ cubic meters
1 cubic foot per second (cfs) $= 0.02832$ cubic meters per second
$= 1.699$ cubic meters per minute
1 acre-in. per hour $= 1.008$ cubic feet per second (cfs)
1 second-foot-day (cfsd) $= 2447$ cubic meters
1 million gallons (mg) $= 3785$ cubic meters
$= 3.785$ million liters
1 million gallons per day (mgd) $= 694.4$ gallons per minute (gpm)
$= 2.629$ cubic meters per minute
$= 3785$ cubic meters per day

TABLE A.2 Properties of Water

Traditional U.S. Units							
Temperature (°F)	Specific gravity	Unit weight (lb/ft³)	Heat of vaporization (Btu/lb)	Kinematic viscosity (ft²/sec)	Vapor pressure		
					mb	psi	in. Hg
32	0.99987	62.416	1073	1.93×10^{-5}	6.11	0.09	0.18
40	0.99999	62.423	1066	1.67×10^{-5}	8.36	0.12	0.25
50	0.99975	62.408	1059	1.41×10^{-5}	12.19	0.18	0.36
60	0.99907	62.366	1054	1.21×10^{-5}	17.51	0.26	0.52
70	0.99802	62.300	1049	1.06×10^{-5}	24.79	0.36	0.74
80	0.99669	62.217	1044	0.929×10^{-5}	34.61	0.51	1.03
90	0.99510	62.118	1039	0.828×10^{-5}	47.68	0.70	1.42
100	0.99318	61.998	1033	0.741×10^{-5}	64.88	0.95	1.94

SI Units							
Temperature (°C)	Specific gravity	Density (g/cm³)	Heat of vaporization (cal/g)	Kinematic viscosity (cs)	Vapor pressure		
					(mm Hg)	(mb)	(g/cm²)
0	0.99987	0.99984	597.3	1.790	4.58	6.11	6.23
5	0.99999	0.99996	594.5	1.520	6.54	8.72	8.89
10	0.99973	0.99970	591.7	1.310	9.20	12.27	12.51
15	0.99913	0.99910	588.9	1.140	12.78	17.04	17.38
20	0.99824	0.99821	586.0	1.000	17.53	23.37	23.83
25	0.99708	0.99705	583.2	0.893	23.76	31.67	32.30
30	0.99568	0.99565	580.4	0.801	31.83	42.43	43.27
35	0.99407	0.99404	577.6	0.723	42.18	56.24	57.34
40	0.99225	0.99222	574.7	0.658	55.34	73.78	75.23
50	0.98807	0.98804	569.0	0.554	92.56	123.40	125.83
60	0.98323	0.98320	563.2	0.474	149.46	199.26	203.19
70	0.97780	0.97777	557.4	0.413	233.79	311.69	317.84
80	0.97182	0.97179	551.4	0.365	355.28	473.67	483.01
90	0.96534	0.96531	545.3	0.326	525.89	701.13	714.95
100	0.95839	0.95836	539.1	0.294	760.00	1013.25	1033.23

APPENDIX: B

TABLE B.1   Areas Under the Normal Curve

$$F(z) = \int_0^z \frac{1}{\sqrt{2\pi}} e^{-z^2/2} dz$$

z	0	.01	.02	.03	.04	.05	.06	.07	.08	.09
0	0	.0040	.0080	.0120	.0159	.0199	.0239	.0279	.0319	.0359
0.1	.0398	.0438	.0478	.0517	.0557	.0596	.0636	.0675	.0714	.0753
0.2	.0793	.0832	.0871	.0910	.0948	.0987	.1026	.1064	.1103	.1141
0.3	.1179	.1217	.1255	.1293	.1331	.1368	.1406	.1443	.1480	.1517
0.4	.1554	.1591	.1628	.1664	.1700	.1736	.1772	.1808	.1844	.1879
0.5	.1915	.1950	.1985	.2019	.2054	.2088	.2123	.2157	.2190	.2224
0.6	.2257	.2291	.2324	.2357	.2389	.2422	.2454	.2486	.2518	.2549
0.7	.2580	.2611	.2642	.2673	.2704	.2734	.2764	.2794	.2823	.2852
0.8	.2881	.2910	.2939	.2967	.2995	.3023	.3051	.3078	.3106	.3133
0.9	.3159	.3186	.3212	.3238	.3264	.3289	.3315	.3340	.3365	.3389
1.0	.3413	.3438	.3461	.3485	.3508	.3531	.3554	.3577	.3599	.3621
1.1	.3643	.3665	.3686	.3708	.3729	.3749	.3770	.3790	.3810	.3830
1.2	.3849	.3869	.3888	.3907	.3925	.3944	.3962	.3980	.3997	.4015
1.3	.4032	.4049	.4066	.4082	.4099	.4115	.4131	.4147	.4162	.4177
1.4	.4192	.4207	.4222	.4236	.4251	.4265	.4279	.4292	.4306	.4319
1.5	.4332	.4345	.4357	.4370	.4382	.4394	.4406	.4418	.4430	.4441
1.6	.4452	.4463	.4474	.4485	.4495	.4505	.4515	.4525	.4535	.4545
1.7	.4554	.4564	.4573	.4582	.4591	.4599	.4608	.4616	.4625	.4633
1.8	.4641	.4649	.4656	.4664	.4671	.4678	.4686	.4693	.4699	.4606
1.9	.4713	.4719	.4726	.4732	.4738	.4744	.4750	.4756	.4762	.4767
2.0	.4772	.4778	.4783	.4788	.4793	.4798	.4803	.4808	.4812	.4817
2.1	.4821	.4826	.4830	.4834	.4838	.4842	.4846	.4850	.4854	.4857
2.2	.4861	.4865	.4868	.4871	.4875	.4878	.4881	.4884	.4887	.4890
2.3	.4893	.4896	.4898	.4901	.4904	.4906	.4909	.4911	.4913	.4916
2.4	.4918	.4920	.4922	.4925	.4927	.4929	.4931	.4932	.4934	.4936
2.5	.4938	.4940	.4941	.4943	.4945	.4946	.4948	.4949	.4951	.4952
2.6	.4953	.4955	.4956	.4957	.4959	.4960	.4961	.4962	.4963	.4964
2.7	.4965	.4966	.4967	.4968	.4969	.4970	.4971	.4972	.4973	.4974
2.8	.4974	.4975	.4976	.4977	.4977	.4978	.4979	.4980	.4980	.4981
2.9	.4981	.4982	.4983	.4983	.4984	.4984	.4985	.4985	.4986	.4986
3.0	.4987	.4987	.4987	.4988	.4988	.4989	.4989	.4989	.4990	.4990
3.1	.4990	.4991	.4991	.4991	.4992	.4992	.4992	.4992	.4993	.4993
3.2	.4993	.4993	.4994	.4994	.4994	.4994	.4994	.4995	.4995	.4995
3.3	.4995	.4995	.4996	.4996	.4996	.4996	.4996	.4996	.4996	.4997
3.4	.4997	.4997	.4997	.4997	.4997	.4997	.4997	.4997	.4998	.4998
⋮	⋮									
4.0	.499968									

Source: After C. E. Weatherburn, *Mathematical Statistics,* London: Cambridge University Press, 1957 (for z = 0 to z = 3.1); C. H. Richardson, *An Introduction to Statistical Analysis*, Orlando, FL: Harcourt Brace Jovanovich, 1994 (for z = 3.2 to z = 3.4); A. H. Bowker and G. J. Lieberman, *Engineering Statistics*, Englewood Cliffs, NJ: Prentice Hall, 1959 (for z = 4.0).

TABLE B.2   *K* Values for Pearson Type III Distribution

Skew coefficient, C_s	Recurrence interval in years										
	1.0101	1.0526	1.1111	1.2500	2	5	10	25	50	100	200
					Percent chance						
	99	95	90	80	50	20	10	4	2	1	0.5
	Positive skew										
3.0	−0.667	−0.665	−0.660	−0.636	−0.396	0.420	1.180	2.278	3.152	4.051	4.970
2.9	−0.690	−0.688	−0.681	−0.651	−0.390	0.440	1.195	2.277	3.134	4.013	4.909
2.8	−0.714	−0.711	−0.702	−0.666	−0.384	0.460	1.210	2.275	3.114	3.973	4.847
2.7	−0.740	−0.736	−0.724	−0.681	−0.376	0.479	1.224	2.272	3.093	3.932	4.783
2.6	−0.769	−0.762	−0.747	−0.696	−0.368	0.499	1.238	2.267	3.071	3.889	4.718
2.5	−0.799	−0.790	−0.771	−0.711	−0.360	0.518	1.250	2.262	3.048	3.845	4.652
2.4	−0.832	−0.819	−0.795	−0.725	−0.351	0.537	1.262	2.256	3.023	3.800	4.584
2.3	−0.867	−0.850	−0.819	−0.739	−0.341	0.555	1.274	2.248	2.997	3.753	4.515
2.2	−0.905	−0.882	−0.844	−0.752	−0.330	0.574	1.284	2.240	2.970	3.705	4.444
2.1	−0.946	−0.914	−0.869	−0.765	−0.319	0.592	1.294	2.230	2.942	3.656	4.372
2.0	−0.990	−0.949	−0.895	−0.777	−0.307	0.609	1.302	2.219	2.912	3.605	4.398
1.9	−1.037	−0.984	−0.920	−0.788	−0.294	0.627	1.310	2.207	2.881	3.553	4.223
1.8	−1.087	−1.020	−0.945	−0.799	−0.282	0.643	1.318	2.193	2.848	3.499	4.147
1.7	−1.140	−1.056	−0.970	−0.808	−0.268	0.660	1.324	2.179	2.815	3.444	4.069
1.6	−1.197	−1.093	−0.994	−0.817	−0.254	0.675	1.329	2.163	2.780	3.388	3.990
1.5	−1.256	−1.131	−1.018	−0.825	−0.240	0.690	1.333	2.146	2.743	3.330	3.910
1.4	−1.318	−1.168	−1.041	−0.832	−0.225	0.705	1.337	2.128	2.706	3.271	3.828
1.3	−1.383	−1.206	−1.064	−0.838	−0.210	0.719	1.339	2.108	2.666	3.211	3.745
1.2	−1.449	−1.243	−1.086	−0.844	−0.195	0.732	1.340	2.087	2.626	3.149	3.661
1.1	−1.518	−1.280	−1.107	−0.848	−0.180	0.745	1.341	2.066	2.585	3.087	3.575
1.0	−1.588	−1.317	−1.128	−0.852	−0.164	0.758	1.340	2.043	2.542	3.022	3.489
0.9	−1.660	−1.353	−1.147	−0.854	−0.148	0.769	1.339	2.018	2.498	2.957	3.401
0.8	−1.733	−1.388	−1.166	−0.856	−0.132	0.780	1.336	1.993	2.453	2.891	3.312
0.7	−1.806	−1.423	−1.183	−0.857	−0.116	0.790	1.333	1.967	2.407	2.824	3.223
0.6	−1.880	−1.458	−1.200	−0.857	−0.099	0.800	1.328	1.939	2.359	2.755	3.132
0.5	−1.955	−1.491	−1.216	−0.856	−0.083	0.808	1.323	1.910	2.311	2.686	3.041
0.4	−2.029	−1.524	−1.231	−0.855	−0.066	0.816	1.317	1.880	2.261	2.615	2.949
0.3	−2.104	−1.555	−1.245	−0.853	−0.050	0.824	1.309	1.849	2.211	2.544	2.856
0.2	−2.178	−1.586	−1.258	−0.850	−0.033	0.830	1.301	1.818	2.159	2.472	2.763
0.1	−2.252	−1.616	−1.270	−0.846	−0.017	0.836	1.292	1.785	2.107	2.400	2.670
0	−2.326	−1.645	−1.282	−0.842	0	0.842	1.282	1.751	2.054	2.326	2.576

TABLE B.2   (*Continued*)

Skew coefficient, C_s	Recurrence interval in years										
	1.0101	1.0526	1.1111	1.2500	2	5	10	25	50	100	200
					Percent chance						
	99	95	90	80	50	20	10	4	2	1	0.5
	Negative skew										
−0.1	−2.400	−1.673	−1.292	−0.836	0.017	0.846	1.270	1.716	2.000	2.252	2.482
−0.2	−2.472	−1.700	−1.301	−0.830	0.033	0.850	1.258	1.680	1.945	2.178	2.388
−0.3	−2.544	−1.726	−1.309	−0.824	0.050	0.853	1.245	1.643	1.890	2.104	2.294
−0.4	−2.615	−1.750	−1.317	−0.816	0.066	0.855	1.231	1.606	1.834	2.029	2.201
−0.5	−2.686	−1.774	−1.323	−0.808	0.083	0.856	1.216	1.567	1.777	1.955	2.108
−0.6	−2.755	−1.797	−1.328	−0.800	0.099	0.857	1.200	1.528	1.720	1.880	2.016
−0.7	−2.824	−1.819	−1.333	−0.790	0.116	0.857	1.183	1.488	1.663	1.806	1.926
−0.8	−2.891	−1.839	−1.336	−0.780	0.132	0.856	1.166	1.448	1.606	1.733	1.837
−0.9	−2.957	−1.858	−1.339	−0.769	0.148	0.854	1.147	1.407	1.549	1.660	1.749
−1.0	−3.022	−1.877	−1.340	−0.758	0.164	0.852	1.128	1.366	1.492	1.588	1.664
−1.1	−3.087	−1.894	−1.341	−0.745	0.180	0.848	1.107	1.324	1.435	1.518	1.581
−1.2	−3.149	−1.910	−1.340	−0.732	0.195	0.844	1.086	1.282	1.379	1.449	1.501
−1.3	−3.211	−1.925	−1.339	−0.719	0.210	0.838	1.064	1.240	1.324	1.383	1.424
−1.4	−3.271	−1.938	−1.337	−0.705	0.225	0.832	1.041	1.198	1.270	1.318	1.351
−1.5	−3.330	−1.951	−1.333	−0.690	0.240	0.825	1.018	1.157	1.217	1.256	1.282
−1.6	−3.388	−1.962	−1.329	−0.675	0.254	0.817	0.994	1.116	1.166	1.197	1.216
−1.7	−3.444	−1.972	−1.324	−0.660	0.268	0.808	0.970	1.075	1.116	1.140	1.155
−1.8	−3.499	−1.981	−1.318	−0.643	0.282	0.799	0.945	1.035	1.069	1.087	1.097
−1.9	−3.553	−1.989	−1.310	−0.627	0.294	0.788	0.920	0.996	1.023	1.037	1.044
−2.0	−3.605	−1.996	−1.302	−0.609	0.307	0.777	0.895	0.959	0.980	0.990	0.995
−2.1	−3.656	−2.001	−1.294	−0.592	0.319	0.765	0.869	0.923	0.939	0.946	0.949
−2.2	−3.705	−2.006	−1.284	−0.574	0.330	0.752	0.844	0.888	0.900	0.905	0.907
−2.3	−3.753	−2.009	−1.274	−0.555	0.341	0.739	0.819	0.855	0.864	0.867	0.869
−2.4	−3.800	−2.011	−1.262	−0.537	0.351	0.725	0.795	0.823	0.830	0.832	0.833
−2.5	−3.845	−2.012	−1.250	−0.518	0.360	0.711	0.771	0.793	0.798	0.799	0.800
−2.6	−3.889	−2.013	−1.238	−0.499	0.368	0.696	0.747	0.764	0.768	0.769	0.769
−2.7	−3.932	−2.012	−1.224	−0.479	0.376	0.681	0.724	0.738	0.740	0.740	0.741
−2.8	−3.973	−2.010	−1.210	−0.460	0.384	0.666	0.702	0.712	0.714	0.714	0.714
−2.9	−4.013	−2.007	−1.195	−0.440	0.390	0.651	0.681	0.683	0.689	0.690	0.690
−3.0	−4.051	−2.003	−1.180	−0.420	0.396	0.636	0.660	0.666	0.666	0.667	0.667

Source: After *Water Resources Council, Bulletin No. 15,* December 1967.

TABLE B.3   Uniformly Distributed Random Numbers

53 74 23 99 67	61 32 28 69 84	94 62 67 86 24	98 33 41 19 95	47 53 53 38 09
63 38 06 86 54	99 00 65 26 94	02 82 90 23 07	79 62 67 80 60	75 91 12 81 19
30 30 58 21 46	06 72 17 10 94	25 21 31 75 96	49 28 24 00 49	55 65 79 78 07
63 43 36 82 69	65 51 18 37 88	61 38 44 12 45	32 92 85 88 65	54 34 81 85 35
98 25 37 55 26	01 91 82 81 46	74 71 12 94 97	24 02 71 37 07	03 92 18 66 75
02 63 21 17 69	71 50 80 89 56	38 15 70 11 48	43 40 45 86 98	00 83 26 91 03
64 55 22 21 82	48 22 28 06 00	61 54 13 43 91	82 78 12 23 29	06 66 24 12 27
85 07 26 13 89	01 10 07 82 04	59 63 69 36 03	69 11 15 83 80	13 29 54 19 28
58 54 16 24 15	51 54 44 82 00	62 61 65 04 69	38 18 65 18 97	85 72 13 49 21
34 85 27 84 87	61 48 64 56 26	90 18 48 13 26	37 70 15 42 57	65 65 80 39 07
03 92 18 27 46	57 99 16 96 56	30 33 72 85 22	84 64 38 56 98	99 01 30 98 64
62 95 30 27 59	37 75 41 66 48	86 97 80 61 45	23 53 04 01 63	45 76 08 64 27
08 45 93 15 22	60 21 54 46 91	98 77 27 85 42	28 88 61 08 94	69 62 03 42 73
07 08 55 18 40	45 44 75 13 90	24 94 96 61 02	57 55 66 83 15	73 42 37 11 61
01 85 89 95 66	51 10 19 34 88	15 84 97 19 75	12 76 39 43 78	64 63 91 08 25
72 84 71 14 35	19 11 58 49 26	50 11 17 17 76	86 31 57 20 18	95 60 78 46 75
88 78 28 16 84	13 52 53 94 53	75 45 69 30 96	73 89 65 70 31	99 17 43 48 76
45 17 75 65 57	28 40 19 72 12	25 12 74 75 67	60 40 60 81 19	24 62 01 61 16
96 76 28 12 54	22 01 11 94 25	71 96 16 16 88	68 64 36 74 45	19 59 50 88 92
43 31 67 72 30	24 02 94 08 63	38 32 36 66 02	69 36 38 25 39	48 03 45 15 22
50 44 66 44 21	66 06 58 05 62	68 15 54 35 02	42 35 48 96 32	14 52 41 52 48
22 66 22 15 86	26 63 75 41 99	58 42 36 72 24	58 37 52 18 51	03 37 18 39 11
96 24 40 14 51	23 22 30 88 57	95 67 47 29 83	94 69 40 06 07	18 16 36 78 86
31 73 91 61 19	60 20 72 93 48	98 57 07 23 69	65 95 39 69 58	56 80 30 19 44
78 60 73 99 84	43 89 94 36 45	56 69 47 07 41	90 22 91 07 12	78 35 34 08 72
84 37 90 61 56	70 10 23 98 05	85 11 34 76 60	76 48 45 34 60	01 64 18 39 96
36 67 10 08 23	98 93 35 08 86	99 29 76 29 81	33 34 91 58 93	63 14 52 32 52
07 28 59 07 48	89 64 58 89 75	83 85 62 27 89	30 14 78 56 27	86 63 59 80 02
10 15 83 87 60	79 24 31 66 56	21 48 24 06 93	91 98 94 05 49	01 47 59 38 00
55 19 68 97 65	03 73 52 16 56	00 53 55 90 27	33 42 29 38 87	22 13 88 83 34
53 81 29 13 39	35 01 20 71 34	62 33 74 82 14	53 73 19 09 03	56 54 29 56 93
51 86 32 68 92	33 98 74 66 99	40 14 71 94 58	45 94 19 38 81	14 44 99 81 07
35 91 70 29 13	80 03 54 07 27	96 94 78 32 66	50 95 52 74 33	13 80 55 62 54
37 71 67 95 13	20 02 44 95 94	64 85 04 05 72	01 32 90 76 14	53 89 74 60 41
93 66 13 83 27	92 76 64 64 72	28 54 96 53 84	48 14 52 98 94	56 07 93 89 30

Source: After L. R. Beard, *Statistical Methods in Hydrology*, U.S. Army Corps of Engineers, 1962.

Index